BEAMS 2002

Related Titles from AIP Conference Proceedings

651 Dense Z-Pinches: 5th International Conference on Dense Z-Pinches
Edited by Christopher Deeney, Jack Davis, and Nino R. Pereira, December 2002, 0-7354-0108-X

648 Beam Instrumentation Workshop 2002: Tenth Workshop
Edited by Gary Smith and Thomas Russo, December 2002, CD-ROM included, 0-7354-0103-9

647 Advanced Accelerator Concepts: Tenth Workshop
Edited by Christopher E. Clayton and Patrick Muggli, December 2002, CD-ROM included, 0-7354-0102-0

625 High Energy Density and High Power RF:
5th Workshop on High Energy Density and High Power RF
Edited by B. E. Carlsten, August 2002, 0-7354-0078-4

592 High Quality Beams: Joint US-CERN-JAPAN-RUSSIA Accelerator School
Edited by S. I. Kurokawa, S. Y. Lee, J. Miles, E. A. Perevedentsev, November 2001, 0-7354-0034-2

581 Physics of, and Science with, the X-Ray Free-Electron Laser:
19th Advanced ICFA Beam Dynamics Workshop
Edited by C. Pellegini, S. Chattopadhyay, M. Cornacchia, and I. Lindau, August 2001, 0-7354-0022-9

576 Application of Accelerators in Research and Industry:
Sixteenth International Conference
Edited by J. L. Duggan and I. L. Morgan, July 2001, 0-7354-0015-6

572 Electron Beam Ion Sources and Traps and Their Applications:
8th International Symposium
Edited by Krsto Prelec, June 2001, 0-7354-0011-3

474 High Energy Density Microwaves
Edited by Robert M. Phillips, May 1999, 1-56396-796-0

To learn more about these titles, or the AIP Conference Proceedings Series, please visit the webpage **http://proceedings.aip.org/proceedings**

BEAMS 2002

14th International Conference on High-Power Particle Beams

Albuquerque, New Mexico 23-28 June 2002

EDITORS
Thomas A. Mehlhorn
Mary Ann Sweeney
Sandia National Laboratories
Albuquerque, New Mexico

SPONSORING ORGANIZATIONS
Department of Energy
Sandia National Laboratories
Defense Threat Reduction Agency
Air Force Research Laboratory

AMERICAN INSTITUTE OF PHYSICS
Melville, New York, 2002
AIP CONFERENCE PROCEEDINGS ■ VOLUME 650

Editors:

Thomas A. Mehlhorn
Target & Z-Pinch Theory Department
Sandia National Laboratories
P.O. Box 5800
Albuquerque, NM 87185-1186
USA

E-mail: tamehlh@sandia.gov

Mary Ann Sweeney
Sandia National Laboratories
MS 1191
P.O. Box 5800
Albuquerque, NM 87185-1191

E-mail: masween@sandia.gov

L.C. Catalog Card No. 2002115712
ISBN 0-7354-0107-1
ISSN 0094-243X

CD-ROM available separately: ISBN 0-7354-0109-8

Printed in the United States of America

CONTENTS

PLENARY TALKS

PULSED POWER TECHNOLOGY

PULSED POWER SWITCHES

RADIOGRAPHY

ELECTRON BEAM APPLICATIONS

ION BEAM PHYSICS AND APPLICATIONS

THEORY AND MODELING

BEAMS FROM ULTRA-INTENSE LASERS

APPENDICES

Preface

The 14th International Conference on High-Power Beams was held from June 23 to June 28, 2002, in Albuquerque, NM. This year, *Beams 2002* was organized and held in conjunction with the 5th International Conference on Dense Z-Pinches (DZP). The combined Conference drew approximately 450 attendees. This volume contains over 110 manuscripts that were submitted in conjunction with the technical program presentations of *Beams 2002*. A companion volume contains the papers from the Dense Z-Pinches portion of the technical program.

Following the pattern of past Beams meetings, this conference began with a series of plenary talks on the status of beams and pulsed power in the U.S., Europe, Japan and Russia and concluded with a summary of the highlights of the meeting by a multi-national panel. The combined technical program committees of Beams and DZP chose to implement a parallel structure where Beams orals overlapped DZP posters, and vice versa. Morning and afternoon sessions began with plenary invited talks that were chosen to have broad technical appeal. This blended program encouraged full participation from both constituencies and provided many opportunities for cross-fertilization.

Beams 2002 demonstrated that pulsed power and high power beams are still areas of fertile technical research and application with presentations ranging from pulsed power technology to the generation of high-brightness ion beams by ultra-intensity laser beams. Approximately 20 papers in this volume deal with pulsed power technology and improve-ments in pulsed power switch technology; over 30 papers deal with electron beam generation and radiography; 15 papers deal with various aspects of high power microwaves; over 30 papers deal with the use of electrons and ion beams for applications such as materials modification and processing; 7 papers describe advances in modeling and simulation techniques; and 3 with laser-related applications.

From this distribution of papers the Editor observes that high-power electron beams remain a key technical area. Almost half of the papers in this volume are related to electron diode physics, radiography applications, or the use of electron beams for emerging commercial applications. A special workshop on radiography was a vital part of the *Beams 2002* meeting and demonstrated the continued importance of radiography within our community. Most of the ion beam papers are concerned with the commercial applications of beams with modest powers to materials modifications and processing. More than 25% of the total papers deal with the use of electron and ion beams to applications such as materials modification and medical sterilization, indicating a continuing maturation of the field beyond traditional defense-related activities.

Finally, although it was only a small portion of *Beams 2002*, the emerging area of the generation of high-power, high-brightness electron and ion beams by ultra-high-intensity laser-matter interactions is an exciting new field at the interface between laser and particle beam physics. This is an area at the forefront of high power beams physics with potential for many exciting applications such as fast ignition, advanced radiography, and high-brightness ion beam sources for accelerators. We can expect to see growing participation in this area in *Beams 2004* to be held in St. Petersburg, Russia.

Thomas A. Mehlhorn
Sandia National Laboratories

Letter from the Chairman of Beams/DZP 2002

The 14th International Conference on High-Power Particle Beams and the 5th International Conference on Dense Z-Pinches (BEAMS/DZP 2002) took place at the Hilton Albuquerque in Albuquerque, New Mexico, on June 23-28, 2002. This year we tried the experiment of combining the programs of these two historically separate conferences because of the significant overlap in subject matter and interests. By most measures the experiment was deemed a success, and we will likely co-locate some future meetings of the two conferences.

As in previous separate Beams and DZP conferences, papers were given as oral (invited and contributed) and poster presentations. The Beams part of the meeting included electron and ion beam physics, free electron lasers, x-ray lasers, high-power microwaves, imploding liners, and radiography, and the DZP topics included magnetohydrodynamics, radiation hydrodynamics, z-pinch physics, and z-pinch diagnostics. The combined topics were pulsed power technology, electrical diagnostics, x-ray spectroscopy, electrodynamic modeling, plasma instabilities, high energy density physics, astrophysical applications, industrial applications, inertial or magnetized target fusion, plasma focus, and x-ray lithography. The 394 participants from 19 countries presented 358 papers, of which 85 were orals and 273 were posters. Most of these papers are included in the conference proceedings.

The 15th International Conference on High-Power Particle Beams will be held in Russia in 2004. This will be the third time the conference has been there. Discussion is underway to determine if the 6th International Conference on Dense Z-Pinches will also meet at this time.

In addition to the outstanding technical aspects of the meeting, the participants took advantage of the unique New Mexico environment, which offers spectacular scenery and a melting pot of Native American, Spanish and Anglo cultures. Social events included a welcome reception on Sunday, a western barbecue on the Sandia Indian Reservation on Tuesday, the conference banquet on Wednesday, and the DZP reception on Thursday.

We would like to acknowledge the technical contributions of all the participants. Furthermore, we thank all the members of the international steering committees, the local steering committee, and the technical program committees for their contributions to the success of the meeting. We would also especially like to thank the many sponsors that financially supported the Beams/DZP 2002 conference.

Jeff Quintenz
Conference Chairman

Dear Colleagues,

It is our great pleasure to welcome you to the 14th International Conference on High-Power Particle Beams and the 5th International Conference on Dense Z-Pinches in Albuquerque, New Mexico.

When Gerry Yonas chaired the first Beams conference in Albuquerque in 1975, the technologies were exciting and young. Now that the Conference returns to Albuquerque 27 years later, with Gerry as its honorary chairman, pulsed power technologies are more mature and applied, but exciting still. The 2002 Conference will be a worthy successor to the 2000 Beams Conference held in Nagaoka, Japan.

With the increased prominence of research on z pinches, the Beams Conference has been co-located in 2002 with the 5th International Conference on Dense Z-Pinches. Since 1988 that Conference has gathered the world's z-pinch researchers approximately every four years. The present Conference skipped one year to be hosted together with Beams 2002 in Albuquerque, home of Sandia National Laboratories' Z accelerator and its z-pinch, which generates the world's most powerful x-ray burst in the laboratory.

The two Conferences received more than 360 abstracts and expect participation from scientists in more than 15 countries. The Technical Chairmen, Tom Mehlhorn and Chris Deeney, combined the two programs in such a manner that it will be possible to attend both Beams and DZP talks and posters during the 4-1/2 day technical program.

Albuquerque is located in the wide valley between the Sandia Mountains and the Rio Grande. It is the largest city in New Mexico and near the center of the Southwest. Albuquerque has become a major research and development center that offers spectacular scenery, casual elegance, and a melting pot of Native American, Spanish, and Anglo cultures. The conference hotel, Hilton Albuquerque, reflects the flavor and history of the Land of Enchantment, with Indian rugs, pueblo art, clay pottery, Kachina dolls, and arched doorways.

The social events include a Western barbecue on the Sandia Indian Reservation on Tuesday evening and the conference banquet at the Hilton on Wednesday evening. Ravi Sudan and Valentin Smirnov will receive Beams Awards during the banquet. In addition, a full schedule of activities for companions is planned in Albuquerque, Santa Fe, and Taos.

We wish all of you a rewarding scientific meeting and an enjoyable time in the Southwest tradition.

Sincerely,

Jeff Quintenz
Conference Chairman

Mary Ann Sweeney
Conference Co-Chairman

Nino Pereira
Jack Davis
Dense Z-Pinches Co-Chairmen

The Organizing Committees thank the following sponsors for their support of Beams/DZP 2002.

Sponsors

Department of Energy

Sandia National Laboratories

Defense Threat Reduction Agency

Air Force Research Lab

12th International Conference on High-Power Particle Beams in Israel

13th International Conference on High-Power Particle Beams in Japan

4th International Conference on Dense Z-Pinches in Vancouver

Ecopulse, Inc.

Ktech

Pulsed Power Conference Inc.

Science Applications International Corporation

Team Specialty Products

Titan Pulse Sciences Division

University of New Mexico

Technical Co-Sponsor

IEEE Electron Devices Society

Organizing Committees

Beams/DZP Local Steering Committee

Mary Ann Sweeney, *Chairman*	Sandia National Laboratories, USA
Lisa Mattox	Sandia National Laboratories, USA
Fran DiMarco	Fran DiMarco Meeting and Event Planning, Albuquerque, NM, USA
Rab Freeman	Albuquerque, NM, USA
Chuck Gilman	Science Applications International Corporation, USA
Judy Gilman	Albuquerque, NM, USA
Pat Helles	Air Force Research Laboratory, USA
Dan Jobe	Ktech, USA
Melissa Murray	Sandia National Laboratories, USA
Steve Nickerson	Sandia National Laboratories, USA
Edl Schamiloglu	University of New Mexico, USA
Bill Shoup	ITT Industries, USA
Dick Siemon	Los Alamos National Laboratory, USA
Marshall Sluyter	Rockville, MD, USA
Al Toepfer	Science Applications International Corporation, USA
Jo Toepfer	Albuquerque, NM, USA

Beams International Advisory Committee

Edek Blaugrund	Weizmann Institute of Science, Israel
Hans Bluhm	Forschungszentrum Karlsruhe, Germany
Donald Cook	Sandia National Laboratories, USA
Gerald Cooperstein	Naval Research Laboratory, USA
Vladimir Fortov	Academy of Sciences, Russia
Timothy Goldsack	Atomic Weapons Establishment, UK
Karel Jungwirth	Czech Academy of Sciences, Czech Republic
Vasili Koidan	Budker Institute of Nuclear Physics, Russia
Alan Kolb	ACK Associates, USA
Gennady Mesyats	Academy of Sciences, Russia
Jeffrey Quintenz	Sandia National Laboratories, USA
Donald Rej	Los Alamos National Laboratory, USA
Joseph Shiloh	Rafael, Israel
Valentin Smirnov	Kurchatov Institute, Russia
Charles Stallings	Stallings and Associates, USA
Ravi Sudan	Cornell University, USA
Roger White	Titan Pulse Sciences Division, USA
Kiyoshi Yatsui	Nagaoka University of Technology, Japan
Gerold Yonas	Sandia National Laboratories, USA

Dense Z-Pinches International Advisory Committee

Hernan Chuaqui	Pontificia Universidad Catolica de Chile, Chile
Jack Davis	Naval Research Laboratory, USA
Malcolm Haines	Imperial College, UK
Thomas Hopkins	Defense Threat Reduction Agency, USA
Hans-Joachim Kunze	Ruhr-Universitat, Germany
Keith Matzen	Sandia National Laboratories, USA
Tetsuo Miyamoto	Nihon University, Japan
Valentin Smirnov	Kurchatov Institute, Russia

Beams Technical Program Committee

Thomas Mehlhorn, *Chairman*	Sandia National Laboratories, USA
Gerold Yonas, *Honorary Chairman*	Sandia National Laboratories, USA
Hans Bluhm	Forschungszentrum Karlsruhe, Germany
Ronald Davidson	Princeton Plasma Physics Lab, USA
Ronald Gilgenbach	University of Michigan, USA
Timothy Goldsack	Atomic Weapons Establishment, UK
Koichi Kasuya	Tokyo Institute of Technology, Japan
Meir Markovits	Rafael, Israel
Craig Olson	Sandia National Laboratories, USA
Paul Ottinger	Naval Research Laboratory, USA
Markus Roth	GSI, Germany
John Sethian	Naval Research Laboratory, USA

Dense Z-Pinches Technical Program Committee

Christopher Deeney, *Chairman*	Sandia National Laboratories, USA
John Apruzese	Naval Research Laboratory, USA
David Bell	Defense Threat Reduction Agency, USA
Melissa Douglas	Los Alamos National Laboratory, USA
Julio Herrera	Inst de Ciencias Nucl, Mexico
Konstantin Koshelev	Institute for Spectroscopy, Russia
Sergei Lebedev	Imperial College, London, UK
Yitzak Maron	Weizmann Institute of Science, Israel
Alexander Shishlov	High Current Electronics Institute, Russia
Marshall Sluyter	Rockville, MD, USA
Alexander Velikovich	Naval Research Laboratory, USA

Brief Histories of the Conferences

The first International Conference on Dense Z-Pinches (DZP), in 1984 in Alexandria, Virginia, was organized by the Naval Research Laboratory's John Sethian in large part to discuss the z pinch prospects for thermonuclear fusion. The meeting turned out to be very lively, with the 30 participants from various countries even discussing optical radiation from z pinches. Shortly thereafter it became possible to discuss production of x rays from z pinches. Hence, Professor Norman Rostoker of the University of California Irvine and Nino Pereira and Jack Davis of NRL decided to add x rays to the conference topics for the second DZP, in Laguna Beach, California, in 1988. The second conference was even more successful, with over 120 participants from all over the world including, for the first time, many from the Soviet Union. Subsequently, the DZP Conference kept this format, in 1993, in London, United Kingdom, and in 1997 in Vancouver, Canada. The fifth DZP Conference maintains the tradition, but is co-located with the Beams Conference in the location of the world's largest z pinch.

Nino Pereira

The first International Conference on High-Power Particle Beams (Beams) was held in Albuquerque, New Mexico, in 1975. The conference, organized by Gerold Yonas of Sandia National Laboratories, attracted scientists from around the world. The success of this first conference prompted the formation of the Beams International Advisory Committee and the organization of subsequent conferences on an approximately biannual basis. In addition to the United States, the Beams conference has been held in Russia, Germany, France, the Czech Republic, Israel, and Japan. As the field has matured, the conference has expanded the list of topics covered. The number of papers on the contributions of z pinches to high energy density physics has grown significantly in recent Beams Conferences, and we are pleased to be co-locating Beams with the Dense Z-Pinches Conference in 2002. Recognizing our beginnings, we are returning to Albuquerque this year and are happy to honor Gerry Yonas' contributions to the Conference by having him serve as Honorary Chairman of Beams 2002.

Jeff Quintenz

Beams Conference Statistics (Location, Dates, Attendance)

Location	Dates	USA	Russia (+ other fSU)	Japan	Germany	France	Israel	Others	Total	Host Country
Albuquerque, NM	Nov. 3-5, 1975	178	4	1	3	4	4	7	201	178
Ithaca, NY	Oct. 3-5, 1977	192	6	2	0	5	3	3	211	192
Novosibirsk, Russia	July 3-6, 1979	38	183	8	5	6	0	3	243	177
Palaiseau, France	June 29-July 3, 1981	81	15	9	14	81	6	16	222	81
San Francisco	Sept. 12-14, 1983	293	0	15	8	14	5	7	342	293
Kobe, Japan	June 9-12, 1986	77	17	155	5	8	5	14	281	155
Karlsruhe, Germany	July 4-8, 1988	99	16	19	75	21	8	21	259	75
Novosibirsk, Russia	July 2-5, 1990	52	149	14	10	6	0	13	244	149
Washington, DC	May 25-29, 1992	286	65	15	10	13	12	22	423	286
San Diego, CA	June 20-24, 1994	173	58	19	13	10	12	10	295	173
Prague, Czech Rep.	June 10-14, 1996	60	113	17	25	19	12	63	309	38
Haifa, Israel	June 7-12, 1998	58	80	17	8	9	62	37	209	62
Nagaoka, Japan	June 25-30, 2000	20	42	161	4	2	9	29	267	161

Dense Z Pinches Conference Statistics (Location, Dates, Attendance)

Location	Dates	USA[2]	Total
Alexandria, VA	March 1984[1]		32
Laguna Beach, CA	April 26-28, 1989	69	120
London	April 19-23, 1993	25	117
Vancouver	May 28-31, 1997	34	129

[1] Precise dates of the 1st conference are unknown.
[2] Number of U.S. attendees is unknown for the 1st conference; all but a few attendees are assumed to be from the USA.

Combined Beams and Dense Z Pinches Conference Statistics (Location, Dates, Attendance)

Location	Dates	USA (Host Country)	Russia (+ other FSU)	Japan	Germany	France	Israel	Others	Total
Albuquerque, NM	June 23-28, 2002	231	72	17	9	6	7	51	394

PLENARY TALKS

Status of Particle Beam and Pulsed Power Research in the US

Gerald Cooperstein

Plasma Physics Division, Naval Research Laboratory, Washington, DC 20375

Abstract. Many exciting papers were presented at BEAMS2002 on the topics of high-power particle beams, pulsed power research, and applications. Taken in their entirety, they represent a snapshot of the world's efforts in these areas of research. This paper highlights major US research efforts and collaborative efforts outside the US.

INTRODUCTION

Pulsed x-radiography of explosively-driven hydrodynamic experiments is presently the major driver for intense beam research and pulsed-power development. Almost 1/3 of the papers presented at the BEAMS2002 are on these topics. Radiography source research includes intense electron beams accelerated by linear induction adders, and the paraxial, magnetically-immersed and rod-pinch electron-beam diodes. New radiography generators include DAHRT and Cygnus at Los Alamos National Laboratory (LANL), the Radiographic Integrated Test Stand (RITS) at Sandia National Laboratories (SNL), the Compact Radiographic Source (CRS) at Lawrence Livermore National Laboratory (LLNL), Mercury at the Naval Research Laboratory (NRL), and the Hydrodynamic Research Facility (HRS) at the Atomic Weapons Establishment (AWE) in Aldermaston, England. Other major new pulsed-power facilities include ZR at SNL, Decade Half at Arnold Engineering Development Center (AEDC), ECF2 at Gramat France, the Spallation Neutron Source (SNS) at Oak Ridge National Laboratory, and Electra at NRL. Other major beams programs include the Heavy Ion Fusion (HIF) and High Power Microwave (HPM) programs. Because of limited space, this paper will not review many other important US particle beam and pulsed power research areas.

NEW RADIOGRAPHY SOURCE RESEARCH

Pulsed hydrodynamic radiography is critical for the US Science Based Stockpile Stewardship program. Point-projection pulsed radiography is used to interpret small-scale experiments such as spall ejecta and large scale dynamic surrogates for sub-critical underground tests. The radiographic probe must be matched to the application. Voltage, which determines penetration power and contrast, ranges from 100's of kV to 10's of MV. Dose, which determines signal level, ranges from 10's of mR to 1000 R at 1 m. Spot sizes, which determine spacial resolution, varies from sub-mm to a few mm. Pulse duration, which determines the ability to freeze motion,

CP650, *Beams 2002: 14th International Conference on High-Power Particle Beams*,
edited by T. A. Mehlhorn and M. A. Sweeney
2002 American Institute of Physics 0-7354-0107-1

varies from 10 to 100 ns. Multiple pulses are desired to follow the objects evolution. A suite of systems must be developed because no one system can meet all these requirements.[1]

Actively-pursued sources are conceptually illustrated in Fig. 1. The accelerated electron beam (Fig. 1a) uses linear-induction-adder (LIA) accelerators to produce 20-MV, 2-kA, mm-diam, multi-pulse beams. They are large, complex and expensive. The newest US program, DAHRT at LANL is well represented at BEAMS2002 with an overview paper, technology papers and beam papers.[2] The paraxial diode (Fig. 1b) was developed at AWE. A large-area, low-current-density electron beam is weakly self-focused onto a gas cell. Beam current and charge neutralization then allows the beam to ballistically transport to a focus. This technique scales to high voltage and dose but, at present, spot size is limited to a few mm. AWE is collaborating with SNL on this diode.[3,4] The magnetically-immersed diode (Fig. 1c) is being developed by SNL. It uses a mm-diam cathode immersed in a 2 - 5 T magnetic field which is high enough to keep the electron gyroradius equal to the cathode radius. The beam then hits a small-diameter anode. Preliminary experiments and extensive theoretical analysis suggest that the technique will scale to high voltage.[1,5]

The most actively pursued source, at present, is the rod-pinch diode (RPD) (Fig. 1d).[6-13] It has produced sub-mm spot sizes and has been chosen for near term sub-critical experiments, but may not scale to very high voltage. As illustrated in Fig. 2, electrons emitted from a hollow cathode self-pinch onto a mm-diam anode rod and deposit their energy at the rod's tip as seen in the x-ray pinhole photo. It has been successfully tested on a number of facilities ranging from 1.5 MV on NRL's Gamble II (5 R at 1 m from sub-mm diam spots)[6,9] to 4 MV on Asterix (over 20 R at 1 m from 1- to 2-mm-diam spots)[10] in Gramat France. It presently is being tested at 6 MV on Asterix in a collaborative experiment with the French DGA and CEA, NRL, and SNL. Extensive theoretical modeling has been performed.[6,11] Preliminary RPD heated-anode results suggest that losses due to low-Z ions can be eliminated and the spot size can be reduced.[12] When the diode is filled with an external plasma, it has operated at very low impedance (1 MV, 500 kA) and has produced over 20 R at 1 m with < 1/2-mm-diam spot sizes.[13]

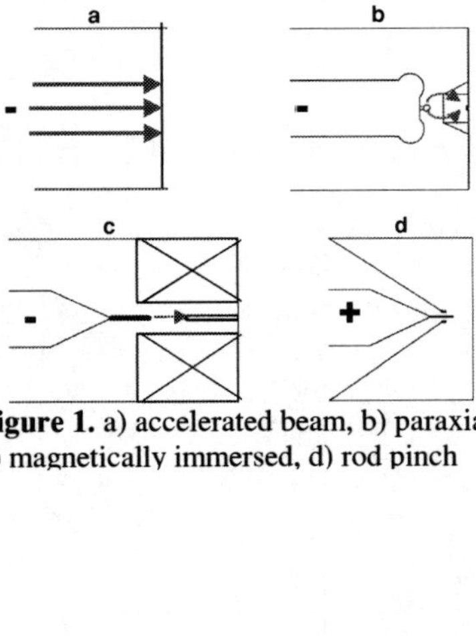

Figure 1. a) accelerated beam, b) paraxial, c) magnetically immersed, d) rod pinch

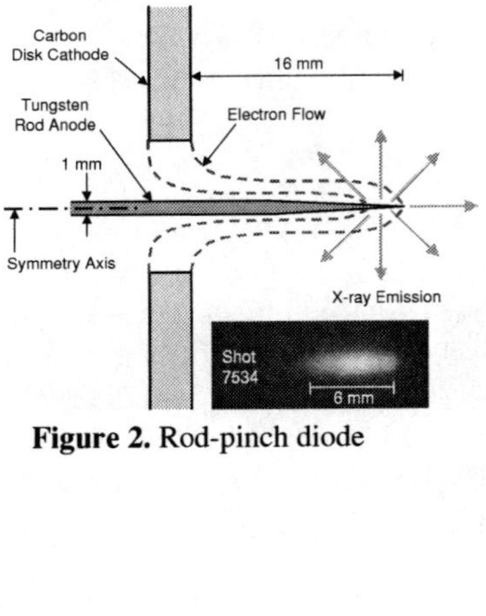

Figure 2. Rod-pinch diode

NEW RADIOGRAPHY GENERATORS

DAHRT: LANL is developing the DAHRT LIA (Fig. 3a) to generate intense x-rays and small spot sizes. Conventional pulsed power drives each cavity which, in turn, accelerates the beam to a higher voltage. The high-voltage electrode of each cavity is inductively isolated from ground by a ferromagnetic core. In phase 1, completed in July 1999, beam parameters of 20 MeV, 2 kA, 60 ns were obtained with a 2 mm spot size (limited by an ion-hose instability caused by counter-streaming low-Z ions from the anode). Phase 2, in installation, has the same voltage, current, and emittance, in a 2-μs pulse. A magnetic chopper will divert all but four 60-ns portions of the pulse that will sequentially hit the converter. An aerial photo of the DAHRT I and II buildings is shown in Fig. 3b. Both machines face a common firing point where high-explosive hydrodynamic experiments can be carried out.[2]

Cygnus: LANL is collaborating with SNL, Titan-PSD and NRL, to produce Cygnus, designed to drive a RPD and to fit into a small underground tunnel at the Nevada test site (Fig. 4). Cygnus uses Inductive Voltage Adder (IVA) architecture, similar to the LIA with the electron beam replaced by a conducting center conductor. All of the voltage is applied to the anode-rod tip, in this case. The Marx generator and low-impedance pulse-forming line are separated from the 3-cell IVA by a long 4-Ω water coax that terminates in an oil manifold. This manifold distributes the voltage to the 12-Ω cells which, in turn, drive a 36-Ω vacuum-transmission line. The prototype module at LANL has met specs and generates a 2.25 MV, 60 kA, 60 ns pulse which produces > 5 R at 1 m with < 1 mm spot size.[7]

RITS: SNL is developing RITS to drive high-intensity x-ray sources. The 3-cavity configuration (Fig. 5), is operational at 4 MV, 160 kA, and 50 ns. RITS is designed to be modularly expandable up to 16 MV. It could then generate multiple pulses at lower voltages by sequentially firing subsets of the cells.[1,4,14]

CRS: LLNL has been developing compact Marx generators for several years. It has recently collaborated with NRL and Bechtel Nevada to couple the Marx to a 50-Ω RPD to generate a 1.5 MV, sub-mm spot with about 0.5 R at 1 m. It is about 1-m diam, 2-m long and will be tested in FY03.[15]

Mercury: NRL recently purchased the Kalif-Helia accelerator[16] from FZK in Germany and will have it operational at NRL as Mercury in Spring 2003. It is a 6-cell IVA which generates a 6 MV, 375 kA, 50 ns pulse. Mercury can be easily converted between positive and negative-polarity operation by reversing the direction of the stalk. It can thus be used to test all radiography diodes including the new low-impedance plasma-filled RPD.

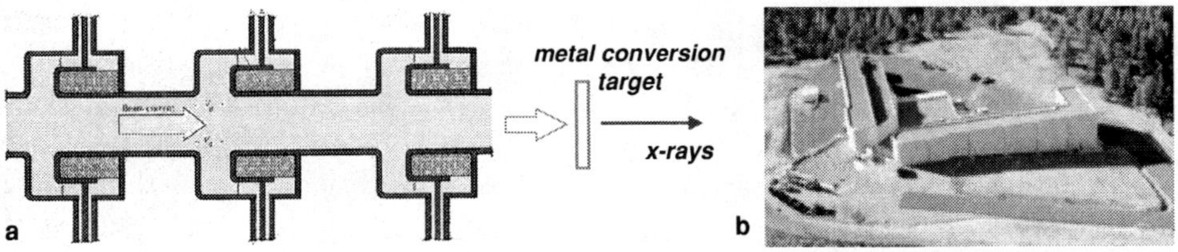

Figure 3. a) Linear Induction Accelerator (LIA), b) Aerial photo of DAHRT I and II

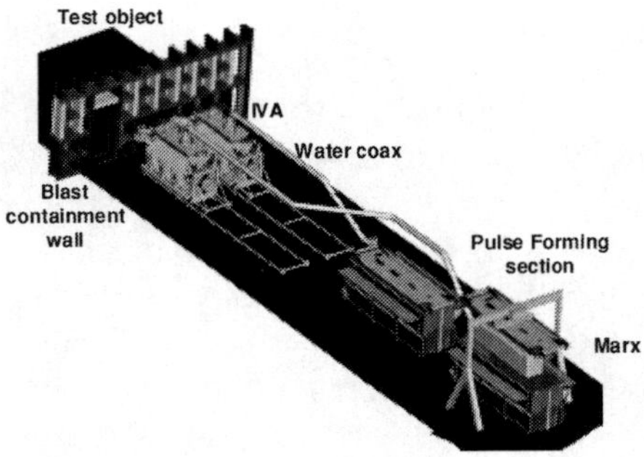

Figure 4. Two Cygnus modules at test site

Figure 5. RITS with 3 cavities

HRF: AWE, in collaboration with SNL and Titan-PSD, is presently developing a multi-MV, multi-axis facility (Fig. 6) that will consist of five 10-stage IVA's producing 14 MV in 100-Ω loads. The goal is to obtain 1000 R at 1 m with a 2 mm spot size. The first IVA is scheduled for operation in 2004. AWE is researching IVA technology and all the usual radiography diodes on their existing facilities.[3,17]

OTHER MAJOR NEW PULSED POWER FACILITIES

ZR: SNL is presently undertaking a major refurbishment of their Z facility called ZR.[18] They are redesigning the Marx charging and triggering circuits, doubling the energy stored in the Marx, optimizing the intermediate store, adding new laser-triggered gas switches, optimizing the water switches and transmission lines, and upgrading the power flow for higher stress. The objective is to obtain more shots, improved precision, and 1.7-times higher current with minimum impact to ongoing programs. The total cost is $60M and ZR is scheduled to be completed in mid FY06. At the projected 26 MA output, the radiated power and energy will be increased to 350 TW and 2.7 MJ.

Decade Half: Titan-PSD, under DTRA sponsorship, is designing a major upgrade to the existing Decade facility located at AEDC. The Decade Quad presently consists of a Marx, four water-transfer capacitors, and four pulse-forming lines coupled to a single z-pinch plasma-radiator-source (PRS) load. The intent is to build an opposing Quad and couple the two together with a low-inductance parallel-plate coupler terminated in a vacuum coupler that drives a PRS. The projected output is 14 MA into a 300 ns implosion-time PRS.[19]

ECF-2: The DGA in Gramat, France is collaborating with Russia and the US to build an inexpensive z-pinch driver (Fig. 7), ECF-2, using magnetic flux compression as a power amplification stage.[20] The generator consists of LTD stages (essentially an IVA cell with built-in low inductance capacitors for pulse forming) based on Russian LTD development at HCEI in Tomsk[21]. There are sixteen 8-stage lines with 12 used to drive the primary generator that compresses the flux created by the other lines. The primary generator stores 3.6 MJ, generates 8 MA in 1 μs, and will be completed in summer 2002. It is a relatively simple modular system with no pulse forming lines, oil, SF_6 or water, only air at atmospheric pressure and vacuum.

Figure 6. Conceptual version of HRF at AWE

Figure 7. ECF-2 at Gramat

SNS: The SNS is included because it is a major new facility that uses plasma, beams and pulsed power. Negative hydrogen ions will be accelerated to 1GeV and stopped in heavy elements to produce the world's most intense neutron source for fundamental research and industrial applications. The project is a $1.4B joint effort between a number of laboratories and will be completed in 2006.[22]

Electra: The purpose of the Electra program at NRL is to develop the technology for an electron-beam-pumped krypton-fluoride laser that can meet fusion energy requirements for rep-rate, efficiency, durability, and cost. At present two 500-KV, 100-kA, 100-ns electron beams are used to produce a 700-J laser pulse. The system can run at 5 Hz for 5 hours.[23]

MAJOR BEAM PROGRAMS

HIF: The US HIF program is run by a Virtual National Laboratory consisting of LBNL, LLNL, and PPPL with support from UMD, NRL, SNL, MRC and MIT. At present, they are attacking important physics issues in short experiments at modest beam powers (1 A, 0.5 – 1 MeV). It is planned to move up to an Integrated Beam Experiment where source-to-target physics can be studied at 5 – 20 MeV. In the longer term, Integrated Research Experiments are planned with multiple beams at 200 - 800 MeV, leading to a $2 - 3 B Engineering Test Facility.[24]

HPM: The US HPM program has a number of civilian applications (such as long-range communication, radar, environmental-waste clean up, bio-toxin neutralization, fusion heating, etc.) in addition to military applications. HPM research consists of sources, pulsed power, antennas, and specific components. The major source research programs are centered at the Air Force Research Laboratory (AFRL).[25] Sources include the Magnetically Insulated Line Oscillator that does not use an external magnetic field and the Relativistic Klystron Oscillator that does. Both generate 1.5 GW. Much research is also being done on carbon velvet cathodes which produce > 300 A/cm^2, for over 850,000 shots with low outgassing. Extensive 3-D modeling is performed on all the sources. Other devices that are being studied at AFRL and elsewhere in collaboration with the US include super PASCO and Relativistic Magnetrons[25], Relativistic Backward Wave Oscillators[26], Marx driven UWB HPM[27], and Reltrons[28].

SUMMARY

The focus of the BEAMS2002 meeting has significantly changed since BEAMS75. Pulsed hydrodynamic radiography is presently the major driver for intense beam research and new pulsed power facilities with almost 1/3 of the US papers presented at BEAMS2002. There are also a number of major new pulsed power facilities and major ongoing beam programs. The exciting papers presented at BEAMS2002 show that our field is moving in many new directions and undertaking challenging science and technology.

ACKNOWLEDGEMENTS

Thanks to J. Maenchen, M. Burns, C. Ekdahl, T. Goldsack, D. McDaniel, P. L'Eplattenier, P. Sincereney, D. Rej, J. Sethian, G. Logan, T. Spencer, D. Mosher, B. Weber, F. Young, and R. Allen for providing materials and/or helping with the paper.

REFERENCES

1. J.E. Maenchen *et al.*, these proceedings.
2. M.J. Burns *et al.*, E. Rose *et al.*, Y.J. Chen *et al.*, C.A. Ekdahl *et al.*, P.J. Christenson *et al.*, J. McCarrick, T.P. Hughes *et al.*, and H.A. Davis *et al.*, these proceedings.
3. T.J. Goldsack *et al.*, and J. O'Malley, these proceedings.
4. K. Hahn *et al.*, these proceedings.
5. D. Rovang *et al.*, and G.M. Cooper *et al.*, these proceedings.
6. G. Cooperstein *et al.*, *Phys. Plasmas* **8**, 4618-4636 (2001).
7. J.R. Smith *et al.*, these proceedings.
8. A.G. Pearce *et al.*, and G. Cooperstein *et al.*, these proceedings.
9. R.J. Commisso *et al.*, *IEEE Trans. Plasma Sci.* **30**, 338-351 (2002).
10. R.J. Commisso *et al.*, these proceedings.
11. B.V. Oliver *et al.*, S.B. Swanekamp *et al.*, P.F. Ottinger at al., and D. Mosher *et al.*, these proceedings.
12. D.D. Hinshelwood *et al.*, these proceedings.
13. B.V. Weber *et al.*, and S.D. Strasburg *et al.*, these proceedings.
14. P. Corcoran *et al.*, V.L. Bailey *et al.*, and D.L. Johnson *et al.*, these proceedings.
15. W. DeHope *et al.*, *2001 IEEE Nuclear Science Symposium* (San Diego, CA, November 2001).
16. P. Hoppe, *et al.*, *Proc. 2001 Pulsed Power Plasma Science Conference*, Las Vegas, NV (2001), IEEE Cat. No. 01CH37251, p. 596.
17. S.G. Clough *et al.*, these proceedings.
18. M.G. Mazarakis *et al.*, S.E. Rosenthal *et al.*, and T.D. Pointon *et al.*, these proceedings.
19. P.W. Spence *et al.*, these proceedings.
20. P. L'Eplattenier *et al.*, these proceedings.
21. A.A. Bastrikov *et al.*, these proceedings.
22. D.J. Rej, these proceedings.
23. F. Hegeler *et al.*, M.C. Myers *et al.*, and D.V. Rose *et al.*, these proceedings.
24. B.G. Logan these proceedings, and I.D. Kaganovich *et al.*, and A. Tauschwitz *et al.*, these proceedings.
25. T.A. Spencer, these proceedings, and K.L. Cartwright *et al.*, these proceedings.
26. N.F. Kovalev *et al.*, these proceedings.
27. J.R. Mayes and W.J. Carey, these proceedings.
28. S.H. Choi and E Schamiloglu these proceedings, and S.D. Putnam *et al.*, these proceedings.

High Power Particle Beams and Pulsed Power for Industrial Applications

Hansjoachim Bluhm, Wladimir An, Vladimir Engelko*, Harald Giese, Wolfgang Frey, Annette Heinzel, Peter Hoppé, Georg Mueller, Christoph Schultheiss, Josef Singer, Ralf Strässner, Dirk Strauß, Alfons Weisenburger, Fritz Zimmermann

Forschungszentrum Karlsruhe, Institut fuer Hochleistungsimpuls- und Mikrowellentechnik
P. O. Box 3640, D-76021 Karlsruhe
**Efremov Institute of Electrophysical Apparatous, 189631 St. Petersburg, Russia*

Abstract. Several industrial scale projects with economic and ecologic potential are presently emanating from research and development in the fields of high power particle beams and pulsed power in Europe. Material surface modifications with large area pulsed electron beams are used to protect high temperature gas turbine blades and steel structures in Pb/Bi cooled accelerator driven nuclear reactor systems against oxidation and corrosion respectively. Channel spark electron beams are applied to deposit bio-compatible or bio-active layers on medical implants. Cell membranes are perforated with strong pulsed electric fields to extract nutritive substances or raw materials from the cells and to kill bacteria for sterilization of liquids. Eletrodynamic fragmentation devices are developed to reutilize concrete aggregates for the production of high quality secondary concrete. All activities have a large potential to contribute to a more sustainable economy.

INTRODUCTION

Several industrial scale projects with economic and ecologic potential are presently emanating from research and development in the fields of high power particle beams and pulsed power in Europe. At Forschungszentrum Karlsruhe most of this work is now conducted in the frame of a novel program on sustainability and technology. One aim of this program is to develop technologies that can contribute to a more sustainable economy. It is obvious that the way of technology use determines how sustainable an economy can be. Future generations will be confronted with the emissions released by our technologies, they will face exhausted deposits and disposed wastes. On the other hand they will also encounter the achieved wealth, the technically gained larger independence from nature and the accumulated technical knowledge. This double face of technology needs careful systems analysis of all socio-economic aspects relevant for sustainability in each special case. However it is generally accepted that sustainability requires drastic changes in many areas of industrial activity. Pulsed power and beam technologies can add to these efforts in certain domains. Here we shall discuss a few selected examples where this has been started.

CP650, *Beams 2002: 14th International Conference on High-Power Particle Beams*, edited by T. A. Mehlhorn and M. A. Sweeney

ELECTRON BEAM SURFACE TREATMENT

Material surface treatment with large area pulsed electron beams offers a wide range of applications for the improvement of material surface properties such as wear, corrosion and oxidation resistance. It is based on the rapid adiabatic melting of a thin surface layer followed by fast quenching of the molten layer through heat conduction into the unaffected cold bulk material [1]. The pulsed electron beams produced with the facilities GESA I and GESA II at Forschungszentrum Karlsruhe can melt any metallic surface of area 50 to 300 cm² with a single pulse. Typical melting depths are up to 40 µm for the GESA I beam and up to 120 µm for the GESA II beam, corresponding to the range of 150 and 400 keV electrons respectively. The main parameters of GESA I and GESA II are listed in Table 1.

TABLE 1. Main parameters of GESA I and GESA II.

Parameters	GESA I	GESA II
Accelerating voltage	50-150 kV	200-400 kV
Power density	up to 2 MW/cm²	up to 6 MW/cm²
Beam diameter on target	5-10 cm	4-6 cm
Pulse duration	4-40 µs	5-250 µs

In both facilities the electron beam is produced with the help of a field emission cathode consisting of 700 carbon fiber bushels. The electrons are extracted through a grid and accelerated in a triode configuration. A combination of ballistic focusing and adiabatic compression with the help of solenoid fields is used to guide the beam to the target which is about 0.5 m distant from the anode. Further details of the facilities can be found elsewhere [2].

Two applications of pulsed electron beams are presently under investigation: Reduction of corrosion attack on steel in Pb or Pb/Bi cooled accelerator driven nuclear reactor systems (ADS) and improvement of protection layers on high temperature gas turbine blades.

Among the most embarrassing wastes that we leave to future generations are long living fission products, plutonium and the higher actinides from nuclear power reactors. Therefore strong efforts are presently undertaken in Europe, USA, Japan, Korea, and other countries to transform these wastes into products with shorter half-life or less radiotoxicity. Because of their hard neutron energy spectra and their inherent safety features accelerator driven subcritical nuclear reactors with Pb or Pb/Bi cooling are presently developed for this task. A key problem in the development of such systems is steel corrosion in the liquid metal. Especially Ni is easily dissolved in liquid Pb or Pb/Bi at temperatures around 600 °C or above. A protective oxide scale can form on the steel surface if a controlled oxygen concentration around $8 \cdot 10^{-6}$ at. % is maintained in the liquid heavy metal. However, this layer consisting of magnetite and FeCr-spinels grows rather fast, and thus can lead to progressing steel oxidation and long term instability. This can be avoided by alloying Al into the steel surface with the help of the pulsed electron beam from GESA. For that purpose a thin Al-layer is deposited onto the steel surface and co-melted with the underlying steel layer of several 10 µm thickness. The Al becomes mixed into the molten steel due to convection and due to the rapid quenching remains frozen in the steel matrix. With oxygen dissolved in the liquid Pb/Bi the alloyed Al leads to the formation of a thin protective alumina layer that grows very slowly and covers the underlying steel surface against corrosion attack. Corrosion experiments in the Pb/Bi test loops at IPPE Obninsk and Promethey, St. Petersburg at flow velocities around 1 m/s showed that only those steels where a layer with 10 % of Al had been alloyed into the surface survived the test without damage after 4000 h (Fig. 1).

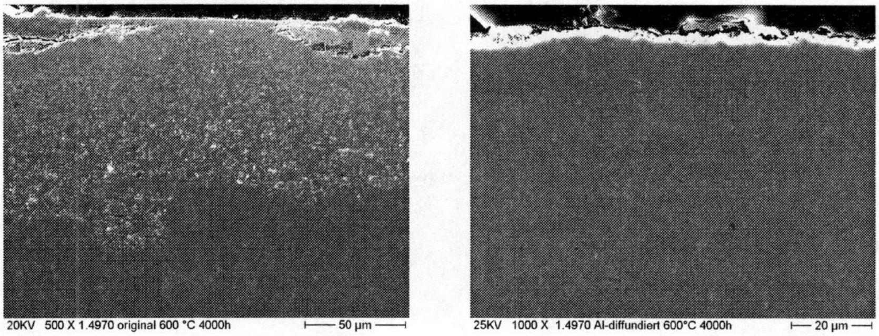

FIGURE 1. Results from tests of 1.4970 steel probes in a Pb/Bi-loop at 600 °C after 4000 h: left, untreated surface, showing strong corrosion; right, Al-alloyed surface showing no attack.

Increased efficiencies in the utilization of energy can make an important contribution to a more sustainable energy production. Gas fired power plants are among those with the highest thermodynamic efficiencies. A further increase could be achieved if a higher gas temperature could be realized in the gas turbine. This requires thermal insulation of the turbine blade surface to keep the temperature of the base material and of the oxidation protection layers within acceptable limits. Actually high temperature gas turbine blades need to be covered with an oxidation resistant alumina former which in most cases is a low pressure plasma sprayed super-alloy MCrAlY. A thermally grown oxide scale (TGO) develops on the surface of this layer, which does not only protect the underlying material against further oxidation but also serves as a bonding layer for a thermal barrier coating (TBC) consisting of yttria stabilized zirconia (YSZO). The lifetime of these systems is limited by failure of the TGO leading to spallation of the TBC. This failure mode depends on the morphology of the TGO. Since the plasma sprayed MCrAlY layer is generally very rough the coating is mechanically polished. Surface treatment with pulsed electron beams also leads to surface smoothing but in addition creates an amorphous or nanocrystalline layer structure. As shown in Fig. 2 both treatments lead to very different oxide scale growth. θ-Al_2O_3 blades appear on the mechanically polished surface while pure α-Al_2O_3 grows on the electron beam treated surface. Depositing the TBC onto this surface results in regularly grown YSZO columns with minimal transverse tensions. Therefore a longer lifetime of these coatings can be expected. Entire turbine blades have been treated at the GESA I facility which are presently undergoing longterm oxidation tests at 950 and 1000 °C. After successful completion of these tests a decision about the construction of an industrial treatment facility will be made. Design studies for such facilities have begun. Fig. 3 shows the conceptual design.

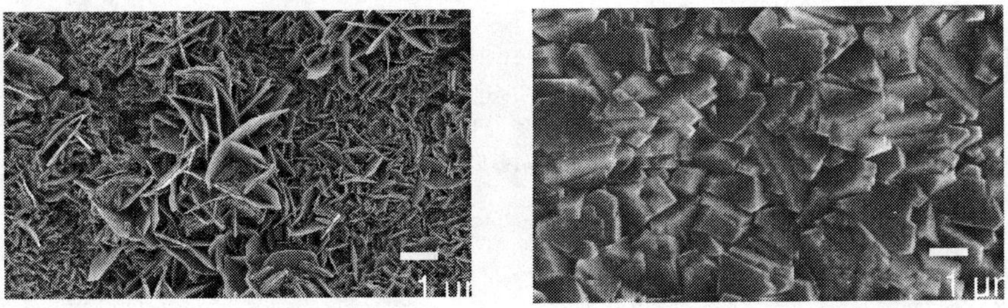

FIGURE 2. Surface morphology of the oxide scale grown on the surface of a NiCrAlY oxidation protection layer after 200 h at 950 °C. Left with conventional polishing of the plasma sprayed layer, right with GESA electron beam treatment.

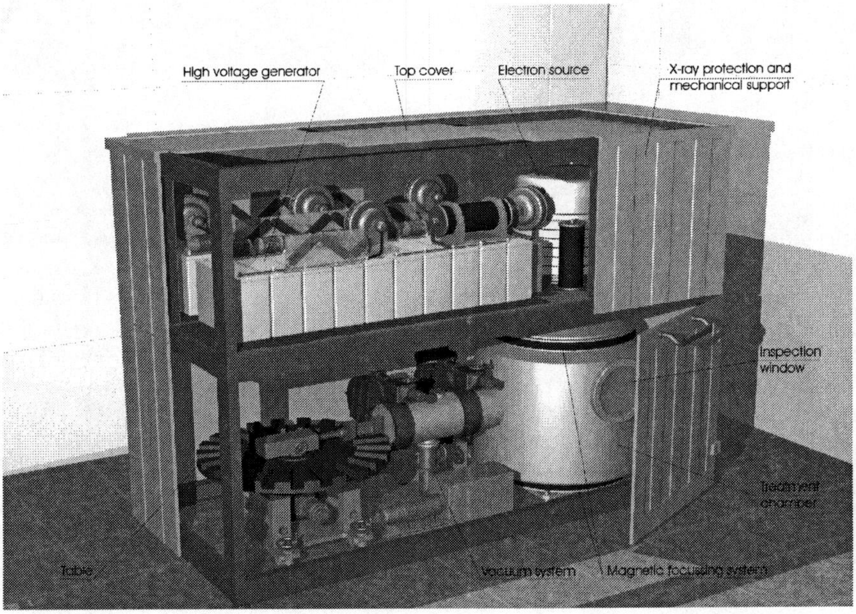

FIGURE 3. Layout of an industrial GESA-facility for the treatment of turbine blades. Daily production rate: 20 blades; pulse rate: 0.1 Hz.

Some applications –especially the treatment of machine parts- require spatially and temporally homogeneous and stable beams. It has been reported previously [3] that the interaction between the beam and the target results in reflected electrons and ions both moving backward into the triode and changing the beam profile from a more rectangular current density distribution to a Gaussian one. Moreover it has been discovered that beam oscillations with frequencies of a few MHz occur whose structure and amplitude depend on the mirror ratio, i.e. the ratio of the magnetic field at the target and at the anode. Two explanations for these oscillations are presently investigated: rotation of the entire beam around a common axis and a two stream instability. Beam rotation can be caused by an azimuthal drift in the radial electric and axial magnetic field if the beam axis does not coincide with the magnetic field axis. A two stream instability can develop due to a counter-stream of ions reflected in the potential well which forms as electrons are extracted from the expanding plasma [4]. A fast diagnostic imaging the x-ray emission from a metallic converter disk with the help of a scintillator and a streak and framing camera is used to distinguish between both hypothesis.

COATING OF MEDICAL IMPLANTS BY CHANNEL SPARK DEPOSITION

Recently, revived interest has occurred to use the electron beam from a channel spark discharge for the coating of temperature sensitive medical implants with thin layers of hydroxylapatite or bioactive glass. Subcutaneous and transcutaneous medical implants frequently lead to inflammation because a tight encapsulation is not achieved. Soft tissue anchoring can be improved by the afore mentioned coatings.

The channel spark is a highly efficient low cost pulsed electron beam system that produces the coatings by ablation from a target and thus maintains their stoichiometric composition. It is a low pressure (~1 Pa) gas system consisting of a hollow cathode and a ceramic tube that separates it from the anode. The electron beam is generated by formation of a plasma in the hollow cathode. It is accelerated through the ceramic channel and space charge neutralized in the background gas. Due to the magnetic selffield the beam is compressed and rather high current densities can be achieved ($\sim 10^5$ A/cm²).

The kind of plasma production influences both the properties of the beam and the lifetime of the system. To increase the ablation efficiency, the reproducibility and the lifetime of the system a Penning discharge has been developed for plasma generation and a tiny gas leak has been introduced in the vicinity of the hollow cathode to maintain a small pressure difference between the hollow cathode and the tube end [5,6].

EFFECTIVE USE AND REGENERATION OF NATURAL RESOURCES

Food production frequently requires large amounts of energy and chemicals to extract substances from plant cells or to sterilize and prepare food products for long term storage.

We have demonstrated that the disintegration of sugar beet cells with the help of strong pulsed electric fields leads to both economic and ecologic advantages [7]. It has been known since several decades that electric fields of 1-100 kV/cm can induce irreversible pores in the membrane of biological cells. This process has been called electroporation. The necessary field strength depends on the cell dimensions and on the pulse duration. To open plant cells field strength of greater than 1 kV/cm are required at pulse durations of a few μs. To destroy the membrane of bacteria up to 100 kV/cm become necessary. Applying an electric field to a cell embedded in a conducting suspension leads to polarization and creates a potential drop across the insulating membrane. Electroporation occurs if the cell membrane potential exceeds 0.7-1 V, corresponding to fields above 1 MV/cm [8].

To demonstrate the potential advantages of plant cell treatment with pulsed electric fields the KEA (Karlsruher Elektroporations-Anlage) facility has been built. It is based on a Marx generator to produce 300 kV pulses with a repetition rate of 20 Hz. These pulses are connected to 4 electrode pairs axially and azimuthally distributed in the wall of a 20 cm plastic tube containing the sugar beets to be treated in a conducting water bath. With this facility we have been able to demonstrate that the common process of thermal denaturalization at 72°C can be replaced by electroporation at much lower temperatures and with negligible additional energy input. Since the extraction of sugar from the treated beets can now occur at lower temperatures less undesirable substances enter into the raw juice. Consequently also the subsequent precipitation of these substances requires smaller amounts of chemicals. Since these advantages were so convincing we are now constructing, together with an industrial partner, a much larger device with a throughput of 2000 t/d that shall confirm the anticipated advantages also on a factory scale. Fig. 4 shows the hoisting drum which has already demonstrated the expected hauling capacity.

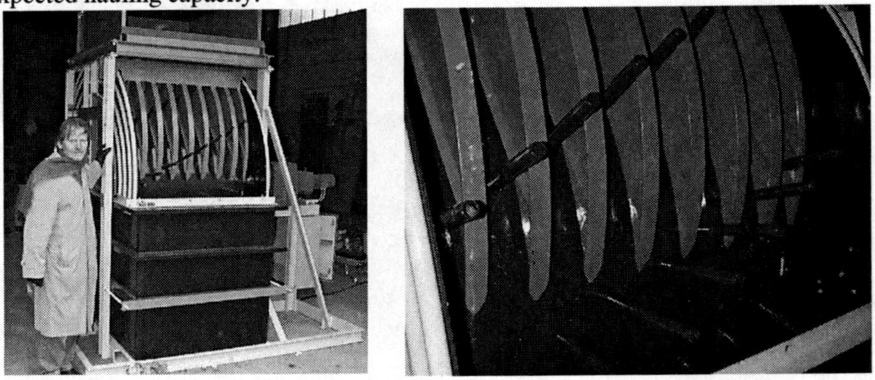

FIGURE 4. Hoisting drum for the transport of sugar beets with a capacity of 80 t/h. Electrodes will be placed in the lower vessel with counter electrodes on the drum. Sugar beet input is on top. The outlet is in the middle part of the apparatus.

Pasteurization of liquid foods is a process with large energy consumption and in addition destroys vitamins and flavour. Strong electric fields can kill microorganisms and thus sterilize liquids in a cold way, preserving vitamins and flavouring. A program has been started to investigate the effectiveness of this process and its scaling to large throughputs. First results achieved with e. coli bacteria indicate that the pulse shape is an important parameter in this treatment.

PHB (polyhydroxybutyrat) is a bio-polymer that can be produced from bacteria with biotechnical methods. It has similar material properties as polyethylene but is biodegradable, biocompatible and resorptive and therefore can be used e.g. with medical implants. Presently it is gained through a chemical enzymatic process which disintegrates the bacterial cell membrane and gives access to the granolas in the cell consisting of PHB [9]. The bacteria are grown in large quantities in bio-reactors filled with a nutrient solution. To enhance the efficiency of disintegration we plan to combine the enzymatic dissolution by electroporation.

Clean water, is one of the basic conditions for the persistence of human life. It is consumed in large quantities and released as waste water both in industrialized and developing countries. A sustainable use requires cleaning of the sewage water in clarification plants before release into rivers. Purified water from such plants may still contain large amounts of pathogenic microorganisms and poisonous chemical substances. Pulsed power technologies have the potential to remedy some of these problems in an ecologically attractive way. A post-treatment of clarified water from sewage plants with pulsed electric fields can reduce the bacterial freight. We are presently investigating the feasibility and the efficiency of such treatments.

RECYCLING OF BUILDING MATERIALS WITH ELECTRODYNAMIC FRAGMENTATION

Building activities are connected with the largest material flows in industry. Approximately 260 million tons of concrete per year are presently produced in Germany (~ 3 t per inhabitant and year). Corresponding quantities are consumed in other industrialized countries. The raw materials gravel, sand and cement are entirely taken from natural resources. On the other hand 58 million tons of rubble are created per year in Germany. Reutilization of these materials is mainly restricted to secondary constructional operations like backfilling, noise protection dams, etc.. Only a very small fraction (less than 3%) is presently reused for building construction. Reutilization without degradation requires an improved separation into sand, gravel and cement. This cannot be achieved with conventional mechanical breakers. In addition these recycling facilities produce large quantities of fine material (20-40 %) with grain sizes < 2mm which cannot be reused as aggregate for concrete fabrication.

Electric impulse fragmentation is however well suited to separate concrete into its initial components. The method is based on the initiation of a pulsed electric discharge inside the solid material. With pulse amplitudes of ~300 kV, material layers of 2-4 cm can be punctured. Specific energy deposition of up to 100 J/cm at a GW power level leads to pressure buildup of up to 10^{10} Pa in the discharge channel. Pressure waves and radially propagating tracks are launched into the solid body, which can lead to the separation of inclusions from the matrix or to detachment at material boundaries. To induce the discharge through the solid material it must be immersed in a dielectric liquid with higher dynamic breakdown strength which for µs-pulses with fast rise times can be water. Concrete is a composite heterogeneous material with a large fraction of gas filled pores and micro-cracks between the aggregates and the cement matrix. Since the acoustic inhomogeneities are rather large in concrete ideal conditions exist for the separation by pressure waves. In addition the pressure pulse mainly creates tensile and shearing forces at the discharge channel, conditions at which the strength of concrete is low. Therefore cracks will originate and spread from the channel. Gravel and sand can be separated from the matrix with relatively small cement adhesions. Thus recycling of these products at a high quality level becomes possible [10].

FIGURE 5. FRANKA-Stein: general view of the facility with reaction chamber and part of the material removal system.

Most industrial applications need large throughputs to become economical. For concrete it has been estimated that a productivity of about 100 t/h is necessary to achieve that goal. It is obvious that the quantity of material that can be treated per arc channel is limited and cannot be increased much above that for an optimized laboratory device. Of course this quantity is also dependent on the requested degree of purification. FRANKA-Stein (300 kV, 500 J, 12 Hz) has been constructed and operated to determine these parameters at continuous treatment and to test different schemes to control the dwell time of the material in the arc zone(Fig. 5). Another purpose of this facility was to develop a water reprocessing system. From the experiments on FRANKA-Stein we can conclude that the throughput per arc is limited to something of the order of 1 t/h. Thus about 100 arcs are necessary to realize a throughput of 100 t/h. This does however not mean that all components need to be multiplied by 100. However, it suggests that appreciable investment is necessary to industrialize this technology.

REFERENCES

1. Engelko, V., et al., Proc. 9th Int. Conf. On high Power Particle Beams, edited by D. Mosher and G. Cooperstein, Washington, DC 1992, Vol. 3, pp. 1935-1941
2. Engelko, V., et al., *Vacuum* **62**, 211-216 (2001).
3. Engelko, V., Mueller, G., *Vacuum* **62**, 97-103 (2001)
4. Dorokhov, I.V., et al., *Sov. J. Plasma Phys.* **13**, 829-833 (1987)
5. Frey, W., et al., "A new trigger device for channelspark systems based on aPenning discharge", these Proceedings
6. Schultheiss, Ch., et al., "Progress in channel spark development and application of pulsed electron beam deposition (PED) in the field of medical coating work", these Proceedings
7. Schultheiss, Ch., et al., "Processing of sugar beets with pulsed electric fields", accepted for publishing in *IEEE Transactions of Plasma Science*, (2002)
8. Weaver, J.C., Chizmadzhev, Yu.A., *Bioelectrochemistry and Bioenergetics* **41**, 135-160 (1996)
9. Babel, W., et al., *Plaste und Kautschuk* **37**, 109-115 (1990)
10. Bluhm, H., et al., *IEEE Transactions on Dielectrics and Electrical Insulation* **7**, 625-636 (2000)

Status of Pulsed Power/Beams/Z-pinch Researches in Japan

Kiyoshi Yatsui

Extreme Energy-Density Research Institute, Nagaoka University of Technology, Nagaoka, Niigata 940-2188, Japan

Abstract. Fruitful progresses have been achieved in pulsed power technology and its applications in Japan in materials science, environmental, medical, biological, etc. At Nagaoka Univ. of Tech., main efforts are devoted in materials science such as the preparation of thin films or ultrafine nanosized powders (UNP) by high-density ablation plasma produced by pulsed ion beam evaporation. Alternatively, UNPs were produced by rapid cooling of the plasma produced by pulsed wire discharge (PWD) as well. The ablation plasma is also very effective to build-in metals (e.g., tungsten, aluminum, copper) in *via holes* of LSIs. Flue-gas treatment of NO_x is studied by pulsed, relativistic electron beam. At Tokyo Inst. of Tech., studies are carried out on pulse discharge using powders, soft X-ray by capillary discharge, and development of pulse power sources for industrial applications. At Kumamoto Univ., studies are concerned on NO_x removal by pulsed power, gas-puff z-pinch, and opening switches. Converting the pulsed power to z-pinches, charged particles beams, or plasma foci have also been carried out in many universities. Systematic studies are carried out under good collaboration among many universities, national laboratories and companies.

INTRODUCTION

Due to rapid growth of pulsed power technology in recent years, it is available easily to achieve huge power. Such a big power can be used directly as direct z-discharge, but there are wider applications if the power could be converted into charged particle beams (pulsed relativistic electron beam or pulsed ion beam) or lasers [1]. In fact, at its early stage, it was mainly directed to pulsed radiation source or energy driver for inertial confinement fusion. Nowadays, however, new applications have been available such as in materials science, flue-gas treatment, biological, and civil engineering, which will be never achieved by conventional technologies. In Japan, many scientists in universities, colleges, national laboratories, and private companies are interested in R&D of pulsed power technology and its applications. Recent activities in this particular field of interests in Japan will be quickly reviewed.

ACTIVITIES

Although there are a lot of activities, they will be introduced briefly according with the field of interests.

CP650, *Beams 2002: 14th International Conference on High-Power Particle Beams,*
edited by T. A. Mehlhorn and M. A. Sweeney

a) Development of pulse power supply/generator

High rep-rate pulsed power supply is energetically developed for actual applications. In Nagaoka Univ. of Technology (NUT), new power supplies have been developed [1]. Using 8-series, 6-parallel MOS-FETs, we have achieved 2-MHz operation with voltage of – 5 kV, current of 75 A, pulse width of 250 ns, rise time of 24 ns, fall time of 26 ns, and jitter of 1 ns. Furthermore, low-jitter, high rep-rate power supply has been developed in NUT by combining IGBT (insulated gate bipolar transistor) and magnetic switch. Very stable operation has been achieved with the output of 16 kV on 5-Ω load, jitter less than 1.5 ns at the rep-rate of 2 kHz. Using SI thyristor, where turn-off is available in addition to turn-on, high rep-rate power supply has been developed for the flue-gas treatment in automobile engines. High rep-rate pulse power generator, "ETIGO-IV", has been operational in NUT [2]. It can deliver to a matched load of the output of 400 kV, 13 kA ,120 ns at rep-rate of 1 Hz (see Fig. 1).

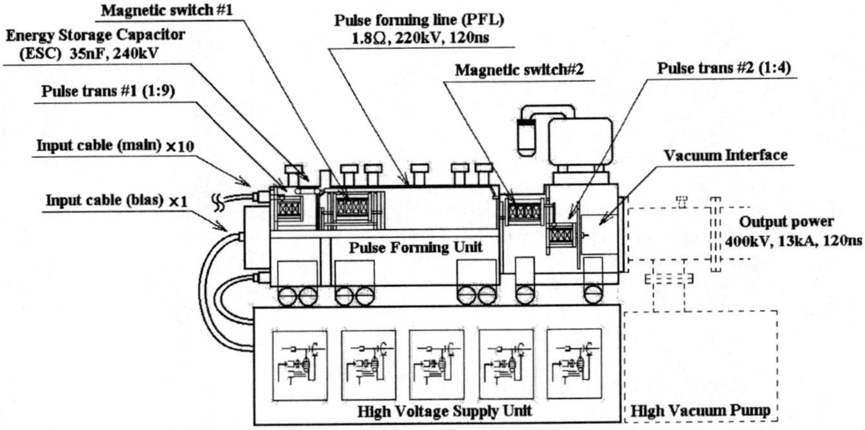

Figure 1. Schematic of "ETIGO-IV".

In Kumamoto Univ., all-solid-state pulse power generator is developed by semiconductor opening switch, producing the output of 150 kV, 60 ns on 300-Ω load.

In Toyama Univ., bipolar pulse accelerator was developed, which consists of two acceleration gaps and a drift tube [3]. Using a bipolar pulse, ion species selected can be accelerated. A Br-type magnetically insulated diode is used with coaxial gas puff plasma gun. Initial plasma of nitrogen (100 A/cm^2) is accelerated by negative pulse in the first gap, and post accelerated by positive pulse in the second gap (see Fig. 2).

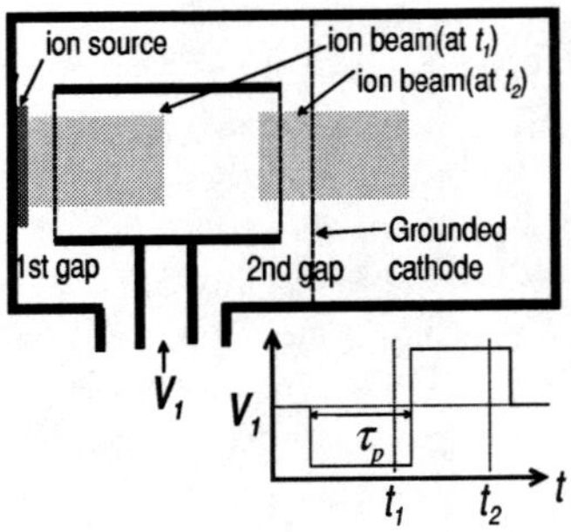

Figure 2. Principle of Bipolar Pulse Accelerator.

Laser ion source is developed on induction adder in Tokyo Inst. of Tech. (TIT) [4]. Low-voltage DC extraction is achieved with low emittance after drifting 65 cm.

b) Z-pinches

A new idea is proposed in TIT to produce high-density plasma by pulsed high current discharge of powders [5]. Powder is transiently injected by an electrostatic powder injection system. It composes of a pair of parallel-plate electrodes, one of which had a hole at the center for ejecting particles. Spherical copper powder (100 μm in diam.) was used. The discharge voltage and current are - 20 kV and 20 kA during 1 μs. Basic process of the plasma was observed by image-converter camera and laser shadowgraph. Powder plasmas can be applied in z-pinches, space, soft X-ray source, high-voltage insulation, and thin film deposition in materials science. Development of radiation source by capillary discharge is carried out by another group at TIT, for imaging, medical treatment and lighography [6].

In Nihon Univ., experimental studies are carried out on compressional z–pinch on SHOTGUN device (Fig. 3) [7]. Soft X-ray is produced by high-density, high temperature plasma. The power supply is 30 kV, 24 μF (10 kJ) with the maximum current of 300 kA. Good correlation was obtained between hot spots of X-ray radiation and spatial pattern of Rayleigh-Taylor instability. The instability wavelength can be explained by taking finite Larmor radius effect into account. Uniform z-pinch can be formed by rapid contraction of the plasma without heating. In argon z-pinch plasma, hot spots were due to K-shell radiation of He-like ions. Contraction of z-pinch was suppressed by a magnetic field, resulting in the K-shell radiation, and enhancing L-shell one. L-shell radiation seems to be promising for lithography or microscopy. In addition, dense z-pinch was proposed for the reduction of growth rate of instabilities [8].

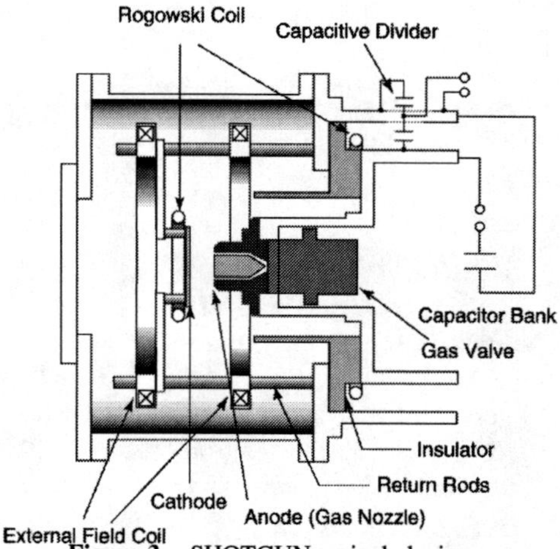

Figure 3. SHOTGUN z-pinch device.

In Kyushu Univ., the influence of voltage polarity is studied on the pinch characteristics of gas-puff z-pinch [9]. Here, a plane or mesh electrode is used as an opposite electrode of the gas-puff electrode. A cone-shaped rod is also used as the opposite electrode to reduce the rebound of the injected gas from the surface. However, electric field distribution between the electrodes is not uniform and the field is concentrated around the tip of rod electrode. Thus, z-pinch characteristics might be affected by the polarity on the cone electrode, which was studied by the measurements of soft X-ray, hot spots formation time, and the pinching behavior.

In Kumamoto Univ., z-pinch X-ray radiation is studied by wire array system [10]. To improve the current distribution, four wires of tungsten (0.1 mm in diam.) were exploded in vacuum. The pulse power was inductive voltage adder giving 160 kA with a quarter period of 1.8 μs. Figure 4 shows the current through and the voltage across the wires. The current is inhomogeneous especially after the wire becomes plasma. Simultaneously, the inhomogeneity remains throughout the discharge.

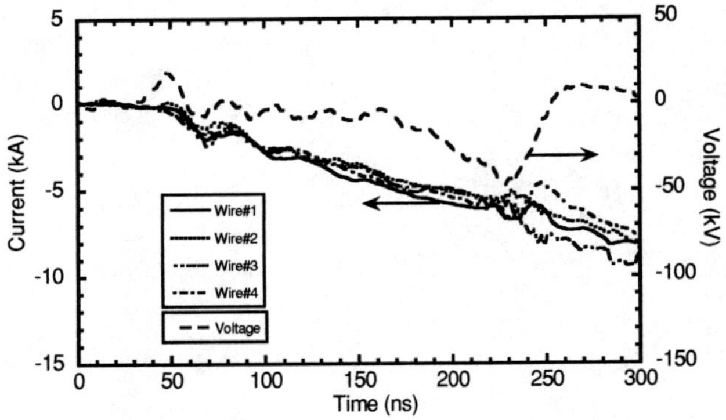

Figure 4. Current and voltage on four wires during explosion.

c) Pulsed electron beam

In Kanazawa Univ., efforts are carried out on the formation of ultrshort relativistic electron beam using multi-stage autoacceleration to study superradiance (SR). Since the SR is driven by an electron bunch whose length is shorter or comparable with interaction length, the beam duration less than 1 ns is needed for mm-waveband SR. Experimentally, the electron beam with 1.1 MeV, 1 kA, 1 ns was obtained by multistage autoacceleration using a series of passive cavities with decreasing length [11].

d) Applications in materials science by ion-beam ablation plasma

In NUT, main efforts are devoted in materials science such as the preparation of thin films or ultrafine nanosized powders (UNP) by high-density ablation plasma produced by pulsed ion beam evaporation (IBE) [12]. Very effective preparation of many kinds of thin films has been demonstrated. The IBE has been found to very useful to synthesize UNP. An interesting result found recently will be introduced on UNP of silicon. Figure 5 shows the *as-deposited* results of (a) size distribution of silicon powders and (b) photoluminescent spectrum, which were obtained at helium gas of 1 Torr, substrate temperature of –10 °C deposited on quartz substrate. We clearly observe blue light emission near 420 nm from silicon powders. Changing the plasma parameters, it was found to change the wavelength of the emission. Alternatively, UNPs were produced by rapid cooling of the plasma produced by pulsed wire discharge (PWD) as well [2]. The ablation plasma is also very effective to build-in metals (e.g., tungsten, aluminum, copper) into *via holes* of LSIs [1].

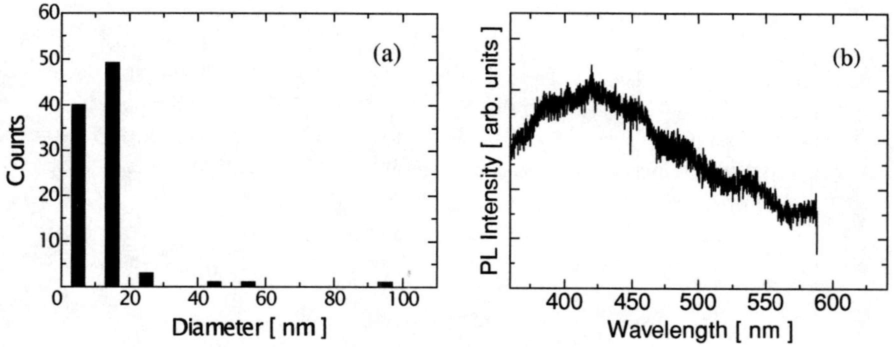

Figure 5. (a) Size distribution and (b) PL spectrum of Si powders.

In Himeji Inst. of Tech., surface modification by pulsed proton beam is studied with 180 keV, 4 J/cm^2, and 65 ns [13]. From EPMA analysis of Al-30%Si target irradiated, they found Si atoms are solved into Al matrix. The result is shown in Fig. 6, where Si content solved in Al determined by XRD is plotted as a function of the number of shots irradiated. After being irradiated by 50 shots, the solution rate attains 5.7 %, which exceeds the solid solubility limit of 1.59 %. Furthermore, the hardness of the layer treated was Hv 120, corresponding to twice the untreated layer.

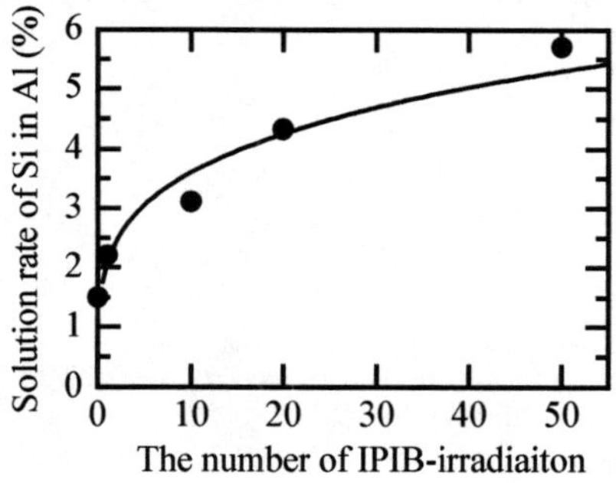

Figure 6. Solution rate of Si in Al.

e) Flue gas treatment

In NUT, flue-gas treatment of NO_x is studied by pulsed, relativistic electron beam (REB) [1,14]. The distant gas chamber is filled up with dry-air-balanced NO gas mixture at 270 kPa, and is irradiated by REB (2 MeV, 30 A, 35 ns) passing through a 1.6-m long atmosphere. With the initial NO concentration of 88 ppm, ~ 70 % of NO_x is successfully removed by firing 10 shots of the beam. The NO_x removal efficiency has been found to be 50 ~ 155 g/kWh. In Osaka City Univ., similar experiments are carried out by low-energy pulsed E-beam [15] (200 keV, 140 A, 2.2 A/cm^2, 500 ns). The beam irradiates gas cell filled at 1 atm N_2 gas with the initial concentration of 200 ppm. The removal efficiency of 190 nmol/J was obtained by the pulsed beam irradiation.

f) Microwave generation

In NUT, microwave generation is carried out on virtual cathode oscillator (vircator). The microwave generation was available by 1-Hz operation, producing the power of ~ 450 MW around the frequency of 2.5 ~ 3.5 GHz [16]. In Niigata Univ., the microwave generation is done by backward wave oscillator [17]. Another group is carrying out the development of gyrotron using intense pulsed electron beam [18].

g) Biological applications by pulsed electric field

Effect of pulsed electric field on Bacilus Stearothermophilus has been studied in Kumamoto Univ. [19]. The survival rate was decreased at higher electric field.

REMARKS

Finally, some information will be added on the activities of pulsed power technology and applications in Japan because we have good network in Japan in this society. We have annual meeting on pulsed power technology and its applications normally in NIFS (National Institute of Fusion Science), which is supported by the NIFS. Normally, 50 scientists participate to discuss approximately 35 papers, whole papers of which are published from NIFS as a NIFS-PROC series. In addition, we have 3 or 4 regular meetings in Japan supported by the Institute of Electrical Engineers of Japan. Furthermore, in three countries in Far East, namely Korea, Japan, and P. R. China, we have International Symposium on Pulsed Power and Plasma Applications once a year. This year, the 3rd Symposium will be held in Mianyang, Sichuan, P. R. China, on Oct. 23 and 24. Scientists who are interested in this field are welcome to participate meetings involved above.

REFERENCES

1. K. Yatsui and W. Jiang: "*Pulsed Electromagnetic Energy Engineering*" [in Japanese] (Inst. Electrical Engineers of Japan, 2002).
2. K. Yatsui et al: *Jpn. J. Appl. Phys.*, **40**, 921 (2001).
3. K. Masugata: *Nucl. Instrum. Methods* **A411**, 205 (1998).
4. J. Hasegawa et al: *Proc. 13th Int'l Conf. High-Power particle Beams*, **II**, 841 (2001).
5. H. Nozawa et al: *Jpn. J. Appl. Phys.*, **40**, 1009 (2001).
6. M. Masnavi et al: *Jpn. J. Appl. Phys.*, **41**, #9 (2002).
7. K. Takasugi et al: *Proc. 13th Int'l Conf. High-Power particle Beams*, **I**, 470 (2001).
8. T. Miyamoto: *J. Phys. Soc. Japan* **68**, 1238 (1998).
9. K. Imasaka et al: *Rev. Sci. Instrum.*, **71**, 4438 (2000).
10. N. Shimomura et al: *Jpn. J. Appl. Phys.*, **39**, 6051 (2000).
11. D. Hasegawa et al: *IEEE Pl. Sci.*, **28**, 2275 (2000).
12. K. Yatsui et al: *Jpn. J. Appl. Phys.*, **36**, 4928 (1997).
13. H. Akamatsu et al: *Vacuum* (in press)(2002).
14. S. Seino et al: *NIFS-PROC-50*, 74 (2001).
15. Nakagawa et al: *Jpn. J. Appl. Phys.*, **37**, L91 (1998).
16. D. Ishida et al: *Proc. Pulsed Power Meeting* [in Japanese], PPT-01-115, 63 (2001).
17. K. Ogura et al: Fusion Eng. & Design **26**, 365 (1995).
18. K. Minami et al: IEEE Trans. Plasma Sci., **27**, 254 (1999).
19. S. Katsuki et al: *IEEE Trans. Plasma Sci.*, **28**, 155 (2000).

PULSED POWER TECHNOLOGY

Study of High Dielectric Constant Materials for use in Compact Transmission Lines for HPM Drivers

Miroslav Joler, Christos Christodoulou, John Gaudet, and Edl Schamiloglu

University of New Mexico, Albuquerque, NM 87131 USA

Abstract. One approach to developing compact pulsed power drivers of High Power Microwave (HPM) sources involves using folded Blumlein transmission lines as the pulse shaping element. In this paper we report progress in the modeling of Blumlein transmission lines, and the advantages of utilizing high dielectric constant ceramics *in lieu* of Kapton™ as the dielectric material. Ceramics with a graded resistivity profile might offer significant advantages in this configuration. Future directions of our research effort will be described.

INTRODUCTION

High Power Microwave (HPM) systems require compact, portable pulsed power in order to be used on mobile platforms [1]. To-date the emphasis in HPM research has been on intense beam-driven sources, with modest attention being paid to the pulsed power. Typical pulsed power systems used tend to be large laboratory facilities. Figure 1 presents a block diagram of a pulsed power-driven HPM source.

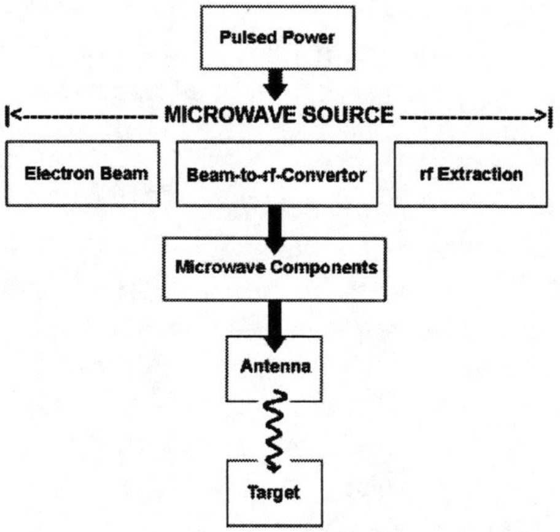

FIGURE 1. A block diagram of a pulsed power-driven HPM source. *(Figure originally appeared in [1], used with the permission of the authors and publisher, © 2001 IEEE.)*

CP650, *Beams 2002: 14th International Conference on High-Power Particle Beams,*
edited by T. A. Mehlhorn and M. A. Sweeney

Capacitive storage-based pulsed power systems are commonly used as drivers for HPM sources in the laboratory. The requirements on such drivers include maintaining a constant impedance (typically of the order of 10-100 Ω) at 0.1-1.0 MV voltages for time scales on the order of 100's ns. Capacitive energy storage-based systems are well-suited to meet these requirements. A block diagram representation of such a system is depicted in Fig. 2. The primary energy source (or "prime power") in the laboratory is the electrical transformer. Our research focuses on the capacitive energy store and pulse forming sections, since these occupy the greatest volume and comprise the largest mass in the system.

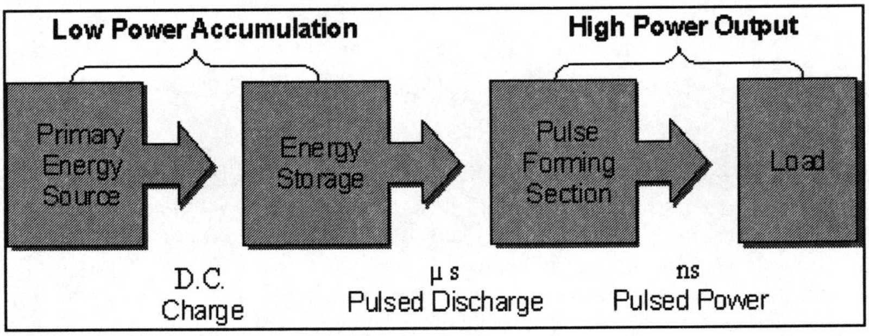

FIGURE 2. A capacitive storage-based pulsed power system. The prime power charges a capacitor bank on a long time scale. This output is then conditioned to meet the needs of the load. (The load of interest to this research is an HPM source.)

BLUMLEIN TRANSMISSION LINE

A Blumlein transmission line consists of two or more coupled transmission lines [2]. Rapidly-shorting switches cause voltage reversal in half of the lines for a time given by the double transit of an electromagnetic pulse. The resulting characteristic feature of a Blumlein transmission line is that the output pulse is roughly equal to the charge voltage of the line. In contrast, traditional transmission lines are characterized by the matched pulse at the load having half the charge voltage at the input to the line.

Researchers at Sandia National Laboratories (Sandia) have been developing a 600 kV, 50 ns, 5 Ω compact pulse-shaping driver that is significantly lighter than comparable coaxial water transmission line systems. Studies by Alexander *et al* [3] demonstrated that propylene carbonate-impregnated Kapton™ sheets are attractive for constructing high average field striplines that can be folded into a compact geometry. Such a compact source can be used to drive low impedance HPM sources (such as a Vircator, MILO, or Radial Acceletron) on a portable platform [1]. This research at the University of New Mexico (UNM) seeks to support this work and advance it through studies of other dielectric materials to be candidates for this configuration transmission line.

Consider the expression for the line impedance of a Blumlein transmission line, shown in Eq. (1):

$$Z \approx \frac{377*d}{w*\sqrt{\varepsilon_r}}\ \Omega,$$

(1)

where d is the thickness of the dielectric medium sandwiched between the metal conductors, w is the width of the line (assuming $d/w < 0.1$), and ε_r is the dielectric constant. In addition, consider the energy density γ (in J/m^3) of a dielectric under the influence of an applied electric field E:

$$\gamma = \varepsilon_0 \varepsilon_r E^2/2$$

(2)

where ε_0 is the permittivity of free space. Whereas the dielectric constant of Kapton™ is about 3.5, the dielectric constant of various ceramics that are being considered in this study (such as TiO_2 and $BaTiO_3$) can be as large as 30,000 for relaxor ferroelectrics, although our focus will be on ceramics with dielectric constants on the order of 100. These ceramics must also be able to hold off electric field stresses on the order of 400 kV/cm. A system utilizing such a large dielectric constant material leads to significantly larger stored energy densities compared with a plastic dielectric system, and a comparable impedance Blumlein transmission line can be made without resorting to folds or bends.

Initial attempts at developing high dielectric constant ceramic-based transmission lines were reported in [4], a collaboration between researchers at Sandia and at the University of Missouri-Rolla (Rolla). The UNM program seeks to continue this work and has begun preliminary discussions with researchers at Rolla [5].

A key aspect of the earlier work that is being studied at UNM (described in the next section) is the issue of the effects of a resistivity grading on the performance of the Blumlein transmission line. One proposal from the Rolla researchers [6] involves a 15-layer composite of 20 mil-thick TiO_2 (that has a dielectric constant of 100). The 5 inner layers of this composite has a resistivity of 3×10^{12} Ω-cm, while the outer 5 layers on both sides have a resistivity that decreases in a step-wise manner down to 10^4 Ω-cm. This resistivity grading can be accomplished by doping the TiO_2 with Nb^{+5} and appropriately annealing the material (which depends on the specifics of the process).

The grading of resisitivity is well-known in the pulsed power and accelerator communities as an effective means of controlling the electric field in a manner that reduces the probability of breakdown occurring (see [7] and references therein).

ELECTROMAGNETIC ANALYSIS

Earlier simulations of the electric field distribution within the dielectric of the Blumlein transmission line were either electrostatic or steady state. The transient nature of the input signal is critical when considering spatially nonuniform complex dielectric constants. The wide spectral bandwidth of the transient pulse on the line leads to a case of multi-mode propagation. Because of this the overall electric field distribution must account for all the significant modes propagating simultaneously.

This aspect of the analysis is particularly well-suited for calculation using CST MICROWAVE STUDIO™, a fully 3-D electromagnetic analysis tool. The procedure we are following is to first identify the power level of interest (based on the wave that is input to the line). This then yields the frequency bandwidth relevant to the problem. Next, we investigate which modes can propagate through the structure and identify their corresponding frequencies. Once a center frequency is selected a calculation can be performed for the electric field distribution. Using a discrete Fourier transform analysis, we sum the power magnitudes that correspond to the frequency intervals for each of the modes. Finally, we superpose the electric fields of each of the modes to obtain the net field distribution. This approach leads to a calculation of the electric field distribution along the line, accounting for multi-mode propagation through the structure, and real magnitudes of power, based on a spectral analysis of the input pulse. This, in turn, allows one to understand the effects of a graded resistivity across the width of the line.

One of the concerns regarding the use of graded high dielectric constant ceramics is the effect of triple points where the step in resistivity meets the ambient dielectric environment. This issue will have to be assessed through experimental testing of the breakdown characteristics of these materials.

ACKNOWLEDGEMENTS

This work is supported by an AFOSR/DoD Grant on Compact, Portable Pulsed Power. The authors would like to thank useful discussions with Prof. Karl Schoenbach, Prof. Ravi Joshi, Dr. Mounir Laroussi (Old Dominion University) and Dr. Ronald Pate and Mr. Jeff Alexander (Sandia National Laboratories).

REFERENCES

1. See, for example, Barker, R.J. and Schamiloglu, E., Eds., *High Power Microwave Sources and Technologies*, IEEE Press, Piscataway, NJ, 2001.
2. Blumlein, A.D., U.S. Patent No. 2,465,840 (1948). Note that the U.S. patent was granted posthumously. For a biographical account of this engineer, see Burns, R., *The Life and Times of A.D. Blumlein*, IEEE, London, United Kingdom, 2000.
3. Alexander, J.A., Shope, S., Pate, R., Rinehart, L., Jojola, J., Ruebush, M., Crowe, W., Lundstrom, J., Smith, T., Zagar, D., and Prestwich, K., "Plastic laminate pulsed power development," *Proceedings, Society of Automotive Engineers,* Paper 00PSC-113, San Diego, CA, October, 2000.
4. Huebner, W., Zhang, S.C., Gilmore, B., Krogh, M.L., Schultz, B.C., Pate, R.C., Rinehart, L.F., and Lundstrom, J.M., "High breakdown strength, multiplayer ceramics for compact pulsed power applications," *Proceedings, 12th IEEE International Pulsed Power Conference*, Monterey, CA, 1999, p. 1242.
5. Dogan, F., University of Missouri-Rolla, private communication, 2002.
6. Gilmore, B.L., *Development of High Energy Density Dielectrics for Pulse Power Applications,* Ph.D. Dissertation, University of Missouri-Rolla, 2002.
7. Peter, W. and Zucker, O., "Resistive-gradient coatings and high-voltage rf breakdown," (submitted to J. Appl. Phys., 2002).

Capacitors for Pulsed Power Systems

N. A. Ratakhin and V. F. Fedushchak

High Current Electronics Institute, Siberian Division of RAS
4, Akademichesky Ave., 634055 Tomsk, Russia

ABSTRACT. Experimental models of capacitors meant for using in directly operated pulsed power systems with energy supply time to load of 500 ns have been designed. It is supposed that the capacitor with energy output of 20 kJ would be rather perspective used as an elementary unit in such systems. At present time, an intermediate model HCEIcap50-4 is being tested and prototype units HCEIcap100-2 have been manufactured. Capacitors consist of plain sections of 2,5 m in length. The sections have a combined insulation, i.e. paper-film insulation with paper and mylar film. The expected characteristics of HCEIcap100-2 are as follows: capacity is 2μF, voltage is 100 kV, current is 1 MA. The results of the first static testing are presented. A specially adapted for the capacitor low-inductance switch to form a unitary high-current module is being developed.

INTRODUCTION

Multimegajoule pulsed generators are progressing taking the path of continuous increase in energy stored. Physical problems solved using such systems are well-known: liners implosion, e-beam production, megagauss fields formation, obtain of ultra-high pressures [1]. Conventionally, operation of such generators could be subdivided as: 1 – energy output from the primary storage (usually it is Marx generator, rarely pulsed transformer); 2 – energy supply to loading (power peaking systems, forming lines, interrupters). Inductance of primary storage can be made rather low. But it is more difficult to have low inductance in circuits of energy supply to load. Analysis of various schemes of energy supply to load suggests that it is possible to achieve total inductance of load and supplying circuits as 50 nH. This means that some of the listed physical problems may be solved using direct action generators. As the energy accumulated is increased, the number of capacitors and switches grows too, and the system reliability falls. Thus, increase in power system capacitance, capacitors reliability and reduction of their total number is issue of the day and noted by the authors of large-scale pulsed systems [3]. Presently, there are no existing nanosecond pulse capacitors with power capacity of 10-20 kJ suitable for use in multimegajoule systems. By means of re-arrangement of output unit and insulator in standard capacitors IK50-3, IK40-5 we have managed to achieve decreased capacitor inductance up to 20 nH. The new-made in such a way capacitors possess a row of attractive peculiarities to be used in creation of multimegajoule generators. They have currents of 400-500 kA in contour with total inductance of 25 nH and off-line high voltage input. Nevertheless, the further inductance reduction is impossible in the frames of techniques of linearly extruded sections due to inductance increase of feedings. In order to solve this problem to create capacitors

CP650, *Beams 2002: 14th International Conference on High-Power Particle Beams*,
edited by T. A. Mehlhorn and M. A. Sweeney

of 10-20 kJ with output power time of up to 500 ns, we selected manual assembled planar sections keeping technological modes without changes.

CAPACITOR

The device has a traditional for power pulsed capacitors arrangement, i.e. capacitor paper, mylar film, aluminum foil, castor oil. Our aim was to decrease inductance maximally, that is why the sections of the capacitor are flat and 2,5 m in length. The other parameters, i.e. the width of paper and covers have been chosen the same as in traditional capacitors are. The paper width is 280 mm, thickness is 10 μm, width of aluminum foil is 260 mm and thickness is 10 μm, thickness of mylar film is 10 μm and width is 280 mm the same as paper. The general arrangement of a section is as follows: foil, two paper layers, a layer of mylar film, a paper layer, a layer of mylar film, a paper layer, a mylar layer, two layers of paper, foil. The total thickness of section insulation is 90 μm. Such type of insulation is used for the most part of pulsed capacitors. The section breakdown voltage on solid electrodes is 12 kV. Capacitor housing is metallic

FIGURE 1. View of capacitor HCEIcap50-2.8, HCEIcap50-4.6.

case to reduce inductance, and in connection with that during its operation its potential does not differ from grounding. Window of capacitor power output, i.e. that real physical size through which commutation in capacitor HCEIcap100-2 is realized, is two times less than in the capacitor HCEIcap50-2.8, that is important for great power supply of to loading.

TABLE 1. HCEIcap100-2 capacitor ratings.

Capacitance	2 mF	Energy density	90 J/l
Voltage	100 kV	Peak current	1 000 kA
Energy	10 kJ	Inductance	4-8 nh

The capacitor HCEIcap100-2 consists of two capacitors HCEIcap50-4. The capacitor HCEIcap100-2 is supposed to be basic in adjustment of manufacture and testing technological regimes. Preliminary works on capacitor energy output have been done. Two ways of energy output were tested, through the end surface and via the side

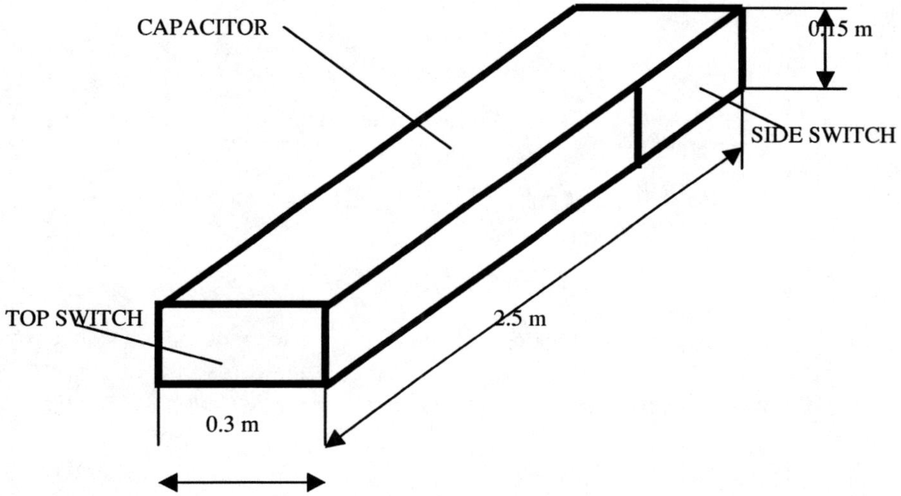

FIGURE 2. View of capasitor HCEIcap 100-2.

capacitor wall. Testing was performed at the level of 30 kV mainly due to absence of switch meant for those aims. In the both cases the circuit inductance did not exceeded 8 nH, hence as from the side wall the inductance was easily reduced to 4 nH. This is related to absence of restrictions on the number of current-carrying lines over the length of the sidewall.Creation of a capacitor for 100 kV is hardly rational due to the more problems arisen regarding electrical insulation of the capacitor itself, as well as high-voltage power supply and triggered circuits.

SWITCH AND CAPACITOR ASSEMBLY

At present time, the works are done to prepare the high-current module presented in Fig. 3 for testing. It is planning to study the parallel operation mode of two capacitors using scheme with unipolar and bipolar charging on one loading. Further, it is planned to perform such works in the mode of voltage reverse in order to have ultimate reducing of discharge loop inductance. It is supposed that the total inductance of discharge loop will not exceed 10 nH. The high current module will be the basic element in the power pulsed generator. Three basic switch versions are being developed for the high-current module: multi-gap switch [2], railgap switch [3], and surface discharge based switch.

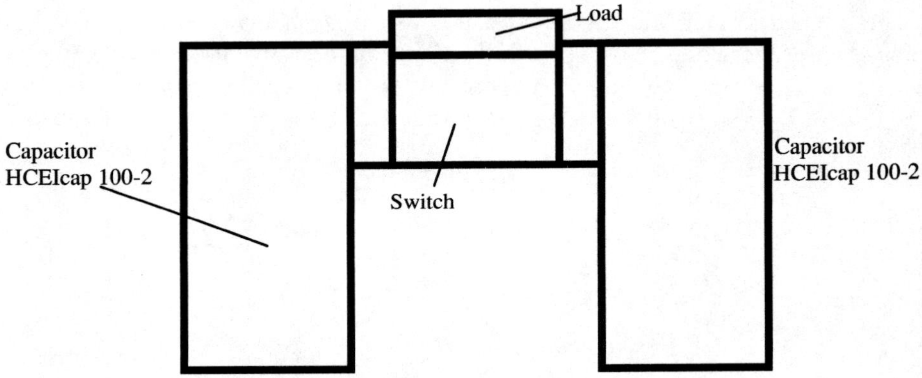

FIGURE 3. Switch, capasitor and load assembly.

CONCLUSION

The early studies results on development and testing of nanosecond capacitors with energy of 10 kJ seem to be rather optimistic. The works continued in this direction will allow development of compact and not expensive multimegajoule generators.

REFERENCES

1. 1. A.V. Louchinsky, N.A. Ratakhin, V.F. Fedushchak, A.N. Shepelev, "Multifunction pulsed transformer-type generator", Izv. VUZov. Fisika.-1997.-№12.-P.67.
2. 2. A.N. Bastrikov, A.A. Kim, B.M. Koval'chuk, et al. "Low-inductance multigap switches", Izv. VUZov, Fisika, 1977, №12, p.5.
3. 3.R.A. Cooper et al.,"High Energy, Low Inductance, High Current Fiberglass Energy Storage Capacitor for the Atlas Machine Marx Modules", Proc. 12th IEEE Int'l. Pulsed Power Conf., June 27-30, 1999, p.122.

Dependence of Flux Losses on MCG Helix Geometry*

Andreas A. Neuber, Thomas Holt,
Juan-Carlos Hernandez, James C. Dickens, and M. Kristiansen

Center for Pulsed Power and Power Electronics Research
Departments of Electrical & Computer Engineering and Physics, Texas Tech University
Lubbock, TX 79409-3102
email: andreas.neuber@ ttu.edu

Abstract. While Magneto Cumulative Generators (MCGs) differ widely in size and operating regimes, it is apparent that the helical flux compression generator is the most promising concept with respect to current amplification and compactness. Though the geometry of the helical generator (dynamically expanding armature in the center of a current carrying helix) seems to be basic, it turns out that the understanding of all involved processes is rather difficult. This fact is apparent from the present lack of a computer model that would be solely based on physical principles and manage without heuristic factors. This means, all programs known to the authors utilize an additional parameter that adjusts for the loss in flux that is currently unexplained. Experimental efforts revealed that the unexplained loss in flux becomes smaller the larger the generator volume is. Specifically, for generators with constant diameter helix and armature, the figure of merit, β, for generator performance exhibits a distinct decline with the angular frequency, ω, of the progressing contact between end-initiated armature and helix. Since ω is proportional to the square root of the ratio of initial MCG inductance to compressed volume, it becomes apparent that the generator performance is limited by size. For large generators on a meter scale, a β of approximately 0.95 has been reported ($\beta = 1$ means no flux loss at all), and own tests indicate that β drops quickly to about 0.7 to 0.6 for generator on the 10 cm scale. Still, even very small generators with about 30 mm outer diameter can have a β of around 0.6. Tapered generators, with tapered helix or tapered armature might perform differently and can be used to further assess the situation. Preliminary results for MCGs with a tapered helix show a performance that is somewhat superior to the straight helix generator.

INTRODUCTION

Since the early 1950's the helical magnetocumulative generator (MCG) has been the subject of extensive study. It is thought that such a device, while capable of amplifying electrical energy a significant amount, has yet to maximize its potential due to intrinsic flux losses. For comparing a variety of helical MCGs, we use the figure of merit, β, defined using the experimental current gain, G_I or the energy gain, G_E:

$$G_I = {I_f}/{I_i} = \left({L_i}/{L_f}\right)^{\beta} \quad \text{or} \quad G_E = \left(\frac{L_i}{L_f}\right)^{2\cdot\beta-1} \qquad (1)$$

CP650, *Beams 2002: 14th International Conference on High-Power Particle Beams*,
edited by T. A. Mehlhorn and M. A. Sweeney

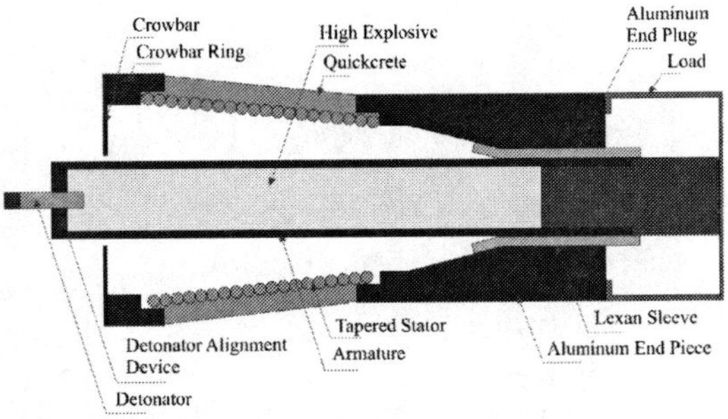

FIGURE 1. Generic diagram of the tapered stator generator, representing TTU VI.

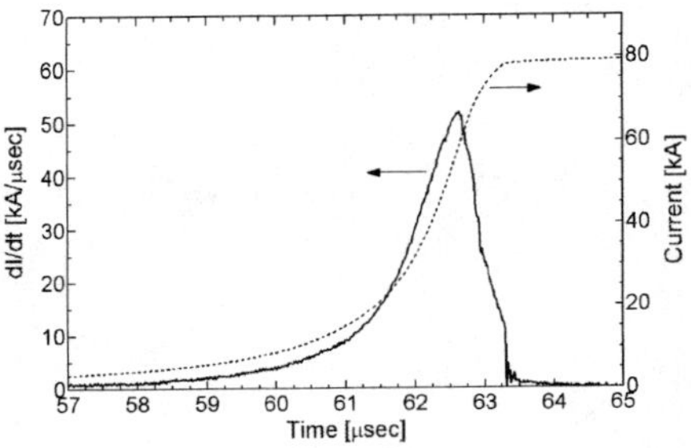

FIGURE 2. dI/dt and current waveform of the tapered stator generator, representing TTU VI.

For a good generator, β should be at least above 0.5 to show some appreciable current amplification. Obviously, a $\beta < 0.5$ is synonymous with the exponent 2β-1 < 0, which means that the final energy produced by the generator is smaller than the initial seed energy. It should be noted that the final current is still larger than the initial even if $\beta <$ 0.5 even as long as $\beta > 0$. The ramifications of this are discussed for a staged system in reference [1].

Our previous experiments with straight stator/armature MCGs revealed that the figure of merit, β, is a strong function of the angular frequency describing the progressing contact point of armature and stator as it is wiping out the stator. Smaller frequencies generally lead to higher β and thus better MCG performance [2]. We present in this study the experimental behavior of an MCG with tapered stator and compare its performance a similar sized straight MCG.

EXPERIMENTAL SETUP

To date, we have tested 10 different types of MCGs. Helix diameters range from as small as 22 mm to as large as 89 mm. The generators presented in this study had the same dimensions at the detonator side: The armature OD was set at 38 mm, the maximum stator diameter was 89 mm, and the amount of explosives used was 250 g. All three presented generators were wound with 32 turns AWG 12 copper wire having a 2 mm core and a 0.5 mm PVC insulation. The Gurney angle of the straight armature MCGs is roughly 15 degrees for all tests.

The straight generator, TTU II, was designed to have a time delay of 8 μs or 21 μs from time of crowbar to the armature impacting the stator's first or last turn, respectively. Its complete stator is wiped out in about 13 μs. The tapered stator generator, TTU VI in Figure 1, was constructed with the same maximum stator diameter as the straight generator at the detonator end and with an expansion ratio of 1.7 at the load end, resulting in a taper angle of 6°. Time from crowbar to the first turn is again about 8 μs with a total burn time of about 17 μs. Hence, its stator is wiped out in about 9 μs, which is distinctly different from the straight generator. A sample of the

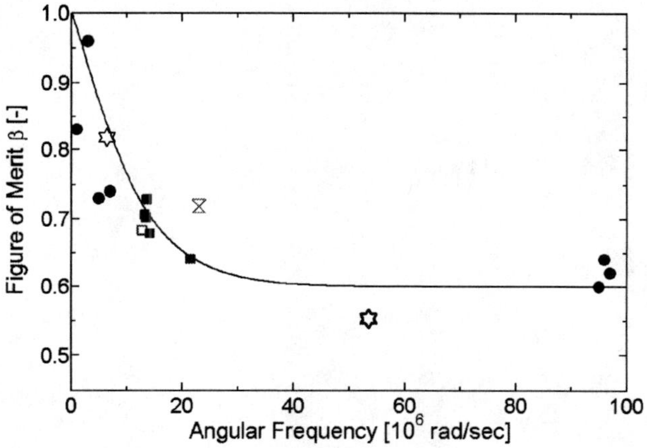

FIGURE 3. Experimental dependence of the figure of merit as a function of the contact point's angular frequency. Data points are calculated from experimental data in references [2], circle symbols, and own tests, all other symbols. TTU II (open square) lies within a group of other straight generators (closed symbols). TTU VI is the upper hourglass. The stars indicate two generators at the extreme of the spectrum, a double helix 89 mm diameter (7 rad-MHz), and single helix with 28 mm diameter (52 rad-MHz). The solid line serves as guide for the eye only.

electric current output of TTU VI is depicted in Figures 2; the output for the straight generator is similar with a somewhat shorter FWHM of the *dI/dt* waveform.

RESULTS AND DISCUSSION

The values listed for the figure of merit, β, reveal that the tapered generator performs better, despite having a 40% larger angular frequency, than the straight generator, see Table 1. This behavior is reflected in the β vs. ω plot, see Figure 3, which shows the improved performance of TTU VI, cf. hourglass symbol in Figure 3. It should be noted that ω scales with the generator volume, $\omega \propto (1/\text{volume})^{1/2}$. Thus smaller angular frequency, meaning larger volume, generally exhibits a better MCG performance.

The detailed physical reason for the observed behavior presented in the previous section is far from being clear. However, all our experiments and simulations [3] point to the existence of an instantaneous flux loss that cannot be explained with resistive losses, turn-skipping, etc. We have yet to find an all-convincing design that would eliminate this instantaneous flux loss; our best attempt is the tapered stator generator or a large generator with low angular frequency, ω, cf. upper star symbol in Figure 3.

Generator	L_{init} µH	L_{load} nH	Current gain	Energy gain	Figure of merit	Gen. vol./cm^3	ω rad-MHz
Straight, TTU II	42	65	83	10.7	0.68	515	16
Tap., TTU VI	53	46	147	18.9	0.71	656	≈ 23

TABLE 1. Generator specifications for generator with straight and tapered stator.

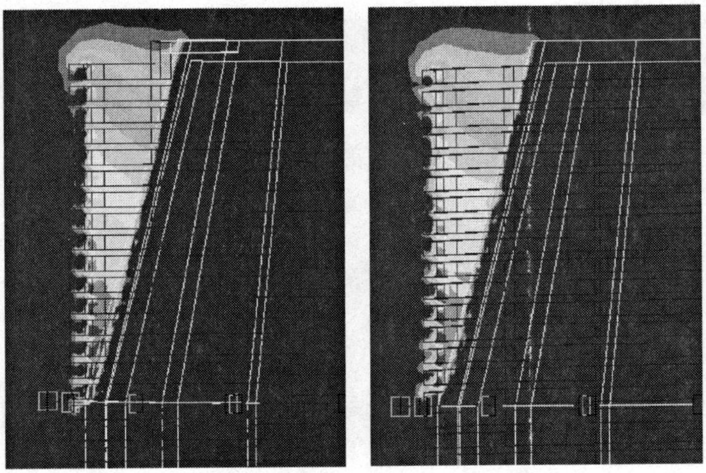

FIGURE 4. Magnetic field magnitude in two cutplanes defined relative to contact point (a) 45 degree, (b) 0 degree (contact point in cut-plane). Armature cone propagates from bottom to top of figure. On a linear scale the field ranges from 1 T (top of figure) to 4 T (volume between stator and armature in the vicinity of the contact point) at 10 kA seed current.

A straightforward approach to understanding the instantaneous flux loss would be the in-depth numerical simulation of an MCG. However, currently available MCG computer software fails to deliver detailed physical results as they lump much of the physics into a less detailed model. Hence, the detailed magnetic field structure, the current distribution, etc. is pretty much unknown especially in the vicinity of the contact point.

Our 3-D field calculations with an AC-current at a frequency of 100 kHz reveal that the magnetic field has a large gradient in the vicinity of the contact point, see Fig. 4. However, a detailed analysis of the calculated fields seems futile since, based on a comparison with experimental data, this adiabatic approach fails to account for current left behind in the stator wires due to the finite current and magnetic field diffusion times.

ACKNOWLEDGEMENT

This work was solely funded by the Explosive-Driven Power Generation MURI program funded by the Director of Defense Research & Engineering (DDR&E) and managed by the Air Force Office of Scientific Research (AFOSR).

REFERENCES

1 A. Neuber, T. Holt, S. Shkuratov, J. Dickens, Y. Talantsev, J. Walter, and M. Kristiansen, "All Explosive Pulsed Power Generator System," to be presented at the IXth International Conference on Megagauss Magnetic Field Generation and Related Topics, Moscow, Russia, 2002.

2 T. Holt, A. Neuber, J. Dickens, and M. Kristiansen, "Impact of Helix Geometry on MCG Flux Losses," Proc. 13th IEEE Intl' Pulsed Power Conf., Las Vegas NV, 2001, pp. 905-908.

3 A. Neuber, J. Dickens, J. B. Cornette, K. Jamison, R. Parkinson, M. Giesselmann, P. Worsey, J. Baird, M. Schmidt, and M. Kristiansen, "Electrical Behavior of A Simple Helical Flux Compression Generator for Code Benchmarking," IEEE Trans. on Plasma Sci. **29**, 573-581, (2001).

New Repetitive X-ray Generator Using Magnetic Field Controlled POS

Georgi I. Dolgachev and Andrey G. Ushakov

Kurchatov Institute Russian Research Center, Moscow 123182, Russia

Abstract. The paper represents new results in repetitive pulsed power production using inductive storage and Plasma Opening Switch operating in applied magnetic field. A design of new compact X-ray production facility capable of producing 4-5 MV pulsed electron beam is proposed. The facility is intended for radiation sterilization technology , food and wastes decontamination.

TECHNOLOGY BACKGROUND

First commercial pulsed power facility using plasma opening switch (POS) technology begun operation in 1992 [1]. This technology produced electron beam with power 1-10 kW at 2-3 MeV for X-ray sterilization requiring high doses [2,3].

The progress in understanding the plasma properties has also been achieved. This resulted in formulating recommendations for both repetitive generators and single pulsed ones employing the POS. Recent generators having utilized repetitive technology were RS-300 facility at 300 kA drive current level [4] and MOL generator (which is a module prototype employed by BAIKAL generator program) [5]. Having invented the repetitive POS controlled with external pulsed magnetic field in 1998 [6] we could create a new compact generator. Table 1 summarizes major results in physics and technology for repetitive systems.

TABLE1. Results of plasma and technology studies to create repetitive POS technology

Results previously reported	Design improvements
Magnetic insulation of POS plasma [6]	More effective POS geometry was proposed capable of producing higher voltages
Charge limit density was obtained [6]	Geometry factors and conduction duration determining effective POS operation were obtained
Switching voltage as a function of energy density spent on ion acceleration [7]	POS voltage limit could be extrapolated, optimum switching voltage may be determined
RS-20 facility exploitation [1,7]	X-ray target at 20 w/cm^2 power density was invented Sterilization dose vs. dose rate effect was studied Repetitive gas triggering switch was invented Plasma gun possessing long life time was designed Positive polarity POS scheme was created
Repetitive e-beam facility [8]	Air electron beam injection was accomplished Negative polarity POS was designed

The above results have demonstrated in particular that a POS system may be realized at relatively low level of current. This may be suitable for continuously

CP650, *Beams 2002: 14th International Conference on High-Power Particle Beams,*
edited by T. A. Mehlhorn and M. A. Sweeney

working applications because low peak energy density prevents materials from destroying and increases their lifetimes.

NEW GENERATOR DESIGN

The purpose of the system was producing repetitively pulsed X-ray and Gamma in the electron beam diode. The tests of main parts have been accomplished. A primary storage system (Pulse Voltage Generator – PVG) is a Marx type. A high-voltage rectifier charges the PVG with 40 kV DC. Then the energy is transformed to 500 ns 1,8 MV pulses with 90% power transformation efficiency. The average power transferred through the capacity is 20 kW. New capacitors (40 kV, 2200 pF 100 Hz) were created and new triggering switches were designed. (see Fig. 1).

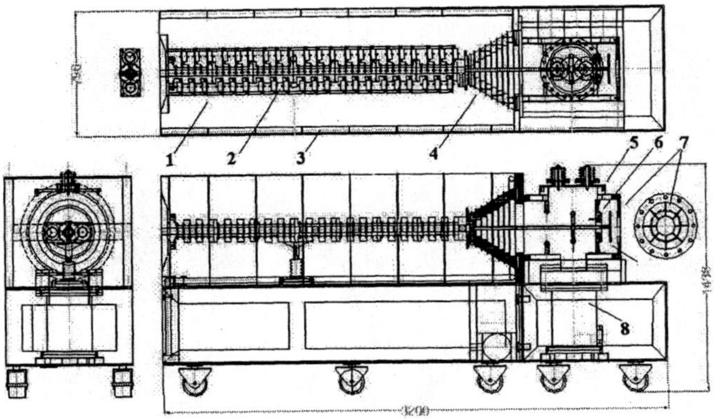

FIGURE 1. Schematic of X-ray generator. Captions: 1 – PVG tank filled with transformer oil, 2 – PVG with caps and gas switches, 3 – PVG tank, 4 – HV insulator, 5 – chamber, 6 – POS, 7 – REB diode and X-ray window, 8 – vacuum system.

POS generator can be considered as a L-C circuit closed through the POS having non-linear resistance. When the PVG is turned on the current through the inductance and the POS. At the conduction phase POS impedance is almost equal to zero and electrical energy of the PVG ($W_e=CU^2/2$) is transformed into the magnetic energy of the vacuum line, which is $W_m=LI^2/2$, where L is inductance of the line, PVG and wires. Then POS impedance increases, so that the energy is transferred to the load. POS resistance increase is called "plasma switching". The switching produces high-voltage pulse and accelerated electron beam. Geometry and magnetic field of the POS and diode allow high-current electron beam formation on the target. The target creates Bremsstrahlung with high dose/dose rate (see Table 2 for the reference).

TABLE 2 Main generator parameters

Parameter, unit	Value
PVG generator output voltage (Marx generator), MV	1,8
REB diode voltage, MV	4.5
PVG generator power transferred, kW	20
REB diode power, kW	10

Pulsed current amplitude kA	3,5
Pulse duration, ns	100-200
Repetition rate, Hz	1-10
Average dose rate in water, kGy*hour	80
Irradiation area, circle diameter, mm	350
Dimensions l-w-h, mm-mm-mm	3200x800x1450

The dose was calculated 30 cm from the target. Simulation results demonstrated that the voltage providing most effective irradiation may be estimated as 4-5 MV.

POS properties depend on geometry, plasma distribution, current amplitude and rise time [7]. A simple formula was used for POS switching voltage:

$$U_{POS} = \alpha U_g^{4/7}, \quad (1)$$

where U_{pos} – POS voltage, U_g – PVG voltage, $\alpha=3.6$ empirical coefficient which has been chosen from a comparison of many facilities. Eq. (1) describes maximum POS voltage as a function of energy density.

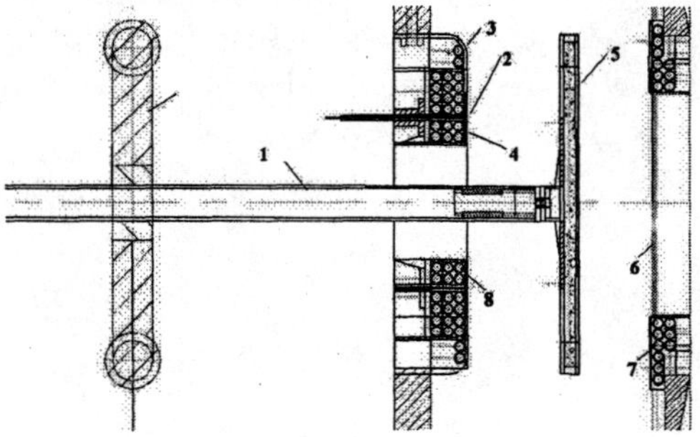

FIGURE 2. POS and diode. 1 – HV electrode, 2 – plasma injection hole, 3 – cathode, 4 – graphite protection, 5 – anode, 6 – x-ray window, 7-8 – magnetic field coil.

To create a system capable of producing high PVG voltage a special attention was paid to 40-stage PVG. Power caps, charge elements and gas triggering switches were combined in one body made from organic glass. The system was placed into the oil tank made from *ss*. The vacuum inductance was connected with the PVG through the insulator, which was designed to provide 2-3 MV within 100-200 ns and 1 MV within tens of μs. 38 kV 100 Hz pulsed caps, PVG gas switches, MF coils and POS "inversed pinch" geometry were tested.

POS AND DIODE DESIGN

POS and diode were combined, so that the drive current coming through the POS after the switching was employed for REB production (see Fig.2). POS uses 5 kGs magnetic field from an additional power supply. Plasma gun surface is hidden from ion bombardment. Cathode is made from graphite to reduce anode sputtering due to heat and ion fluxes. The pulsed magnetic field used anti-cusp configuration and provided magnetic insulation between cathode and anode and, as well, between the diode anode and the x-ray emission window. 4.5 MeV e-beam hits the *W*-target producing Bremsstrahlung [9]. Target melting and sputtering was avoided by applying temperature resistant graphite plates with *W* wire arrays between them (see Fig. 3). The target mass thickness was optimized for maximum dose output. Graphite thickness was minimized providing both mechanical strength and low electron absorption. The target was stable at 20-30 W/cm^2.

FIGURE 3. Multi-layer target design. Left – target composition 1,3 – protective and supporting carbon layers, 2 – tungsten wire array. Right – tungsten wire array photograph.

Fig. 4 illustrates the effectiveness of X-ray conversion of the target. Anode life-time exceeds 10^3 hours. Tests were done at the RS-20 facility. The effective power transformation makes POS system cost-effective in comparison with other existing accelerators.

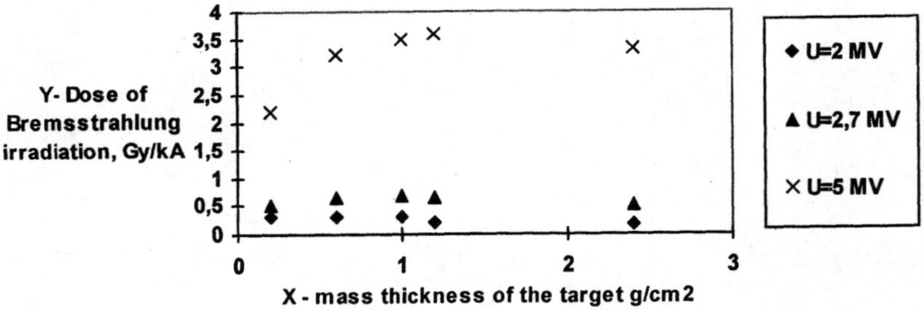

FIGURE 4. Effectiveness of electron beam conversion in X-ray vs. effective target thickness.

The generator is planned for using in programs requiring high dose levels such as sterilization, organic compounds utilization and electronic equipment testing.

REFERENCES

1. Babykin V. M. et al., 9th Int. Conf. on High-Power Particle Beams, Proc. V. 1, 512-517 (1992).
2. Dolgachev G. I. et al. 11th IEEE Int. Pulsed Power Conf. Proc., V.1, 281-286 (1997)
3. Dolgachev G. I. et al 12th Int. Conf. on High -Power Particles Beams, Proc. V.2, 977-980 (1998).
4. Barinov N.U. et al., Instrum. Exp. Tech. V.1, 1-8 (2002).
5. V.V.Alexandrob et al., 13th Int. Conf. on High -Power Particles Beams, Proc. V.1, 147-150 (2000).
6. Dolgachev G. I. et al., Sov. J. Plasma Phys. V.17, 1171-1182 (1991).
7. Dolgachev G. I. and Ushakov A. G., Plasma Phys. Reports V. 27, 110-118 (2001).
8. Barinov N. U. et al. IEEE Trans. Plasma Science, V. 23 (6), 945-948 (1995).
9. Dolgachev G. I. et al. IEEE Trans. Plasma Science, V. 26 (10), 1410-19, (1998).

Use of a Tapered Transmission Line as an Ideal Transformer

Robert J. Vidmar

University of Nevada, Reno, 5625 Fox Ave, Reno, NV 89506 USA

Abstract. A tapered transmission line [1, 2] has an impedance that slowly varies spatially from source end to load end. Use of a tapered line yields a nearly ideal transformer for frequencies above the transmission-line cutoff frequency, which is related to the taper rate and phase velocity in the transmission medium. Use of a material with high permeability and high permittivity reduces the wavelength in a transmission line, and so extends the use to lower frequencies. Calculations for an exponential line utilizing a high-permeability high-permittivity material include the range of impedance transformation, transformer ratio, phase velocity, input impedance, power reflection coefficient, power deposition, and thermal loading.

INTRODUCTION

Tapered transmission lines have been investigated and used as a high-pass impedance matching component for some time, [1-4]. Because of the high-pass characteristic, use gravitated toward high frequencies or short pulse applications, where the physical design is rather compact. For example, an impedance match [5] from 1.2 Ω to 50 Ω over a frequency band of 700 MHz to 1.4 GHz is achieved with a dielectric filled ($\varepsilon_r = 9.9$) exponential line with a length of 9 cm and a cutoff frequency of 318 MHz. For a cutoff frequency of 1 MHz, the line would increase in length to 28 m. In this paper, basic theory is cited, and an emphasis is placed on extending the use of tapered transmission lines to lower frequencies, longer pulses, and high power. The metrics for comparison are the input impedance, the voltage reflection coefficient, power rating, volumetric power deposition, and thermal loading.

THEORY

The analytic treatment of an exponential transmission line has been treated by [1, 2] and is based on the lossy transmission line equations for voltage, V, and current I:

$$\frac{dV}{dx} = -L\frac{dI}{dt} - IR \quad \text{and} \quad \frac{dI}{dx} = -C\frac{dV}{dt} - GV \tag{1}$$

where per unit length quantities are used for resistance R, inductance L, capacitance C, and conductance G: $L = L_0 \exp(+\eta x)$, $R = R_0 \exp(+\eta x)$, $C = C_0 \exp(+\eta x)$, $G = G_0 \exp(+\eta x)$, η is the exponential taper rate, and the subscript 0 denotes the parameter value for $x = 0$. Wave propagation occurs in a complex media with conductivity σ, permittivity $\varepsilon = \varepsilon_o(\varepsilon_r - i\varepsilon_i)$, and permeability $\mu = \mu_o(\mu_r - i\mu_i)$, where ε_o and μ_o are the permittivity and permeability of vacuum, $i^2 = -1$, and the subscripts r and i denote real and

CP650, *Beams 2002: 14th International Conference on High-Power Particle Beams*, edited by T. A. Mehlhorn and M. A. Sweeney

complex quantities. Solutions to the line equations are obtained for waves proportional to $\exp[+i\omega t + \eta x/2 \pm kx]$, where ω (= $2\pi f$) is angular frequency, t is time, k is complex wave number, x is distance, and ℓ is line length. Transmission line parameters are expressed as a network analysis ABCD matrix [3, 4] normalized to impedance Z_{nor}:

$$\begin{bmatrix} A & B \\ C & D \end{bmatrix} = \frac{1}{k} \begin{bmatrix} \exp(-\eta\ell/2)\left[k\cosh(+k\ell) + \frac{\eta}{2}\sinh(+k\ell)\right] & \frac{\gamma Z_{01}}{Z_{nor}}\exp(+\eta\ell/2)\sinh(+k\ell) \\ \frac{\gamma Z_{nor}}{Z_{01}}\exp(-\eta\ell/2)\sinh(+k\ell) & \exp(+\eta\ell/2)\left[k\cosh(+k\ell) - \frac{\eta}{2}\sinh(+k\ell)\right] \end{bmatrix} \quad (2)$$

with $\gamma^2 = (i\omega L_0 + R_0)(i\omega C_0 + G_0)$, $k = \sqrt{\gamma^2 + \eta^2/4}$, $Z_{01} = \sqrt{(i\omega L_0 + R_0)/(i\omega C_0 + G_0)}$, and Z_{01} is the line impedance for x = 0. The condition of k = 0 defines a high-pass cutoff frequency, ω_c, which is computed from $\gamma^2(\omega_c) = -\eta^2/4$. The general solution is

$$\omega_c = +i\frac{(R_0C_0 + L_0G_0)}{2L_0C_0} \pm \sqrt{+\frac{(R_0G_0 + \eta^2/4)}{L_0C_0} - \frac{1}{4}\left(\frac{R_0}{L_0} + \frac{G_0}{C_0}\right)^2} \quad (3)$$

The cutoff frequency for a lossless system ($R_0 = 0$ and $G_0 = 0$) is $\omega_c = \eta/\left(2\sqrt{L_0C_0}\right)$. For a complex medium with $R_0 \neq 0$ and $G_0 \neq 0$, the cutoff frequency becomes complex and its interpretation difficult. One can use ω_c as a guide and deduce the cutoff frequency by sweeping the power reflection coefficient for a matched load.

For an exponential line terminated in a complex load, Z_L, the input impedance, Z_{in}, and voltage reflection scattering coefficient, S_{11} are

$$Z_{in} = \frac{AZ_L + B}{CZ_L + D} = \frac{\exp(-\eta\ell)Z_L\left[k\cosh(+k\ell) + (\eta/2)\sinh(+k\ell)\right] + \gamma Z_{01}\sinh(+k\ell)}{\gamma(Z_L/Z_{01})\exp(-\eta\ell)\sinh(+k\ell) + k\cosh(+k\ell) - (\eta/2)\sinh(+k\ell)} \quad (4)$$

$$S_{11} = \frac{Z_L\left[k\cosh(+k\ell) + \left(\frac{\eta}{2} - \frac{\gamma Z_m}{Z_{01}}\right)\sinh(+k\ell)\right] - Z_m\exp(+\eta\ell)\left[k\cosh(+k\ell) - \left(\frac{\gamma Z_{01}}{Z_m} + \frac{\eta}{2}\right)\sinh(+k\ell)\right]}{Z_L\left[k\cosh(+k\ell) + \left(\frac{\eta}{2} + \frac{\gamma Z_m}{Z_{01}}\right)\sinh(+k\ell)\right] + Z_m\exp(+\eta\ell)\left[k\cosh(+k\ell) + \left(\frac{\gamma Z_{01}}{Z_m} - \frac{\eta}{2}\right)\sinh(+k\ell)\right]} \quad (5)$$

where Z_m is the impedance of the source that is connected to the exponential line.

Power flow is quantified under the assumption that power into the line equals the power out. The power into the line is $P_o = I_oV_o$ and $V(x) = Z_o(x)I(x)$. For an exponential line, the impedance, impedance ratio, voltage, and current are

$$Z_o(x) = Z_{01}\exp(+\eta x), \quad N = Z_o(\ell)/Z_o(0), \quad V(x) = V_o\exp(+\eta x/2), \quad I(x) = I_o\exp(-\eta x/2) \quad (6)$$

where $V_o = \sqrt{P_oZ_{01}}$, $I_o = \sqrt{P_o/Z_{01}}$; and $V(\ell) = V_o\sqrt{N}$ and $I(\ell) = I_o/\sqrt{N}$. The cross sectional area for power propagation decreases exponentially from the low to the high impedance end of a line. This produces maximum values of B, E, and power dissipation in the media at the high impedance end. For a planar exponential line as in Fig. 1, which has a maximum width b_o and height a, the maximum fields are

$$B_{max} = \frac{\mu_o\mu_r I_o}{b_o}\exp(+\eta\ell/2), \quad E_{max} = \frac{V_o}{a}\exp(+\eta\ell/2) \quad (7)$$

EXPONENTIAL LINE EXAMPLE

The exponential line in Fig. 1 has a uniform height *a* but a width $b(x) = b_o \exp(-\eta x)$ that varies exponentially with x. Other geometries, such as tapered coaxial and three dimensional shapes, are possible, but the complex media is expensive to fabricate.

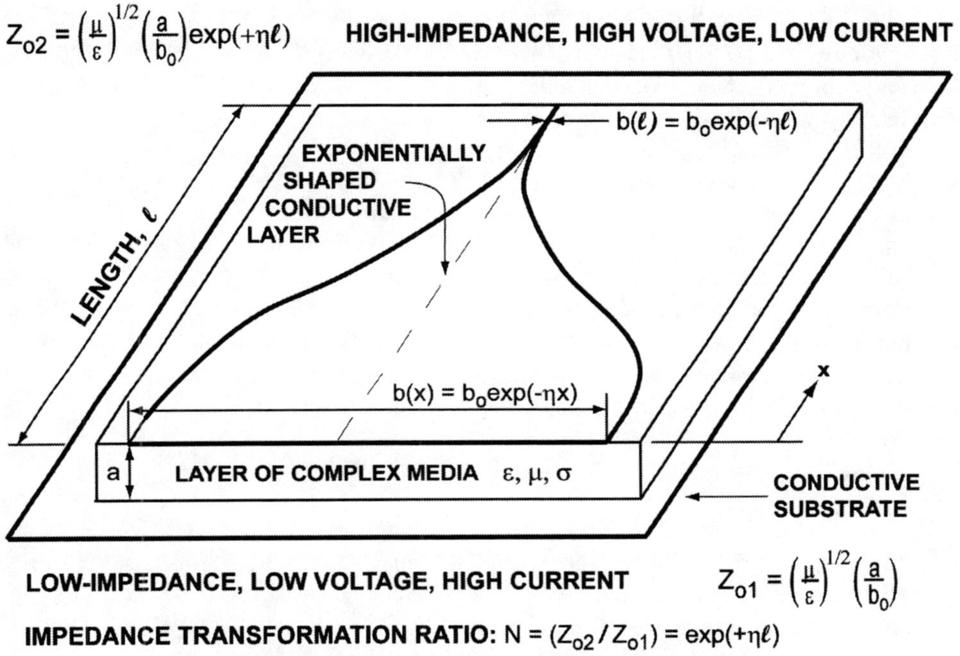

FIGURE 1. Planar exponential transmission line.

The values of C and L are $C = \varepsilon_o \varepsilon_r (b_o / a)\exp(-\eta x)$ and $L = \mu_o \mu_r (a / b_o)\exp(+\eta x)$. The value of R is of the sum of the magnetic dissipation loss plus the conductive layer loss, $R = [\omega \mu_o \mu_i (a / b_o) + 1/(\sigma_s \delta b_o)]\exp(+\eta x)$, where $\delta = (2 / \omega \sigma_s \mu_s)^{1/2}$ is skin depth, σ_s is conductor conductivity, and μ_s is relative conductor permeability. The value of G is the sum of dielectric dissipation loss plus bulk conductivity loss in the media: $G = [\omega \varepsilon_o \varepsilon_i (b_o / a) + \sigma (b_o / a)]\exp(-\eta x)$. Although power dissipation per unit length is constant along the transmission line, the power per unit volume varies exponentially:

$$\frac{P_{media}(x)}{Volume} = \left[\frac{\omega \mu_o \mu_i I_o^2}{b_o^2} + \frac{\omega \varepsilon_o \varepsilon_i V_o^2}{a^2} + \frac{\sigma V_o^2}{a^2}\right]\exp(+\eta x), \quad \frac{P_s(x)}{Volume} = \frac{\omega \mu_s I_o^2}{2b_o^2}\exp(+\eta x) \quad (8)$$

Temperature extremes occur at the high impedance end, x = ℓ. For a system with no external cooling and a pulse of duration Δt, the temperature increase per unit time is

$$\Delta T(x) / \Delta t = [P(x) / V]/(\rho\, c_s) \quad (9)$$

where ρ and c_s are material density and specific heat, respectively

As an example, consider an exponential line as in Fig. 1 with a = 4 mm, b_o = 1 m, ℓ = 1 m, η = 4.828 m^{-1}, $b(\ell)$ = 8 mm, and impedance ratio of N = 125, made with Fair-Rite material 67 and aluminum conductors. Material 67 is a NiZn ferrite with a density of 4600 kg/m^3, a specific heat of 700 J/kg-°K, and B_{max} = 0.23 T. For f < 80 MHz, $\mu_r = 40$, $\mu_i \leq 0.70$, $\varepsilon_r = 12$, $\varepsilon_i = 0$ (unknown), and $\sigma = 1.0 \times 10^{-5}$ $(\Omega\text{-m})^{-1}$. The aluminum layers have $\sigma_s = 3.54 \times 10^7$ $(\Omega\text{-m})^{-1}$, a density of 2,700 kg/m^3, and a specific heat of 896 kJ/kg-°K. The transmission line parameters are C = 26.5 nF/m, L = 201 nH/m, R = 287 mΩ/m, and G = 2.50×10^{-3} $(\Omega\text{-m})^{-1}$, Z_{o1} = (2.75, 0.00366) Ω, Z_{o2} = (343.9, 0.456) Ω, a high-pass cutoff frequency of 5.25 MHz, and a phase velocity of 1.37×10^7 m/s. For f > 10 MHz, S_{11} is less than -17 dB into a matched load.

From Eq. (7) and B_{max}, the value of I_o is = 409 A, which implies P = 460 kW and V_o = 1.125 kV. At the high impedance end of the line $I(\ell)$ = 36.6 A and $V(\ell)$ = 12.6 kV. Power deposition is 40 kW into the ferrite (anomalous magnetic and dielectric losses were not included) and 22 kW into the conductor. Ferrite operation near B_{max} exhibits waveform dependent nonlinear and anomalous behavior. Linear operation with reduced distortion requires a value of B much less than B_{max}. For a power of 460 kW, the maximum temperature rise per unit time is 390 °K/s for the ferrite and 340 °K/s for the aluminum conductor. A conservative maximum temperature increase of 50 °K for the ferrite limits the duration of a waveform to less than 125 ms.

SUMMARY

The example of a 460 kW exponential line using Fair-Rite material 67 quantifies a transformation ratio of 125 for impedance, 11 for current and voltage, and a voltage reflection coefficient of less than -17 dB at frequencies above 10 MHz. At this power level, nonlinear effects and heating of the ferrite both at the high impedance (narrow) end of the transmission line limit the duration of a waveform to 125 ms.

ACKNOWLEDGMENT AND DISCLAIMER

This material is based on research sponsored by the Air Force Research Laboratory, under agreement number F49620-01-1-0354 and a subaward from the University of New Mexico as part of an AFOSR/DOD MURI grant. The U.S. Government is authorized to reproduce and distribute reprints for Governmental purposes notwithstanding any copyright notation thereon. The views and conclusions contained herein are those of the author and should not be interpreted as necessarily representing the official policies or endorsements, either expressed or implied, of the Air Force Research Laboratory or the U.S. Government.

REFERENCES

1. Wheeler, H.A., *Proc. I.R.E.* **27**, 65-71 (1939).
2. Johnson, W.C., *Transmission Lines and Networks,* McGraw-Hill, New York, 1950, pp 201-208.
3. Ramachandran, V., and Nair, K.K., *IRE Trans. on Circuit Theory* **CT-7**, 71-74 (1960).
4. Dutta Roy, S.C., *Proc. of the IEE* **67**, 1162-1163 (1979).
5. Arnold, R.P., and Bailey, W.L., *Electronic Design*, **22**, 136-139 (1974).

High-Voltage Characterization for the Prototype Induction Cells

Wang Huacen, Zhang Kaizhi, Wen Long, Lai Qinggui, Zhang Linwen,

Deng Jianjun

Institute of Fluid Physics, CAEP, P.O.Box 919,Mianyang ,Sichuan, 621900, P.R.China

Abstract. Two linear induction prototype cells expected to work at 250kV, 3kA,with accelerating voltage flattop ($\pm 1\%$) $\geq$70ns, have been tested to determine their high-voltage characteristics. Each cell is composed of a ferrite core immersed in oil, a gap with curved stainless steel electrodes, a solenoid magnet, and a insulator. The experiments were carried out with full-scale cells. The high voltage pulses were applied to two cells using a 100ns, 12Ω pulse Blumlein. The tests were performed at various high-voltage levels ranging from – 250kV to –350kV. No breakdown was observed during the test at vacuum level (7-10) $\cdot 10^{-4}$ Pa. The cell schematic, the experimental set up, and the measured voltage waveforms are presented in this paper.

INTRODUCTION

Two linear induction prototype cells and their related pulsed power system have been designed and fabricated at our laboratory. Each cell is composed of 11 ferrite toroids immersed in oil, 19.2mm wide accelerating gap with curved stainless steel electrodes, a solenoid magnet, and a 42mm high insulator made of cast nylon, slanted between the electrodes. The cell schematic diagram can be seen in Fig 1.

The breakdown electric field of the insulator separating the oil filled ferrite core from the vacuum potion of the cell was maximized by holding 42^0 angle between the insulator and the calculated electric potential line and carefully treating the"triple point".

CP650, *Beams 2002: 14th International Conference on High-Power Particle Beams,*
edited by T. A. Mehlhorn and M. A. Sweeney

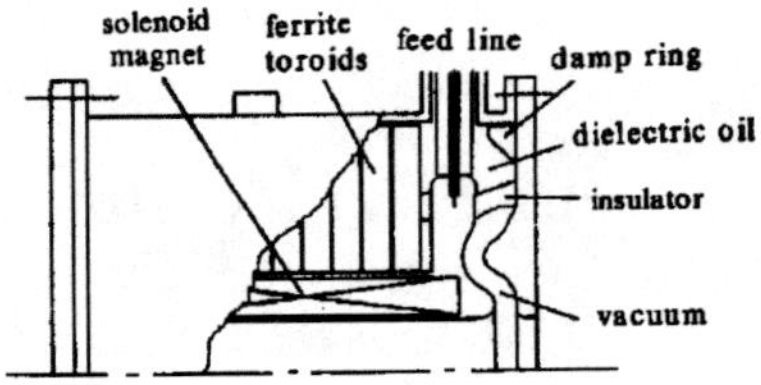

FIGURE 1. Prototype Cell Schematic

The insulator was also completely shield from direct line-of-sight with the beam to prevent it charging and UV-induced breakdown in a way that reduces the cell transverse impedance. The maximum field on the electrode surface calculated by LWE2D and ELECAF2D is 185kV/cm and 70kV/cm in the insulator [1].

The pulsed power system is consists of a Marx generator and a 100ns, 12Ω Blumlein capable of generating –300kv on a matched impedance.

The high-voltage characteristics of the cells were examined with this pulsed power system.

Experimental Set-Up

The cells were driven by the pulse Blumlein during the test. The Blumlein output pulse has about 75ns flattop and 24ns risetime and can be adjusted continuously from –140kV to –300kV. The output of Blumlein was connected to four 48Ω high-voltage solid dielectric cables in parallel. Each cell was driven by two of the four cables. Each cable feed into the cell was terminated in the load formed by a 67Ω compensation resistor and the cell impedance in parallel. The equivalent load impedance for the two cables connected to one cell would be matched if there were a 3kA beam through the cell. In fact, it was mismatched, since there was no beam at all. The actual voltage on the cell gap was higher than that of Blumlein output. The gap voltage was measured by a Capacitive Voltage Probe (CVP) and two Resistive Voltage Probe (RVP), as shown in Fig 2.

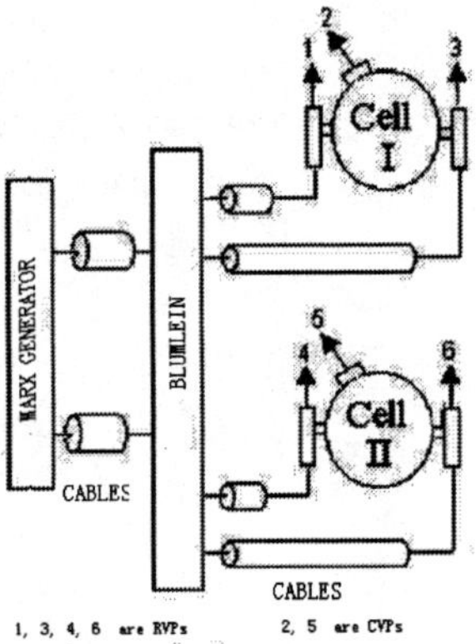

FIGURE 2. Experimental Set-up

In order to examine the influence on the cell insulator from voltage reflected at the mismatched points, the cables connected to the cell I were once disconnected when the gap voltage was raised to –250kV. 500 shots have been taken only for cell II at this voltage level.

Experimental Results

The cells were pulsed 1000 shots respectively at the voltage –250kV and –295kV. The high-voltage pulse applied to the cell gap, then, was raised in 10kV steps till to –330kV. 50 shots were taken at each voltage level while additional 50 shots at –330kV. Some of high-voltage waveforms on the cell gap recorded by the oscilloscopes are shown in Fig 3.

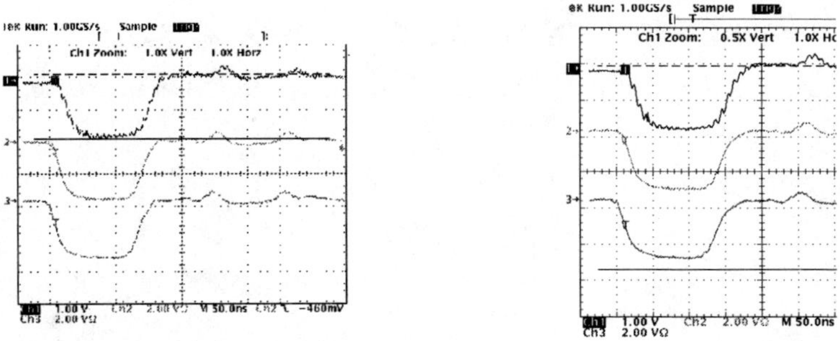

a) 325kV b) 295kV

(ch1: CVP; ch2 & ch3: RVP)

FIGURE 3. Recorded High-Voltage Waveforms across the Cell Gap

The voltage flattop is more than 70ns at the voltage of –325kv, according to Fig 3.

No breakdown was observed for the whole test process at the vacuum level (7 - 10)·10^{-4} Pa. However, as shown in Fig 4, the voltage waveforms would be distorted at the voltage beyond –330kV, because of the saturation of ferrite core. The deduced average saturation flux density of the ferrite core from Fig 4 is higher than 420mT.

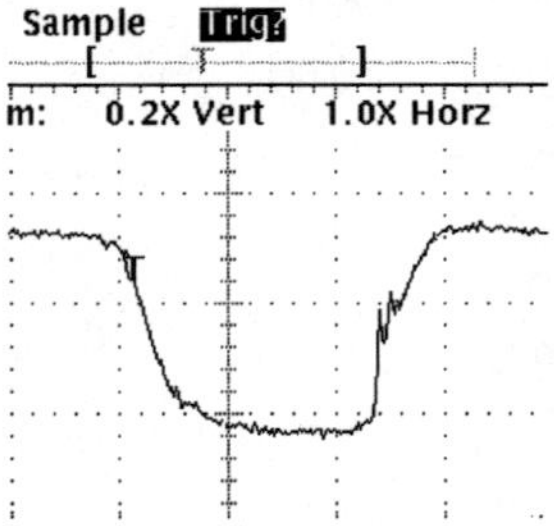

50ns/div 2.0V/div

FIGURE 4. Gap voltage wave of –338kV

The test with an electron beam through the cell vacuum gap is under way. The preliminary results are satisfactory.

Summary

The results of high-voltage test for the prototype induction cells indicated that the cells are adequate for –250kV operation, although the test with beam through the gap is still in progress.

The previous measurements for the cells transverse impedance [2] and magnet alignment [3] show that the prototype cell meets the demands for accelerating 3kA electron beam. In another words, the prototype cell is suitable for constructing an induction linac.

ACKNOWLEDGEMENTS

The authors would like to thank Prof. Li Xianwen and Prof. Cheng Nianan for their help and fruitful discuss.

REFERENCES

1. Wang Huacen, Zhang Kaizhi, Long Wen,et al., Physical Design Study On Linear Induction Accelerator Cell. High Power Laser and Particle Beams, 1996, 8 (3): 313
2. Zhang Kaizhi, Wang Huacen, Zhang Wenwei, et al., Transverse Impedance Measurement Of Induction Accelerating Cavity, High Power Laser and Particle Beams, 2000, 12 (2): 221
3. Zhang Wenwei, Li Hong, Wang Huacen et al. Pulsed Wire Magnetic Field Alignment Measurements On A High Current Induction Cell, High Power Laser and Particle Beams, 1996, 8 (2): 204

A Low Cost, New Method To Generate Double Pulses For Linear Induction Accelerator

Jianjun Deng[a,b], Zhiping Huang[b], Huacheng Wang[b]

[a] *Engineering Physics Dept., Tsinghua University*
[b] *Institute of Fluid Physics, China Academy of Engineering Physics P.O.Box 919-150,MianyangSichuan, 621900,P.R.China*

Abstract. A new idea to generate double pulse for linear induction accelerator by use of cable delay was proposed. Experimental results on generating double pulse are presented in the paper. And the results show that the new method works well.

INTRODUCTION

Pulsed power technology, especially repetitive pulsed power technology, has many applications in military and civil economy such as high power microwave, flash X-ray radiography and environment protection and so on. The rep-rate was limited to several kHz due to available high power switches. Recently, many efforts have been made to achieve very high rep-rate up to MHz pulsed power technology by use of solid-state switches [1]. In this paper, a new idea to generate double pulses time interval to several hundred nanoseconds was proposed for linear induction accelerator based on traditional spark gap switch technology. In the first section, a layout of the single pulse system to drive two induction accelerating cavities is introduced. In the second section, a new method to generate double pulses based on the same pulsed power system to drive the same cavities is described. Experimental results of double pulses are given in the third section. The discussion and conclusion on the results are carried out in the final section.

LAYOUT OF THE SINGLE PULSE INDUCTION ACCELERATOR MODULE

In the pulsed power system of the single pulse induction module, a Marx generator was used to charge a Blumlein PFL with impedance of 12 ohm and a cylindrical field-distortion spark gap switch. The spark gap switch is triggered when the Blumlein is charged to the peak value, and a high voltage pulse with amplitude of 250kV, pulse width of 100 ns is generated and feed into two induction cavities through two pairs of HV cables symmetrically. Two cables of each cavity have the same length and there is a time delay between the cables of adjacent cavities to synchronize the electron beam. Each cable is terminated by a $CuSO_4$ resistor with ~50 ohm to match the load. In each induction cavity there is 11 pieces of ferrite cores with B_s =0.4T and

CP650, *Beams 2002: 14th International Conference on High-Power Particle Beams,*
edited by T. A. Mehlhorn and M. A. Sweeney

B_r =0.3T which can induce 250kV pulse between the accelerating gap to accelerate the electron beam. Figure 2 is the output pulses of the cavities.

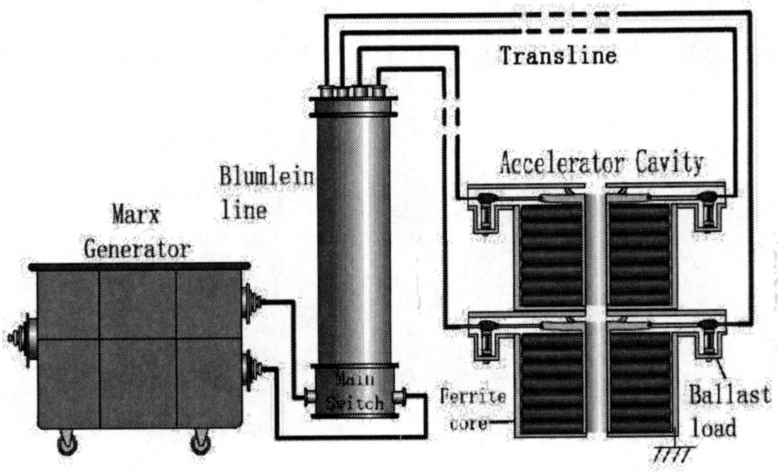

FIGURE 1. Layout of one Blumlein, single pulse to drive two induction cavities

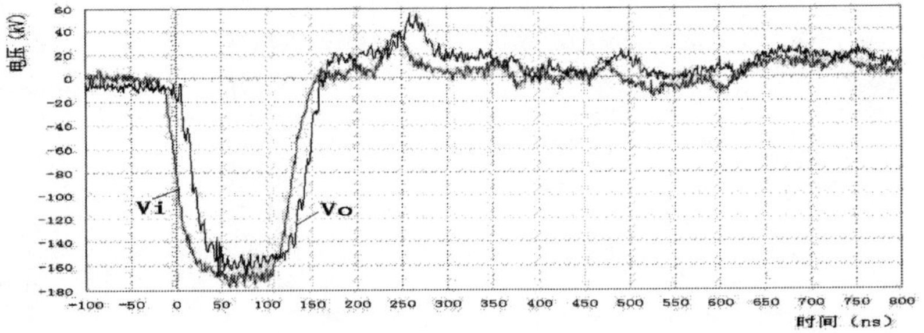

FIGURE 2. Output voltage waveform of the induction cavity at single pulse mode V_i is the input voltage and V_o the output voltage

IDEA TO GENERATE DOUBLE PULSES FOR THE INDUCTION CAVITY

For the linear induction accelerator, multi-pulse capability is of very importance. Therefore, how to generate multi-pulse in which time interval is in the range of several hundred nanoseconds to microseconds is a key problem and challenge for the pulsed power technology now. As we know, it's very difficult to achieve MHz rep-rate, high power, and high voltage up to several hundred kV pulses. The spark gap switch can be only operated in up to several kHz regime even working with blowing-gas. By use of solid state switches such high demands could be satisfied by connecting thousands of high power devices in serial and parallel, for instance high power MOSFETs, but it's very complicated and also very expensive. In this paper a low cost, new method to generate double pulses by cable delay is proposed and tested based on the traditional spark gap switch. The idea is so simple that you just need to change the length of one cable of each induction cavity as you want based on the same pulsed power system above. This will provide both two HV pulses for the induction cavities with time interval as you want and reset current for the ferrite toroids of the cavity.

EXPERIMENTAL RESULTS OF DOUBLE PULSES WITH THE INDUCTION MODULE

We have done a bench-test with a small ferrite core to check this idea whether the double pulses could be generated and the induction cavity could induce proper voltage at such high rep-rate condition. [2] After that the experiment was carried out with the real induction module at different time separations. Fig. 3 is the corresponding output voltage waveforms (V_o).

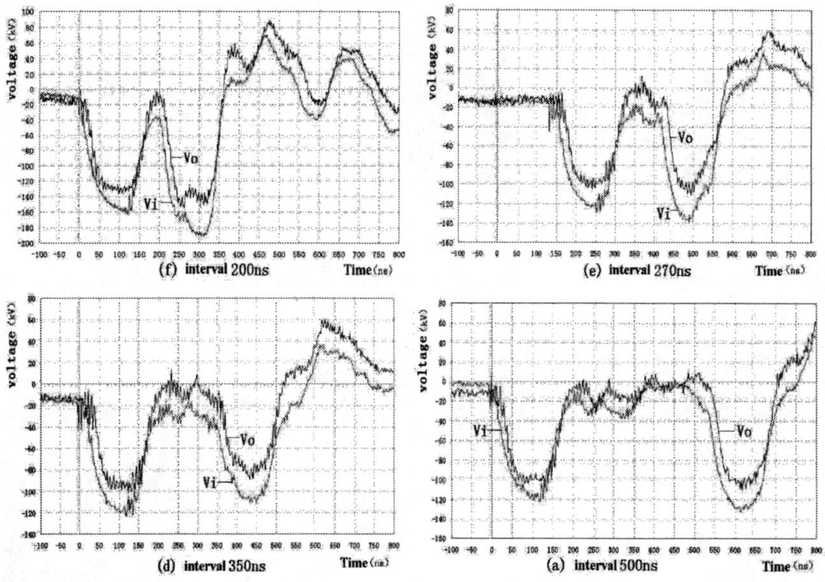

FIGURE 3. Output voltage waveforms of the induction cavity at double-pulse mode with different separations

DISCUSSIONS AND CONCLUSIONS

As shown above we have obtained double pulses with separation varied from 200ns to 500ns even longer by means of cable delay. The second output waveform changed a little due to mismatch and contribution of reflection of the first pulse to the second one. Good match could improve this. From the results above we can come to the conclusions that double pulses can be generated with short time separation up to 200 ns based on the traditional spark-gap technology by means of cable delay, the ferrite core has good high frequency characteristics and could be used for high rep-rate condition up to several MHz regime. Multi-pulses more than two could be generated based on this idea. Also double pulses could be achieved by triggering-delay of the Blumlein switch so that one Blumlein could be used to drive-four cavities (See Fig. 4).

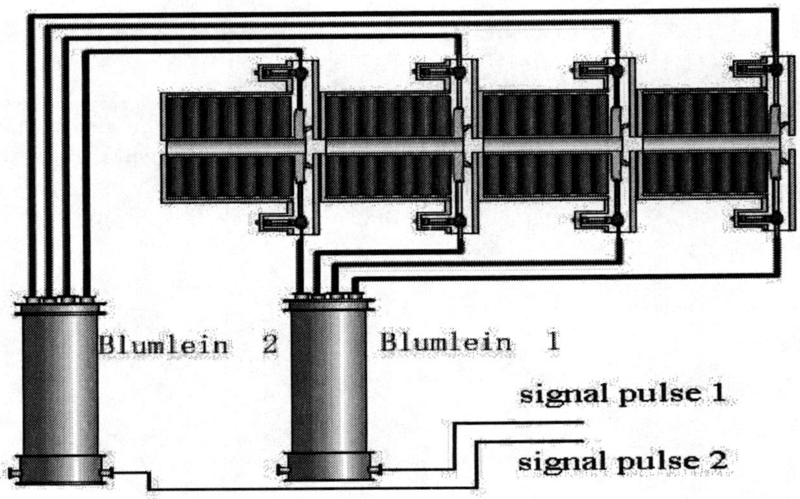

FIGURE 4. Schematic of one Blumlein-drive-four induction-cavities at double-pulse mode

REFERENCES

1. Deng, J.J ,Wang, J.G., Reiser, M., Godlove, T., Proc. Of International Particle Accelerator Conference, Vancouver, 1997.
2. Huang, Z.P., Master thesis, Graduate School, China Academy of Engineering Physics, 2002

Electrical Explosion of Conductors as a Source of High Electrical Field Strength Pulses

V.A. Burtsev, A.B. Andrezen, A.A. Drozdov, N.V. Kalinin

Efremov Scientific Research Institute of Electrophysical Apparatus
196641, St.-Petersburg, Metallostroy, NIIEFA, Russia

Abstract. Results of experimental, analytic and numerical researches of "idle" electrical explosions of conductors at the wide range of changing of current density j_m in maximum of stored magnetic energy are presented. Experimentally founded out monotonous dependence of locally optimized values of peaked field intensities versus maximum current density $E_m^* = kj_m$ is analyzed.

INTRODUCTION

One of applications of electrical explosions of conductors is the receiving of high-voltage pulses, first of all, for accelerating engineering and radiating sources[1]. For suppression of ionizing phenomena resulting in ignition of the shunting discharge, researchers usually apply different arc-quenching media. The comparative experiments, carried out by us early have allowed to choose powdery artificial quartz as working media for electroexplosive current breakers [2]. In the present work, the main attention is given to the analysis of questions connected with receiving of high intensity electrical fields. Used here experimental results have been received at electrical explosions of aluminum foils with length up to 1м in artificial powdery quartz on installation "Iren" with the inductive-capacitive storage parameters $C_0 = 15{,}3$ μF, $L_0 = 0{,}09$ μH, $U_0 = 80$ кV [3].

EXPERIMENTAL OPTIMIZATION OF FOIL EXPLOSIONS

The analysis of experimental results, obtained at research of electrical explosions of aluminum foils in powdery quartz has shown, that observed single-humped dependence of the maximal values of pulsed voltage versus the foil length $U_m = f(l)$ [3] can be explained by influence of two factors [3]. At the large lengths of samples the stored specific magnetic energy in current maximum does not sufficient for complete the process of explosion, that prevents the achievement of high peaked voltages, though in these cases the complete switching off the circuit with a zero current pause was observed. At small lengths the stored magnetic energy begins to be spent for development of the shunting discharge; peaked voltages in these cases are also much below observed in optimum modes.

CP650, *Beams 2002: 14th International Conference on High-Power Particle Beams,*
edited by T. A. Mehlhorn and M. A. Sweeney

On Fig.1 results of two series of optimization experiments [3] are presented as dependencies $E_m = f(j_m)$. The values of maximal current density j_m, i.e. the current density in a maximum of stored magnetic energy, is chosen as the characteristic of the rate of energy input in metal. The maximal values of peaked field strength in these series represent the locally optimized quantities E^*_m. It is necessary to note, that electrical conductivity of explosion products, corresponding to the maxima of field strength in both series reaches identical value $\sigma_{mU} = 0{,}26.10^4$ (Ohm.cm)$^{-1}$.

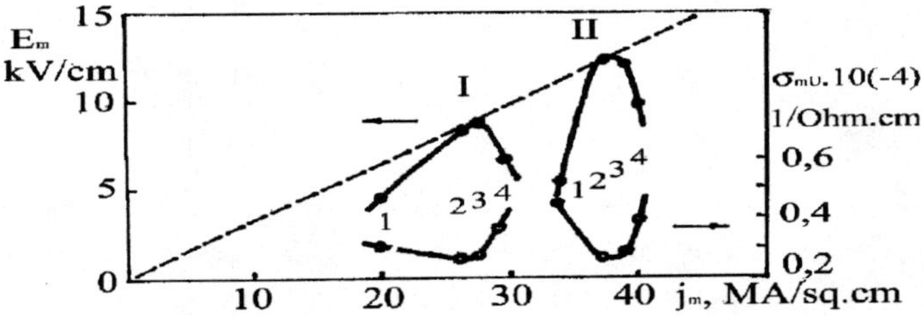

Figure 1. Dependencies $E_m = f(j_m)$ and $\sigma_{mU} = f(j_m)$ at different lengths and cross-sections of foils. l (mm): 400 - 1; 600 - 2; 670 - 3; 930 - 4. s (mm^2): 1,5 - I; 0,4 - II.

As it is possible to see on Fig.2a, locally optimized values for two considered series of experiments (dark circles), and also optimized data, received in work [4] at explosions of copper foils in powdery quartz (light circle), lie on the linear graph $E^*_m = g j_m$, where = 0,33.10^{-3} Ohm.cm.

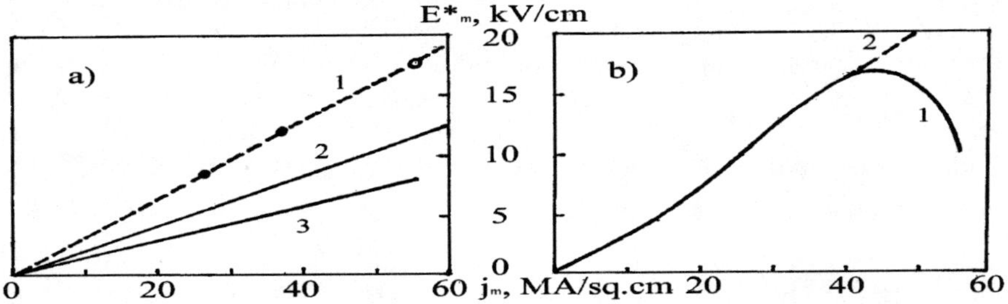

FIGURE 2. Dependencies $E^*_m = f(j_m)$, built:
a) - on data of experiments [3, 4] - graph 1 and calculation [5] - graph 2, [6] - graph 3;
b) - on data of calculations with using (7) - graph 1 and corrected (7) for $j_m \geq j_m^{cr}$ - graph 2.

On the other hand, one can write down for optimized regimes:

$$E^*_m = \frac{j_{mU}}{\sigma_{mU}} = \frac{I_{mU}}{s_{mU}} . \frac{1}{\sigma^*_{mU}} . \frac{s_{mI}}{s_{mI}} = \alpha . \frac{s_{mI}}{s_{mU}} . \frac{1}{\sigma^*_{mU}} . j_m = g j_m \, , \qquad (1)$$

where $\alpha = I_{mU} / I_m$ - quantity, near to 1. As we see, factor g is defined mainly by expansion degree and electrical conductivity in the voltage peaks. There is a lot of physical models, with help of which the authors try to describe transport processes at transitions metal - nonmetal and metal - nonmetal-plasma. If to use these models and numerically to define minimum conductivity $\sigma_{\min}$ at the density of liquid metal, it is possible to estimate g by calculation way. Among the considered models, the wide-range models [5,6] give the closest results to experimental date: $\sigma_{\min} \approx 0{,}4.10^4$ and $0{,}6.10^4$ (Ohm.cm)$^{-1}$ accordingly, which define the graphs 2 and 3 on Fig.2a.

APPROXIMATE ANALYSIS OF EXPLOSION PROPER

Mechanical extrapolation of the graph 1 on Fig.2a to area of higher current density shows an opportunity of the achievement larger peak field strengths. This important conclusion requires the additional analysis. The point is that in experiments described in [3], a displacement of an initial point of the stage of natural explosion has been revealed on the curve $R/R_0 = f(w_{inp})$ in the direction of larger values of $R_{\exp}/R_0$ and $w_{\exp}$ at increasing of j_m. Besides during this stage, the acceleration of the growth of relative resistance versus specific input energy was observed. These experimental facts have been to take into account at studying the perspectives of achievement high field strengths by increasing of j_m.

The analyze of the stage of natural explosion was carried out with help of equation for LR- circuit, the solution of which relatively $E(t)$ at any $R(t)$ has view:

$$E(t) = \frac{R(t)}{l} I_0 \exp\left(-\int R(t)/L \, dt \right) \qquad (2)$$

Here, $I_0(0)$ - current in the initial point of natural explosion. The made estimations have shown, that its resistance can be written down in the following kind $R(t) = R_{\exp} + kt$, where $R_{\exp}$ and k - values, independent versus t.

The analysis of expression (2) shows, that the intensity of electric field reaches the maximum value at the moment of time $t_m = \sqrt{L/k} - R_{\exp}/k = \sqrt{\tau_0 \tau_R} - \tau_R$, where $\tau_0 = L/R_{\exp}$ and $\tau_R = R_{\exp}/k$. As we see, maximum of electrical field E_m, interesting for us, will take place at the initial constant circuit τ_0, exceeding the time constant of resistance growth τ_R. Magnitude of E_m^* is defined by expression

$$E_m = \frac{I_0 R_{\exp}}{\sqrt{e}l} \sqrt{\frac{\tau_0}{\tau_R}} \exp\left(\frac{\tau_R}{2\tau_0}\right), \text{ at } \tau_0/\tau_R \to \infty \text{ we have } E_m \cong \frac{I_0 R_{\exp}}{\sqrt{e}l} \sqrt{\frac{\tau_0}{\tau_R}} \qquad (3)$$

Now let use approximation formulas in system of units CI, reflecting indicated above phenomena, connected with the initial point of explosion [3].

$$R_{\exp}/R_0 = R^0_{\exp}/R_0 + 0{,}3.10^{-33} j_m^3 = 14 + 0{,}3.10^{-33} j_m^3 \qquad (4)$$

$$w_{\exp} = w^0_{\exp} + 0{,}06.10^{-33} j_m^3 = 2{,}5.10^6 + 6.10^4.10^{-33} j_m^3 \qquad (5)$$

In contrast to [3] we shall use linear approximation for $R/R_0 = f(w)$

$$R/R_0 = R_{\exp}/R_0 + \delta(w - w_{\exp}) \quad \text{in range of } w \geq w_{\exp}$$

At transition to veriable t in chosen system of coordinates we have

$$R/R_0 = R_{\exp}/R_0 + a\delta j_m^2 t \text{, where } \delta = 0{,}7 + 5.10^{-22} j_m^2, \; a = 0{,}24 j_m^{-1} \qquad (6)$$

The substitution of formulas (5, 6) in the expression (3) gives

$$E_m^* = \left(\frac{2\rho_0 a\delta}{e\sigma_0}(w_{vap} - \Delta w_{\exp}) \right)^{1/2} .\exp\left(\frac{(R_{\exp}/R_0)^2}{4\rho_0\sigma_0 a\delta(w_{vap} - \Delta w_{\exp})} \right) j_m \qquad (7)$$

In carried out transformations, the condition of locally optimized regimes was used: $LI_o^2/2 = m_0(w_{vap} - \Delta w_{\exp})$, где w_{vap} - specific heat of vaporization, $\Delta w_{\exp} = w_{\exp} - w_{\exp}^0$.

On Fig.2b, the dependence $E_m^* = f(j_m)$ is presented, which was obtained with using of the expression (7). As one can see after $j_m = j_m^{cr} = 4.10^7$ A/см^2, a sharp fall of E_m^* is observed. However, if to assume, that dependencies (4, 5) and the function $\delta(j_m)$ reach the saturation at j_m^{cr}, in result of this the quantities $R_{\exp}/R_0$, $w_{\exp}$ and δ in area $j_m \geq j_m^{cr}$ become independent versus j_m, one can obtain practically linear dependence, well agreeing with experimental data: $E_m^* = 4{,}1.10^{-4} j_m$.

In works [3] the discharge stage, preceding to the stage of natural explosion of foils in powdery quartz was treated as a stage of the overheat liquid metal relatively temperature of quasi-stationary boiling at atmospheric pressure (2450 °C). According to estimations, limiting parameters $R_{\exp}, w_{\exp}$ at $j_m \approx j_m^{cr}$ correspond to temperature of liquid metal of 4000-5000 °C. At using of dry powdery quartz it is very difficult to present external pressure, which could cause such increase of boiling temperature. It gives the base to assume, that metal state gets in metastable area and in a limiting case reaches the liquid spinodal. Then the stage of natural explosion at $j_m \geq j_m^{cr}$ could be understood as process of vaporization in the field of two-phase states.

CONCLUSION

1. Locally optimized values of field strength E_m^* are defined by minimum electrical conductivity at transition of metal-nonmetal.
2. Effect of displacement of explosion initial point and amplification of dependence $R/R_0 = f(w_{inp})$ at increasing of j_m develops in the range of (1-4).10^7 A/cm^2.
3. Observed linear dependence $E_m^* = gj_m$ speaks about good perspectives for obtaining of pulsed field strength of ~ 100 кV/cm at transition to super-high current densities (≥ 10^9 A/cm^2).
4. The further researches of similar regimes would have not only large practical, but also fundamental importance.

REFERENCES

1. V.A. Burtsev, N.V. Kalinin, A.V. Luchinsky. Electrical explosion of conductors and its application in electrophysical installations.-M.: EnergoAtomIzdat, 1990 (In Russian).
2. V.A. Burtsev et al. Zhurnal Techn. Fiziki, 1977, т.47, № 8, p. 1653-1661 (In Russian).
3. A.B. Andrezen, V.A. Burtsev et al. Zhurnal Techn. Fiziki , 1980, т. 5, № 11, p. 2283-2294; Preprints of Efremov Institute К-0463, 1980 and К-0501 , 1984 (In Russian).
4. V.A. Demidov, V.I. Skokov. In Proceed. of 7-th Intern. Conf. on generation of megagaus magnetic fields and related experiments (Sarov, 1996), v. 1, p. 381-384.
5. A.Ja. Polishuk. Teplofizika visokikh temperatur. 1990, т. 28, № 5, p. 877-886 (In Russian).
6. Y.T. Lee, R.M. More. Phys. Fluids. V. 27, No 5, May 1984, p.1273-1285.

Measuring Multi-Megavolt Diode Voltages*

N. R. Pereira[a], S. B. Swanekamp[b], B.V. Weber[c], R. J. Commisso[c], D. D. Hinshelwood[c], S. J. Stephanakis[c]

[a]*Ecopulse, Inc., PO Box 528, Springfield, VA 22150,*
[b]*Titan/Jaycor, McLean, VA 22102*
[c]*Plasma Physics Division, Naval Research Laboratory, Washington DC 20375*

Abstract. The voltage in high-power diodes can be determined by measuring the Compton electrons generated by the diode's bremsstrahlung radiation. This technique is implemented with a Compton-Hall (C-H) voltmeter that collimates the bremsstrahlung onto a Compton target and bends the emitted Compton electron orbits off to the side with an applied magnetic field off to Si *pin* diode detectors. Voltage is determined from the ratio of the Compton electron dose to the forward x-ray dose. The instrument's calibration and response are determined from coupled electron/photon transport calculations. The applicable voltage range is tuned by adjusting the position of the electron detector relative to the Compton target or by varying the magnetic field strength. The instrument was used to obtain time-dependent voltage measurements for a pinched-beam diode whose voltage is enhanced by an upstream opening switch. In this case, plasmas and vacuum electron flow from the opening switch make it difficult to determine the voltage accurately from electrical measurements. The C-H voltmeter gives voltages that are significantly higher than those obtained from electrical measurements but are consistent with measurements of peak voltage based on nuclear activation of boron-nitride targets.

It is difficult to measure the voltage across a high-power (TW) bremsstrahlung diode, especially when this voltage is in the multi-megavolt range. Conventional electrical techniques, such as capacitive monitors, do not work well at high power because the diode contains plasma. In the plasma-free, liquid-filled pulse-forming line capacitive monitors work well, but they are usually so far electrically from the diode that they require a substantial inductive correction. Radiation-based methods [1] work for sub-MV voltages and perhaps when the radiation is tens of MeV, but do not work well in the intermediate range from 1 MV to 10 MV. The C-H voltmeter is designed for this intermediate range.

Current and voltage in a diode determine how much radiation the diode produces. Electrons accelerated by a voltage, V, interact with a radiation target to produce a spectrum of photons. The spectrum's highest photon energy $h\nu_e=eV$ is the electron energy corresponding to the voltage, but most of the photons have energies ¼ to 1/3 of the diode voltage. Therefore, it is possible to determine the voltage by measuring the hardness of the radiation spectrum.

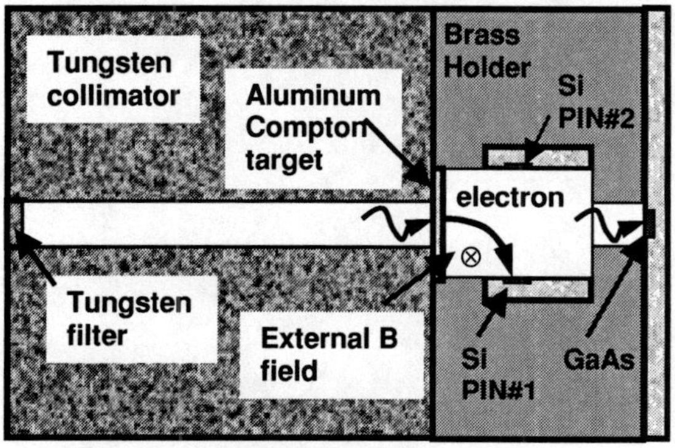

Figure 1. The Compton-Hall voltmeter geometry.

For photon energies less than 0.5 MeV, transmission through suitable filters increases rapidly with increasing voltage, and the voltage can

CP650, *Beams 2002: 14th International Conference on High-Power Particle Beams,* edited by T. A. Mehlhorn and M. A. Sweeney

be inferred from the ratio of photon intensities transmitted though the different filters. Instruments that implement this differential filter technique use multiple filters. Good results have been obtained with 7 filters and the photon energy spectrum can be determined with still more filters.[2,3]

The differential filter technique is limited by the atomic number of the filter materials. The energy at which photoelectric absorption dominates Compton scattering increases with atomic number Z roughly as $h\nu_k \approx Z^2 \times Ry$ (where Ry = 13.6 eV is Rydberg's constant), so that, even for filters with the highest atomic number material (U, with Z = 92 and $h\nu_k \approx$ 100 keV), the ratio of transmitted photon energies does not vary sufficiently with voltage to measure diode voltages much higher than 1 MeV. The Compton-Hall (C-H) voltmeter replaces differential filtering with separating the radiation-produced Compton electrons according to energies.[4] Our paper extends the C-H voltmeter range to the 2-5 MV range, addresses the sensitivity of the voltmeter to the electron angle of incidence on the converter, and presents time-dependent data.

Figure 1 is a sketch of the C-H voltmeter. Photons incident from the left pass through a tungsten filter into a collimator and strike an aluminum Compton target. Most of the photons pass through the target and reach a GaAs photoconductive detector (PCD). Some of the photons generate Compton electrons in the Compton target and a portion of these Compton electrons are bent by a magnetic field toward a silicon *pin* detector (# 1) on the bottom wall. Since the Compton electrons' radius of curvature varies as $r_e \propto \sqrt{\gamma^2 - 1}/B$, the signal recorded by *pin* diode #1 is sensitive to both the electron energy and the magnetic field. The voltage is determined from a ratio of the electron dose in *pin* #1 to the direct x-ray dose in the GaAs PCD. A second Si *pin* detector (# 2) on the top wall evaluates any background, e.g., from scattered or insufficiently shielded direct x-rays. The instrument is called the Compton-Hall voltmeter because the technique uses Compton electrons whose current is affected by a magnetic field.

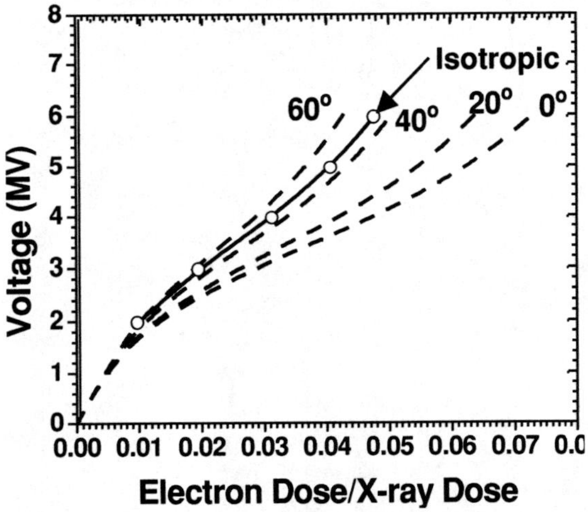

Figure 2. Calculated calibration curves in the forward direction for the CH-voltmeter for different incident electron angular distributions.

How the C-H voltmeter works can be qualitatively understood by comparing r_e to the cavity's geometrical scale length. However, the response can not be calculated analytically but must be computed with a radiation transport code such as Sandia's Integrated Tiger Series of codes.[5] The computations are done in three basic steps. First the instrument's response to mono-energetic photons is computed by directing them onto the Compton target and calculating both the forward x-ray dose per photon flux in the PCD and the Compton electron dose per photon flux in the *pin* diodes. All doses are determined as a function of photon energy by repeating these calculations for many photon energies. The next step is to compute the x-ray source spectrum from the interaction of an assumed diode electron distribution with the bremsstrahlung target used on Gamble II. The bremsstrahlung target consisted of a 100-μm thick tantalum foil followed by a 3-cm thick aluminum beam stop. These computations create a spectrum of x-rays with different directions and energies. The last step is to take the x-ray spectrum in the direction of the voltmeter, apply any filters, and then convolve the spectrum with the instrument's photon response.

Figure 2 shows the dependence of the voltage on the ratio of the electron dose in *pin* diode # 1 to the x-ray dose in the GaAs PCD for several angles of incidence of the primary electron beam on the bremsstrahlung target. In these calculations, the C-H voltmeter is positioned on the diode axis approximately 1 m from the source. This figure shows that the voltage is relatively insensitive to electron angle of incidence for voltages less than 2 MV. Therefore, for V< 2 MV, the voltage can be obtained to within 10% accuracy directly from the dose ratio without any additional information about the electron angular distribution on the radiation target. The V < 2 MV results in reference [4] are therefore still valid even though the role of electron incidence angle was not considered.

For V > 2 MV, figure 2 shows that voltage derived from the voltmeter depends on the electron angle of incidence. In this case, additional information about the electron angle of incidence on the bremsstrahlung target is required to determine the voltage. This sensitivity to electron angle of incidence occurs because the photon spectrum is significantly harder along the direction of the high-energy electrons. For comparison, Figure 2 also shows the dependence of the voltage on dose ratio for an isotropic distribution of angles in the forward 2π radians ($0^o < \theta < 90^o$). It is interesting to note that the calibration curve for an isotropic angular distribution is similar to that for a 45° angle of incidence.

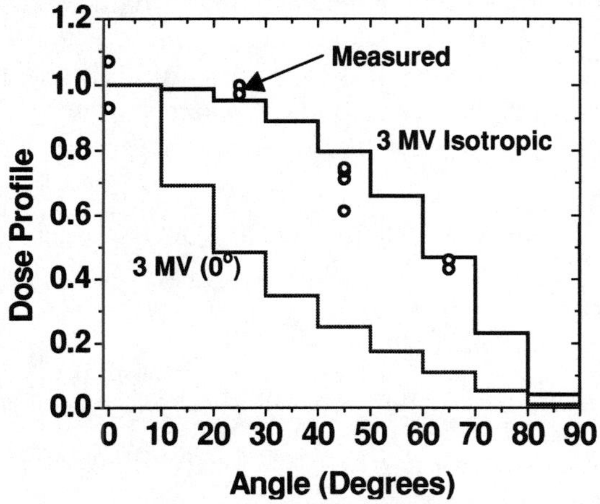

Figure 3. Variation of dose with angle for perpendicular and isotropic incidence of the electrons on the converter.

Time averaged dose measurements provide a convenient way to estimate the electron angle of incidence on the converter. Figure 3 compares the angular dose profile for different electron angular distributions on the converter. For 0° angle of incidence (lower line) the dose is peaked in the forward direction and falls off rapidly with increasing angle. In contrast, for isotropically incident electrons (top line) the dose shows less variation with viewing angle. The dose profile measured on Gamble II with time-integrated thermoluminescent dosimeters (TLDs) is plotted as open circles. Since the measured angular dose pattern agrees well with an isotropic electron incidence, we use the isotropic calibration curve of Figure 2 to infer the diode voltage from the C-H dose ratio.

The performance of the C-H voltmeter was tested on the Gamble II accelerator at the Naval Research Laboratory. A plasma-opening switch was used as a power-conditioning element to produce voltages of 2 to 5 MV across the diode. Figure 4 shows the current measured upstream of the plasma opening switch (I_G), and two diode voltage measurements. The voltage V_{cor} is determined from electrical measurements and includes a correction for inductive effects. The voltage V_{CH} is the voltage obtained from the C-H voltmeter.

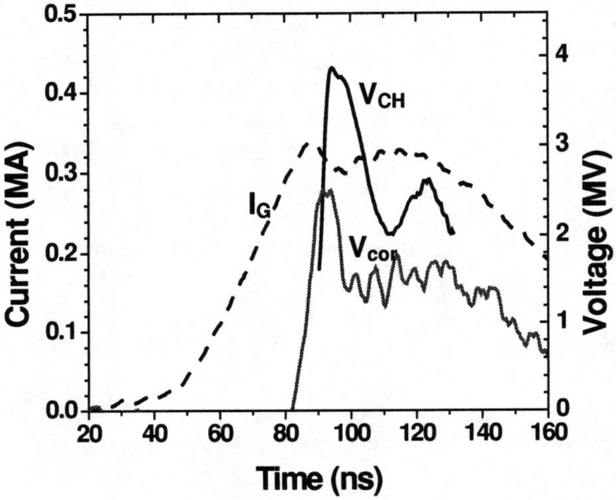

Figure 4. The voltage from the CH voltmeter (V_{CH}) compared to an inductively corrected voltage from electrical measurements (V_{cor}).

The electrically measured voltage with its 2.5 MV peak is much lower than the 3.9 MV peak obtained with the C-H voltmeter. An independent measurement of the endpoint voltage using proton activation of boron-11 in boronnitride targets gives 3.6±0.2 MV.[6] Therefore, the voltage shape from the C-H voltmeter is consistent with the measured x-ray dose rate and the peak voltage is consistent with the peak voltage obtained from proton activation. In conclusion, the C-H voltmeter is an exciting new way to measure the voltage in high-power diodes.

[*] NRL's portion of this work was supported by Defense Threat Reduction Agency.

1. as implemented by e.g., J. Rauch at Maxwell Laboratories and by J. Riordan and others at Physics International in the 1980's. See also Reference 3.

2. S. G. Gorbics and N. R. Pereira, Rev. Sci. Instrum. **64**, 1835 (1993).

3. J.C. Riordan, J.E. Faulkner, J.R. Goyer, D. Kortbawi, J.S. Meachum, R.S. Menderhall, I.S. Roth, and B.A. Whitton, Rev. Sci, Instrum. **63**, 4792. And most recently by J. Riordan, S. Chantrenne, et al., these proceedings.

4. N. R. Pereira, S. G. Gorbics, and D. M. Weidenheimer, Proc. 11th Intl. Conf. Particle Beams, Prague, p. 1071 (1996).

5. J.A. Halbleib, R.P. Kensek, G.D. Valdez, S.M. Seltzer, and M.J. Berger, IEEE Trans. Nucl. Sci. **41**, 1025 (1992).

6. Dr. Frank Young obtained the proton activation measurements.

On the Beam Loading of Induction Voltage Modulator

K. Horioka, M. Nakajima, M. Watanabe, M. Honda, M. Shiho
K. Takayama* and J. Kishiro*

Department of Energy Sciences, Tokyo Institute of Technology
*High Energy Accelerator Research Organization (KEK)**

Abstract Effects of beam loading on the induction voltage modulator have been experimentally evaluated using a FET (field effect transistor) based driving pulser and a resistive secondary circuit. The core materials used in this study, are made of a nano-crystalline alloy (Finemet) and a Co-amorphous alloy. The voltage modulation is evaluated as a function of magnetization rate dB/dt, magnetization level of the magnetic material and the current rise-rate of the secondary loop (dI/dt). Experimental results indicate that the voltage modification depend not only on the magnetization rate but also on the initial state of the magetic materials and also flux swing B from it. These results indicate that the response of the modulator against the loading depends on the transient magnetic-domain motion in the core materials.

INTRODUCTION

Recently, interests in a new concept of accelerator system based on repetitive and controllable inductive voltage modulators are rapidly growing[1,2]. In the concept of Induction Synchrotron (IS)[3,4], the acceleration and the longitudinal modulation are independently provided to accelerate a high intensity proton beam bunch. Unlike conventional RF synchrotrons, two types of voltage pulses are excited by two independent induction modulators in the IS. One generates a so-called barrier bucket bipolar waveform with short pulses (40-100nsec), which captures the injected beam bunches and/or stacks up them to a super-bunch[5]. The other makes a flat-top, long pulse (500-1500nsec) waveform followed by a short reset pulse, which accelerates the super-bunch. As the beam is expected to circulate about 10^4 times in the acceleration ring in the IS system at KEK 12GeV-PS, precise control of the waveform is required.

On the other hand, a high current accelerator system is a key component for heavy ion inertial confinement fusion (HIF)[6]. Especially an accelerator system based on the induction recirculator appears to be an economically attractive option [7,8,9,10]. In the HIF system, a carefully programmed voltage waveform is required under the heavy beam loading condition

When a steep beam bunch comes into the induction unit, it conceivably causes unacceptable voltage modification and/or beam instability. For the ion acceleration, the beam loading effect is one of the important issues because the careful control of the acceleration voltage waveform is required to accelerate and modulate the beam shape of the ions. Induction accelerators intrinsically have a capability of high current acceleration than RF accelerators. However the driving circuit has finite impedance and the beam bunches have to be recirculated many times in the accelerator ring under a variable loading condition. For example, stacked FETs are employed as the switching device of high rep-rate induction modules in the Induction recirculator system [11,12]. In this case, the module has a finite impedance composed of stray inductance of the driving circuit and the ON-resistance of the stacked FETs, which may cause a voltage modulation of intolerable level against the beam loading. Thus, when we use the induction modulator for acceleration of a high intensity beam bunch, the evaluation of the voltage modulation by the beam load is important from the point view of the beam dynamics.

EXPERIMENTAL SET UP

As mentioned in the previous section, the induction modulator is considered to be equivalent to a "non-

CP650, *Beams 2002: 14th International Conference on High-Power Particle Beams,*
edited by T. A. Mehlhorn and M. A. Sweeney

ideal transformer circuit" with the beam acting as a single-turn secondary load. Figure.1 shows a schematic of the beam loading. Typical waveforms of a beam current and a modulated acceleration voltage by the beam loading, are schematically expressed. In Fig,1, Vm shows the modulation voltage induced by the beam loading. As the return current of the beam flows around the induction module, not only the impedance of the driving circuit but also the non-linearity of the core material would modify the acceleration voltage and those effects are difficult to estimate. Then the effect of beam loading on the acceleration voltage of the induction modulator should be experimentally estimated.

Typical ferromagnetic cores were employed in this study; a nano-crystalline alloy (Finemet FT-1H) [13,14], and a cobalt-based amorphous alloy(ACO-5H). The wound-tape-type cores have a toroidal shape with inner radius Ri of 30mm, outer radius Ro of 77.5mm and width of 26mm which are placed in a coaxial enclosing structure. The tape with 20μm thickness is insulated by SiO2 coating with packing factor of 0.7. At first the magnetic core is completely demagnetized by a damped oscillation waveform of a demagnetization circuit.[15] Before the application of secondary current pulse, the primary circuit drives an acceleration voltage to bias the magnetization levels of the cores and the product defines progresion of the magnetization. The beam current load was simulated by a resistive secondary circuit.

EXPERIMENTAL RESULTS

We expected that not only the dI/dt, but the magnetization rate dB/dt, its time scale t, and the initial shape of domain in the magnetic core B, should affect the response against the pulsed loading.

Namely, we predicted that Vm is a function of dI/dt, dB/dt, t and B, as

$$Vm = f(dI/dt, dB/dt, t, B) \qquad (1).$$

So, before the measurements, the core was completely demagnetized. Thereafter, progression of mgnetic domain was controlled by the voltage level and the delay from the start of the primary pulsing. By using this procedure, we could define the initial shape of magnetic domain in the core before the pulsed loading. The non-linear response is expected to depend on the domain wall shape. So we investigated core responses in typical operational region of magnetization levels; one is in a low magnetization rate region which can be described with bar-domain model and the other is in a region with saturation wave model[15].

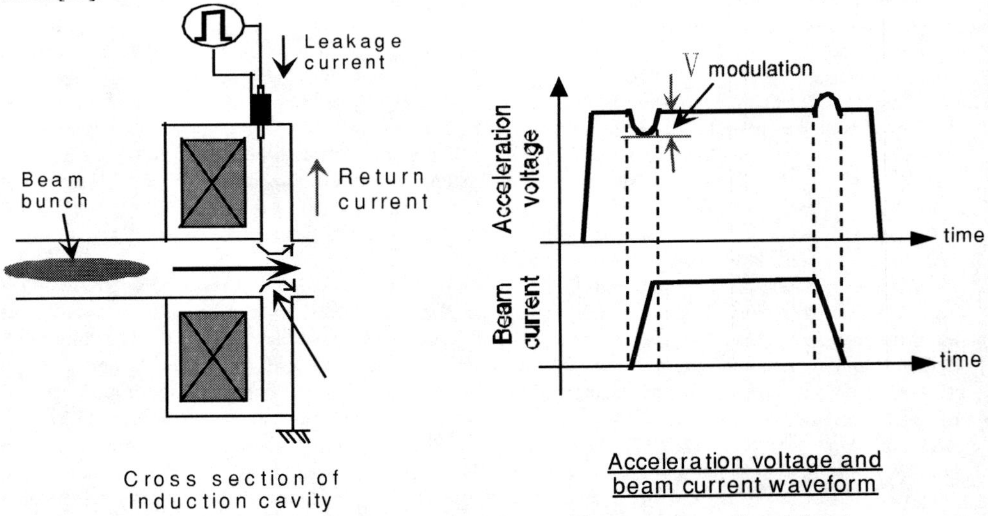

Fig.1 Schematic illustration of beam loading, typical shapes of beam current, and modulated acceleration waveform

Modified voltage values versus pulsed loading were investigated as a function of the progression of magnetization B at its rate of dB/dt=0.12 T/μs. The biased magnetization level before the pulsed loading is adjusted by controlling the delay time. This magnetization rate correponds to the magnetization based on the bar-domain mode for both materials.

Changes of the acceleration voltage against the biased-magnetization level are shown in Fig 2(a). The modulated values were almost constant as long as the core was not saturated (the value of saturation induction is 1.35T for Finemet and 0.62T for Co-amorphous).

Changes of the acceleration voltage against the pulsed loading at high magnetization rate are shown in Fig.2 (b). In this magnetization rate level, both magnetic core is expected to operate with magnetization corresponding to the saturation wave mode (dB/dt=4.3T/μs for Finemet, dB/dt=2.8T/μs for Co-amorphous). As shown in the figure, the values of voltage modulation gradually decreased against the progression of magnetization (B).

The difference would be explained by the domain mode motion in the ferromagnetic cores. If we operate the core at bar-domain mode region, the domain wall grows at different place but with the same wall shape. Then the core response is considered to be constant in spite of the magnetization levels, as long as far from the saturation region, as shown in Fig.2 (a). On the other hand, when the induction module is operated at saturation wave mode region, the boundary of the domain wall becomes gradually smaller with progression of the magnetization. Then, the core response could be gradually weakened with the magnetization progress, as in Fig.2 (b).

To make clear the effect of domain motion in the material, we have reinvestigated precise magnetization curves as a function of initial magnetization state. Fig.3 shows the typical magnetization curves of Finemet (a) and Co-amorphous (b). In the figure, two groups of B-H curves corresponds to the magnetization process from different initial state; from demagnetized state B=0 and from the remanence condition B=Br. As Fig.3 shows, magnetization processes from the different initial state have different traces on the B-H plane especially at high magnetization rate, and this is the cause of the different response against the loading.

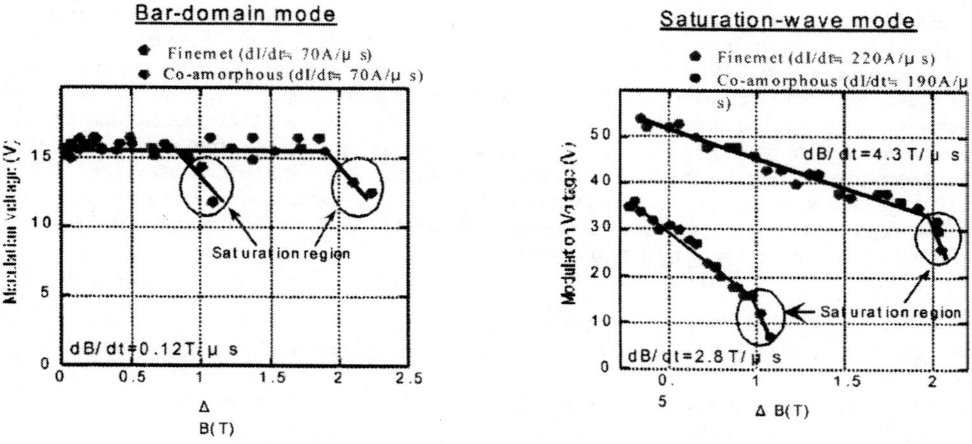

(a) (b)

Fig.2 Modulation voltages versus flux swing of magnetic materials for low magnetization rate (a) and high magnetization rate (b)

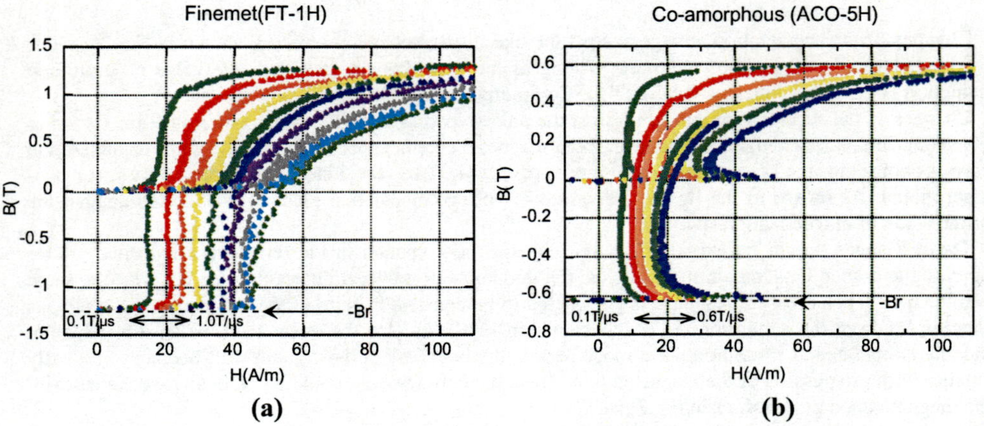

Fig.3 Magnetization curves as a function of dB/dt for Finemet (a) and Co-Amorphous (b)

CONCLUSIONS

Effects of beam loading on induction voltage modulators have been experimentally evaluated. It is clarified that; (1) the voltage is modulated differently for short and long loading pulses, (2) the voltage modifications are almost the same level against B when the cores are operated in bar-domain mode region, (3) on the other hand, the voltage modulation gradually decreases against B at the regime of magnetization rate corresponding to the saturation wave mode. These results indicate that the core response against the modulation of beam load current, is dependent not only on the magnetization rate ; dB/dt, and rise rate of the secondary current ; dI/dt, but on the initial shape of magnetic domain and the progression of magnetization levels. All of the experimental results denote that the modification levels of acceleration voltage depend on the transient domain motion in the magnetic core material and those effects can be predictable by investigating the precise characterization of the magnetic materials.

The data base of magnetic response of the modulator [16,17,18], together with the beam physics issue and the engineering issues of repetitive and controllable modulator, allow us to evaluate the concept of accelerator systems based on inductive voltage modulator.

REFERENCES

[1] M.A.Newton, F.J.Deadrick, R.L.Hanks, et.al., Fusion Engineering and Design 32-33 (1996) 259-266

[2] C.Burkhart, L.Ahle, J.R.Bayless, R.Hanks, E.Kelm, Nucl. Instr. Meth., A464 (2001) 452-456.

[3] K. Takayama and J. Kishiro, Nucl. Instr. and Meth. A451/1, 304(2000).

[4] J.Kishiro, K.Takayama, K.Horioka, M.Watanabe et.al., Proc. of EPAC, 1966(2000).

[5] K.Takayama et.al., Proc. of European Particle Accelerator Conference, 382-384(2000).

[6] Edward. P.Lee and J.Hovingh, AIP Conference Proc. No249. PT.2. (1992) 1713-1724.

[7] J.Barnard et.al., Nucl. Instr. Meth., A415 (1998) 218-228.

[8] M. Watanabe, M. Nakajima and K. Horioka, Nucl. Instr. and Meth. A464, (2001)440-444.

[10] K.Horioka. M.Nakajima, M.Shiho, Nucl. Instr. Meth., A415 (1998) 291-295.

[11] H. Kirbie et.al., 10th IEEE Int. Pulsed power Conf. (1994)441-447.

[12] H. Kirbie, W. R. Cravey, S. Hawkins et.al., 9th IEEE Int. Pulsed power Conf. (1993)415-418.

[13] S. Nakajima, S. Arakawa, Y. Yamashita and M. Shiho, Nucl. Instr. and Meth. A331, 318-322(1993).

[14] Y. Yoshizawa, K. Yamauchi, T. Yamane and H. Sugihara, J. Appl. Phys. 64, 10(1988) 6047-6049.

[15] M.Watanabe, M.Nakajima, M.Shiho, K.Horioka et.al., Rev. Sci. Instr., Vol.73, 4 (2002) 1756-1760.

[16] A.Molvic, A.Faltens, L.Reginato, M.Blaszkiewicz, R.Wood, Nucl. Instr. Meth., A415 (1998) 315-319.

[17] C. H. Smith, J. Appl. Phys. 67, 9, 5556(1990).

[18] A.W.Molvik, A.Foltens, Nucl. Instr. Meth., A464 (2001) 445-451.

Material dependence of ferroelectric emission under strong accelerating field

Yasushi Hayashi and Eiki Hotta

Department of Energy Sciences, Tokyo Institute of Technology
4295 Nagatuta, Midori-ku, Yokohama 226-8502, Japan

Abstract, We have studied emission characteristics of a PZT ferroelectric cathode under influence of a strong accelerating field using two kinds of PZT materials, LZT-2 and APC850, with the same cathode geometry. The perveances of the electron guns differ by more than 100%. The difference of the emitted electron is found to be related to the difference of frequency characteristics of the materials, which implies that the electron emission is possibly caused by the local domain oscillation in the close vicinity of vacuum-ferroelectric-metal triple-junction. The experimental results provide a key-point, which may lead to a more complete understanding of electron emission from ferroelectric materials.

Introduction

Strong electron emission from a ferroelectric material was reported in 1988, and has been studied extensively in the recent decade[1]. In our research, efforts have been made to use the electron beam as a source suitable for high power microwave generation[2,3]. However it seems that the mechanism of the electron emission from a ferroelectric material still has not been thoroughly understood. Until now, it is commonly recognized that there are two kinds of electron emission: a weak and a strong emission. In the case of the weak emission, the electrons are emitted due to the tunneling effect associated with "domain switching" of the ferroelectric material, where the emission reveals strong material dependence. This emission process has been basically accepted by the researchers around world. In the case of strong electron emission, the mechanism has not been thoroughly understood yet and many problems are still left unsolved. Some researchers have suggested that the emission is caused by surface plasma formed on the cathode. This is true under some experimental conditions, and when this happens, emission should not show any dependence on the material properties other than the dielectric permittivity. However, in our recent studies, it is found that the formation of surface plasma depends on the electric field configuration[4]. When a strong accelerating field applies, the emission reveals field emission characteristics. In the present study, we have conducted electron emission experiments using two kinds of PZT materials with different material properties. It is found that the emitted current is also related to the frequency characteristics of the material, namely to the frequency of the domain oscillation. This suggests that the domain movement also give a contribution to the strong electron emission under our experimental conditions, as it does in the weak emission. In the following sections, the details of the observed results and discussions will be presented.

CP650, *Beams 2002: 14th International Conference on High-Power Particle Beams,*
edited by T. A. Mehlhorn and M. A. Sweeney

Experiment

The electron gun and pulsed power system used in the present study are the same with that described in reference 2 and 3. Two kinds of cathode material (LZT-2[5] APC850[6]) with the same geometry have been used to examine the emission characteristics of the electron gun. The material properties, as specified by the manufacturers, are given in Table 1.

The first experiment uses the LZT-2 cathode. Figure 1 plots a time evolution of the diode voltage and beam currents, where t=0 is defined as the time that the voltage starts to rise. The peak value of the diode voltage is approximately 430 kV. A positive trigger pulse is applied on the back electrode at 110 ns to start the electron emission at 180 ns. The beam current shows two phases during the diode operation: a rising phase in which the current increases from 0 to a steady value; and a steady phase in which the current basically follows a three-halves power rule of the diode voltage. The peak values of the beam current, as shown in curves 5-8, are approximately 200 A, and the emitted current density is 28.2 A/cm^2. The averaged perveance of the gun using LZT-2 cathode is 0.77 μP. This suggests that in the steady phase: 1) the beam current is "space-charge limited", here " " indicates that the space-charge limited current is similar to but not the Child-Langmuir current. and 2) the current amplitude is not related to the amplitude of the trigger voltage. The Child-Langmuir current limit is found to be 185 A from a simulation using a thermionic emission of the electron gun with the same geometry.

The second experiment is performed using a cathode of APC850 material. The diode voltage and the beam current are shown in Fig. 2. In this case, the beam current is 440 A for a diode voltage of 380 kV. This yields a perveance of 1.88 μP, which is more than twice that obtained using the LTZ-2 cathode. The Child-Langmuir current limit is 140 A for the diode voltage. Because the current is quite high in this case, we are unable to apply a diode voltage above 380 kV due to the internal impedance of the power supply. The emitted current density is 62.1 A/cm^2, which is a factor of two higher than that of the previous case.

Some studies have been conducted to find out the cause of the difference in emitted current. It is found that the difference is not caused by either shot to shot variation, prepoling conditions, fabrication error and surface plasma[4]. For these reasons, we are considering that the difference of the emitted current may be caused by the difference of the material properties. From Table I, It is found that these two materials have surprisingly similar properties, and the only significant difference is the frequency constant in radial direction. In the case of our cathode, this quantity approximately applies to the oscillation in parallel direction to the ferroelectric surface. This also implies that the domain oscillation may also contribute to the strong electron emission, under our experimental conditions, as it does in the case of weak emission. The difference of the contributions is that in a weak emission, the domain switching is occurred in perpendicular direction to the ferroelectric surface; while in the case of the strong emission, the direction of the domain oscillation is parallel to the ferroelectric surface and perpendicular to the silver stripe, and the oscillation is limited in the close vicinity of the vacuum-ferroelectric-metal triple point.

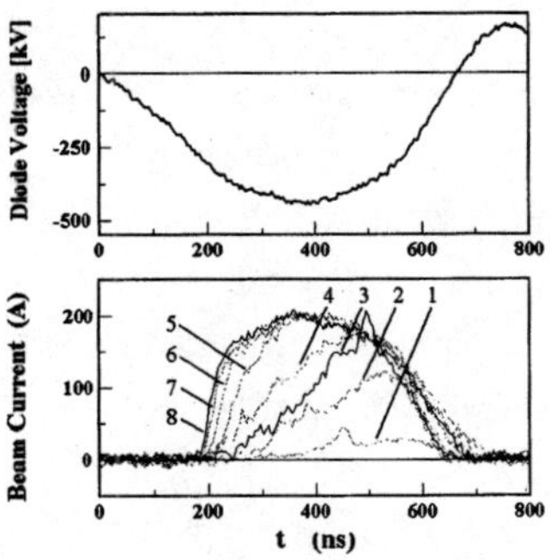

Fig. 1 Beam current obtained from LZT-2 cathode. Each trace corresponds to the trigger voltage as: 1) 1.12; 2) 1.28; 3) 1.44; 4) 1.76; 5) 2.08; 6) 2.56; 7) 2.88; and 8) 3.2 kV, respectively.

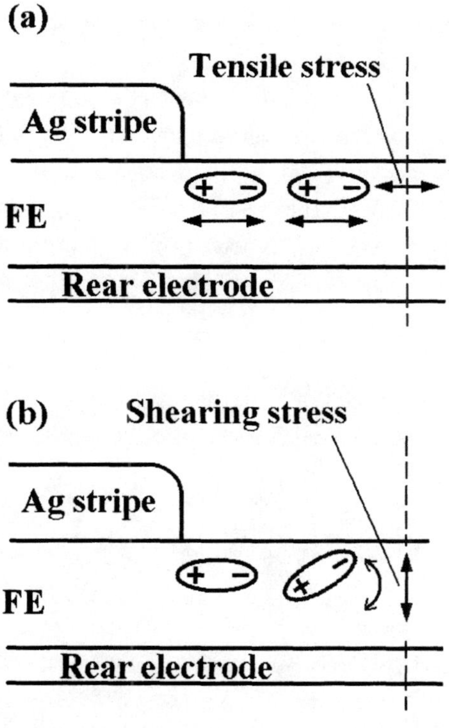

Fig. 3 Domain movement that may cause the electron emission and crack of the cathode.

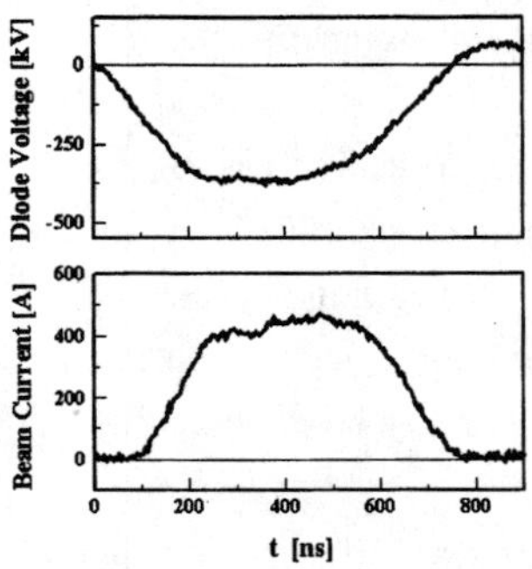

Fig. 2 Diode voltage and beam current obtained in the experiment using APC850 cathode.

Table I. Material Properties of LZT-2 and APC850

Item/material	LZT-2	APC850
Dielectric Constant	1900	1750
Loss Tangent	0.015	0.014
Density (g/cm^3)	7.5	7.7
Curie Temperature (C)	360	360
Coupling Coefficient		
k_{33}	0.74	0.72
k_{31}	0.36	0.36
k_p	0.65	0.63
Piezo Constants (x10^{12} m/V)		
d_{33}	405	400
d_{31}	-179	-175
Piezo Constants (x10^3 mV/N)		
g_{33}	25.5	26
g_{31}	-11.3	-12.4
Young Modulii (x10^{-10}N/m^2)		
Y_{11}^E	6.10	6.3
Y_{33}^E	5.32	5.4
Mechanical Quality factor		
	75	80
Frequency Constants (Hz·m)		
N_L	1422	1500
N_T	1930	2032
N_R	**1016**	**1980**

The domain oscillations are induced by the trigger voltage. Because the cathode material is working in its ferroelectric state and each domain is polarized, therefore the domain oscillations are accompanied with electric field oscillations. It is likely that the emitted electrons are pumped out from the triple junction by this oscillation of the electric field, and then accelerated into the A-K gap. Assuming that the width of the triple-point region is of an order of micrometer, the oscillation frequency can reach to a very high value, thus the emission appears to be continuous in time, even if the emission may have a temporal dependency on the electric field configuration, which is related to the domain oscillation.

Observation of a cathode after a few thousands of discharges shows that the front side of the cathode is surprisingly clean and burn-mark free. The lifetime of the cathode is terminated by a crack along the ferroelectric stripe, not by accumulation of burn-mark due to the plasma formation. This provides a clear evidence of the existence of the domain oscillation perpendicular to the ferroelectric stripes, and also indicates that the mechanical stress should be addressed as an important issue for the long-lifetime operation of the ferroelectric electron gun. Processes that may contribute to the cracking are illustrated in Fig. 3. In the case (a), the cracking of the cathode is caused by a tensile-stress due to the domain oscillation parallel to the surface of the cathode, and in the case (b), by a shearing stress due to the domain rotation.

Conclusions

We have performed electron emission experiments using PZT emitters with different material properties. An electron current emission of 62.1 A/cm^2 has been obtained. The beam currents obtained by LTZ-2 and APC850 differ by more than 100 %. This difference is found to be related to the difference in the frequency characteristics of the materials, which suggests that the electron emission may be caused by the domain oscillation. This new discovery may provide a key point leading to a more complete explanation of electron emission from ferroelectric materials.

Acknowledgement

The experimental work described in this paper was part of a program carried out at High Voltage Laboratory at Cornell University by one of the authors YH and other HVL personnel while YH was with Cornell University.

Reference

[1] H. Gundel, H. Riege, E. J. N. Wilson, J. Handerek and K. Zioutas: Nucl. Instrum. Methods Phys. Res. A **280** (1989) 1.

[2] Y. Hayashi, D. Flechtner, Cz. Golkowski, J. Ivers and J. A. Nation: Jpn. J. Appl. Phys. **40** (2001) p. 397.

[3] J. D. Ivers, D. Flechtner, Cz. Golkowski, G. S. Kerslick, and J. A. Nation: Proc. Particle Accelerator Conf. 1998, **1** 1275.

[4]Yasushi Hayashi, Donald Flechtner and Eiki Hotta: Journal of Physics D: Applied Physics, Vol. 35, 2002, p.281.

[5] Transducer material specifications.

[6] Piezoelectric ceramic materials characteristics: see APC home page at http://americanpiezo.com/.

Measurements of High Voltage across the Vacuum Diode of a High-Current Generator by Using Photoneutron Reactions

Seregej A.Dan'ko, Georgy I.Dolgachev, Yury G.Kalinin, Dmitry D.Maslennikov

Russian research Centre "Kurchatov Institute" ,NFI,!,Kurchatov Sq., 123182,Moscow, Russia.

Abstract. A new method is proposed for measuring the voltage across the vacuum diode of a high-current generator by determining the high-energy edge of bremsstrahlung from an electron beam with the help of photoneutron reactions. The method is approved in the RS-20 facility with a plasma opening switch in the voltage range2.5÷4.5 MV. The results obtained by using this technique are in good agreement with the data from electrotechnical measurements.

RS-20 FACILITY AND ELECTROTECHNICAL MEASUREMENTS

The pulsed voltage amplitude in the diodes of megavolt pulsed generators is rather difficult to measure because of the specific design of these systems. A typical example is a generator with a plasma opening switch (POS). The POS is a segment of a vacuum coaxial line filled with plasma, which, in the open-circuit state, behaves like a vacuum diode. The RS-20 facility is a POS-based pulsed high-current generator consisting of an inductive-capacitive circuit closed through a POS. The facility operation and the measurement techniques are described in detail in [1]. In experiments, we employed conventional electrotechnical diagnostic techniques, namely, shunts, voltage dividers, and loops measuring the voltage associated with the magnetic field derivative. The electrotechnical methods satisfactorily reproduce the time behavior of the voltage; however, the accuracy of determining the voltage amplitude turns out to be insufficient. This inaccuracy is related to the ambiguity in determining the inductance of the vacuum component of the electric circuit before and after switching the POS. Moreover, the electrotechnical methods fail to provide the direct measurements of the POS voltage.

MEASUREMENTS OF THE ELECTRON ENERGY WITH THE HELP OF PHOTONEUTRON REACTIONS

To determine the voltage across the accelerator diode, we measured the high-energy edge of the bremsstrahlung γ-spectrum from the electrons accelerated in the diode gap. The filter method, which is usually used to determine the electron energy in the range <1 MeV, is inapplicable to measuring the γ-spectrum under our experimental conditions. The reason is that, in the range 1÷5 MeV, the mass

CP650, *Beams 2002: 14th International Conference on High-Power Particle Beams,*
edited by T. A. Mehlhorn and M. A. Sweeney

absorption coefficients of all substances depend only slightly on the photon energy; frequently, this dependence is nonmonotonic [2]. Within this range, the high-energy edge of the γ-spectrum can be measured by recording photoneutrons produced in the reactions $Be^9(\gamma,n)Be^8$ and $D^2(\gamma,n)p$, which have the lowest thresholds, namely, E_{Be} = 1.6 MeV and E_D = 2.25 MeV. The cross sections $\sigma(\gamma,n)$ for these reactions are shown in Fig.1 [3]. Indeed, at voltages close to those corresponding to the threshold energies, the photoneutron yields in these reactions, as well as their ratios, should strongly depend on the maximum energy of γ-photons. The expected photoneutron yield was calculated numerically. The intensity and spectrum of hard γ-radiation from the accelerator diode were calculated by the Monte Carlo method using the KASKAD numerical code [4] with allowance for the actual geometry of the RS-20 facility. The neutron yield due to

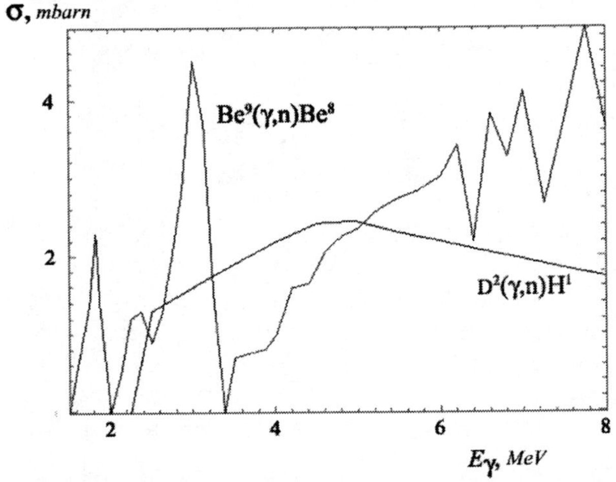

FIGURE 1. Cross sections for the photoneutron reactions $Be^9(\gamma,n)Be^8$ and $D^2(\gamma,n)p$.

the interaction of γ-radiation with matter (Be, D_2O) can be represented in the form of convolution $N_n = \int_{E_0}^{\infty} \sigma(\gamma,n)\cdot N_\gamma(h\nu)\cdot N_{Be,D} d(h\nu)$, where $N_\gamma(h\nu)$-is the number of photons arriving at the target and $N_\gamma(h\nu)$-is the surface density of beryllium or deuterium nuclei. The targets made of beryllium or heavy water were placed outside of the vacuum chamber of the generator at a distance of 1.2 m from the POS. The neutrons produced in each reaction were recorded by two identical neutron detectors based on the activation technique with delayed recording. First, the fast neutrons were decelerated in a polyethylene cube; then, the newly created thermal neutrons interacted with six indium plates placed on the cube faces, which resulted in the production of β-active nuclei in the $In^{115}(n, \beta^-)Sn^{116}$ reaction. The emitted β-particles were recorded with six Geiger counters located on the rear sides of the indium plates. Schematic of the experiment is shown in Fig.2.

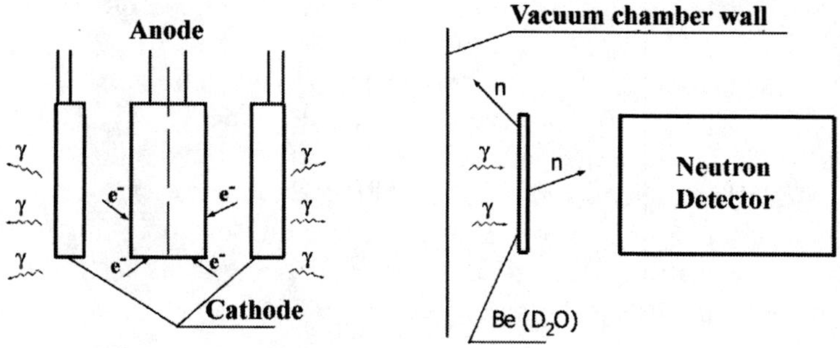

FIGURE 2. Schematic of the accelerator diode and the locations of the measuring detectors used to calculate the photoneutron reaction yields.

It should be noted that this measurement scheme completely eliminates the effect of both electromagnetic interference and X radiation on the measuring apparatus, which is important for the stable operation on high-power pulsed generators. The absolute calibration of the entire neutron tract was not carried out; only the ratio between the readings from the two recording channels was determined. The number of counts in each recording channel was calculated for the peak voltages U_d = 2.5±4.5 MV. The shape of the voltage pulse was taken from the dB_{POS}/dt signals, and the amplitude was varied in the range 2.5÷4.5 MV. Computations were performed for two models in which electrons were assumed to fall onto either the side or end surface of the anode (Fig.4). The ratio between the yields of the two photoneutron reactions, Y_D/Y_{Be}, as a function of the peak voltage U_d is shown in Fig.3. The efficiency of this method is

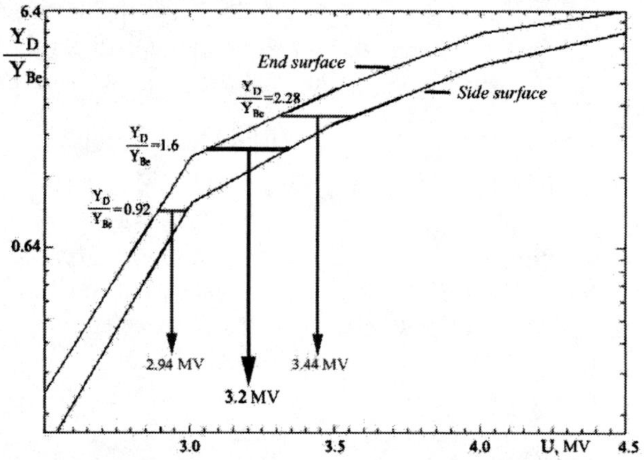

FIGURE 3. Ratio between the photoneutron reaction yields, Y_D/Y_{Be}, vs. the maximum voltage for two models in which electrons are assumed to fall onto either the side or end surface of the cylindrical anode. The vertical arrows indicate the voltages for three different Y_D/Y_{Be} ratios, and the horizontal segments between the curves correspond to the measurement errors noted in the table.

mainly determined by the fact that the above ratio very strongly depends on the peak voltage. As the voltage increases from 2.5 to 4.5 MV, the yield ratio increases by almost two orders of magnitude. It is this sharp dependence that ensures the high accuracy in determining the voltage regardless of the possible measurement and/or calculation errors.

EXPERIMENTAL RESULTS AND DISCUSSION

The readings from the two channels, the ratio between the readings, and the diode voltage determined from the computed dependence (Fig.3) are listed in the Table 1. To increase the γ-radiation yield, the end and side surfaces of the cylindrical anode were covered with Ta plates, which somewhat deteriorated the POS operation in this set of experiments consisting of about 20 shots. In five shots, the signals exceeded standard deviation. For the shots shown in the table, the noise (background) was about

TABLE 1. Photoneutron reaction yields.

Shot no.	Be-channel, counts	D-channel, counts	YD/YBe ratio	U, MV
1	34(±12)	53(±12)	1.6(±0.68)	3.2±0.15*
2	32(±12)	33(±12)	1.(±0.51)	2.95±0.1*

Note: Asterisk denotes the error related to an uncertainty in the current distribution over the anode surface.

90 counts and the excess over this level amounted to 3÷5 standard deviations. At such a low useful signal, the measurement error is mainly statistical in character. Thus, for shot no.1, the ratio between the reaction yields can vary from 0.92 to 2.28 within the standard deviation (Fig.3). Correspondingly, the maximum voltage is within the range 2.94÷3.44 MV. It is this shot which signals presented in Fig.3 were measured. The maximum voltage measured with the help of electrotechnical detectors, U_M = 3.5 ± 0.2 MV, falls within the voltage range calculated from the standard deviation. As the useful signal increases, the statistical error becomes negligible and the error related to an uncertainty in the current distribution over the anode surface (about 5%) starts to dominate (Fig.3). It is this error (marked by the asterisk) that is listed in the last column of the Table 1.

ACKNOWLEDGMENTS

This study was supported by the Russian Foundation for Basic Research (project no. 01-02-17259), INTAS (grant no. 97-0021), and the U.S. Civilian Research and Development Foundation for the Independent States of the Former Soviet Union (CRDF) (grant no. RP1-2113).

REFERENCES

1. N. U. Barinov, S. A. Budkov, G. I. Dolgachev, et al., Fiz. Plazmy 28, 202 (2002) [Plasma Phys. Rep. 28, 177 (2002)].
2. Tables of Physical Quantities, Ed. by I. K. Kikoin (Atomizdat, Moscow, 1976), p. 962.
3. R. J. Hughes, in Proceedings of the International Conference on Photonuclear Reactions and Applications, Asilomar, CA, 1973, p. 151; G. Breit, in Proceedings of International Conference on Photonuclear Reactions and Applications, Asilomar, CA, 1973, p. 323.
4. G. S. Ageev, Report of Chelyabinsk State University for 1982, Inv. No. 02825090754, Governmental Registration no. 80049937 (Chelyabinsk, 1982).

High Voltage, Sub Nanosecond Feedthrough Design for Liquid Breakdown Studies

Michael Cevallos, James Dickens, Andreas Neuber, Herman Krompholz

Center for Pulsed Power and Power Electronics Research
Departments of Electrical & Computer Engineering and Physics, Texas Tech University
Lubbock, TX 79409-3102
email: andreas.neuber@ ttu.edu

Abstract. Experiments in self-breakdown mode and pulsed breakdown at high over-voltages in standard electrode geometries are performed for liquids to gain a better understanding of their fundamental breakdown physics. Different liquids of interest include liquids such as super-cooled liquid nitrogen, oils, glycerols and water. A typical setup employs a discharge chamber with a cable discharge into a coaxial system with axial discharge, and a load line to simulate a matched terminating impedance, thus providing a sub-nanosecond response. This study is focused on the feed-through design of the coaxial cable into this type of discharge chamber, with the feed-through being the critical element with respect to maximum hold-off voltage. Diverse feedthroughs were designed and simulated using Maxwell 3-D Field Simulator Version 5. Several geometrically shaped feed-through transitions were simulated, including linearly and exponentially tapered, to minimize electrostatic fields, thus ensuring that the discharge occurs in the volume of interest and not between the inner and outer conductor at the transition from the insulation of the coaxial cable to the liquid. All feedthroughs are designed to match the incoming impedance of the coaxial cable. The size of the feedthroughs will vary from liquid to liquid in order to match the coaxial cable impedance of 50Ω. The discharge chamber has two main ports where the feed-through will enter the chamber. Each feed-through is built through a flange that covers the two main ports. This allows the use of the same discharge chamber for various liquids by changing the flanges on the main ports to match the particular liquid. The feedthroughs were designed and built to withstand voltages of up to 200 kV. The feedthroughs are also fitted with transmission line type current sensors and capacitive voltage dividers with fast amplifiers/attenuators in order to attain a complete range of information from amplitudes of 0.1mA to 1 kA with a temporal resolution of 300 ps.

INTRODUCTION

Experiments in self-breakdown mode and pulsed breakdown at high over-voltages in standard electrode geometries are preformed in liquid to test various liquid switching media for use in high voltage high energy density compact pulse power devices. Often ultra high power switching is desired with closure times in the sub-nanosecond regime. Thus it becomes more and more necessary to design feedthroughs that will not distort data taken during the experiments. This calls for feedthroughs that do not cause reflections in the coax system and can maintain a constant impedance through all transitions.

In this paper we will design feedthroughs based on a coaxial system. We try to address two main locations that can cause distortion to signals. The junction created

CP650, *Beams 2002: 14th International Conference on High-Power Particle Beams,*
edited by T. A. Mehlhorn and M. A. Sweeney

when the coax cable enters the discharge chamber and the transition of the solid insulator to the liquid insulator.

EXPERIMENTAL SETUP

The setup includes a pair a rigid coax cable made to electrically resemble an RG-19 transmission lines with an impedance of 52Ω. They are connected through electrical feedthroughs on opposite sides of the discharge chamber. The discharge chamber is a six-way stainless steel cross where two 4-inch ports at the top and bottom are used to fill and drain the chamber. Of the remaining four 8-inch ports, two are quartz view ports that allow for ultraviolet emissions to be recorded. Test gap electrodes are connected to either transmission line center conductor inside the discharge chamber. For these experiments needle electrodes will be used. A cylindrical grid inside the discharge chamber connects the outer conductors to maintain a constant impedance.

THEORETICAL DESIGN AND DISCUSSION

The first area to be addressed was the junction where the rigid coax enters the discharge chamber. In many cases this junction is not continuous and therefore can cause a mismatch in impedance and therefore be a source of reflections. The simplest method to solve this problem is to have the rigid coax enter the chamber as one continuous piece. This can be accomplished by using a modified Kajon fitting. The modification can be seen in the figure below. This modification allows the rigid coax system to enter the chamber to enter the chamber and still remain sealed under low pressures. With further modifications it will be able to adjust the width of the spark gap while the chamber is full.

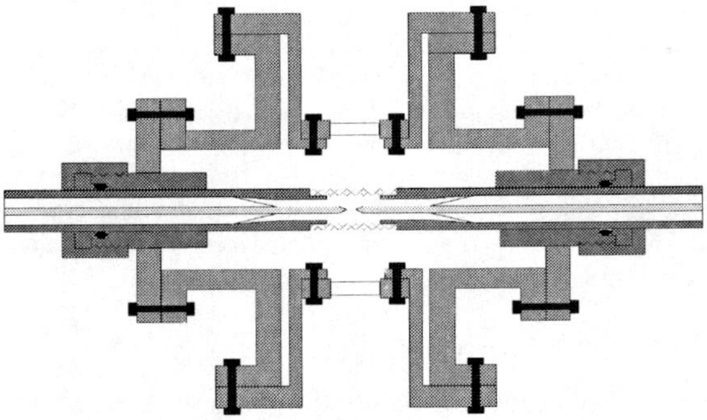

FIGURE 1. Basic Set up which may include double polarity Charging.

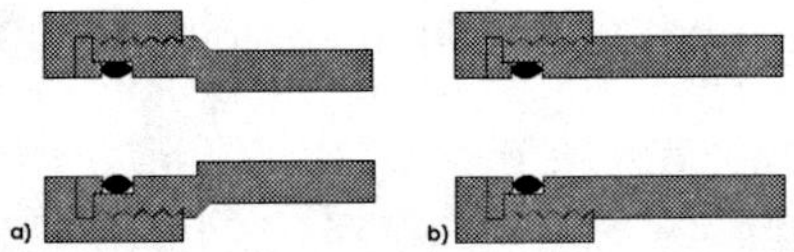

FIGURE 2. a) Unmodified Kajon Fitting b) Modified Kajon Fitting.

The second are of interest is the transition from the insulator of the rigid coax to the insulator of the coax system inside the discharge chamber. The transition in this experiment is from Lexan to oil where the dielectric constants are 2.7 and 2.2 respectively with an inner conductor diameter of 0.25 inches. In order for the impedance to match the 52Ω of the coax cable the following equation was used to find the outer diameter of the oil coax system inside the discharge chamber.

$$R_o = \sqrt{\frac{\mu}{\varepsilon}} \cdot \frac{\ln\left(\frac{b}{a}\right)}{2 \cdot \pi} \qquad R_o = \sqrt{\frac{L}{C}} = \sqrt{\left(\frac{\ln\left(\frac{r_2}{r_1}\right)}{2 \cdot \pi \cdot \varepsilon_{Lexan}} + \frac{\ln\left(\frac{r_3}{r_2}\right)}{2 \cdot \pi \cdot \varepsilon_{Oil}}\right) \cdot \frac{\mu_{Lexan.Oil} \cdot \ln\left(\frac{r_3}{r_1}\right)}{2 \cdot \pi}} \tag{1}$$

$$b = \exp\left[2 \cdot R_o \cdot \frac{\pi}{\left(\frac{\mu}{\varepsilon}\right)^{\frac{1}{2}}}\right] \cdot a \tag{2}$$

The outer diameter of the oil coax system was found to be 0.479 inches. Now the transition from the outer conductor of the rigid coax cable with Lexan to the oil coax system inside the chamber is known. For these experiments the transition distance between a 100% Lexan coax to a 100% oil coax will be taken to be 1 inch. Now it is necessary to determine the radii of the insulators through out the transition phase.

$$\frac{\ln\left(\frac{r_2}{r_1}\right)}{2 \cdot \pi \cdot \varepsilon_{Lexan}} + \frac{\ln\left(\frac{r_3}{r_2}\right)}{2 \cdot \pi \cdot \varepsilon_{Oil}} = \frac{1}{C} \tag{3}$$

$$\frac{\mu_{Lexan} \cdot \ln\left(\frac{r_2}{r_1}\right)}{2 \cdot \pi} + \frac{\mu_{Oil} \cdot \ln\left(\frac{r_3}{r_2}\right)}{2 \cdot \pi} = L \tag{4}$$

$$r_2 = \exp\left[-\left(\ln\left(\frac{r_1}{r_3}\right) + R_o^{\,2} \cdot \frac{4 \cdot \pi^2}{\mu \cdot \ln\left(\frac{r_3}{r_1}\right)} \cdot \varepsilon_{Oil}\right) \cdot \frac{\varepsilon_{Lexan}}{\left(\varepsilon_{Lexan} - \varepsilon_{Oil}\right)}\right] \cdot r_1 \tag{5}$$

$$r_3 = \exp\left[\frac{\ln\left(\frac{r_2}{r_1}\right) \cdot \left(-\mu \cdot \varepsilon_{Oil} - \mu \cdot \varepsilon_{Lexan}\right) + (a - b + c + d)^{\frac{1}{2}}}{2 \cdot \mu \cdot \varepsilon_{Lexan}}\right] \cdot r_2 \tag{6}$$

$$a = \ln\left(\frac{r_2}{r_1}\right)^2 \cdot \mu^2 \cdot \varepsilon_{Oil}^{\,2} \qquad b = 2 \cdot \ln\left(\frac{r_2}{r_1}\right)^2 \cdot \mu^2 \cdot \varepsilon_{Oil} \cdot \varepsilon_{Lexan}$$

$$c = 16 \cdot \mu \cdot \varepsilon_{Lexan}^{\,2} \cdot R_o^{\,2} \cdot \pi^2 \cdot \varepsilon_{Oil} \qquad d = \mu^2 \cdot \varepsilon_{Lexan}^{\,2} \cdot \ln\left(\frac{r_2}{r_1}\right)^2$$

If one of the insulator transition geometry is known the other can be found using the above equations. For this discussion a stair step, linear taper and an arctangent taper have been selected.

RESULTS

In Figure 3b) the linear taper was first applied to the outer conductor, r1, and the insulator transition, r2, was found. In Figure 3b) the linear taper was applied to the insulator transition and the outer conductor geometry was found. This process was used again in Figures 3d) and 3e) using an arctangent function in order to remove all abrupt changes in geometry and reduce any field enhancement points. After the geometries were found they were placed back into the equation for characteristic impedance and every point was found to have a characteristic impedance of 52Ω. Figure 3 also shows the electric field within the coax cable. As can be seen the electric field for the different transitions are similar

It should also be noted that simulations at high frequencies to find reflection coefficients at various points in the coax system are planned as well as finding the electric field vector along the surface of the insulator transition as this location is the most likely to breakdown. Simulations will also be run to obtain the best transition length at high frequencies.

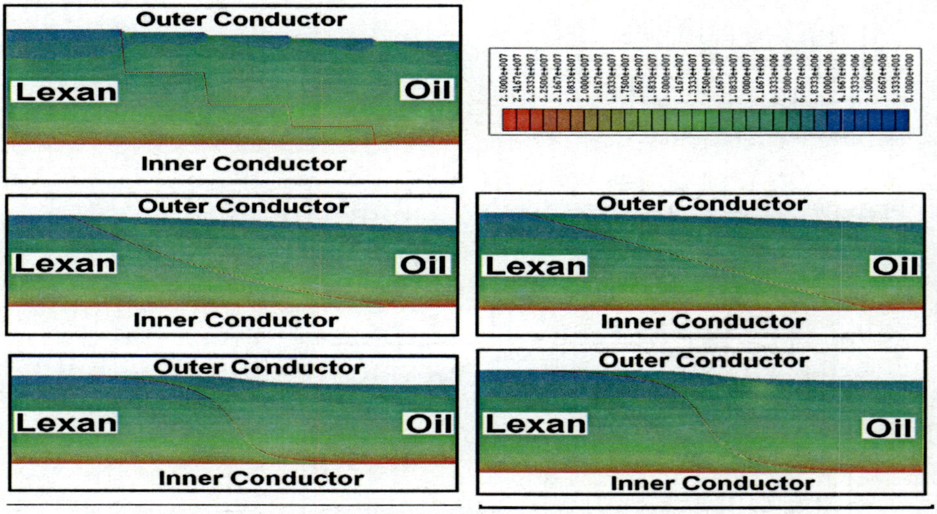

FIGURE 3. Insulator transitions and electric field inside the coaxial system.

ACKNOWLEDGEMENT

This work was solely funded by the Compact Pulsed Power MURI Program funded by the Director of Defense Research & Engineering (DDR&E) and Managed by the Air Force Office of Scientific Research (AFOSR).

Fast Discharge Energy Storage Development for Improving X-ray Simulators*

Peter Sincerny,[a] Vic Carboni,[a] Kendall Childers,[a] Pat Corcoran,[a] Jud Hammon,[a] S. K. Lam,[a] Richard Miller,[a] Tom Naff,[a] Ian Smith,[a] Terry Tucker,[a]
Joel Ennis,[b] Robert Cooper[b], David Bell,[c] Randy Davis[c]

[a]*Titan Pulse Sciences Division, 2700 Merced Street, San Leandro, CA 94577 USA*
[b]*Sorrento Electronics, 4949 Greencraig Lane, San Diego, CA 92123 USA*
[c]*Defense Threat Reduction Agency, 6801 Telegraph Rd., Alexandria, VA 22310 USA*

Abstract. Over the last two years there have been design studies to investigate the impact of improvements in fast energy storage systems on the design of simulator upgrades (Double-EAGLE and Decade Quad) and on larger future simulators (40-MA to 60-MA PRS machine). The fast energy storage systems investigated in these design studies included Fast Marx Generators (FMG with $\sqrt{LC}$ = 200 ns and $\sqrt{LC}$ = 300 ns) and Linear Transformer Drivers (LTD). A design sketch of a compact 20-MA PRS driver and a potential upgrade of Double-EAGLE using FMG technology will be presented. The first concept that will be discussed is a 16-MA driver for PRS (plasma radiation source) loads. This generator would consist of 48 eight-stage FMG units and 13 m diameter and would drive the PRS directly without further pulse compression. The second concept that will be presented is a potential upgrade of an operational simulator, Double-EAGLE. This concept would utilize the FMG to replace the existing slower Marx generator, transfer capacitor and triggered gas switch.

The basic building blocks for these future FMG driven machines are a low- inductance Marx switch and a low-inductance capacitor designed to be integrated with the new switch. These components are configured in a low-inductance FMG stage and then stacked in series to form a unit for the voltage required and a number of units in parallel for the required system inductance and stored energy. A review of the FMG component requirements and the status of the FMG component testing in a single-stage FMG configuration will also be presented. A four-stage FMG unit is being built and tested to demonstrate the required stage voltage and inductance. Results of these initial tests will be presented.

SYSTEM DESIGNS FOR A LARGE PLASMA RADIATION SOURCE (PRS) MACHINE

Present Plasma Radiation Source (PRS) drivers, Saturn (Ref. 1), Z (Ref. 2), Double-EAGLE (Ref. 3), and Decade Quad (Ref. 4) are capable of delivering current up to 20 MA (Z at Sandia). In a previous system study pulsed power options for driving PRS with currents up to 50 MA were investigated. (Ref. 5-7) The results of this system study provided

* Work sponsored by the Defense Threat Reduction Agency

CP650, *Beams 2002: 14th International Conference on High-Power Particle Beams,*
edited by T. A. Mehlhorn and M. A. Sweeney

the motivation and design requirements for the fast Marx generator (FMG) component development presented in this paper. The system study focused on simulator concepts capable of delivering 400 kJ of 13 keV radiation from a PRS load. Some of the concepts investigated are shown in Figure 1. The concepts included: a fast LTD driver, a relatively slow conventional Marx generator driving a transfer capacitor, an FMG driving a peaking capacitor and a very fast Marx generator directly driving a PRS load. All four designs produced the required 400 kJ of radiation with a 250 ns implosion time. The chart shown in Figure 2 is a comparison of the four designs. Note the LTD system actually had 448 parallel circuits since there were seven sub-modules for each of the 64 modules. The other disadvantage of the LTD is the requirement to trigger all stages while the Marx concepts require only trigger on the first few stages of each module. The last two concepts in the chart (FMG with peaking capacitor and the very FMG) require Marx generators with $\sqrt{LC}$ = 300 ns and 175. The goal of the present FMG development program is to develop low- inductance capacitors and switches capable of achieving these $\sqrt{LC}$ parameters. Once these components are demonstrated at a significant level of energy storage and at an affordable cost, this new FMG technology will provide the capability to build large PRS machines in a significantly more compact configuration. The new FMG technology minimizes or eliminates the need for storing the energy in a large transfer capacitor.

NEAR-TERM APPLICATIONS OF THE FMG TECHNOLOGY

The new FMG technology can also be utilized to build smaller PRS machines or to upgrade existing machines such as Double-EAGLE. Figure 3 shows a sketch of a 16-MA PRS machine that utilizes the $\sqrt{LC}$ = 300 ns FMG as the prime energy storage system. The system consists of 48 FMG sub-modules driving a water coupler and finally a four-level (two triplates) vacuum insulator and MITL. The total energy stored is 5.7 MJ. A sketch of a sub-Marx module is shown in Figure 4. This sub-Marx is; .75 μF per stage, ± 100 kV with eight stages for a total voltage of 1.6 MV. The circuit model predicts a peak current of 15.8 MA with 397 ns implosion time into a PRS load.

Figure 5 shows the potential improvement to an existing machine, Double-EAGLE (DE), if the new FMG technology was utilized to replace the aging DE Marx generator. Figure 5a shows the existing DE machine that includes a slow Marx, transfer capacitor, triggered gas switch, two pulse-forming lines, and the final transformer and vacuum section. Figure 5b is a sketch of the DE machine with the Marx, transfer capacitor, and gas switch all replaced by a new FMG. This new FMG technology will reduce the overall size of the machine by 25% and will eliminate the transfer capacitor and triggered high voltage switch. The FMG for this upgrade would require output voltage of 4.0 MV and will store 480 kJ (compared with the existing DE Marx that stores 560 kJ).

STATUS OF FMG COMPONENT DEVELOPMENT

New low inductance Marx switches and capacitors have been built and are presently being tested in a single-stage configuration of the FMG. The near-term goal is to develop and demonstrate a four-stage FMG with a $\sqrt{LC}$ of 300 ns. The four-stage system will be capable of storing 60 kJ at ± 100 kV (43 kJ at ± 85kV). A faster version of this FMG can be realized by simply building capacitors of the same size with half the capacitance.

The Marx switches are rail gaps with the following specifications: ± 100 kV, 300-kA peak current. The initial test switch housings were fabricated with Rexolite. The main electrodes are brass while the trigger electrode is stainless steel. The plastic-case capacitors have the following dimensions: 7 cm thick, 61 cm wide and 97 cm tall.

Low Capacitance Test Results

Two of the rails switches were initially tested with low capacitance. This test was performed to assure the voltage integrity of the switch prior to using the full energy stored to mitigate catastrophic faults. The switches performed well for more than 300 shots at charge voltages ≥ 80 kV. No switch prefire was recorded below 100 kV charge. There were three prefires in the 200 shots at and above 100 kV. The switches showed reliable multi-channeling with low (2 ns) jitter in the closure time.

Full Capacitance (Full Energy) Testing

After completing the high voltage, low-capacitance tests a full energy stage was assembled and checked out with two of the new low inductance capacitors. The closure time jitter remained below 2 ns, at ± 85kV charge voltage. The peak current was 155 kA into 0.7 ohm. The switch was operated at 50% to 70% of self-break pressure with no prefires.

Inductance $\sqrt{LC}$ Measurement

The single stage inductance was measured by ringing the stage into a very low inductance short circuit at ± 50kV. The measured stage inductance was 103 nH, which corresponds to a $\sqrt{LC}$ of 264 ns since the capacitance of the as-built capacitors was 675 nF/stage (two capacitors in series). This inductance does not include the return inductance when the stage is used in a multi-stage Marx generator. The calculated ground return inductance for a four-stage FMG is 13 nH per stage on average. Therefore the total average stage inductance for a four-stage system will be 116 nH for a $\sqrt{LC}$ of 280 ns. This inductance will increase with the number of stages in the Marx and will be somewhat higher (5%) if the next batch of capacitors is built closer to the goal specification of 1.5 μF rather than the as-built value of 1.35μF. The projected total inductance per stage of 116 nH is close to the original program goal of 120 nH ($\sqrt{LC}$ = 300 ns for a four-stage Marx with 0.75 μF per stage capacitance).

Future Testing & Plans

Figure 6 is a picture of the four-stage FMG prototype. This system will be tested at low voltage in July and will be up to 85 kV in August.

The focus of the next phase of this program will be to demonstrate the reliability and lifetime of the new FMG technology and to reduce the component costs. A cast version of the switch and a cost-effective method to build the plastic-case capacitors will be key to lowering the cost of future.

References

1. R. B. Spielman, "A Double Post-hole Vacuum Convolute Diode for Z-pinch Experiments on Saturn," 7th IEEE Pulsed Power Conference, 1989.
2. R. B. Spielman et al., "Pulsed Power Performance of PBFA-Z," 11th IEEE International Pulsed Power Conference, 1997, p. 709.
3. P. Sincerny et al., "Performance on Double-EAGLE," 5th IEEE Pulsed Power Conference, 1985, p. 151.
4. J. Douglas et al., "Decade Quad Water Coupler – Electrical Design and Performance," 13th IEEE Pulsed Power Conference, 2001.
5. P. Corcoran et al., "Pulsed Power for Future X-ray Simulators," 13th IEEE Pulsed Power Conference.

6. I. Smith et al., "Pulsed Power for Future (and Past) X-ray Simulators," Special Issue of IEEE Transactions on Plasma Sciences to be published in 2002.
7. P. Sincerny et al., "Concepts for an Affordable High Current Imploding Plasma Generator," 12th IEEE Conference, 1999.

FIGURES

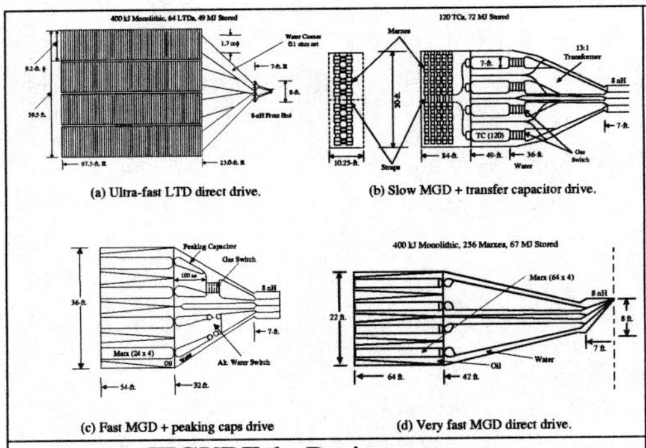

FIGURE 1. Design concepts.

	Very Fast LTD (125 ns)	Slow Marx (.500 ns) + Transfer Cap (150 ns)	Faster Marx (300 ns) + Peaker (150 ns)	Very Fast Marx (175 ns)
Overall Size	620 k ft^3	390 k ft^3	330 k ft^3	280 k ft^3
DC Stored Energy	49 MJ	72 MJ	55 MJ	67 MJ
Energy Stored in Water	None	All	< 1/2	None
# of Modules for 400 kJ	64	60	96	256

FIGURE 2. Comparison of design concepts.

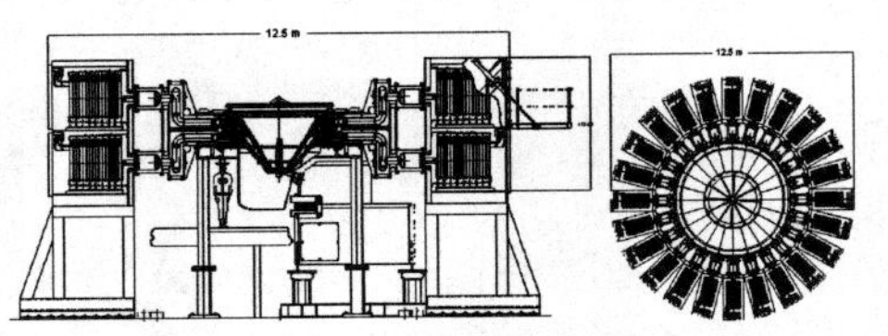

FIGURE 3. Fast Marx, 16-MA, Z-pinch simulator – side (a) and top (b) view.

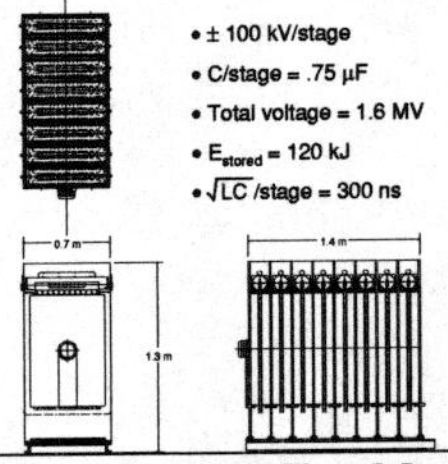

FIGURE 4. Eight-stage Fast Marx submodule building block for the 16-MA simulator.

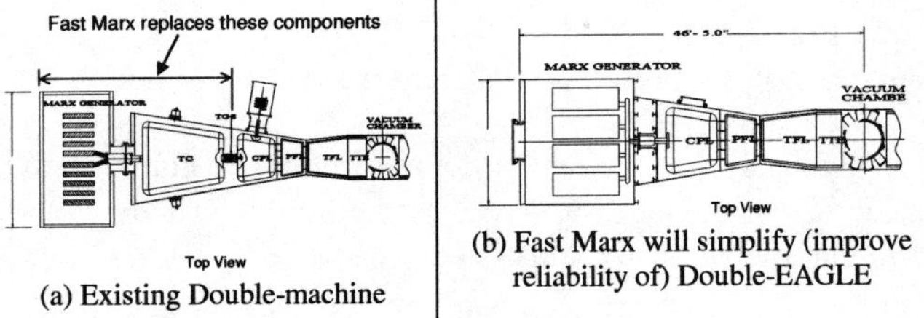

FIGURE 5. New Fast Marx technology can be utilized to upgrade Double-EAGLE for increased output.

FIGURE 6. Photograph of the four-stage Fast Marx.

Development of the Ultra-Fast LTD Stage

Alexandre A. Kim, Alexandre N. Bastrikov, Sergey N. Volkov,
Vjacheslav G. Durakov, Boris M. Kovalchuk, Vadim A. Sinebryukhov

High Current Electronics Institute, Academichesky Ave. 4, 634055 Tomsk, Russia

Abstract. The technology of the fast LTD stages is proposed for 1 MV driver with the goal to get a 60 ns pulse width at 80 % of peak voltage in the diode. For this driver a new LTD-100R stage is designed consisting of 20 GA 35389 capacitors. A 1 MV driver will include 7 such stages in series and provide 1 MV, 125 kA pulse in a critically matched load. The footprint of the whole system is ~1.75 m^2. The report includes the design of the new stage and results of the tests.

INTRODUCTION

Compared to previous fast LTD stages (LTD-100 and LTD-100Z) [1] consisting of Maxwell 31165 capacitors (100 kV, 40 nF) the output pulse of the new stage has to be ~1.5 times sharper. In RLC circuit which is mainly the electric scheme of the LTD stages such pulse can be obtained if only the product LsCs is reduced ~ 2 times, where Ls and Cs are the inductance and the capacitance of the stage. The inductance of the LTD-100 stages can not be reduces sufficiently, therefore the new stage, LTD-100R, was build of capacitors with 2 times reduced capacitance (GA 35389: 100 kV, 21 nF, 25 nH).

DESIGN OF THE LTD-100R STAGE

The design of the LTD-100R stage is presented in Fig. 1. Basically this is the same design as that of LTD-100 [1], the difference is the new GA 35389 capacitors that are 150 mm height/50 mm header compared to 277 mm height/75 mm header for Maxwell 31165, and the total number of capacitors in the stage which is 20. The outer diameter of the stage is 1250 mm, the length along the axis is 200 mm. So, the whole length of serial assembly of 7 stages that are needed to get 1 MV on the diode, is ~1400 mm resulting in the footprint of the 1 MV module of ~1.75 m^2.

The core cross-section is $5\cdot10^{-3}$ m^2, it consists of several rings made of ET-3425 iron. This iron needs ~1.5 kA premagnetization current and allows passive premagnetizing. In last case the voltsecond integral of the stage core reduces from ~18 mV*s to ~15 mV*s.

There are 2 charge cables (±100 kV) entering the stage and 1 trigger cable. The stage in Fig. 1 includes also 2 inputs for premagnetizing cables, but in case of passive premagnetizing only 1 input in vicinity of the diode is needed to premagnetize all the stages in the module.

CP650, *Beams 2002: 14th International Conference on High-Power Particle Beams,*
edited by T. A. Mehlhorn and M. A. Sweeney

For ±100 kV operation the stage cavity is filled with transformer oil. The switches operate in dry air at a pressure of 3.8 ata (±100 kV charge).

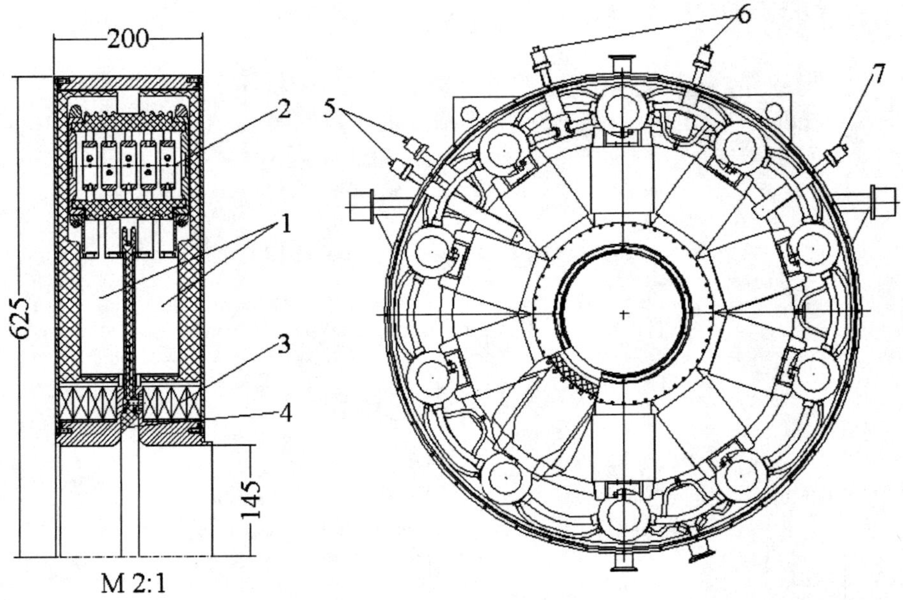

FIGURE 1. LTD-100R with 20 GA 35389 capacitors: 1-storage capacitors, 2-switch, 3-core, 4-vacuum interface, 5-premagnetizing cables, 6-charge cables, 7-trigger cable.

TEST RESULTS

The stage was tested at ±100 kV charge at Rload varied from ~0.3 Ohm to ~1.5 Ohm in order to get detailed information on the stage parameters. Preliminary results are summarized in Figs. 2, 3 by solid circles.

In Fig. 2a the peak load power is given. It reaches maximum of ~19 GW at Rload ~0.5 Ohm. If we consider the stage circuit as a simple RLC circuit where the peak output power is reached at Rload ~ $1.1(Ls/Cs)^{0.5}$, at Cs = 105 nF this means the inductance of the stage of Ls ~ 22 nH. As soon as the stage consists of 10 identical parallel circuits (bricks) including two capacitors, the switch and the output buses, each brick inductance becomes equal to L ~220 nH. Detailed Pspice simulation that takes into account the resistance of the capacitors and the switches as well as energy losses in the core is given in Figs. 2, 3 by solid line for L = 236 nH. It fits rather good the experimental data in Fig. 2a at Rload = 0.5-0.8 Ohms, goes above the experiment at larger resistance and below the experiment at lower resistance. Fig. 2b shows the peak output voltage. For a 1 MV module consisting of 7 stages in series the output voltage of each stage has to be ~140 kV. This voltage is reached at Rload ~1.3 Ohm.

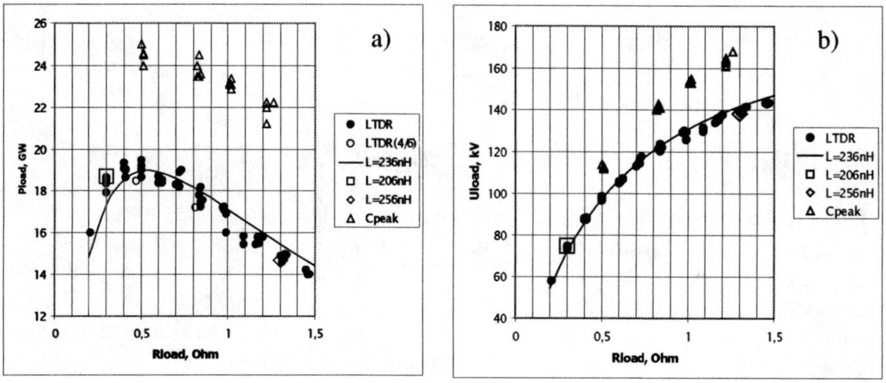

FIGURE 2. Stage performance depending on load resistance: a – output power, b – output voltage.

Fig. 3 gives the width of the load voltage pulse at 80% of peak amplitude. It exceeds the limit of 60 ns being equal to ~64 ns at small Rload and increasing to ~75 ns at large Rload. As soon as the width of the pulse should be proportional to $\sim(LC)^{0.5}$, the increase of the width in Fig. 3 indicates that the inductance of the brick increases with Rload. The only inductance in the brick that may depend on Rload is the inductance of the switch, it could rise as the current through the switch reduces. In Fig. 3 open squares are simulation results at L=206 nH at Rload = 0.3 Ohm and at L=256 nH at Rload = 1.3 Ohm. They fit the experiment much better. Same simulation runs are presented in Fig. 2 also with open squares indicating better agreement with observed load power and voltage.

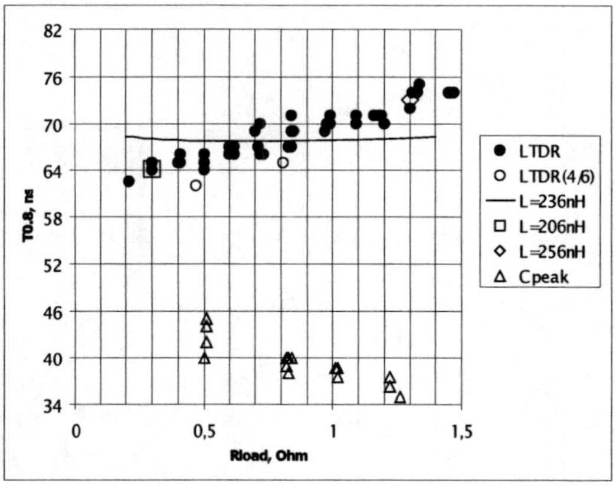

FIGURE 3. Width of the voltage pulse at 80 % of peak amplitude.

In order to increase the output voltage and power and reduce the pulse width to required 60 ns level the stage was equipped with five 1.4 nF water peaking capacitors that were installed in parallel to the output line between the switches evenly around the axis. The results of this modification are presented in Figs. 2, 3 with open triangles. One can see that these capacitors provide ~25 % increase of the output power and reduce the pulse width to less than ~45 ns.

SIMULATION OF THE 1 MV MODULE PERFORMANCE

Fig. 4 presents the simulated voltage pulse across the 8 Ohm resistive load at the output of the module consisting of 7 stages in series (with peaking capacitors). The output line is assumed to be a cone with the outer diameter of 274 mm in the first stage, reducing evenly to 214 mm in the output stage (total inductance ~ 47 nH). The diode inductance is assumed to be 100 nH. The voltage rises to 1 MV corresponding to diode current of 125 kA, the pulse width at 80% of peak amplitude is ~ 45 ns.

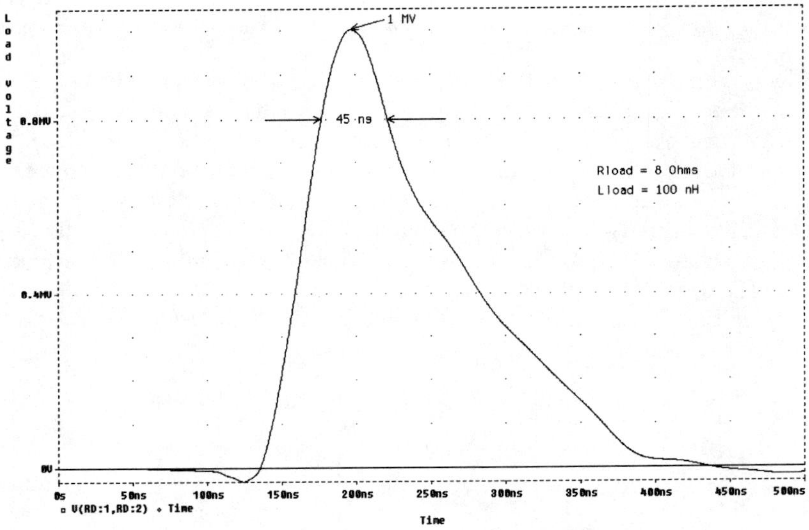

Figure 4. Pspice simulation of the 1 MV module consisting of 7 stages in series.

ACKNOWLEDGMENTS

The authors would like to thank Dr. John E. Maenchen and Dr. David L. Johnson for numerous fruitful discussions. This work is supported by the contract BE-6073/ITHPP/HCEI/99-06.

REFERENCES

1. Kim A. A., Kovalchuk B. M., Bastrikov A. N., etc., *Proc. 13th IEEE Intern. Pulsed Power Conf.*, June 17-22, Las Vegas, Nevada, USA, 1491-1494 (2001).

PULSED POWER SWITCHES

A New Trigger Device for Channel Spark Systems Based on a Penning Discharge

Wolfgang Frey, Christoph Schultheiss, Hansjoachim Bluhm

Forschungszentrum Karlsruhe, Institut für Hochleistungsimpuls- und Mikrowellentechnik P. O. Box 3640, D-76021 Karlsruhe, Germany

Abstract. The ablation efficiency of a channel spark system is mainly influenced by its trigger plasma source. Compared to the application of a sliding discharge trigger plasma source a channel spark system triggered by a Penning discharge exhibits high rates of ablated bulk material per shot, a high reproducibility of the ablated mass per shot and an increased lifetime.

INTRODUCTION

Gas-filled channel spark (CS) systems are high efficiency and low cost electron beam sources used for thin film production by means of pulsed electron beam ablation [1]. An electron beam (≈15 keV, 10^5 A·cm^{-2}, pulse duration: 100 ns) is directed onto a target, consisting of the desired film material. The beam penetrates a few hundreds of nanometers into the target material, ablating bulk material, which finally deposits on the substrate.

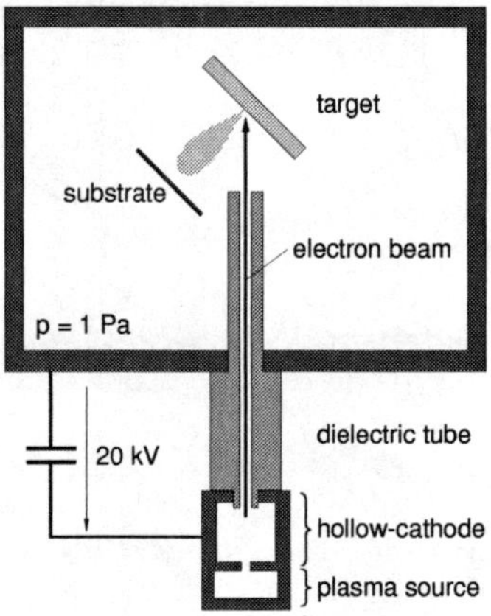

FIGURE 1. Components of a channel spark system for thin film production.

The electron beam is generated by injecting a plasma into the hollow cathode. While the trigger source delivers plasma to the hollow cathode, the electrons escape through the dielectric channel and are accelerated towards the target. The electron

CP650, *Beams 2002: 14th International Conference on High-Power Particle Beams,* edited by T. A. Mehlhorn and M. A. Sweeney

beam propagates confined by its own magnetic field (z-pinch). The beam current density is about 10^5 A·cm^{-2}. Thus, the neutral gas density at a working pressure of 1 Pa is sufficient to achieve space charge neutralisation [1].

The merits of this method of thin film production are its simplicity, the low heat stress for substrate materials, low costs and the conservation of stoichiometry. Contrary to UV-laser ablation, the pulsed electron beam ablation is not restricted to non-transparent materials.

SLIDING DISCHARGE TRIGGER PLASMA SOURCE

At first a plasma source based on a sliding discharge was used, Fig. 2. To ignite the electron beam the energy storage capacitor $C_E = 7.5$ nF is charged. When the charging voltage amounts to $V_0 = 20$ kV the spark gap closes. This causes a voltage drop along the ceramic tube located between the electrodes of a conventional spark plug, Fig. 2, resulting in a discharge along the ceramic surface, generating the plasma for electron beam ignition. After beam generation the current rises up to 1.5 kA discharging C_E.

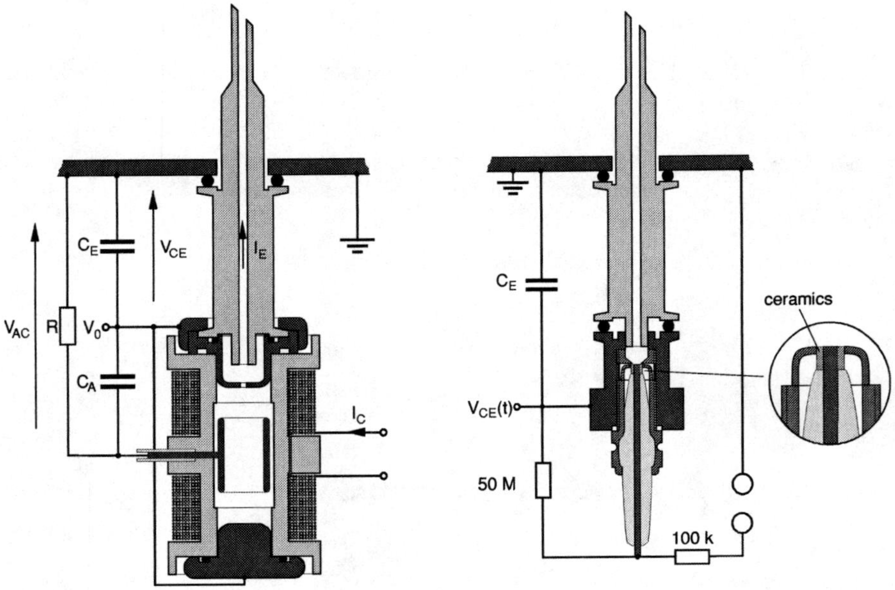

FIGURE 2. Channel spark system with sliding discharge trigger plasma source, right, and with Penning discharge trigger plasma source, left.

The plasma produced by the surface discharge mainly consists of ceramic material. Due to the erosive nature of this process of plasma generation, the lifetime of the plasma source is limited to a maximum of $1.5 \cdot 10^7$ shot. The trigger lifetime is only sufficient for a 46 hour continuous operation at a repetition frequency of 60 Hz. On the other hand, this trigger plasma source is cheap and easy to replace.

PENNING DISCHARGE TRIGGER PLASMA SOURCE

In order to increase lifetime and to improve ablation performance a new trigger plasma source based on a non-erosive process was developed. A reflex discharge, also known as Penning discharge, is used for trigger plasma generation. The source consists of the anode cylinder, two cathodes and a Helmholtz coil to provide a magnetic field directed parallel to the electron beam axis, Fig. 2.

For trigger plasma generation and subsequent electron beam ignition the energy storage capacity C_E and the auxiliary capacity C_A are charged to $V_0 = 20$ kV providing positive potential at the anode cylinder and negative potential at both cathodes.

The trigger plasma generation is started by the application of a pulsed magnetic field to the Penning tube. The pulse duration is 240 µs, Fig. 3. The magnetic field amplitude amounts to B = 0.2 T.

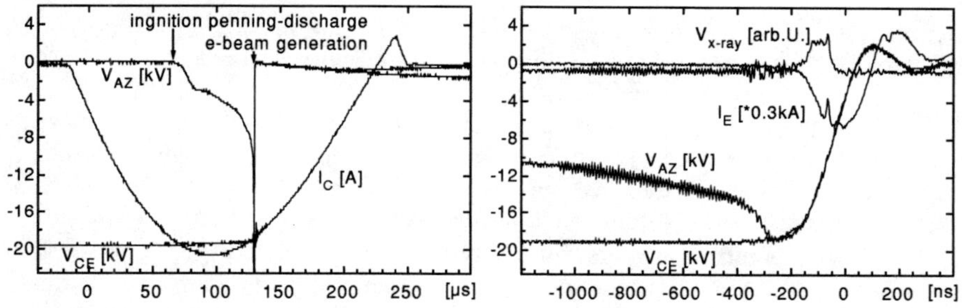

FIGURE 3. Ignition of the Penning discharge and generation of the electron beam during the transient magnetic field pulse. The diagram at right shows the beam generation at a higher time resolution.

Due to the axial magnetic field seed electrons are forced to move azimuthally. They are confined in the penning discharge tube and multiplied by electron impact ionisation. Ions colliding with the cathodes's surfaces further produce electrons and increase the plasma density. At a pressure of p = 1 Pa the mean free path length of electrons for impact ionisation is approximately 4.0 cm. Due to the dimensions of the Penning tube (diameter: 20 mm, length: 44 mm) charge carrier multiplication by impact ionisation without magnetic field is not possible.

Due to the increasing conductivity of the penning discharge plasma, the potential at the anode cylinder V_{AC} approaches the hollow cathode potential V_{CE} within 45 µs, Fig. 3. At a sufficient high density, provided by discharging the auxiliary capacity, the plasma passes through the bore hole in the upper cathode into the hollow cathode volume and finally ignites the electron beam. The beam duration is 160 ns indicated by Bremsstrahlung measurements, Fig. 3.

RESULTS

In order to quantify the ablation efficiency of the channel spark system applying different trigger plasma sources, ablation tests with float glass targets (AR-glass) were done. Depending on the distance d from the end of the dielectric tube to the target the average values of ablated material vary from 1.5 µg per shot to less than 0.25 µg per shot when the sliding discharge trigger and an inner tube diameter of 3 mm are used, Fig. 4, left.

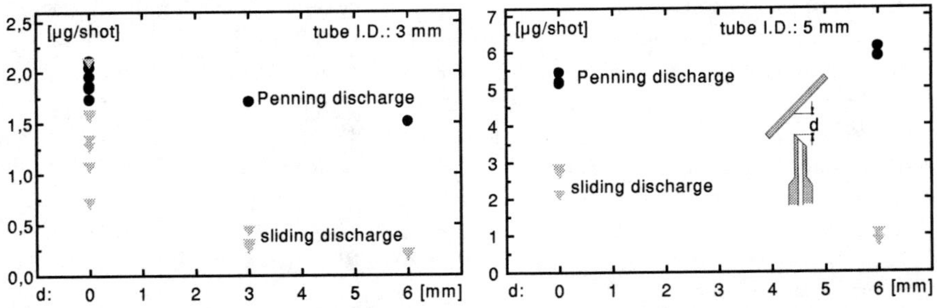

FIGURE 4. Rates of ablated bulk material per shot from float glass targets depending on the distance d from the end of the dielectric tube to the target, the inner tube diameter and the trigger plasma source.

Applying the sliding discharge trigger the reduction of ablation efficiency with increasing tube to target distance a is suspected to be caused by an insufficient space charge neutralisation of the electron beam outside the dielectric tube. Thus, the electron beam diverges, resulting in a lower current density on the target surface.

For both inner tube diameters the CS system with penning trigger plasma source exhibits larger ablation values and an improved beam transport. The plasma generation in the penning tube starts several 10 µs before electron beam ignition. Compared to the delay time of the CS system with surface discharge trigger of about $T_D = 1$ µs, the time for plasma expansion into the dielectric tube is 50 times longer, causing a higher partial gas and plasma density in the dielectric tube which results in a better space charge neutralisation of the electron beam.

Especially at larger tube to target distances a, the rate of ablated mass increases if the inner tube diameter is chosen to be 5 mm. Tube erosion patterns observed with tubes having an inner diameter of 3 mm clearly indicate a loss of beam energy by beam/wall interactions resulting in lower rates of ablated material. The lifetime of a 3 mm channel is about 10^6 shot. A 5 mm channel still works after $50 \cdot 10^6$ shot.

The CS system with penning discharge trigger further exhibits a better reproducibility of ablated bulk material proved by the low scattering of successive ablation rate measurements at d = 0 mm, Fig. 4, left. Up to now the system has run at a repetition frequency of 60 Hz for $1.3 \cdot 10^7$ shot without failure. The developed system promises to be a high-efficiency and long-lifetime electron beam source for industrial applications [2].

REFERENCES

[1] Müller, G., "Thin film deposition by means of pulsed electron beam ablation", in *Science and Technology of Thin Films*, edited by F.C. Mattacotta, World Scientific Publishing Co., Singapore, 1995, pp. 89-119.

[2] Schultheiss, C., et al, Progress on channel spark development, this conference.

Development of Photoconductive Switches for Stacked Blumlein Pulsers*

Farzin Davanloo[a], Carl B. Collins[a] and Forrest J. Agee[b]

[a]*University of Texas at Dallas, P. O. Box 830688, NB11, Richardson, TX 75083-0688*
[b]Air Force Office of Scientific Research, 801 N. Randolph St. Arlington, VA 22203-1977

Abstract. The stacked Blumlein pulse generators consist of several triaxial Blumleins stacked in series at one end. The lines are charged in parallel and synchronously commutated with a single switching element at the other end using a conventional thyratron, spark gap, or photoconductive switch. Several intense photoconductively-switched stacked Blumlein pulsers producing high power nanosecond pulses with risetimes as fast as 200 ps have been demonstrated. These devices have compact geometries and are commutated by avalanche type GaAs photoconductive switches triggered by low-power laser diodes. This report presents the progress in the development and use of these novel pulsers. Special attention is placed on testing a diamond-treated switch that drastically improves high gain operation and lifetime in our Blumlein pulsers.

INTRODUCTION

During the avalanche-mode switching of a pulser, the current is concentrated in filaments that extend from the cathode to the anode across the insulating region of the GaAs switch [1]. Carrier recombination results in the emission of characteristic band gap photons in the near infrared region that can be seen by an infrared viewer. We have observed these effects in experiments where laser diodes provided trigger photons for the avalanche commutation of the stacked Blumlein pulsers.

As soon as the avalanche is initiated, a single filament can be observed which approximately follows the collimated laser beam that was focused in a line from the cathode to the anode. Filamentary currents with densities of several MA/cm^2 and diameters of 15-300 μm passing through a narrow channel can cause switch damage, especially at the contacts points. A greater number of filaments during each cycle of commutation reduce the stress on the switch, thereby increasing its lifetime.

Our recent efforts have been directed to study and implement the broadening of the current channels in the avalanche photoconductive switch in order to improve lifetime and increase switching peak power. This is being performed by use of amorphic diamond coatings at the switch electrode to enhance the operation and lifetime of a GaAs photoconductive semiconductor switch (PCSS) in our stacked Blumlein pulse generators.

* This work was supported by the Air Force Office of Scientific Research (AFOSR) under Grant F49620-00-1-0296.

CP650, *Beams 2002: 14th International Conference on High-Power Particle Beams,*
edited by T. A. Mehlhorn and M. A. Sweeney

STACKED BLUMLEIN PULSERS

Stacked Blumlein pulsers have been extensively characterized at the University of Texas at Dallas (UTD) and their versatility has been demonstrated [2]. Our recent efforts have resulted in demonstration of several intense photoconductively-switched stacked Blumlein pulsers. Presently, these devices operate with a switch peak power in the range of 50-80 MW and activating laser pulse energies as low as 300 nJ [3].

Examinations of output waveforms have indicated pulse durations in the range of 1-5 ns and risetimes as fast as 200 ps. Examples of the output voltage generated by a 2-line stacked Blumlein prototype with Blumlein length of 11 and 17 cm together with the photograph of the pulser are given in Fig. 1. This pulser was constructed by connecting the corresponding high voltage and ground copper plates for each of the Blumleins to the switch electrode assembly. Blumleins with an impedance of about 100 Ω were extended to the proximity of the stacking location, 11 or 17 cm from the switch electrodes, where the center, high-voltage conductor was terminated. Then the lines angled toward each other along with the dielectric, where they were stacked directly on top of one another.

The lines were joined in series for about 2.5 cm, and the top and bottom plates were connected to a 200 Ω non-inductive resistive load as shown in the top and bottom photographs of Fig. 1. At the switch assembly, the pulse-forming line was screwed to the electrodes in a low-profile geometry as shown in the middle photograph of Fig. 1. A semi-insulating liquid encapsulated Czochralski (LEC) type GaAs switch was used to commutate the pulser. The switch assembly was enclosed in a SF_6 pressurized chamber. Pulse durations of voltage waveforms generated were proportional to the Blumlein lengths corresponding approximately to the two-way transit time of Blumleins. Peak voltage values for pulses presented in Fig. 1 were 98 kV, and corresponded to a voltage gain of 1.96.

PHOTOCONDUCTIVE SWITCH LIFETIME ENHANCEMENT

Basic research at UTD has described a conformal coating that has the hardness of natural diamond. Deposited at room temperature, it forms a strong bond to any material onto which it is applied. It has been demonstrated that only a 1-3 μm coating of amorphic diamond could protect fragile substrates against erosive environments [4].

In this work, an amorphic diamond coating was deposited on one side of semi-insulating LEC grown GaAs substrates. Electrode copper foils were attached to either side using conducting silver paint and epoxy. Electrodes were placed on opposite sides of the switch, and conduction was through the bulk of GaAs with the switch gap setting of 5 mm. Typical current-voltage characteristics measured for 0.5-μm and 1.0-μm amorphic diamond films on the GaAs are shown in Fig. 2. For comparison, the I-V plot for an uncoated similar GaAs substrate is included in this figure. The rectifying behavior for the coated sample seen in Fig. 2 was attributed to the amorphic diamond/GaAs heterojunction because the I-V plot for the uncoated sample showed symmetrical and ohmic character with change in the voltage polarity.

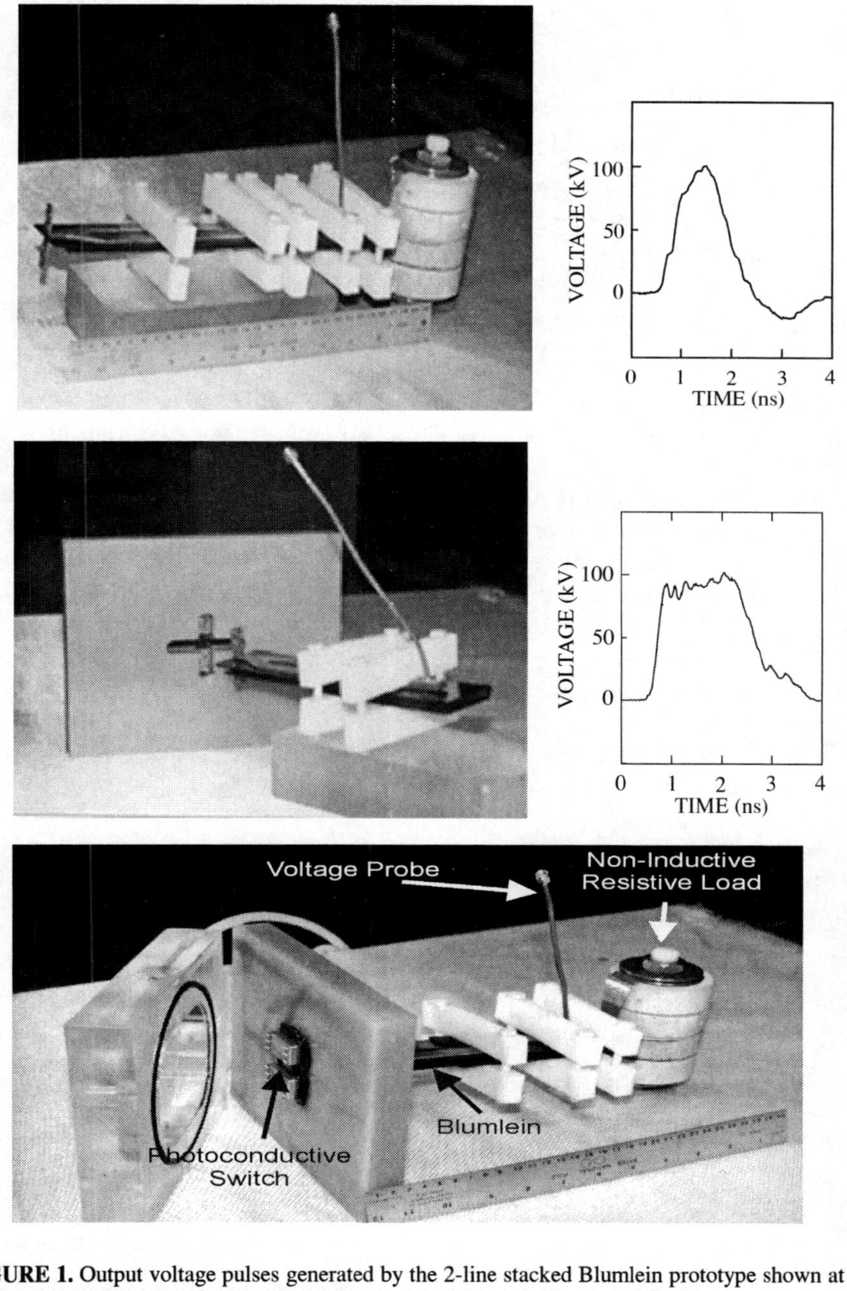

FIGURE 1. Output voltage pulses generated by the 2-line stacked Blumlein prototype shown at the bottom of this figure with pulse-forming lengths of 11 and 17 cm. Device was operated with a charging voltage of 50 kV. Top photograph shows the pulse-forming assembly connected to a resistive load. The middle photograph shows the pulse-forming section attached to the switch assembly.

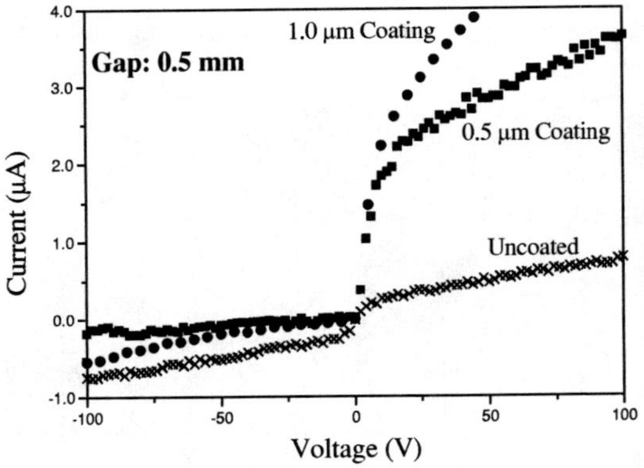

FIGURE 2. Current-Voltage characteristics measured in the dark for an uncoated sample and two samples of diamond-coated GaAs switch substrates at a gap setting of 5 mm.

The rapid current increase in the forward direction for the coated sample has been attributed to tunneling of electrons from amorphic diamond to the conduction band of GaAs, a process similar to Fowler-Nordheim tunneling. Thus, it is anticipated that, by depositing films of amorphic diamond near the PCSS cathode electrode contact, the number of carriers and avalanche sites increase, thereby improving the switch lifetime and performance.

In this work, we studied the lifetime of a GaAs switch with a 1-cm strip of 0.5 µm amorphic diamond coating deposited on the switch at the cathode location. The switch was installed in an opposed configuration and a switch gap of 5 mm was chosen for this study. The switch was tested at 18 MW (0.6 kA current and 30 kV voltage) for 10^5 shots, after which it failed. Under similar conditions of experiments and switch configuration, we performed the switch longevity experiment with a bare GaAs switch. The uncoated switch lifetime was found to be 1×10^4 commutation cycles. With the application of amorphic diamond, not only was the switch lifetime increased by an order of magnitude, but also the damage at the cathode contact was found to be less than that found for the anode contact. This indicated that the diamond coating protected and hardened the cathode side.

REFERENCES

1. Zutavern, F. J., Loubriel, G. M., Helgeson, W. D., O'Malley, M. W., Gallegos, R. R., Baca, A. G., Plut, T. A., and Hjalmarson, H. P., Fiber-optic Control of Current Filaments in High Gain Photoconductive Semiconductor Switches, in *Conference Record of the 1994 Twenty-First Power Modulator Symposium*, IEEE, Piscataway, 1994, pp. 116-119.
2. Davanloo, F., Collins, C. B., and Agee, F. J., *IEEE Trans. Plasma Sciences*, **26**, 1463-1475 (1998).
3. Davanloo, F., Dussart, R., Iosif, M. C., Collins, C. B., and Agee, F. J., Photoconductive Switch Enhancements and Lifetime Studies for Use in Stacked Blumlein Pulsers, in *Proceedings of the 12th International Pulsed Power Conference*, edited by C. Stallings and H. Kirbie, IEEE, Piscataway, 1999, pp. 320-323.
4. Davanloo, F., Collins, C. B., and Koivusaari, K.J., in *Adhesion Measurements of Films & Coatings, Vol. 2*, edited by K.L. Mittal, VSP, Netherlands, 2001, pp. 141-157.

Liquid Nitrogen as Fast High Voltage Switching Medium

J. Dickens, A. Neuber, M. Haustein, J. Krile, H. Krompholz

Center for Pulsed Power and Power Electronics
Departments of Electrical and Computer Engineering & Physics, Texas Tech University
Lubbock. TX 79409-3102

Abstract. Compact pulsed power systems require new switching technologies. For high voltages, liquid nitrogen seems to be a suitable switching medium, with high hold-off voltage, low dielectric constant, and no need for pressurized systems as in high pressure gas switches. The discharge behavior in liquid nitrogen, such as breakdown voltages, formative times, current rise as function of voltage, recovery, etc. are virtually unknown, however. The phenomenology of breakdown in liquid nitrogen is investigated with high speed (temporal resolution < 1 ns) electrical and optical diagnostics, in a coaxial system with 50-Ohm impedance. Discharge current and voltage are determined with transmission line type current sensors and capacitive voltage dividers. The discharge luminosity is measured with photomultiplier tubes. Preliminary results of self-breakdown investigations (gap 1 mm, breakdown voltage 44 kV, non-boiling supercooled nitrogen) show a fast (2 ns) transition from an unknown current level to several mA, a long-duration (100 ns) phase with constant current superimposed by ns-spikes, and a final fast transition to the impedance limited current during several nanoseconds. The optical measurements will be expanded toward spectroscopy and high speed photography with the aim of clarifying the overall breakdown mechanisms, including electronic initiation, bubble formation, bubble dynamics, and their role in breakdown, for different electrode geometries (different macroscopic field enhancements).

INTRODUCTION

Liquid switching media for the use at high voltages and high energy density for compact pulsed power devices gain more and more importance. For special applications, and for the use of semiconductor components, cryogenic systems might be of further advantage. The first choice for the switching medium for cryogenic operation would be liquid nitrogen. For ultrafast high power switching with gap closure times in the sub-nanosecond regime, high pressure spark gaps have been used, with pressures approaching several 100 atmospheres. Here, liquid nitrogen as a switching medium, with rather high hold-off voltages, low dielectric constants, and relatively simple handling as compared to high pressure systems, might be advantageous as well.

The phenomena involved in liquid breakdown, and especially in liquid nitrogen, are not completely clear at this time, due to the complex interaction of electronic and hydrodynamic processes. Most widely accepted ideas on breakdown in liquid nitrogen include field emission of electrons from the cathode, local heating, density reduction,

CP650, *Beams 2002: 14th International Conference on High-Power Particle Beams*,
edited by T. A. Mehlhorn and M. A. Sweeney

and the development of electron avalanches. Basic investigations of breakdown initiation are described in [2]. The role of thermal and electrically induced bubbles influencing or preceding breakdown has been investigated in numerous papers, see e.g. [3, 4, 5, 6, 7].

In this paper, we try to characterize breakdown in LN2 for the first time on a sub-nanosecond time scale, with emphasis on high speed high resolution electrical diagnostics (measurement of current and voltage), supplemented by optical diagnostics (measurement of the discharge luminosity). It is planned to expand optical diagnostic methods with high speed photography, spectroscopy, and active optical diagnostics, such as Schlieren photography.

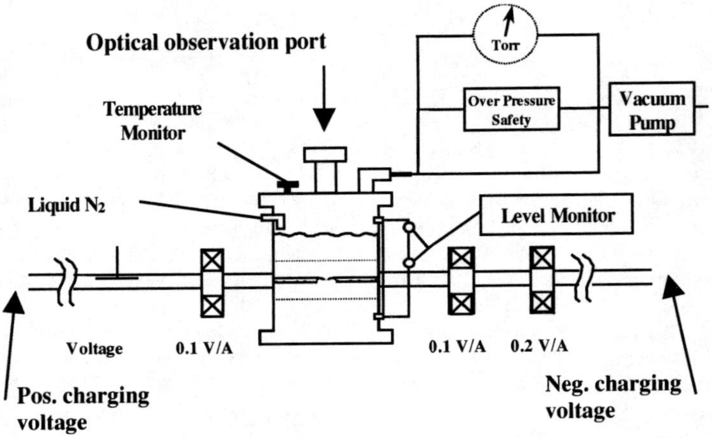

FIGURE 1. Basic setup with double-polarity charging and diagnostic elements

EXPERIMENTAL SETUP

The setup consists of two RG-19 transmission lines with an impedance of 52 Ω and a transit time of 150 ns. They are connected trough electrical feedthroughs on either side of a thermally insulated cylindrical stainless steel container, with the test gap electrodes connected to either transmission line center conductor. A cylindrical grid inside the container connects the outer conductors, maintaining constant impedance. The electrodes are hemispherical with a radius of curvature at the tip of 0.3 mm. For some experiments, needle electrodes with a radius in the micrometer regime have been used. All measurements have been performed in supercooled nitrogen without thermal bubbles, by lowering the container pressure to less than 34 kPa and returning to atmospheric pressure, which results in a LN2 temperature of about 65 K.

The current was measured at the positions along the transmission lines indicated in Fig. 1 with transmission-line-type current sensors [8]. In combination with fast clipping diodes and fast amplifiers, the range of the current measurement was 10^{-3} to several 10^{2} A. Overall temporal resolution was limited by the Hewlett-Packard Infinium oscilloscopes used, with a sampling rate of 2 Gs/s, a record length of 32768 points, and a risetime of 500 ps. Luminosity measurements were performed using Hamamatsu H6780-04 photosensors, with a risetime of 780 ps and a sensitivity of $4x10^4$ A/W.

RESULTS

Fig. 2 shows an example for the measured voltage and current during breakdown. The voltage drop occurs rather fast, it reaches half the charging voltage after about 2 ns. As a consequence, the current amplification decreases rather fast as well. It starts with an e-folding time of about 1 ns at an estimated spatially averaged field amplitude of 200 kV/cm. The macroscopic field enhancement due to the cathode radius of curvature is about 10. Observed craters and protrusions on the electrode surface let us expect microscopic field enhancements of at least 100. The luminosity signal has the same waveform as the current for the main breakdown.

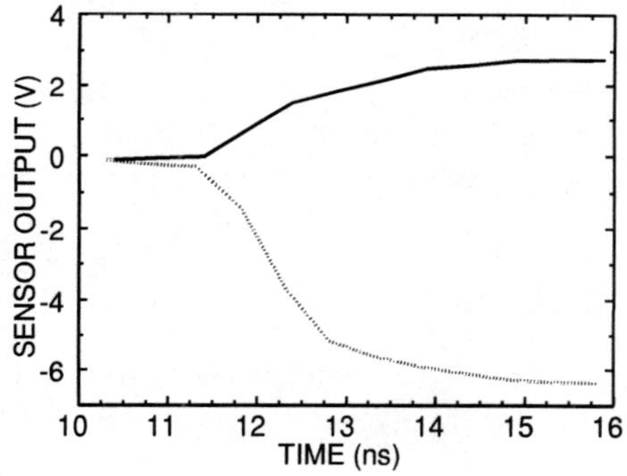

Fig. 2 Current (upper trace) with sensitivity of 100 A/V, and voltage (lower trace) with sensitivity 4 kV/V during breakdown. The capacitive sensor acts as highpass, i.e. the starting voltage of 0 in this diagram corresponds to a positive breakdown voltage of 56 kV, and the voltage at –1.1e-7 s corresponds to zero voltage across the gap. The slow voltage rise after this time is due to the droop of the sensor, the actual voltage remains at zero.

Fig. 3 shows, on a much longer time scale, an example for observed predischarges, i.e. current and luminosity signals. Only the first pulse in the current pulse trains (simultaneously to the luminosity spikes) represents a "true" current signal, the rest of the spikes in each decaying pulse train is caused by reflections at the charging sides of the cable, and reflections at the (capacitve) gap associated with the fast polarity change of each reflected pulse. The behavior of the pre-breakdown pulses is of stochastic nature, and varies from discharge to discharge. In some cases, more than 10 pulses are observed, in others there is no prebreakdown. Additionally, we observed the tendency that boiling LN2 exhibits distinctly less prebreakdown than the supercooled, non-boiling liquid.

DISCUSSION AND CONCLUSIONS

The results presented in this paper cover only electrical and some luminosity measurements and have to be completed by near-future high-speed photography etc. They show the existence of pre-breakdown phenomena, which consist of individual current/luminosity spikes which are most likely associated with bubble formation due to local heating by electron emission, as discussed in [7]. We expect explosive electron emission from micrometer sized asperities on the cathode, causing local

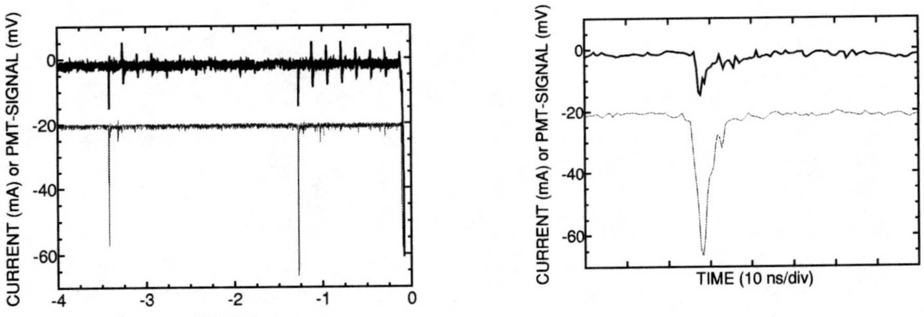

Fig. 3. Left: Prebreakdown pulses. Upper trace: current (first negative spike indicates current in the gap, following "bipolar" signals are due to reflections, lower trace: luminosity. Right: Second pre-breakdown pulse with expanded time scale

heating and vaporization. Below a breakdown threshold, the carrier multiplication does not seem to reach densities sufficient for the development of streamers. This process is repeated until one site reaches conditions with sufficient amplification leading to breakdown.

ACKNOWLEDGEMENT

This work was solely funded by the Compact-Pulsed Power MURI program funded by the Director of Defense Research & Engineering (DDR&E) and managed by the Air Force Office of Scientific research (AFOSR).

REFERENCES

[1] Giesselmann, M., Mahmud, Z., Carson, S., "Investigation of Power MOSFET switching at cryogenic (LN_2) temperatures", Proceedings of the 22nd International Power Modulator Symposium, Boca Raton, Florida, June 24-27, 1996.

[2] Jones, H.., Kunhardt, E.E "Development of pulsed dielectric breakdown in liquids", J. Phys. D, Appl. Phys. 26 (1995) 178

[3] Bernoual, A. et. al. "Propagation and structure of streamers in dielectric liquids", IEEE Electrical Insulation Magazine, **14** (1998)6

[4] Haykawa, N., et.al., "Breakdown mechanism of liquid nitrogen viewed from area and volume effects", IEEE Trans. on Dielectrics and Electrical Insulation, **4** (1997) 127

[5] Mc Cluskey, F.M.J., Denat, A., "The behavior of small bubbles generated by electrical current impulses over a wide range of applied pressure", J. Appl. Phys, **80** (1996) 2049

[6] Jefferies, M.J., Mathes, K.N. "Dielectric loss and voltage breakdown in liquid nitrogen and hydrogen", IEEE Trans. on Electrical Insulation, **5** (1970)83

[7] A. Denat, F. Jomni, F. Aitken, N, Bonifaci, IEEE Transactions on Plasma Science, **9**, (2002) 17

[8] H. Krompholz, H., Schoenbach, K., Schaefer, G., "Transmission Line Current Sensor", IEEE Instrumentation and Measurement Technology Conference, Tampa, Florida, USA 1985, p. 224 (IEEE catalog number 85CH2159-2)

Plasma Opening Switch Synchronization Facility

Aleksey A. Altuhov, Petr I. Blinov, Georgi I. Dolgachev,
Dmitry D. Maslennikov and Ivan A. Khodeev

Kurchatov Institute Russian Research Center, Moscow 123182, Russia

Abstract. The Baikal program provides a superpower generator creation which uses a plasma opening switch for output pulse production. The MOL facility is being developed within the frames of this program to test main units and systems of the planned generator at 4-6 MV, 3 MA and 100 ns. The MOL facility is planned to use 7 parallel plasma opening switch modules placed in the relatively compact vacuum chamber of 140 cm diameter. This aims at solving two main technical issued: plasma opening switch module development with external applied magnetic field and plasma switch modules synchronization. The paper describes plasma switch module design and low inductance load synchronization experimental tests results.

POS FOR BAIKAL PROGRAM

The "Baikal" generator is planned to achieve output parameters of 10 MV, 50 MA and 150 ns using plasma opening switch (POS) at the final stage of power intensifying. The MOL facility, which is under the construction now, models main units of the generator at 4-6 MV, 3 MA and 100 ns [1]. Effective plasma switching is achievable when limiting the linear density of the charge coming through the POS ($q < 5$ mK/sm) and using longitudinal magnetic field [2]. Meeting a condition of non-exceeding the charge limit results in POS geometry with a large diameter which can not be manufactured. E.g., for 50 MA generator and 2 μs conduction time a POS with an outer electrode diameter (d_0) of near 30 m and total length about 100 m (l_0) is required.

The authors suggested reducing the total space of the facility applying several modules with a diameter ($d=d_0/n$). The total length of the outer electrode in this case is the same as in the case of one module of d_0 diameter. But all n modules could be placed in a cylinder with a diameter of $d_0/n^{0.5}$, which is less than d_0. For the above example, if one uses 100 modules of 30 cm diameter they could be placed inside a cylinder of 3 m diameter. This approach is utilized in a design of the "Baikal" generator [1]. This task comprises of several stages: POS module creation, two-module test-bench creation and two module synchronization experiments using RS-20 facility (see [3] for a description).

MODULE DESIGN

First successful results on two modules synchronization were obtained in [4], for two parallel anodes placed in the common cylindrical cathode with a solenoid producing external magnetic field. The test-bench included two independent anodes operating in parallel POS modules (see Fig. 1). A quasi-stationary external longitudinal magnetic field was produced by a solenoid placed inside the internal POS electrode

CP650, *Beams 2002: 14th International Conference on High-Power Particle Beams*,
edited by T. A. Mehlhorn and M. A. Sweeney

(anode). An outer electrode of the POS is made from 4-mm thickness Copper plate (which is thicker than skin depth) so that external magnetic field was located inside the module.

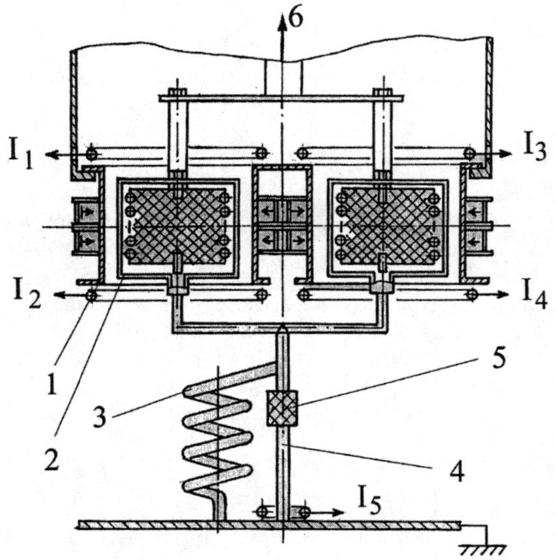

FIGURE 1. Two module synchronization test-bench. Figure captions: 1 – POS cathode with plasma injectors, 2 – POS anode with magnetic field solenoid, 3 – line connecting the magnetic field solenoid with a power supply, 4 –low-inductance load, 5 – switch. I_1, I_2, I_3, I_4, I_5 - Rogovsky coils.

This design allows providing module packaging inside a cylindrical chamber without any interaction of magnetic fields of the adjacent modules. Plasma guns with plasma storing volume [5] were placed outside the Copper electrodes. Plasma filled the inter-electrode gap through several 2 mm holes in the cathode. Power supply of solenoids was provided using (10 μG) inductance. Experiments used 2 modules of 4-stage Marx generator having 150 kA total current and 160 kV output voltage. Several electrical parameters were measured - total generator current - I_0, input currents from both POS modules - I_1 и I_2 , as well as output POS modules currents - I_3 и I_4 and low inductance load (1 μG) - I_5. The load was connected with the POS modules through the non-triggering switch with a dielectric surface breakdown.

EXPERIMENTAL RESULTS

Fig. 2 demonstrates waveforms of typical signals without turning on plasma injectors and with shorting load circuit closing switch. The pulsed voltage generator goes through the low-inductance load. The current is divided between two POS modules on two equal parts. Fig. 3 demonstrates same signals with plasma injectors turned on and give high plasma concentration (short-circuit mode). Output POS currents and load current (fig.1: I_2,I_4,I_5) are equal to zero.

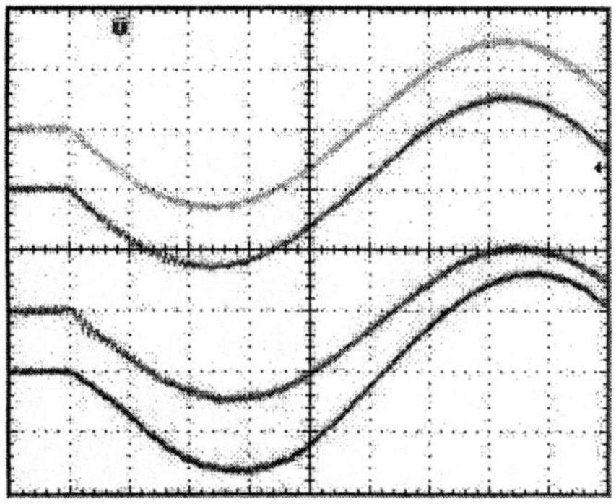

FIGURE 2. Two modules voltage and current waveforms without plasma injectors.

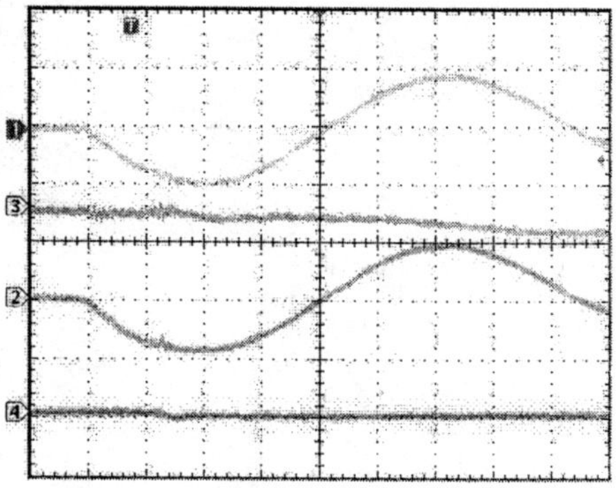

FIGURE 3. Short circuit mode of operation.

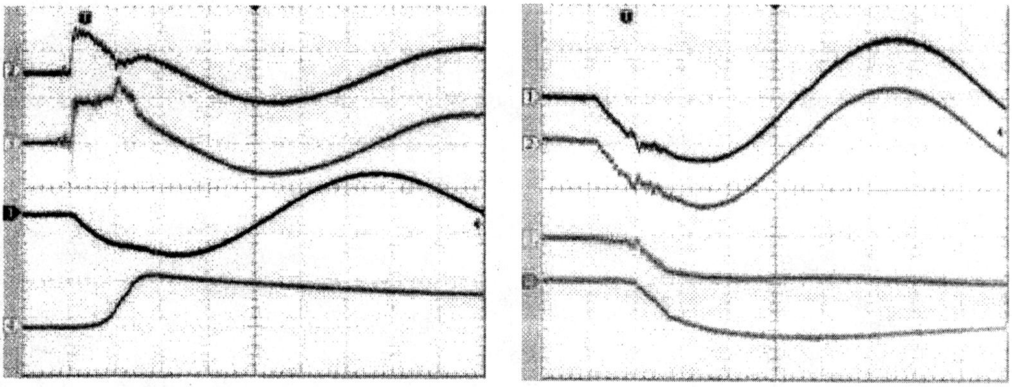

FIGURE 4. Current switching to the load indicated as a set of oscillograms counting from the top: current derivative measured by loop, voltage divider signal, total current, load current I_5 (left); signals obtained from the Rogovsky coils I_1, I_3, I_2, I_4 (right).

Fig. 4 represents the same signals with optimum plasma concentration. The switch 5 (see Fig. 1) is short. The module #1 POS (I_1) switched faster, its resistance was increased and current dropped. Simultaneously, the current through the module #2 (I_3) was also being increased. It also switched the current to the load. Total switching time was 400 ns. We have the same operation mode with low plasma concentration for the module #1 (50% of plasma injectors were turned off). The current from the first POS went to the second one from two sides and resulted in switching. The conduction time was reduced due to POS modules ion concentration drop.

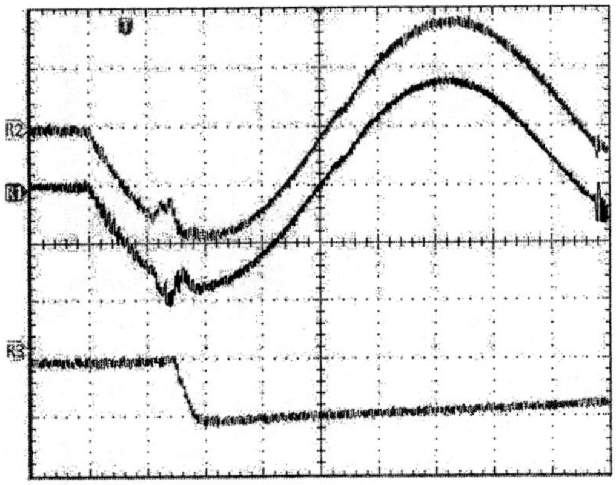

FIGURE 5. Synchronization in conditions when switch 5 (Fig.1) is used. Signals obtained from the Rogovsky coils I_1, I_3, I_5 .

Fig. 5 illustrate module operation mode when the load was connected through the separating closing switch. At the first stage both modules were switched and voltage over them was increased. Then the separating closing switch was closed due to overvoltage and current switched to the load in 200 ns. When electric strength of the closing switch gap was increased, sharper current switching took place. Time interval before reclosure was increased and load current pulse amplitude was reduced.

CONCLUSION

Plasma opening switch module with external applied magnetic field is development. This design allows providing module packaging inside a cylindrical chamber without any interaction of magnetic fields of the adjacent modules. Accurate synchronization of two high-current POS modules was observed with a low-impedance load. Even with asymmetry in plasma conditions a full transfer of the drive current to the load was observed. The separating closing switch improves POS module synchronization, increases voltage amplitude during switching and sharpens a rising front of the load current pulse. Those parameters may be adjusted using various electric strength of the separating closing switch.

REFERENCES

1. Azizov, E.A., Aleksandrov, V.V., Alikhanov, S.G., et al, Proc. 13th IEEE Int. Pulsed Power Conf., **1**, 773-776 (2001).
2. Barinov, N.U., Budkov, S.A., Dolgachev, G.I. et al., Plasma Dynamics, **28**, 202-207 (2002).
3. Barinov, N.U., Budkov, S.A., Dan'ko, S.A., et al., Inst. and Exp. Tech., **45**, 112-119 (2002)
4. Golovanov, Yu.P., dolgachev, G.I., Zakatov, L.P., et al, Sov. VANT Topics of Atomic Science and Tech. , Series Thermonuc. Fusion, **2**,. 35-37 (1987).
5. Barinov, N.U., Dolgachev, G.I., Maslennikov, D.D., et al, Proc. 13th Int. Conf. on High -Power Particle Beams, **1**, 595-598, (2000).

External Applied Magnetic Field Influence on Plasma Opening Switch Performance

Georgy I. Dolgachev, Dmitry D. Maslennikov, Ivan A. Khodeev

Applied Physics Department, Institute of Nuclear Fusion,
Russian Research Center "Kurchatov Institute", Moscow, 123182, Russia.

Abstract. The results of experiments on the extrinsic magnetic field influence on high-impedance plasma opening switch (POS) operation are reported. It has been demonstrated that POS voltage could be increased by the factor of 1.5 using an extrinsic magnetic field at erosion regime. The requirements for the magnetic field value and POS parameters to achieve maximum POS voltage have been determined. The experimentally tested dependency of the POS switching voltage as a function of the Marx one is presented.

INTRODUCTION

Plasma opening switches (POS) being effective and relatively inexpensive power intensifying elements have got wide recognition in pulsed power systems. The programs to create super-power generators for inertial fusion experiments have been recently launched [1,2]. Within the frames of the "BAIKAL" program it is expected to achieve a current pulse at approximately 50 MA with 150 ns duration and voltage amplitude - up to 10 MV. The POS is expected to be placed in the output cascade. According to [3], the POS voltage U_{POS} is determined by the power density w_I spent for plasma erosion, i.e. ion acceleration: $U_{POS} \sim \beta w_I^{4/7}$. At most of generators the density of the transmitted charge (q) is constant, i.e. $w_I = qU_{MARX}$ and

$$U_{POS}(MV) = \alpha (U_{Marx}(MV))^{4/7} \qquad (1)$$

To suppress the shunting electron current and to increase the POS voltage it was proposed to apply an extrinsic magnetic field [4]. This work aims at clearing up the role and influence of the extrinsic magnetic field and charge of the transmitted through the POS, and to define the voltage scaling more exactly.

SCHEME AND PARAMETERS OF GENERATOR

RS-20 generator was built in 1991 [5] with a purpose radiation-biology tests. Recent years the generator has been modified [6] :

1. Marx power output was multiplied by 12.5 and reached 180 kJ with Marx charging voltage (U_{MARX}) 0.85 MV. The current rise time was increased to 2 μs, current amplitude - to 300 kA.

CP650, *Beams 2002: 14th International Conference on High-Power Particle Beams,*
edited by T. A. Mehlhorn and M. A. Sweeney

2. The POS arrangement in the vacuum volume was changed (Fig. 1).

3. The POS (Fig.2) includes 78 plasma guns. The maximum linear charge density (along the circle of the outer POS electrode) is ~9 mC/cm which is twice as much as it had been determined in experiments [4] with a limited density (2.5-5 mC/cm).

4. The plasma guns form 6 sections. The guns of each section are placed uniformly along the circle. To provide programmed filling POS gap with plasma a separate operation of different sections is supposed [7].

5. The capability to change energy and power contribution to POS is provided. It could be reached either by varying the number of Marx's cascades, i.e. voltage under the fixed charge density or by varying the number of Marx's modules, i.e. charge density under the fixed voltage.

EXPERIMENTAL RESULTS

We used standard electrotechnical diagnostics: low-Ohmic electrical shunts, bleeders at the Marx's outlet, loops. Besides we used a method measuring the Bremsstrahlung of electrons accelerated in POS by photoneutron activation. This method employes two reactions: $Be^9(\gamma,n)Be^8$ and $D^2(\gamma,n)p$ with reaction thresholds E_{Be}=1,65 MeV and E_D=2,25 MeV correspondingly.

According to the results of experiments carried out with a four-stage Marx at an input voltage of 32-48 kV, output voltage of 128-192 kV, a current 200-320 kA, and a linear charge density (5-9) mC/cm, the coefficient in Eq. (1) is equal to α=(3.2-3.6)$MV^{3/7}$ [6]. With 20 stage and four Marx modules (i.e. at q_l = 9 mC/cm) the

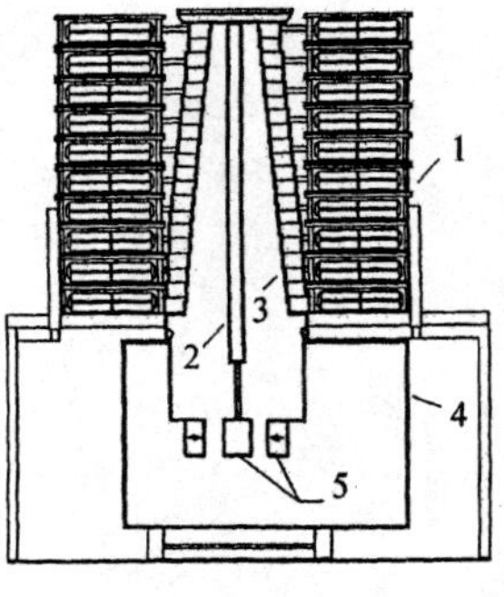

FIGURE 1. Scheme of generator: 1- Marx, 2- High voltage drive, 3- insulator, 4- vacuum chamber, 5- POS.

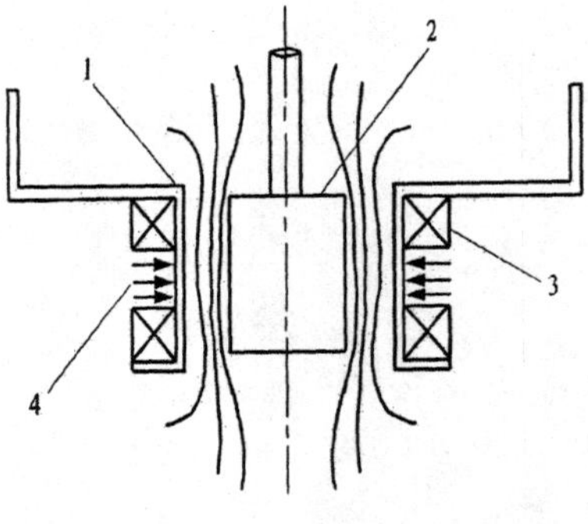

FIGURE 2. POS scheme: 1- cathode, 2- anode, 3- solenoid, 4- plasma guns.

maximum output voltage is no higher then 3 MV. The reason is that, to increase the charge value through the POS it is mandatory that the plasma concentration should be increased. This means that erosion regime conditions are no longer met. That is why the idea of programmed POS gap filling with plasma has been utilized [7]: plasma guns were switched once more within ~ 6-8 μs of the first plasma pulse and after ~ 3 more μs the Marx generator starts. In this case at the beginning the current is transferred by the portions of plasma from both the first and the second plasma shots. A second portion of plasma comes before the opening. This regime requires special attention to be drawn to the initial parameters constancy. So the repetitiveness of the results was rather poor. In this mode of operation we managed to transport all charge through the POS and in rare cases we could increase the switching voltage to the interval of 3-3.5 MV (fig.3a).

Using only 2 Marx modules (the corresponding charge density was: q_l=4.5 mC/cm) it was possible to get 3-3.5 MV (Fig. 3b) in near ~ 50 % of total number of shots. In this experiment the activation method has been employed and it has clearly demonstrated that the POS voltage exceeds 3 MV. This corresponds with electrical probe method. The longitudinal extrinsic magnetic field affects not only on the gap magnetic insulation. In order to reach "quality" opening (when the current drops down to zero) it is necessary to exceed the drive current magnetic field value. In this case electron drift trajectories in crossing **ExB** fields are scarcely closed.

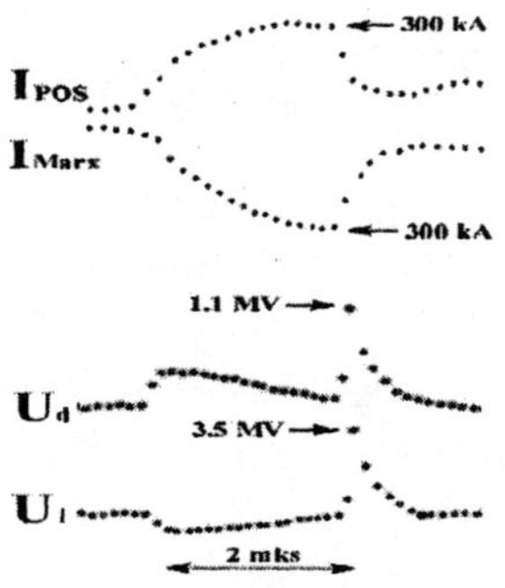

FIGURE 3-a). I_{POS} – current through POS, I_{Marx} – Marx generator current, U_d – voltage monitor signal, U_l – inductive voltage probe signal Voltage amplitude at POS during switching is 3.5 MV.

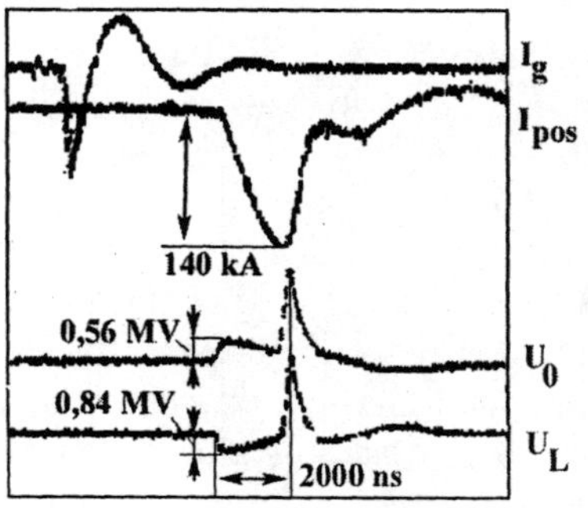

FIGURE 3-b). I_g – plasma guns current, I_{POS} – current through POS, U_o – voltage monitor signal, U_l – inductive voltage probe signal. Voltage amplitude at POS during switching is 3.4 MV which corresponds with nuclear activation voltage measurements.

In the self azimuthal magnetic field directed the axial electron drift takes place. And even if we apply additional external azimuthal magnetic field this does not help much: the switched current drops only to 50% of its amplitude value [8]. The axial drift results in space charge storing in the area of butt-ends of POS electrodes and, consequently, destruction of magnetic insulation.

Fig. 4 illustrate the dependance of POS voltage (U_{POS}) as a function of capacity storage initial voltage (U_{Marx}). Those curves were first obtained in [3] on the basis of analysis of the most known experiments. Now some new points have been added to the graph (figure 4).

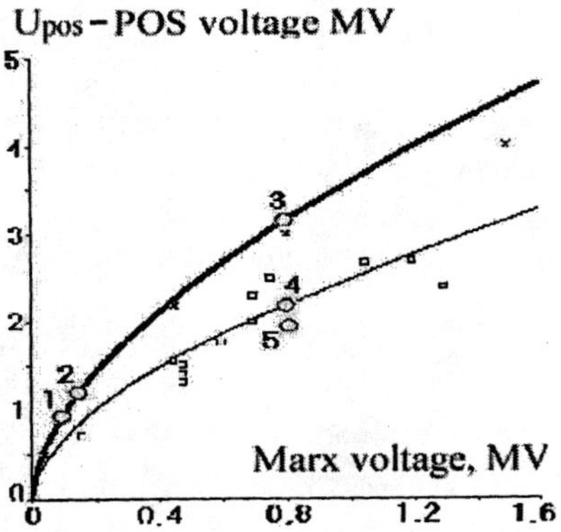

FIGURE 4.
The circles show the values of $U_{POS}(U_{MG})$ obtained in the present study: (1) RS-20 with four Marx modules with four stages in each (U_{MG}=0.128 MV, B=12 kG), (2) the facility with four modules with four stages in each (U_{MG}=0.16 MV, B=10 kG), (3) RS-20 with two modules (B=12 kG) or with four modules and the controlled plasma filling of the gap (B=16 kG), (4) RS-20 with two modules (B=0), and (5) RS-20 with four modules (B=0).

CONCLUSION

1. POS voltage could be determined by a proposed scaling $U_{POS}=\alpha U_{Marx}^{4/7}$. Where α is 3.6($MV^{3/7}$) for the erosion POS with applied extrinsic magnetic field regime and 2.5($MV^{3/7}$) for it without a magnetic field

2. External applied magnetic field strength should exceed the self one. Providing this condition will help to obtain the current switching down to zero.

3. The programming filling the POS gap with plasma may be used to provide the charge density through the POS. The programmed filling the gap with plasma allows to maintain plasma concentration relatively low. This allows to maintain POS operation in erosion regime which is the most effective from the viewpoint of highest voltage achieved.

REFERENCES

[1] V. A. Glukhikhet al. // Proc. Intern. Conference on High-Power Particle Beams, #12, 1998, vol.1, p.71.
[2] V. V. Alexandrov et al. //Abstracts of the 13th Int. Conf. on High-Power Particle Beams, BEAMS-2000, Nagaoka, Japan, June 25-30, 2000, p. 214.
[3] G. I. Dolgachev. and A. G. Ushakov // Plasma Physics Reports, , vol. 27, p.121, Feb. 2001.
[4] G. I. Dolgachev et al.// Sov. Journal of Plasma Physics, vol. 17, pp.1171-1183, Sept. 1991.
[5] V. M. Babykin et al. //Proc. International Conference on High-Power Particle Beams, BEAMS'92, #9, 1992, vol.1, pp. 517-522.
[6] N. U. Barinov et al.//Proc. International Congress on Radiation Physics, High current Electronics and Modification of Materials, #1, 2000, vol.1, p.274.
[7] N. U. Barinov et al.//13th intern. conference on High -Power particles beams, Nagaoka, Japan, June 25-30, 2000, Proceeding., p. 595
[8] Yu. P. Golovanov et al.// Sov. Topics on Atomic Science and Technology, Series Thermonuclear Fusion,1990, issue #3., p. 28.

Plasma Opening Switch Conduction Phase Adjustment

Georgy I. Dolgachev, Dmitry D. Maslennikov, Dmitry V. Ryzhov

Kurchatov Institute Russian Research Center, Moscow 123182, Russia

Abstract. Plasma Opening Switch conduction phase duration adjustment has a practical importance for several POS applications. They are POS synchronization and increase of energy and current density coming through the POS. The report represents the results of both experiments and simulations on POS conduction phase duration on experimental test-bench.

INTRODUCTION

The work has been initiated by the "Baykal" program which requires multiple parallel POS synchronization in a multi-module system and long-duration prepulse through the POS at the conduction phase [1]. The programmed fill of the POS gap is suggested to enhance the conduction phase duration. In this case plasma concentration decrease due to erosion process is compensated by additional plasma injection [2] and during the whole of conduction phase plasma concentration is not above the limit determined by the erosion regime. The goal of the work is the development of plasma gun for programmed fill of POS gap and experimental validation of the suggestion in the real POS scheme with extrinsic axial magnetic field for conduction phase duration adjustment.

EXPERIMENT

The experiments were made on streamlined machine [3] with double-stage power source (fig. 1). The first stage – capacitor battery with quarter of period ~ 40 mks, the second – Marx generator with quarter of period ~1 mks. POS operation was investigated both with one of two stages and with all two stages of the power source. The purpose of using the first stage was to study the possibility to enhance conduction phase duration. Switching on the second stage would increase the rate of plasma erosion and thus shortening of conduction phase duration due to sharp rise of current through POS. For programmed fill of POS gap with plasma the next features were used: multiple plasma gun switching with time interval of ~10 mks, collecting volume with outlets through which plasma flows out uniformly into the POS gap and the

CP650, *Beams 2002: 14th International Conference on High-Power Particle Beams*, edited by T. A. Mehlhorn and M. A. Sweeney

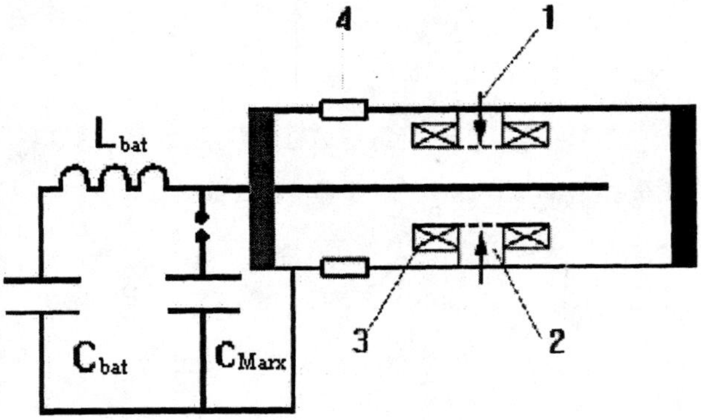

FIGURE 1. 1 – plasma guns, 2 – collective volume, 4 – shunt, 3 – magnetic field coil.

combination of these two ways. As the type of plasma gun we chose "flash-board" with collecting volume (fig. 2a).

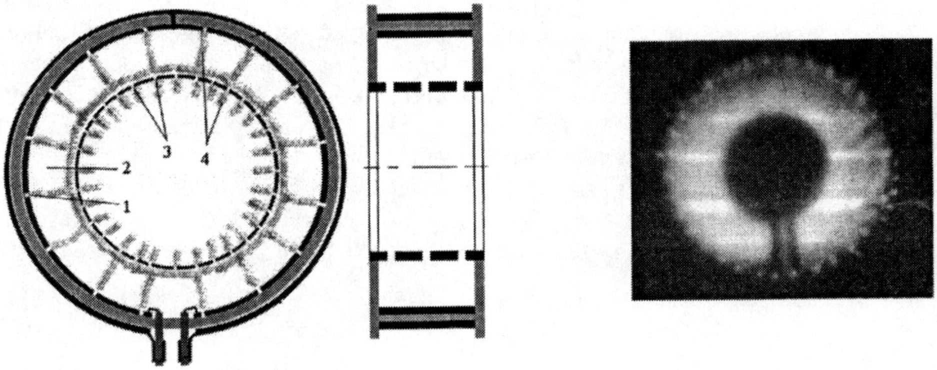

Fig. 2a. Flash-board plasma gun. 1 – plasma gaps, 2 – collective volume, 3 – outlets, 4 – plasma.

Fig. 2b. Image converter photo.

Such plasma gun allows to reduce greatly the number of supply cables which is of great importance for high-power generators. The plasma gun circuit inductance is minimal when current flows through all gaps and reverse wireway – this allows to avoid shunting to collective volume side of plasma gun gaps by plasma. Image converter photo of plasma flowing out the outlets of collecting volume shows that all the gaps of plasma gun are working properly and that there is no plasma closing (fig. 2b). The second purpose of collecting volume is electron-shielding of dielectric surface during the conduction phase that eliminates secondary plasma generation. Secondary plasma expands even in the case of metallic surface with velocity of 10^6

cm/s [4]. Its density is determined by power density. Critical power density from the point of view of secondary plasma generation is ~ 0.2-20 J/cm^2 and depends on material [5]. Principle generator scheme without capacitance store ("Baykal" program) is more simple if POS also plays a role of circuit closer: POS electrodes are upon voltage (~ 10 kV) and after closure of the POS gap with plasma current starts. But in this case there is an "active phase". When generated plasma moves across the gap electrons are emitted from the plasma surface and bombard anode generating secondary plasma. In our conditions electron power density is about critical. Applying extrinsic axial magnetic field allows to suppress the electron current and to prevent generation of secondary plasma. Basic experimental results are presented at fig. 3. Fig. 3a shows the influence of the active phase with no magnetic field, which causes secondary plasma generation and the absence of break.

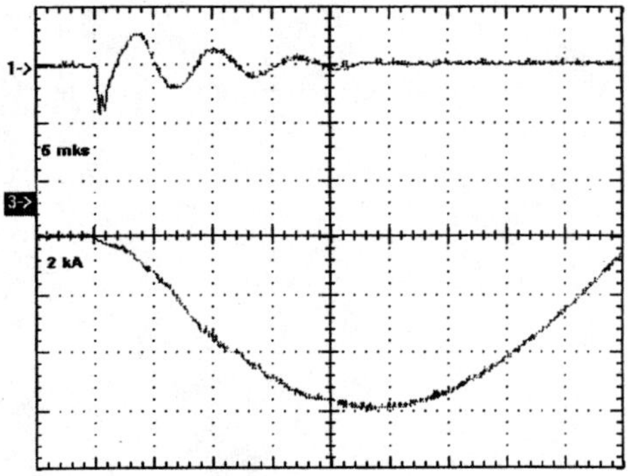

Fig. 3a. The influence of the active phase.

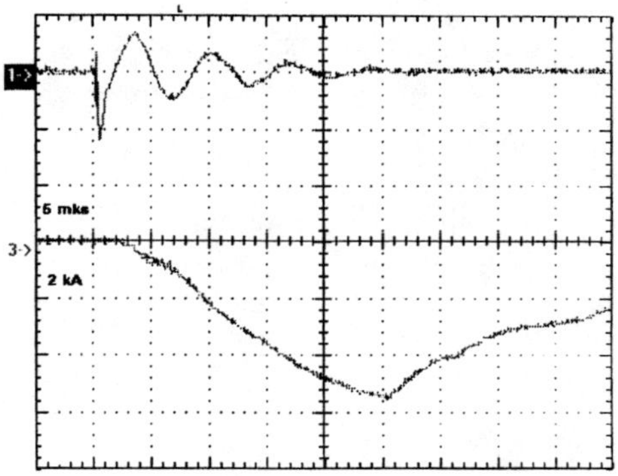

Fig. 3b. Axial magnetic field suppresses active phase

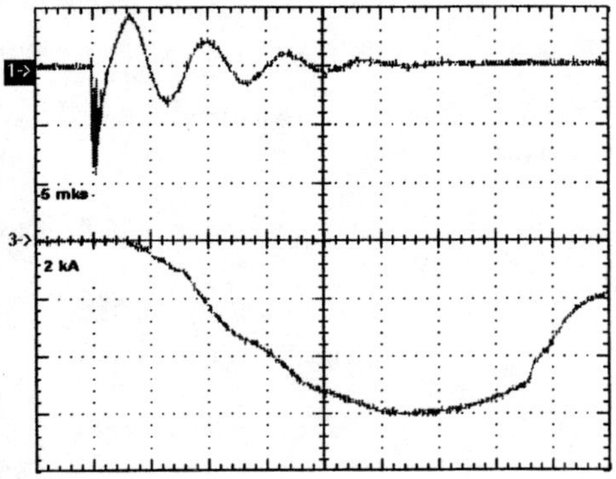

Fig. 3c. Enhanced conduction phase duration

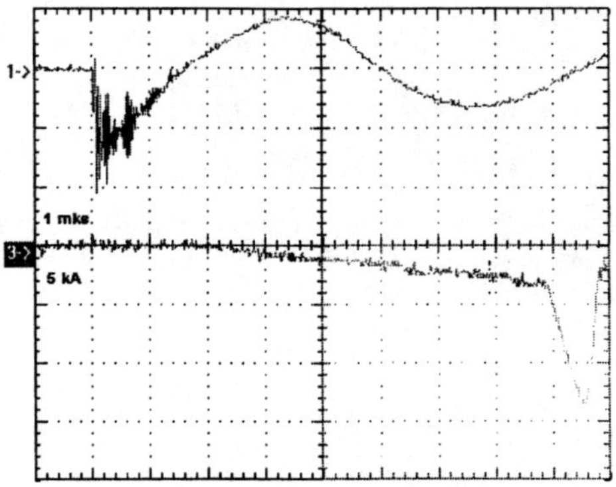

Fig. 3d. The influence of sharp current rise – shortening of conduction phase duration

The suppression of the electron current during the active phase by magnetic field allows to control plasma density more precisely and achieve break (Fig. 3b). Having increased plasma gun voltage we enhanced conduction phase duration up to ~ 40 mks (Fig. 3c). Fig. 3d shows the influence of sharp current rise (and transferring charge) due to switching on Marx generator. It shortens conduction phase duration. This fact could be used for parallel POSes synchronization, when the break of one module causes current rise in the others and thus shortening of their conduction phase duration.

CONCLUSION

1) Flash-board plasma guns has been developed with collective volume permitting to control plasma injection and to enhance conduction phase time up to 40 mks.
2) POS could be used as a circuit closer. Active phase could be suppressed using axial magnetic field.
3) Shortening of conduction phase duration due to the current rise shows the possibility of POS modules synchronization.

REFERENCES

[1] E.V.Grabosky, E.A.Azizov, S.G.Alikhanov, A.N.Gribov et al."Development of X-ray facility "Baikal" based on 900MJ indactive store and related problems" Proceeding of 13th International Pulsed Power Conferece, 2001, Las Vegas, USA, v.1, p.773-776

[2] Barinov N.U., Dolgachev G.I., Maslennikov D.D., "The increase of the time duration of plasma opening switch conduction phase"//13th Intern. Conference on High -Power particles beams, Nagaoka, Japan, June 25-30, 2000, Proceeding., p.595-598.

[3] Dolgachev G.I.,Ushakov A.G., Nitishinsky M.S., "New repetitive POS electron accelerator for air electron beam output" Proceedings of 11th IEEE International Pulsed Power Conference, Baltimore, USA, June 29- July 2, 1997, v.1, pp. 281-286.

[4] Ya. E. Krasik et all., "Spectroscopic investigation of plasma opening switch using a novel gaseous plasma source", 9th international conference on high-power particle beams, 1992. V. 1, p. 633.

[5] Blinov P.I., Dolgachev G.I., "Heavy ion acceleration in magnetic-isolated diode", Plasma physics, 1982, v.8, 5, p. 958-962

Electrical Breakdown of Submillimeter Water Gaps

K.H. Schoenbach,[a] J. Cooper,[b] A. Garner,[a] B. Goan,[a] R.P. Joshi,[a] J. Kolb,[a] S. Katsuki,[c] S. Kono,[c] M. Laroussi,[a] F. Leipold,[a] X. Lu,[c] C. Mallot,[a] J. Quian,[a] S. Xiao[a]

[a]*Electrical and Computer Engineering Department,* [b]*Chemistry and Biochemistry Department*
Old Dominion University, Norfolk, VA 23529, USA
[c]*on leave from Kumamoto University, Japan, and Ariake National College of Technology, Japan*

Abstract. Electrical breakdown and recovery processes in water have been studied using electrical and nanosecond optical diagnostics. The breakdown electric field in submillimeter gaps with 200 ns voltage pulses applied has been measured as 1 MV/cm, the rate of current rise during breakdown reaches $4 \cdot 10^{11}$ A/s. The switch recovery time is determined by expansion and decay of a vapor bubble. The experimental results, together with the results of a model with a percolative approach, provide design criteria for compact, high power, high repetition rate, liquid-switch pulse generators.

INTRODUCTION

Because of their high dielectric constant and high dielectric strength polar liquids, particularly water, have from the early beginnings of pulsed power technology been considered as dielectrics and as switch media. Water has a relative dielectric constant of 81 up into the GHz range, and electric breakdown fields of MV/cm have been reported [1]. These numbers indicate that energy densities of more than 3 J/cm^3 should be achievable in water lines. The high dielectric strength of liquids allows also reducing the size of high power switches considerably compared to gaseous switches. Besides the advantage of lower switch inductance, small switch volumes allow us to remove the liquid quickly after each shot, and to operate them at high repetition rates.

We have focused in the first phase of a project on energy storage and breakdown in liquids on water as a baseline dielectric and have studied the electrical characteristics of water at high electrical stress, up to values of 1 MV/cm between sphere-plane electrodes with gaps in the submillimeter range. The initial phase of the electrical breakdown, the prebreakdown phase, has been studied, and the recovery phase of water switches has been explored. The experimental studies are correlated with modeling.

EXPERIMENTAL SETUP

The experiments have been performed between a spherical and a planar electrode made of stainless steel (Fig. 1a). The gap distance was varied between 50 μm

CP650, *Beams 2002: 14th International Conference on High-Power Particle Beams,*
edited by T. A. Mehlhorn and M. A. Sweeney

and 1 mm. Optical diagnostics used to study the temporal and spatial development of the discharge includes high-speed photography, shadowgraphy, Schlieren diagnostics and interferometry. The temporal resolution for these diagnostic techniques was 1 ns. A trapezoidal voltage pulse of up to 40 kV and 200 ns FWHM, generated by means of a Blumlein pulse generator with an impedance of 100 Ω was applied to the electrodes. At gap distances of less than 400 μm this allowed us to generate average electric fields in the water between of electrodes in excess of 1 MV/cm. The liquid which was used was mainly distilled water with a resistivity of 100 kΩcm, but in some cases also tap water with a resistivity on the order of 1 kΩcm measured at room temperature.

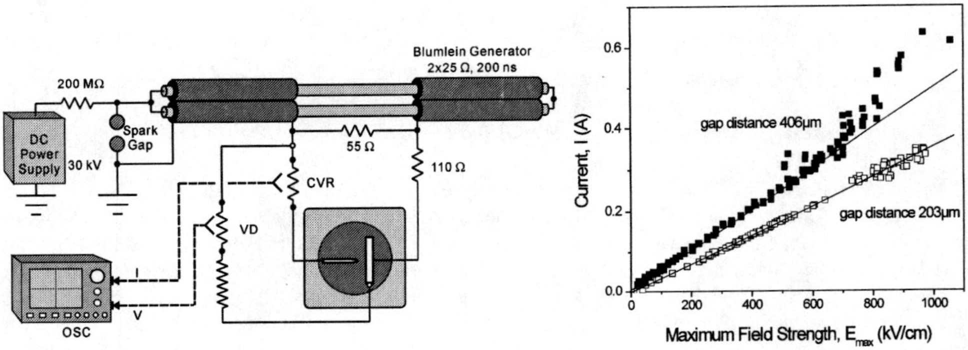

Figure 1a. Schematics of the experimental setup including electrical diagnostics; **1b**. current versus electric field on the electrode axis.

EXPERIMENTAL RESULTS

Current-voltage measurements in distilled water at two gap distances of 203 μm and 406 μm show a linear increase of current with voltage up to electric fields of approximately 1 MV/cm for the 200 ns pulse (Fig. 1b). Calculations of the resistivity of the water based on a field mapping code proved that the two curves in Fig. 1b become identical if plotted in a current density versus electric field diagram. This indicates that the electrical characteristics of the switch before breakdown is determined only by the bulk properties of the water, and the effect of boundary layers on the electric characteristic can be neglected.

For electric field values approaching 1 MV/cm the current increases superlinearly with voltage. High-speed photographs of the gap taken during this phase, the prebreakdown phase, show the development of a plasma at the spherical electrode. The plasma expands, and bridges the gap in approximately 10 ns. For the spherical electrode being the anode, 100% of the streamers develop at this electrode. If the polarity is reversed, only 60% of the discharges originate from the higher electric field electrode, approximately 30% originate from the plane electrode, and a few discharges seem to originate from fluctuations between the electrodes. When the gap is closed the current increases with a rate of rise of $4{\cdot}10^{11}$ A/s for distilled water, approximately $2.5{\cdot}10^{11}$ A/s for tap water.

The energy deposited into the water spark causes the water to transit from the liquid state to the vapor phase. The vapor bubble, which is created, expands nonlinearly with time ($t^{0.4}$) [2] until it reaches an equilibrium state with the compressed surrounding liquid, and then it decays. This process, which takes in our experiments approximately 1 ms, is shown in a sequence of photographs in Fig. 2. The bubble development and decay time determines the maximum switch repetition rate. This was confirmed by experiments with double-pulses, which showed that full recovery of the switch was achieved after approximately 1 ms.

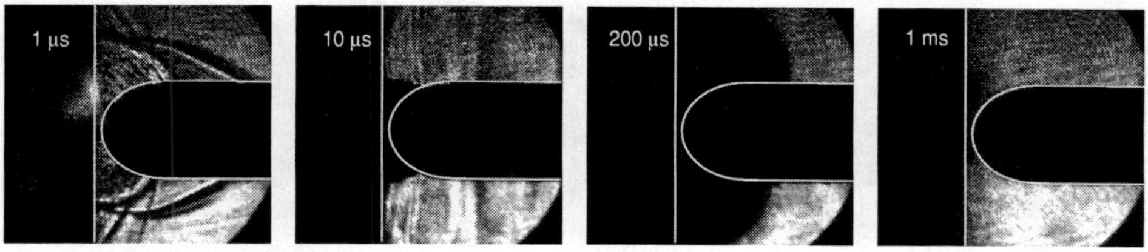

Figure 2. Sequence of Schlieren photographs, showing shockwaves (1 µs) and the development and decay of a vapor bubble.

MODELING

A two-dimensional, percolative approach to model dielectric breakdown, based on a network of parallel resistor-capacitor elements having random values, has been developed [3]. A set of two breakdown criteria has been used. One relies on a threshold electric field, and the other, on energy dissipation relative to the heat of vaporization. By carrying out this time-dependent analysis of a two-dimensional electrical network, the development and propagation of streamers and pre-breakdown dynamical evolution have been obtained directly. This model simulation also provides streamer shape and velocity, the pre-breakdown delay time, and the relationship between breakdown times and applied electric fields.

It has been shown that despite its simplicity, the model can successfully characterize breakdown and it's transient behavior in water. Fig. 3 shows the breakdown delay times versus applied electric field strength for water in a 100 µm electrode gap. The simulation results match the available experimental data very well over the entire 15 ns - 0.185 µs range. The current amplitudes obtained also matched measurements, with large swings in both the negative and positive directions at the onset of breakdown. The model demonstrates that the current is fashioned not only by dynamic changes in local resistance, but also by capacitive modifications associated with vaporization.

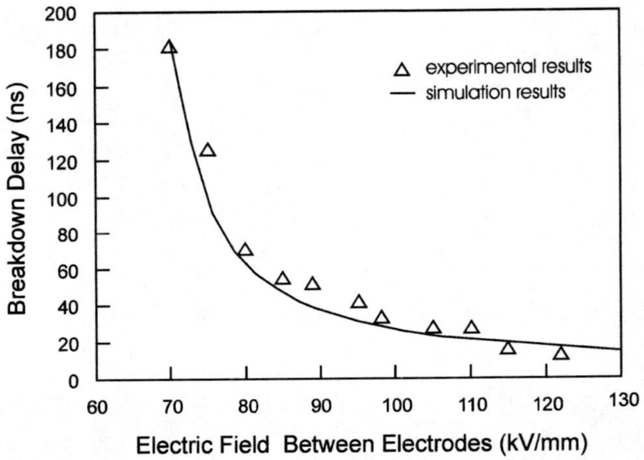

Figure 3 Predicted breakdown delay times versus applied electric field strength for water in a 100 μm electrode gap. Available experimental data are also shown.

TOWARDS ALL-WATER PULSE POWER GENERATORS

Based on the results of basic studies on water switches an all all-water pulse power system has been designed, where water is used both as dielectric in the pulse forming network (PFN) and as switch medium. First experiments with flowing water through the volume of the vapor bubble have shown that a reduction in recovery time and a consequent increase in repetition rate can be achieved. Also, by using a trigatron configuration in the liquid switch geometry we were able to control the switching event, so far however with relatively large jitter (> μs). The research and development results promise to lead to a simple, compact, triggerable pulse power generator with multi kHz repetition rate.

ACKNOWLEDGEMENTS

This research has been supported by an AFOSR/DOD MURI grant on Compact, Portable Pulsed Power, administered through the University of New Mexico.

REFERENCES

[1] R.E. Tobazeon, "Streamers in Liquids," in The Liquid State and its Electrical Properties, E.E. Kunhardt, H.L. Christophourou, and H.L. Luessen, eds., Plenum Press, 1988, p. 466.

[2] A. Larsson, A. Sunesson, J.Garmer and S. Kroll,"Laser-Triggered Electrical Breakdown in Liquid Dielectrics," IEEE Trans on Dielectrics and Electrical Insulation, Vol 8, No.2, April, 2002.

[3] J. Qian, R. P. Joshi, K. H. Schoenbach, M. Laroussi, E. Schamiloglu, and C. Christodoulou, "A Percolation Approach to Modeling Electric Breakdown in Water", to appear in IEEE Trans. Plasma Science, August 2002.

RADIOGRAPHY

Intense Electron Beam Sources for Flash Radiography

J.E. Maenchen[a], K. Hahn[a], M, Kincy[a], D. Kitterman[a], R. Lucero[a], P.R. Menge[a], I. Molina[a], C. Olson[a], D.C. Rovang[a], R.D. Fulton[b], R. Carlson[b], J. Smith[b], D. Martinson[c], D. Droemer[c], R. Gignac[c], T. Helvin[c], E. Ormand[c], F. Wilkins[c], D.R. Welch[d], B.V. Oliver[d], D.V. Rose[d], V. Bailey[e], P. Corcoran[e], D.L. Johnson[e], I.D. Smith[e], D. Weidenheimer[e], G. Cooperstein[f], R. Commisso[f], D. Mosher[f], S. Stephanakis[f], J. Schumer[f], S. Swanekamp[f], F. Young[f], T.J. Goldsack[g], G.M. Cooper[g], A.G. Pearce[g], M.A. Phillips[g], M.A. Sinclair[g], K.J. Thomas[g], M. Williamson[g], S. Cordova[h], R. Woodring[h] and E. Schamiloglu[i]

a. Sandia National Laboratories, Albuquerque, New Mexico, 87185-1193
b. Los Alamos National Laboratory, Los Alamos, New Mexico, 87545
c. Bechtel Nevada Corporation, Las Vegas, Nevada 89030
d. Mission Research Corporation, Albuquerque, New Mexico, 87106
e. Titan Pulse Sciences Division, San Leandro, California 94577
f. Naval Research Laboratory, Washington D.C. 20375
g. Atomic Weapons Establishment, Aldermaston, Reading, Berkshire, RG7 4PR, UK
h. Ktech Corporation, Albuquerque, New Mexico, 87106
i. University of New Mexico, Albuquerque, New Mexico, 87131

Abstract. High intensity pulsed electron beams are used to create bremsstrahlung x-ray sources for flash radiographic interrogation of dynamic experiments. Typical industrial sources operate below 200 GW/cm^2 intensities, while experimental requirements can demand above 50 TW/cm^2. Recent developments in pulsed power-driven high intensity electron beam systems have significantly increased these operating regimes, demonstrating 20 TW/cm^2, and computations predict successful extrapolation to higher intensities. Detailed studies of electron beam configurations, both theoretical and experimental, and the prognosis for each to increase to the required levels is discussed.

INTRODUCTION

Pulsed power-driven flash radiography, where a high intensity pulsed electron beam stops in a high-atomic-number bremsstrahlung converter producing a dose shown below, is a standard method to interrogate the interior mass distributions of dynamic experiments [1]. In this equation V is the applied voltage in megavolts and $\beta_\perp$ is the electron beam relativistic velocity component transverse to the radiographic axis. The x-ray

$$\frac{\text{rads@1 - m}}{\text{Coulomb}} = 1290\, V^{2.8} \exp\left[-\frac{(V+0.5)\beta_\perp}{0.67\pi}\right]$$

CP650, *Beams 2002: 14th International Conference on High-Power Particle Beams,*
edited by T. A. Mehlhorn and M. A. Sweeney

source intensity, the radiographic contrast, the geometry, and the detector resolution and efficiency all must be optimized together to maximize the radiographic information content for a specific experiment [2]. This paper discusses research on understanding and improving radiographic source brightness. The next sections discuss electron beam diode approaches leading to a description of a proposed flash radiographic source which should operate at unprecedented intensities.

INDUSTRIAL RADIOGRAPHIC X-RAY SOURCE

Industrial radiographic sources are typically positive polarity conical anodes irradiated by an annular electron beam drawn from a surrounding grounded cathode ring (Figure 1). The electrons spread across the anode surface producing a diffuse bremsstrahlung x-ray source. The electron current is [3] I_{SCL} (kA) = 14.7 (L/k^2R_a) $V^{3/2}$ where L is the cathode emission length, R_a is the anode radius, and k is a geometric term in this non-relativistic formulation, corresponding to a few kiloamperes. The low intensity nature of this geometry produces minimal anode damage, allowing multiple shot operation without maintenance. These industrial diodes typically operate below 2 MV and 300 GW/cm^2.

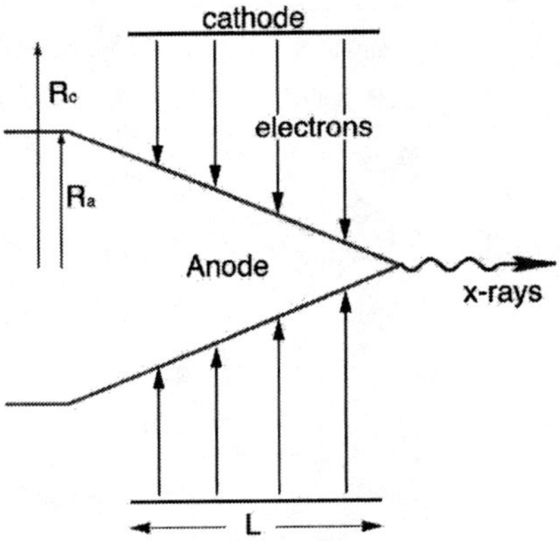

FIGURE 1. Industrial radiographic diode

The remainder of this paper will discuss high brightness x-ray sources which differ in that they operate at very high electron beam intensities where the deposited energy density in the anode bremsstrahlung converter exceeds vaporization thresholds, limiting them to single shot operation between maintenance.

PINCHED BEAM X-RAY SOURCE

A significant intensity increase can be obtained by operating above the critical current I_{crit} (kA) = 8.5 $(\gamma^2 - 1)^{1/2}$ (R_c/d) where R_c is the cathode radius and d is an axial vacuum gap spacing (Figure 2). Here the current produces a strong azimuthal magnetic field which, near the planar conducting boundary, bends the electrons inward. Formation of a counter-streaming ion flow further neutralizes the electron space charge, enabling a high intensity focus [5]. In negative polarity electrons sweep onto an axial focal spot with typically a 30° mean incidence angle and emit a broad x-ray cone.

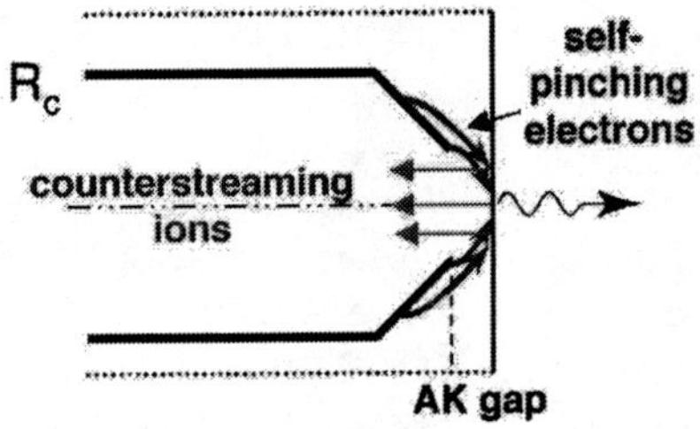

FIGURE 2. Pinched beam x-ray source

The small vacuum gap in pinched beam diodes increases their sensitivity to prepulse, and the requirement for counter-streaming ions limits the impedance lifetime as the anode plasma evolves across the vacuum gap during the pulse. The sweeping electron beam produces a radiographic halo during the pinch phase and the core typically wanders, making post-diode x-ray collimation difficult. These diodes can couple (for example) 30 kA onto about a 2-mm diameter focal spot, operating at a peak focal intensity of about 1 TW/cm^2.

PARAXIAL DIODE X-RAY SOURCE

A more robust high focal intensity diode launches a large area electron beam, such that the energy deposition on the anode is insufficient to form a plasma. This beam passes through a thin foil and is compressed during transport through a gas cell before irradiating the bremsstrahlung converter (Figure 3). Gas ionization occurs within a few nanoseconds and fractionally neutralizes the electron beam space charge. Depending on the plasma magnetic decay time, the electron beam either slowly sweeps through a focus at the converter or rapidly collapses into fixed betatron oscillations. An innovative option includes a short vacuum transport region between the gas and target, which trades spot diameter for beam temperature [1]. Figure 4 shows electron orbits calculated with the hybrid PIC-Fluid LSP code [6].

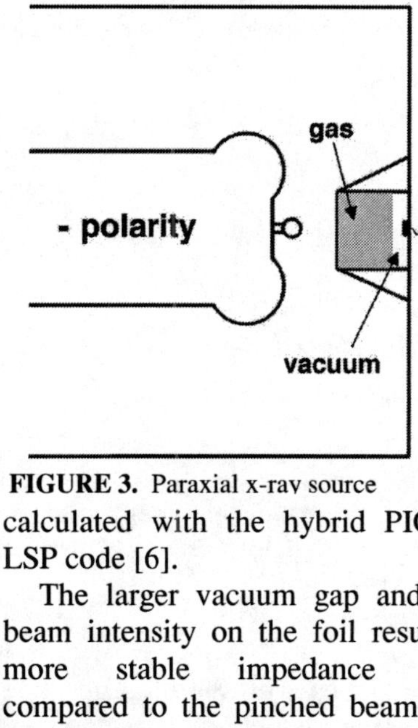

FIGURE 3. Paraxial x-ray source

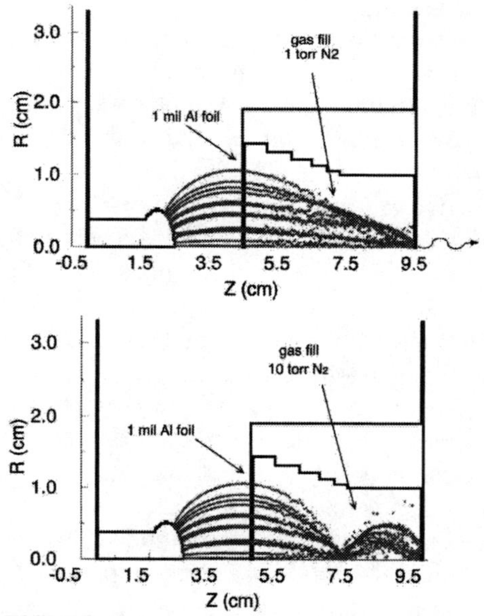

FIGURE 4. Paraxial diode orbits. top: Ballistic, bottom: Betatron

The larger vacuum gap and lower beam intensity on the foil result in a more stable impedance history compared to the pinched beam diode, although this system is also sensitive to prepulse (foil turn-on) and pulse risetime (gas breakdown). Radiation produced while the beam is pinching similarly produces a radiographic halo and spot wander precludes tight collimation. These paraxial diodes are very reliable radiographic sources which operate from 5 to 10 MV, coupling typically 35 kA into foil scattering limited spot diameters of 3-mm at peak amplitudes of about 1 TW/cm^2 [1].

ROD PINCH X-RAY SOURCE

The rod pinch diode [7] (Figure 5) combines the geometry of the positive polarity industrial source with self-field pinched beam focusing, operating above critical current
$I = \alpha * I_{crit} \sim 17\, V^{0.136} (\gamma^2 - 1)^{1/2} / \ln(R_c/R_a)$ kA
Simulations show this diode rapidly transitions from the industrial source Langmuir-Blodgett radial flow into magnetic pinching along the anode rod which extends through the cathode, and then stabilizes with electrons impinging only near the rod tip. The spot size and position are therefore dictated only by the rod diameter and placement, making this a very convenient radiographic x-ray source.

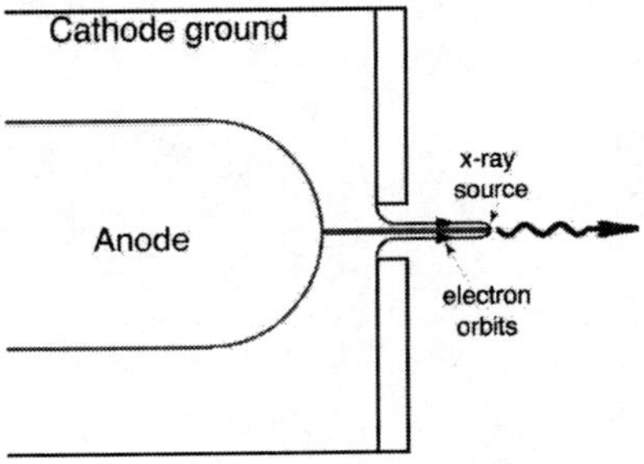

FIGURE 5. Rod Pinch x-ray source

The converter rod thickness may be made comparable to or smaller than the electron range leading to electrons which scatter through the anode to re-enter the diode and reflex, significantly modifying the electron phase space. The very high electric fields and beam intensities form plasmas which evolve, affecting the electromagnetic profiles and the electron orbits and reducing the impedance in time. The rod is observed to hydrodynamically expand during the pulse, limiting the ultimate spot size (Figure 6). Angular radiation measurements suggest that the electron beam strikes the rod predominantly radially, producing less dose in the desired forward cone than the other advanced x-ray sources discussed. For a 0.5-mm gold rod the useful filtered dose rate is [8] Dose = 7e-4 $I(t)*V(t)^{1.11}$ rads/sec at 1-m. For comparison, a 1-mm rod scales as $V^{1.25}$. The rod pinch diode operation is therefore quite subtle while the geometry appears simple.

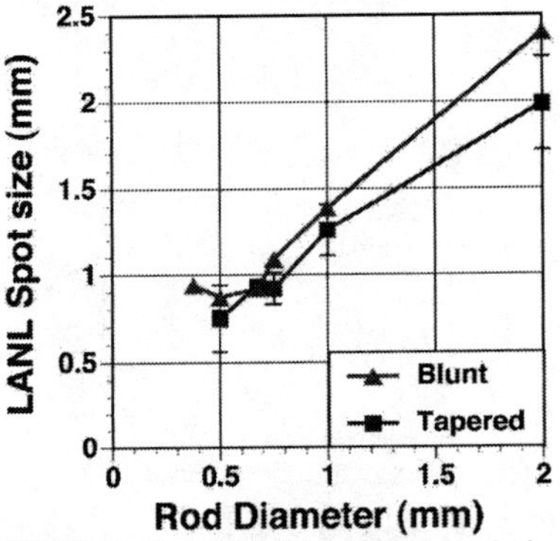

FIGURE 6. Rod Pinch spot variation measurements on 2.2 MV SABRE

This complex physics has been been successfully modeled with the modified LSP hybrid (PIC, Monte Carlo, and fluid) code. PIC electrons which enter the rod are followed with Monte Carlo techniques (a step which also serves to provide the source bremsstrahlung distribution). Slowed and scattered electrons (and secondaries) which exit the rod are re-initialized into the PIC simulation. Surface heating calculated in the Monte Carlo phase is used to prescribe gas release and ionization processes are followed to predict plasma formation and evolution.

These simulations provide impedance history and radiation production relations which are in excellent agreement with numerous experiments [9] spanning 1 to 4 MV with 0.5- to 2-mm diameter anode rods. This diode is found to be insensitive to pulse shape or misalignment, operating reliably in modest vacuums (<50 mPa) with a falling impedance well predicted with the above equation using 1–2 cm/μs electrode plasma closure velocities. The rod pinch diode has been shown to provide a reliable, high intensity radiographic x-ray source up to 4 MV from 20 to 40 ohms at 20 TW/cm^2, and experiments are underway to explore performance up to 6 MV.

MAGNETICALLY IMMERSED X-RAY SOURCE

The magnetically immersed diode utilizes a very small cathode emitter and augments the self-magnetic field with a solenoidal applied magnetic field to contain the electron orbits to a tight spot (Figure 7). Coupling a beam current of (kA)

$$I_{beam} = \frac{17\ \text{kA}\left[\left(1+\frac{V}{511\ \text{kV}}\right)\right]}{\left(1-\frac{Zn_i}{n_e}\right)\left(1+2\ln\left(\frac{R_w}{r_b}\right)\right)}$$

The magnetically immersed diode has demonstrated few-mm-diameter radiographic spots from 3 to 9 MV, operating at 10 to 20 TW/cm^2. These very high intensities promptly produce energetic anode plasmas which enable a 3D electromagnetic ion hose instability. This instability saturates when the beam offset approaches the electron gyro-radius, and the spot size can be expressed as [10]

$$\text{spot} = 1.3\sqrt{\text{offset}^2 + \text{needle}^2}$$

$$\text{offset} = \sqrt{\frac{2\gamma I}{17\ \text{kA}}}\ \frac{0.17\ \text{Tesla}}{B_z}\ \text{cm}$$

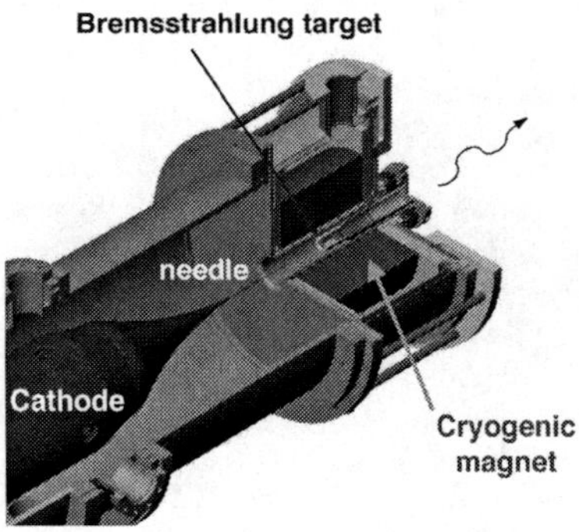

FIGURE 7. Magnetically Immersed x-ray

Experiments [11] have shown this diode is capable of providing either a benign ion hose instability limited spot size described above or a deleterious enhanced instability mode, where the spot grows abruptly to many times the analytic prediction and the impedance collapses. Numerical simulations indicate this enhanced mode is caused by multi-species ions entering into the beam and ionizing in-flight, resulting in a feedback loop of mismatched space charge limited anode and cathode boundary conditions. It is possible that these may be controlled by surface preparation, adequate magnetic insulation of the wall ions, and magnetic mirroring of the target ions. A magnetically immersed diode is under development to apply these techniques.

RADIOGRAPHIC SOURCE SUMMARY

The radiographic diodes discussed above can be summarized in terms of their electron beam intensity as a function of operating voltage, as shown in Figure 8. Industrial radiographic sources from a variety of manufacturers all operate below 300 GW/cm^2 and typically below 2 MV. The pinched and paraxial diodes developed at

AWE in the United Kingdom operate quite reliably near 1 TW/cm^2 on a variety of drivers from 1 to 10 MV. The rod pinch diode operates above 10 TW/cm^2 from 1 to 4 MV but may not scale effectively to the 10 MV range due to the large electron temperature. The magnetically immersed diode has operated above 10 TW/cm^2 from 1 to 9 MV. Improvements to the paraxial and magnetically immersed diodes will be investigated on the Radiographic Integrated Test Stand (RITS [12]) family of accelerators to continue development of high intensity flash radiographic x-ray sources.

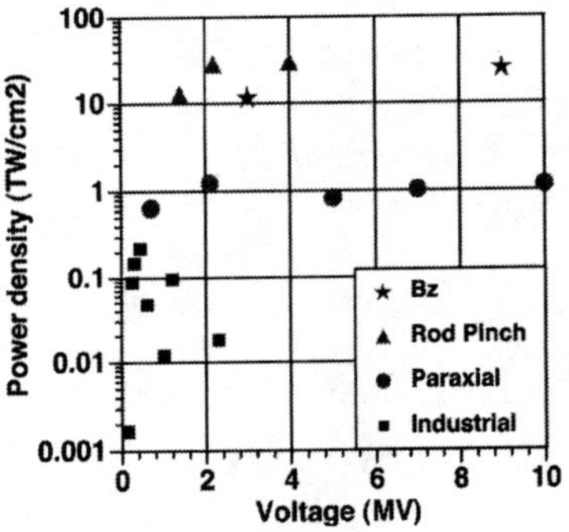

FIGURE 8. Radiographic diode comparison

REFERENCES

[1] T. J. Goldsack, T. F. Bryant, P. F. Beech, S. A. Clough, G. M. Cooper, R. Davitt, R. D. Edwards, N. Kenna, J. McLean, A. G. Pierce, M. J. Phillips, K. P. Pullinger, D. J. Short, M. A. Sinclair, K. J. Thomas, J. R. Threadgold, M. C. Williamson, and K. Krushelnick, IEEE Transactions on Plasma Science **30**, 239 (2002).

[2] S. Watson, "A Closed-Form, Fellget-Linfoot Radiographic Figure of Merit", DARHT Technical Note 94, Los Alamos National Laboratory, Los Alamos, New Mexico, 87545.

[3] I. Langmuir and K. Blodgett, *Phys. Rev.*, **22**, 347 (1923).

[4] J. Creedon, *J. Appl. Phys.*, **48**, 1071 (1977), and C. W. Mendel Jr. and D. B. Seidel, *Physics of Plasmas* **6**, 4791 (1999).

[5] A. E. Blaugrund and G. Cooperstein, *Phys. Rev. Lett.*, **34**, 461 (1975).

[6] T. P. Hughes, S. S. Yu, and R. E. Clark, Phys Rev. **2**, 11041 (1999); D. R. Welch, D. V. Rose, B. V. Oliver, and R. E. Clark, Nucl. Instrum. Methods Phys. Res. A. **464**, 134 (2001).

[7] R. A. Mahaffey, J. Golden, S. A. Goldstein, and G. Cooperstein, *Appl. Phys. Lett.*, **33**, 795 (1978); G. Cooperstein, R. J. Commisso, D. D. Hinshelwood, D. Mosher, P. F. Ottinger, S. J. Stephanakis, J. W. Schumer, S. B. Swanekamp, B. V. Weber, F. C. Young, J. E. Maenchen, P. R. Menge, C. L. Olson, B. V. Oliver, D. V. Rose, D. R. Welch, R. Carlson, D. Oro, D. Fulton, E. Hunt, and D. Droemer Proceedings of the 13th Int'l Conf. on High Power Beams, Nagaoka, Japan, (2000).

[8] D. V. Rose, D. R. Welch, B. V. Oliver, R. E. Clark, D. L. Johnson, J. E. Maenchen, P. R. Menge, C. L. Olson, and D. C. Rovang, Journal of Applied Physics 91, 3328, (2002).

[9] P. R. Menge, J, Gustwiller, D. L. Johnson, J. E. Maenchen, I. Molina, D. C. Rovang, R. Shear, S. Cordova, D. Droemer, E. Hunt, E. Ormand, L. Woo, R. J. Commisso, G. Cooperstein, S. B. Swanekamp, R. V. Oliver, D. V. Rose, and D. R. Welch, IEEE Int'l Conf. on Plasma Science, June 4-7, 2000, New Orleans, Louisiana, IEEE Conf. Record 00CH37087, 232 (2000);
R. J. Commisso, G. Cooperstein, D. D. Hinshelwood, D. Mosher, P. F. Ottinger, S. J. Stephanakis, B. V. Weber, and F. C. Young, IEEE Transactions on Plasma Science **30**, 338 (2002); and

[10] D. R. Welch and T. P. Hughes. *Laser and Particle Beams*, **16**, 285 (1998).

[11] M. G. Mazarakis, J. W. Poukey, D. C. Rovang, J. E. Maenchen, S. R. Cordova, P. R. Menge, R. Pepping, L. Bennett, K. Mikkelson, D. L. Smith, J. Halbleib, W. A. Stygar, and D. Welch., *Appl. Phys. Lett.*, **70**, 832, (1997).

[12] I. D. Smith, .V. L. Bailey, P. Corcoran, J. Kishi, J. Fockler, J. Gustwiller, D. L. Johnson, J. E. Maenchen, and D. Droemer, "Design Of A Radiographic Integrated Test Stand (RITS) Based On A Voltage Adder, To Drive A Diode Immersed In A High Magnetic Field", *12th IEEE International Pulsed Power Conference*, (IEEE, New York, 1999) p. 403

Magnetic Insulation, Power Flow, and Pulse Power Results on RITS-3

David L. Johnson[a], Ian Smith[a], Patrick Corcoran[a], Vernon Bailey[a], John Maenchen[b], Dean Rovang[b], Isidro Molina[b], Kelly Hahn[b], Robert Lucero[b], Mark Kincy[b], David Kitterman[b], Bryan Oliver[c], Dale Welch[c], David Rose[c], Timothy J Goldsack[d], Martin A Phillips[d], Mark A Sinclair[d], Kenneth J Thomas[d]

a. Titan – Pulse Sciences Division, San Leandro, CA
b. Sandia National Laboratories, Albuquerque, NM
c. Mission Research Corporation, Albuquerque, NM
d. Atomic Weapons Establishment, Aldermaston, England

Abstract. RITS (Radiographic Integrated Test Stand) is an induction voltage adder designed by Sandia and PSD to provide 16-MV, 150-kA electron beams and other capabilities. Previous publications have reported on tests of a single pulse forming line and adder cell, including initial results of the effects of various degrees of non-uniform injection of current into the adder bore on magnetic insulation and power flow in the downstream MITL. Now RITS-3 has been constructed, consisting of three pfls driven by a common intermediate store; three induction cells, one driven by each pfl; a three-stage, 4-MV, 150-kA vacuum voltage adder; and an output MITL and diode. Here we report on (1) simulations of the three-stage adder using the MRC 3-D particle-in-cell code LSP that address the effects of injected current non-uniformities on magnetic insulation and power-flow both upstream and downstream in a multi-cell adder; (2) experimental results compared with simulations; and (3) initial performance of the RITS-3 pulse power.

INTRODUCTION

The Radiographic Integrated Test Stand (RITS) [1] is a modular, multi-megavolt, 150-kA, 70-ns, induction voltage adder. RITS-3 is shown in Figure 1. RITS-3 has three 1.35-MV cavities in series on a common, magnetically insulated transmission line. Each cavity is driven by a single water switched pulse forming line (PFL). An azimuthal transmission line, shown in Figure 2, in each cavity provides symmetrization of the current from the one point PFL feed to the MITL. The azimuthal line is connected to the cathode of the cavity at the bottom 180° of the cathode and at two tabs at 45° on either side of the top of the catode. Results of tests on a single RITS cavity are detailed in reference [2].

CP650, *Beams 2002: 14th International Conference on High-Power Particle Beams,*
edited by T. A. Mehlhorn and M. A. Sweeney

The RITS-3 tests studied the effects of injected current non-uniformities on magnetic insulation and power flow in the MITL, compared the results to 3-D PIC code simulations, and evaluated the pulsed power performance of the system.

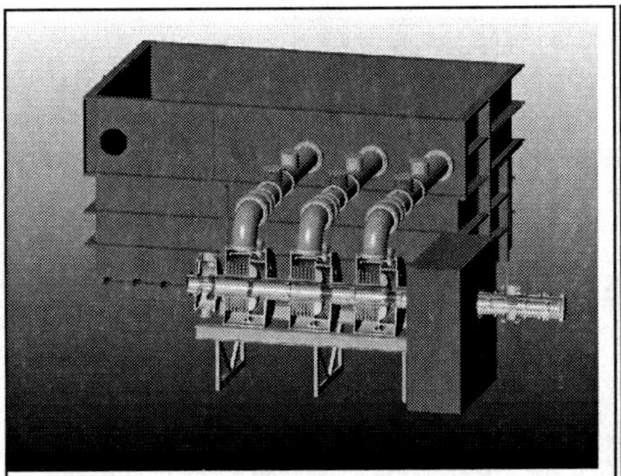

FIGURE 1. Cut away view of RITS-3.

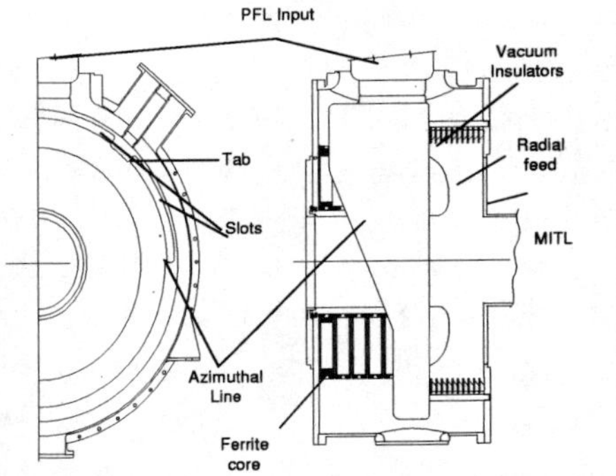

FIGURE 2. Cavity azimuthal transmission line.

Pulsed Power Results

The PFL voltages before and after the prepulse/sharpening switch are shown in Figure 3. Figure 3a is an overlay of output voltages from the three pulse forming lines. Jitter measurements taken on repeated shots indicate a water switch jitter of 4 to 5 ns for a 3 inch water gap charged to 2.7 MV in 200 ns. Figure 3b is the forward going voltage after the prepulse/sharpening switch. The prepulse has been reduced from ~200 kV to less than 50 kV; a second prepulse switch in the pulse forming section further reduces the prepulse to less than 10 kV.

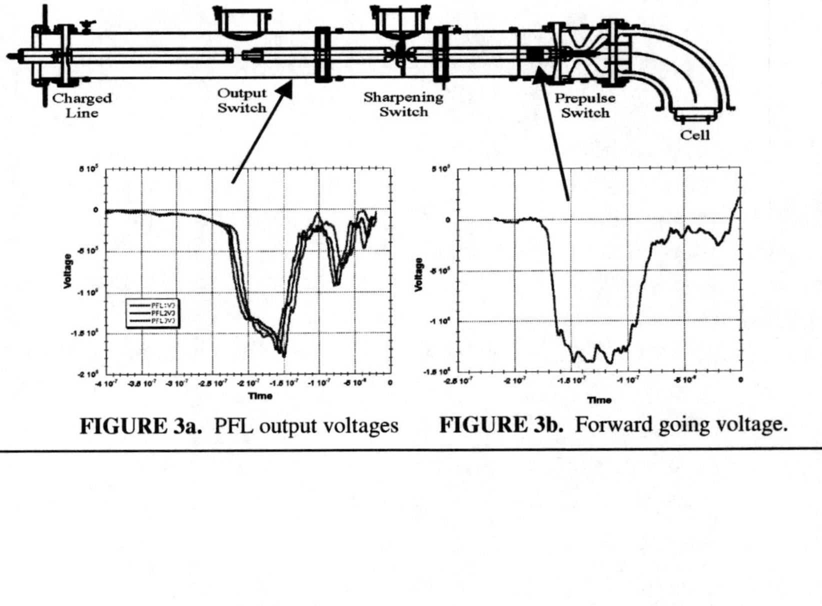

FIGURE 3a. PFL output voltages **FIGURE 3b.** Forward going voltage.

Current Symmetry and MITL Results

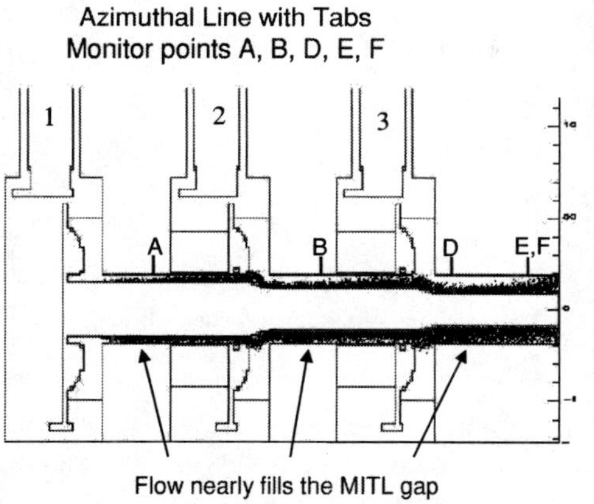

FIGURE 4. Geometry of the RITS cavities and MITL used in the LSP 3-D simulation.

The particle in cell code, LSP [3], was used to model the RITS-3 cavities and MITL. Figure 4 shows a cross section of the geometry used in the 3-D simulation; electrons are also shown in MITL gap. A voltage pulse was applied to the input to each cavity. Power flowed in the azimuthal lines to the gap where the MITL is connected to the cavity. Simulations and tests for two connections of the azimuthal lines were done: one without the upper tabs connected to the cathode and the other with the tabs connected. In both the modeling and the experiment a more uniform current distribution was achieved for the case without the tabs connected (±20% with tabs compared to ±9% without tabs). This injected current asymmetry caused non-uniform sheath current in the MITL. The lower portion of the MITL is more nearly filled by the electron sheath. B-dot monitors were azimuthally placed every 45° from 0° to 180° at monitor locations A, B, D, E, and F. At position D, just beyond the third cavity, it was found that the anode currents were more uniform than the cathode currents both experimentally (Figure 5) and in the simulations (Figure 6). At position F, 1.8 m beyond position D, both the anode currents and cathode currents, shown in Figure 7, become azimuthally symmetric.

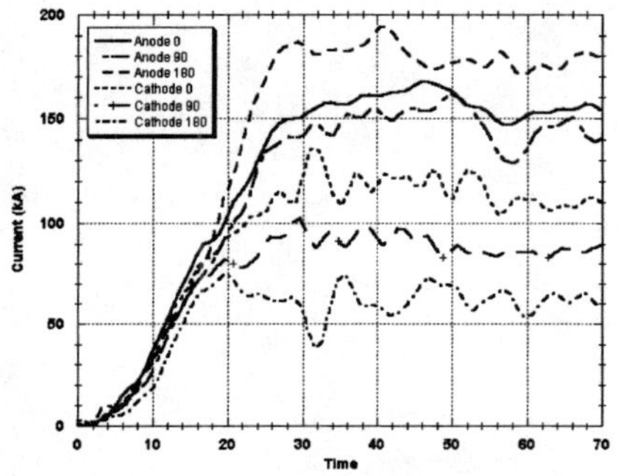

FIGURE 5. Anode and cathode current measurements at position D.

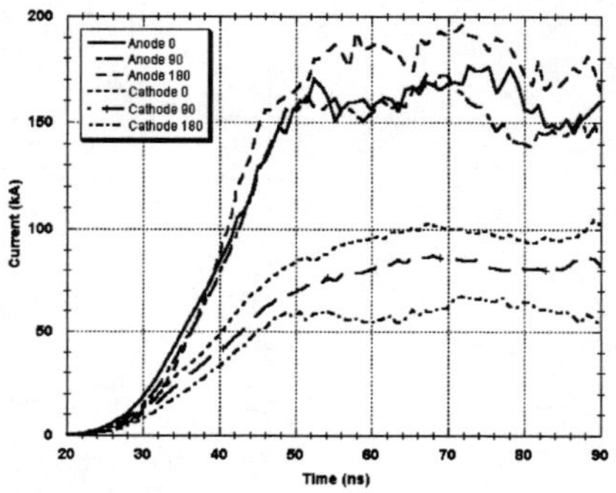

FIGURE 6. Anode and cathode current simulations at position D.

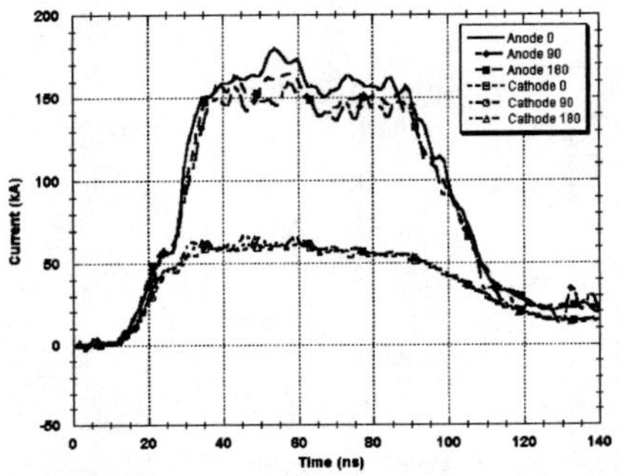

FIGURE 7. Anode and cathode current measured at position F.

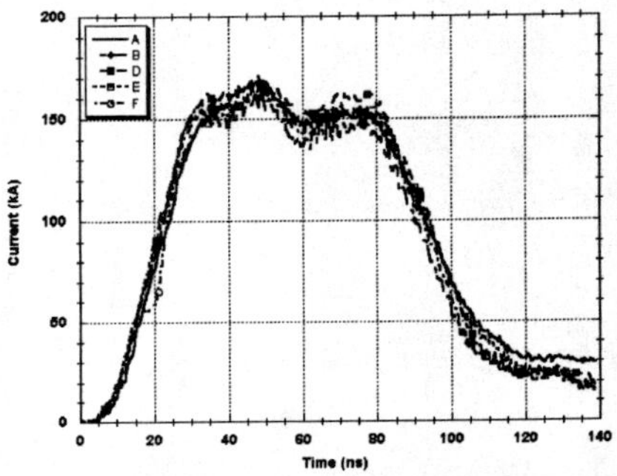

FIGURE 8. Average anode currents measured at positions A through F.

Dispite the azimuthal variations in the injected currents the average anode currents at each of the monitor positions, shown in Figure 8, were very nearly equal. No current loss down the length of the MITL was measured.

Tests are planned to remove the azimuthal lines from the cavities. With this configuration a 2:1 top to bottom current variation is expected. If there is no or minimal current loss in the MITL, smaller diameter, less expensive cavities can be used in single point feed IVA systems.

REFERENCES

1. Smith, I. D., et al, "Design Of A Radiographic Integrated Test Stand (RITS) Based On A Voltage Adder, To Drive A Diode Immersed In A High Magnetic Field", *12th IEEE International Pulsed Power Conference*, (IEEE, New York, 1999) pp. 403-407.
2. Johnson, D. L., Smith, I. D., et al, "Tests of the first RITS PFL and Cell", *13th IEEE International Pulsed Power Conference*, (IEEE, New York, 2001) pp. 600-603.
3. Hughes, T. P., Yu, S. S., and Clark, R. E., Phys. Rev. ST Accel. Beams 2, 110401 (1999).

Status And Plans For The Next Generation Magnetically Immersed Diodes On RITS

Dean Rovang [a], Dale Welch [b], Harry Ives [c], David L Johnson [d], Mark Kincy [a], Benny Lesch [e], John Maenchen [a], Peter Menge [a], Isidro Molina [a], Bryan Oliver [b], Craig Olson [a], Chuck Swenson [e], and David VanDeValde [c]

a. Sandia National Laboratories[1], Albuquerque, NM
b. Mission Research Corporation, Albuquerque, NM
c. EG&G, Albuquerque, NM
d. Titan Pulse Sciences Division, San Leandro, CA
e. National High Magnetic Field Laboratory, Tallahassee, FL

Abstract. Sandia National Laboratories is investigating and developing high-dose, high-brightness flash radiographic sources. We are in the process of designing; fabricating and conducting engineering tests on the next-generation magnetically immersed electron diodes. These diodes employ unique, large-bore (80-110 mm), high-field (28-45 T), cryogenically-cooled solenoid magnets to help produce an intense electron beam from a needle-like cathode "immersed" in the strong B_z field of the magnet. The diode designs and status of the engineering development are described. Later this year we plan to conduct experiments with these sources on the new Radiographic Integrated Test Stand (RITS) [1], now in operation at Sandia. In its present three-stage configuration, RITS provides a 4-MV, 150-kA, 70-ns pulse to the diode. Fully three-dimensional particle in cell LSP code [2] simulations are used to investigate relevant physics issues and the expected radiographic performance (spot size and dose) of this system. Preliminary results from these simulations are described.

IMMERSED DIODES PAST AND PRESENT

The magnetically immersed diode has been used to produce mm-scale radiographic spot sizes on Sabre, Hermes III, TriMeV in the United States [3,4] and on Superswarf [5] in the United Kingdom. A high voltage (3-9 MV), high current (10s of kA) pulse (10s of ns) is used to accelerate electrons from a needle-like cathode as shown in Figure 1 toward the anode. The anode is typically a high-Z Bremsstrahlung converter such as tantalum, gold or tungsten. In the past, these diodes have been unable to achieve a high dose from a small spot. This limitation is shown in Figure 2 where a correlation in spot and dose is seen: either small spots producing low dose or higher dose from larger spots. The other interesting feature illustrated in Figure 2 is that most of the data lie on or near a constant line of beam current density, 100 kA/cm^2. A similar limit current density limitation was observed on TriMeV. In Figure 3 the time resolved beam current density for three different cathode sizes (0.5, 1 and 2) is seen to

[1] Sandia is a multiprogram laboratory operated by Sandia Corporation, a Lockheed Martin Company, for the United States Department of Energy under contract DE-AC04-94AL85000.

CP650, *Beams 2002: 14th International Conference on High-Power Particle Beams,*
edited by T. A. Mehlhorn and M. A. Sweeney

converge to 100 kA/cm^2 in less than 10 ns. This limitation is believed caused by on the onset of stripping of non-protonic ions. Numerical simulations that include models for stripping of non-protonic ions exhibit many of the same trends or tendencies observed in experiment [6]. In particular these simulations have demonstrated the following behaviors when compared to proton-only simulations: larger overall impedances, larger spot sizes, increased impedance lifetimes with larger A-K gaps and more rapid loss of impedance with smaller cathodes. Thus the design goals for the next generation immersed diodes are to improve understanding of this non-ideal diode behavior and to delay or avoid the onset of this behavior.

NEXT GENERATION IMMERSED DIODES

To improve the understanding of this non-ideal diode behavior it will be necessary to diagnose the formation and evolution of cathode and anode plasmas and to determine their densities and species. To this end, a split-magnet diode as shown in Figure 4, has been designed to allow direct line of sight access to the A-K gap region. In addition to the split magnet diode, two other diodes, 30 T and 45 T, are under construction with monolithic magnets. All of these diodes require liquid nitrogen cooling to operate at their design fields and the split-magnet and 30-T diode are designed to accommodate either a 85 mm room temperature anode drift tube or a 100-mm liquid nitrogen cooled anode drift tube. The 45-T diode will accommodate a 55-mm room temperature or 70-mm liquid nitrogen anode drift tube. The cold bore diameters are larger since they do not require thermal breaks (double wall cryostats) to separate the liquid nitrogen bathing the magnet from the anode drift tube wall.

To delay or to avoid the onset of the non-ideal diode behavior three strategies will be pursued: heating of the anode, stepped-fields, and a solid hydrogen anode. Heating of the anode surface will eliminate surface and bulk contaminants, as has been demonstrated [7]. PIC simulations show stepped-fields will decouple the anode ions from influencing the cathode emission by imposing a magnetic field axial gradient between the cathode and anode. Lower magnetic fields at the cathode relative to the anode should cause the ions to diverge away from the cathode decreasing the detrimental effect on the diode impedance. Other advantages offered by stepped-field include improved magnetic insulation of wall ions, lower inherent impedance (larger ratio of r_a/r_k) and possible reduction in magnetic energy requirements.

Perhaps the strategy with the most promise is the solid hydrogen anode concept shown in Figure 5. Numerical simulations of realistic immersed diode geometries and voltage pulses with only protons emitted from the anode have demonstrated stable impedance and spot size behavior [6]. The premise is to make the experiment as close as possible to this ideal modeled diode. This requires the formation of a pure source of protons (or deuterons). It is believed that a thick layer (100s of microns) of solid hydrogen or deuterium on the surface of the high-Z converter should provide such a source. Although this approach is unprecedented in intense beam diode engineering, independent work by other researchers suggest that this concept is reasonable. At Sandia in the late 1980's Hanson et. al. produced a pure proton beam from an ion diode using a thin layer of hydrogen frozen unto a liquid helium cooled copper surface

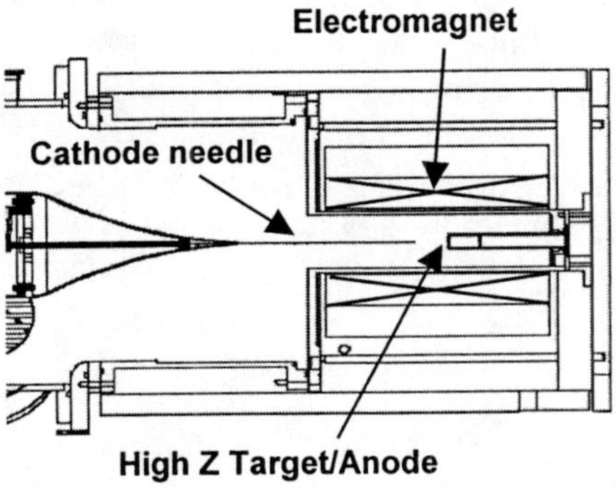

FIGURE 1. Magnetically immersed diode

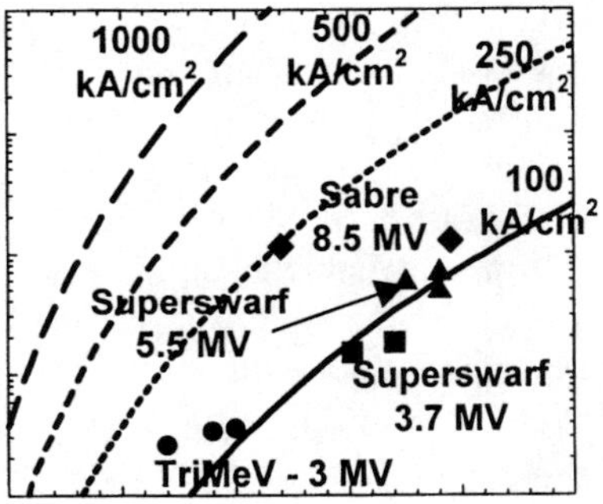

FIGURE 2. Operational limit of present diodes

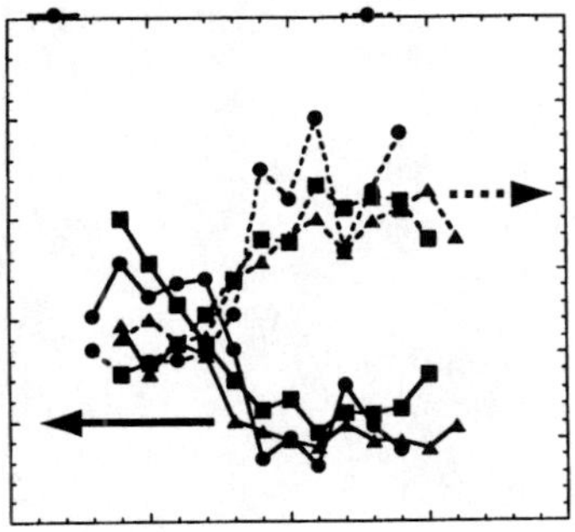

FIGURE 3. Current density & spot size on TriMeV

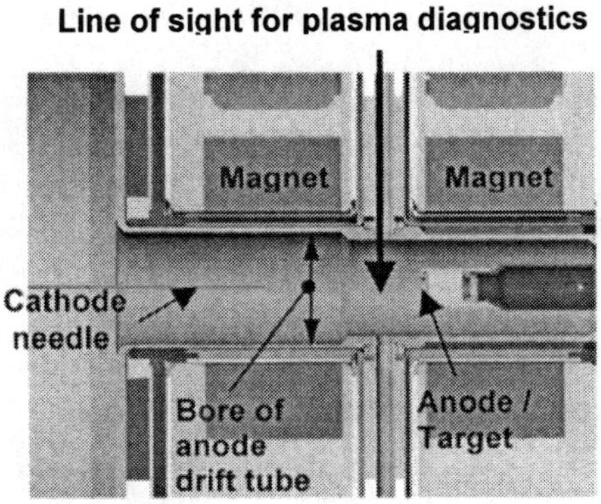

FIGURE 4. 28-T split magnet diode

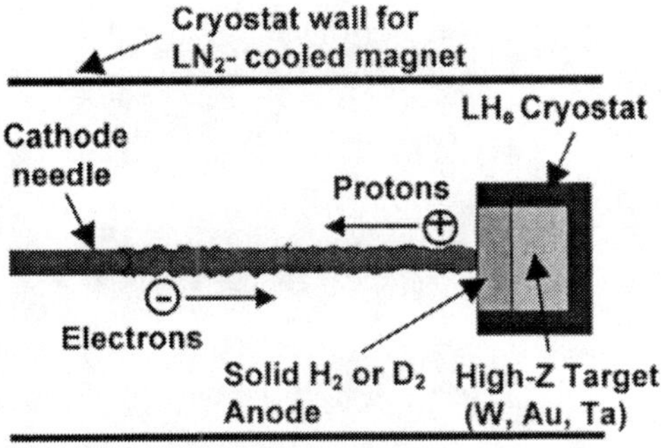

FIGURE 5. Solid hydrogen anode concept

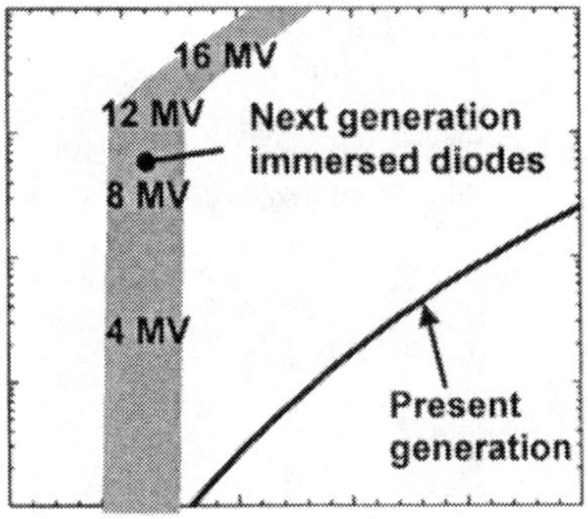

FIGURE 6. Next generation performance scaling

[8]. Solid deuterium foils have been created and used in pulsed power systems as debris shields on z-pinch soft x-ray sources [9]. Hanson et. al. now regularly uses liquid deuterium targets on the Z pulsed power accelerator at Sandia [10]. This work demonstrates the use of complicated cryosystems in harsh and difficult environments. In Japan S. Ishimoto et. al. successfully produced windowless solid hydrogen targets 25 mm in diameter and 5 – 10 mm thick [11]. Numerical simulations and analytic models of the spot size, the beam current and dose have been developed for an ideal immersed diode with a pure proton source. These models are used to perform system studies and scaling of the expected radiographic performance. As seen in Figure 6 there is considerable potential for improvement in achieving smaller spots and higher doses with the next generation immersed diodes. The projections for the next generation assume a 50 ns voltage pulse and the use of 0.5 – 1 mm cathodes. They invoke a linear scaling of applied magnetic field scales with voltage, $B_z = 5$ T /MV up to 60 T, the present engineering limit of large bore pulsed electromagnets. This performance scaling appears adequate to address anticipated radiographic missions, and will be explored on the RITS family of accelerators over the next few years.

REFERENCES

1. John Maenchen, et al. "Intense Electron Beams for Radiography", 12th Symposium on High Current Electronics, Tomsk, Russia, Sept. 24, 2000
2. T. P. Hughes, S. S. Yu, and R. E. Clark, Phys. Rev. ST-AB, 110401 (1999)
3. M. G. Mazarakis, J. W. Poukey, D. C. Rovang, J. E. Maenchen, S. R. Cordova, P. R. Menge, R. Pepping, L. Bennett, K. Mikkelson, D. L. Smith, J. Halbleib, W. A. Stygar, and D. Welch., Appl. Phys. Lett., 70, 832, (1997).
4. P. R. Menge, D. L. Johnson, J. E. Maenchen, C. L. Olson, and D. C. Rovang, SAND2002-0082, January 2002
5. G. Cooper, private communication, AWE, Aldermaston, England
6. D. Welch, private communication, MRC, Albuquerque, NM
7. R. D. Genuario and V. L. Bailey, Appl. Phys. Lett., Vol. 33, No. 8, 15, October 1978; B. V. Weber, et al, 13th IEEE International Pulsed Power Conference, (Las Vegas, 2001), pp. 1348-1351
8. D. Hanson, et al., J. Appl. Phys. 70 (6), September, 1991, 2925 – 2938
9. Pereira, private communication, Ecopulse, Springfield VA, 1993.
10. D. Hanson, et al., SAND2001-1975C
11. S. Ishimoto, et al., Nuclear Instrs. & Mthds. In Phys Res. A 480 (2002) 304-314

Initial Powerflow Studies on RITS

K. Hahn, [a,b] D. Rose, [c] D. Rovang, [a] V. Bailey, [d] S. Cordova, [e] R. Gignac, [f] D. L. Johnson, [d] M. Kincy, [a] R. Lucero, [a] J. Maenchen, [a] I. Molina, [a] B. Oliver, [c] D. Welch, [c] F. Wilkins, [f] R. Woodring [e]

a. Sandia National Laboratories, Albuquerque, NM*
b. University of New Mexico, Albuquerque, NM
c. Mission Research Corporation, Albuquerque,NM
d. Titan Pulse Sciences Division, San Leandro, CA B
e. Ktech Corporation, Albuquerque, NM
f. Bechtel Nevada, Las Vegas, NV

Abstract. Sandia National Laboratories is investigating and developing high-dose and high-brightness flash radiographic sources. In collaboration with colleagues from MRC, AWE and Titan Pulsed Sciences Division, Sandia is conducting powerflow studies into paraxial diodes on the Radiographic Integrated Test Stand (RITS) [1]. Historically high-impedance accelerators have driven these diodes with little excess powerflow into the diode. However, a number of facilities now under design will use lower impedance Inductive Voltage Adder (IVA) technology to drive these diodes resulting in additional excess powerflow into the diode. The goals of these recent experiments conducted at Sandia were to assess these powerflow issues and to help benchmark LSP [2], the Particle-in-Cell code being used as a design tool for studying the coupling of powerflow from the accelerator into the diode. An integral part of the AWE-style paraxial diode is the non-emitting "knob". LSP was used early in the design of the experiments to study the effects of different size knobs on the powerflow into the diode region for both a hollow cathode load and a radiographic diode. These simulation results suggested that larger knobs are more effective in retrapping the upstream MITL bound current while keeping the excess power flow out of the diode region. Initial experiments were performed with a hollow cathode to diagnose how well each size knob works in diverting the excess powerflow. Preliminary results of these experiments and comparisons to LSP simulations are presented.

INTRODUCTION

Initial studies utilizing a hollow cathode conducted on RITS-3, an intense radiographic x-ray source at Sandia National Laboratories (SNL), have led to an increased understanding of the issues involved with basic powerflow and retrapping of upstream MITL bound current at the diode. RITS-3 is a three stage Inductive Voltage Adder (IVA) capable of generating a 4 MV (negative polarity), 60 ns pulse with 150 kA total current along a MITL into a load. RITS will ultimately characterize three radiographic diodes: the paraxial, the immersed-B, and the rod-pinch diodes. Since these diodes require between 20-60 kA, most of the excessive ~150 kA of supplied

---

* Sandia is a multi-program laboratory operated by Sandia Corporation, a Lockheed Martin Company, for the United States Department of Energy under Contract DE-AC04-94-AL85000.

CP650, *Beams 2002: 14th International Conference on High-Power Particle Beams*, edited by T. A. Mehlhorn and M. A. Sweeney

current must be redirected. To accomplish this, an AWE-inspired knob was placed just before the diode region to redirect the excess current to the chamber walls (dustbin) without interfering with the diode characteristics.

Figure 1 shows the RITS diode section with the knob and hollow cathode configuration. Three different sized knobs (7", 9", and 14") can be placed just before the diode section to effectively divert most of the excessive 150 kA current to the dustbin section. The MITL is outfitted with several B-dot style current monitors placed in six azimuth positions at several positions along the MITL. The inner conductor of the MITL at the diode section is tapered from 8" to 2" in diameter upstream of the diode. To measure the current of the diode, B-dots are placed inside the knob section and on the outer knob to measure the beam current.

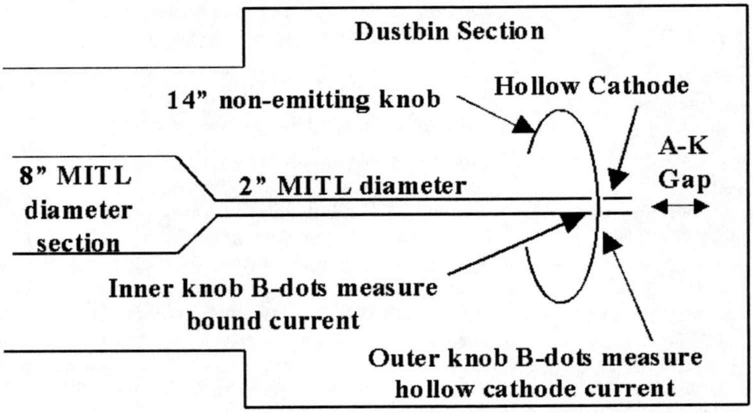

FIGURE 1. RITS experimental layout for diode section.

COMPUTER SIMULATIONS

LSP [2] was initially used to characterize the effect of different size knobs on the powerflow into the diode region for both a hollow cathode load and a radiographic diode. The ultimate goal of the powerflow studies is to drive a radiographic diode, which requires ~20-60 kA, without the ~150 kA excess current wrapping around the knob into the diode section. A hollow cathode was tested first simply to identify how the excess flow is directed, i.e. to the chamber walls or into the diode. These simulation results suggested that larger knobs (14") are more effective in keeping the excess powerflow out of the diode region. Figure 2 shows the results for the two AWE-style knobs at an A-K gap of 7.5 cm. Clearly, the 14" knob is the most effective in redirecting the powerflow. For the smaller knobs (i.e. the 9"), excess flow electrons essentially wrap around the knob and interfere with the beam electrons thus affecting the desired diode current. This consequence has serious ramifications for a true radiographic diode, which is the aim of these studies. Other measures will have to be taken to prevent the excess flow from entering the diode region.

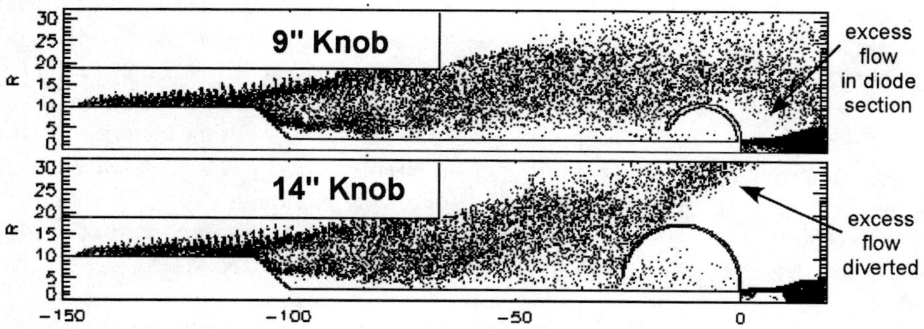

FIGURE 2. LSP simulations of the two (9" and 14") knobs show how the excess powerflow is redirected from the diode section.

EXPERIMENTAL RESULTS

Utilizing the results obtained by LSP, experimental data was taken at several A-K gaps for the 14" knob because it should be the most effective in separating the total current (~150 kA) from the beam current. Figure 3 shows the results for the 14" knob at five A-K gaps along with the predictions made by LSP. The beam current is measured along with the current on the inside of the knob (I_{bound}). At A-K gaps greater than 5 cm, the data is in excellent agreement with LSP's prediction. At 2.5 cm, the experimentally obtained currents are much higher than LSP calculations. Based on these initial experimental findings, it appears that LSP is an extremely useful tool for characterizing diode physics.

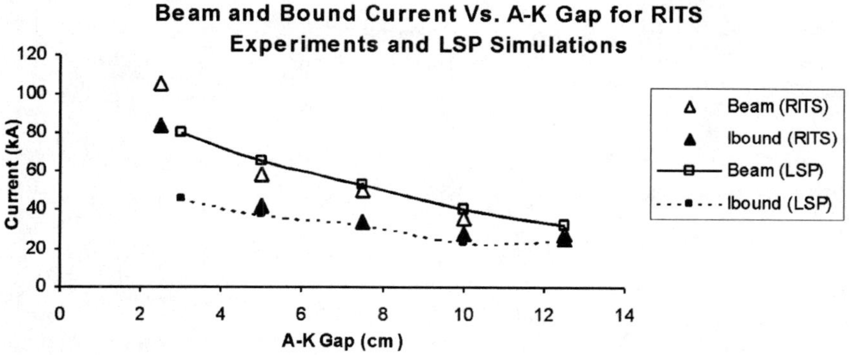

FIGURE 3. Hollow cathode and bound (inside knob) currents are plotted for various A-K gaps from LSP simulations and RITS experimental results.

CONCLUSION

An important consequence of the data obtained thus far is the issue of retrapping. Just before the MITL tapers down from 8" to 2" in diameter, the measured current remained fairly consistent at ~60-65 kA. For the 7.5 cm A-K gap, the beam current was ~50 kA. Much of the current was essentially retrapped from the 8" MITL bound current and the additional excess flow. At A-K gaps less than 7.5 cm, LSP predicts that electrons will wrap around the knob and contribute to the beam current. At smaller A-K gaps, the beam current exceeded the bound current measured at the 8" diameter upstream indicating significant retrapping. However, this increase in current in the diode region comes at the expense of voltage. Although voltage is not measured directly, the PIN diode signal (which measures dose in rads which is proportional to voltage) decreased by a factor of ~10 at the 2.5 cm gap versus the 7.5 cm gap, which received the highest dose. Figure 4 shows the ratio of retrapped electrons (I_{bound}/I_{beam} where I_{bound} is measured at the 8" diameter before the taper) to the normalized dose measured by the PIN diode. At the 5 cm gap, the amount of retrapped current is nearly identical to the current at the last 8" current measurement yet there is a notable decrease in voltage. LSP predicted an A-K gap of ~7-8 cm would maximize bound current retrapping without the excess flow interfering with the diode section. The experimental results agree quite well with LSP for optimizing retrapping and redirecting excess flow. It is therefore concluded based on experiment and LSP that for the 14" knob, an A-K gap of 7.5 cm provides the most retrapping without significantly decreasing the voltage.

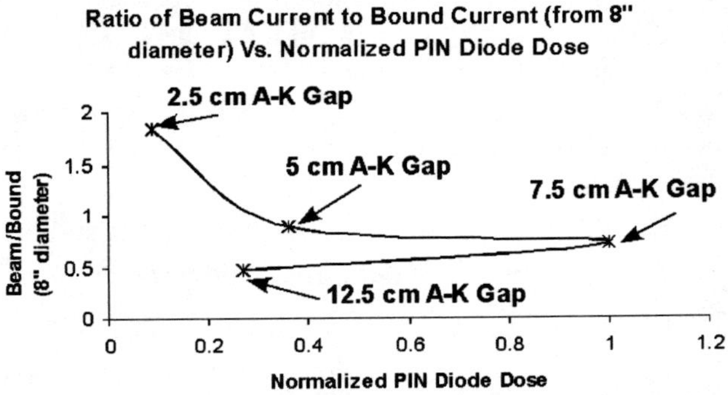

Figure 4. Ratio of retrapped electrons (I_{bound}/I_{beam} where I_{bound} is measured at the 8" diameter before the taper) to the normalized dose measured by the PIN diode.

ACKNOWLEDGMENTS

The authors would like to thank AWE, Atomics Weapons Division, for motivating this study as well as Pat Corcoran, Craig Olson, and Ian Smith for additional support in accomplishing these studies.

REFERENCES

1. I. Smith, et al. 12th IEEE Pulsed Power Conf. Proceedings, (IEEE, New York, p. 403, 1999).
2. T. P. Hughes, S. S. Yu, and R. E. Clark, Phys. Rev. ST-AB, 110401 (1999).

Performance of the Cygnus X-ray Source

John R. Smith[a], Randolph Carlson[a], Robert D. Fulton[a], R. Altes[b], V. Carboni[b], Jacob R. Chavez[a], P. Corcoran[b], William L. Coulter[a], J. Douglas[b], D. Droemer[c], William A. Gibson[a], Thomas B. Helvin[c], David J. Henderson[c], David L. Johnson[b], John E. Maenchen[d], Charlas V. Mitton[c], Isidro Molina[d], H. Nishimoto[b], Eugene C. Ormond[c], Paul A. Ortega[a], Robert J. Quicksilver[a], Rae N. Ridlon[a], Evan A. Rose[a], David W. Scholfield[a], I. Smith[b], Antonio R. Valerio[a], and R. White[b]

[a]*Los Alamos National Laboratory, P. O. Box 1663, Los Alamos, NM 87545 USA*
[b]*Titan Pulse Sciences Division, San Leandro, CA 94577 USA*
[c]*Bechtel Nevada Corporation, Las Vegas, NV 89030 USA*
[d]*Sandia National Laboratories, Albuquerque, NM 87185 USA*

Abstract. Cygnus is a radiographic x-ray source developed for support of the Sub-Critical Experiments Program at the Nevada Test Site. Major requirements for this application are: a dramatically reduced spot size as compared to both Government Laboratory and existing commercial alternatives, layout flexibility, and reliability. Cygnus incorporates proven pulsed power technology (Marx Generator, Pulse Forming Line, Water Transmission Line, and Inductive Voltage Adder sub-components) to drive a high voltage vacuum diode. In the case of Cygnus, a relatively new approach (the rod pinch diode [1]) is employed to achieve a small source diameter. Design specifications are: 2.25 MeV endpoint energy, < 1 mm source diameter, and >3 rads dose at 1 meter. The pulsed power and system architecture design plan has been previously presented [2]. The first set of Cygnus shots were geared to verification of electrical parameters and, therefore, used a large area diode configuration offering increased shot rate as compared to that of the rod pinch diode. In this paper we present results of initial rod pinch operation in terms of electrical and radiation parameters.

CYGNUS

Cygnus was designed to provide an x-ray source of specific parameters with exceptional reliability and fit in an environment limited in spatial dimensions and accessibility. Toward this goal there are three key design elements (small spot size, layout flexibility, and reliability) which dictated the final design and layout. Although all the individual Cygnus elements are proven technology, any newly built system will require a test phase where unforeseen problems are discovered and solved. This phase is addressed in this paper. Details of the Cygnus pulse power components are found in Reference 2. In this work we present the end result, which is the diode electrical characteristics and radiation output. Results are broadly categorized by two different diode configurations: 0.5 mm diameter gold rod and 0.75 mm diameter tungsten rod.

CP650, *Beams 2002: 14th International Conference on High-Power Particle Beams,*
edited by T. A. Mehlhorn and M. A. Sweeney

VOLTAGE AND CURRENT MEASUREMENTS

Diode voltage is measured using two D-dot probes placed in the center stalk, at an axial position 30 cm upstream of the cathode plate. The D-dot signals are summed, integrated, and inductively corrected to obtain the voltage at the diode gap. Diode current is measured using three B-dot probes located in the cathode plate. These three signals are summed and numerically integrated to obtain diode current.

Results of two different diode geometries are presented below. The first geometry uses a 0.5 mm diameter gold rod with an 8 mm diameter cathode aperture. A ten shot overlay of diode voltage and current is shown in Fig. 1.

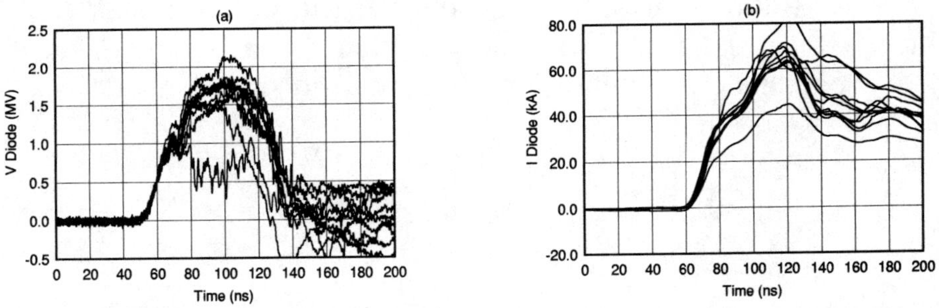

FIGURE 1. Ten shot overlay of diode (a) voltage and (b) current using a 0.50 mm diameter rod.

Diode behavior for this case is erratic with some shots exhibiting characteristics of a shorted diode. The second geometry uses a 0.75 mm diameter tungsten rod with cathode apertures ranging from 8.5 to 11 mm. Also for some shots the rod was tapered in an attempt to reduce source spot size. A ten shot overlay of results is given in Fig. 2.

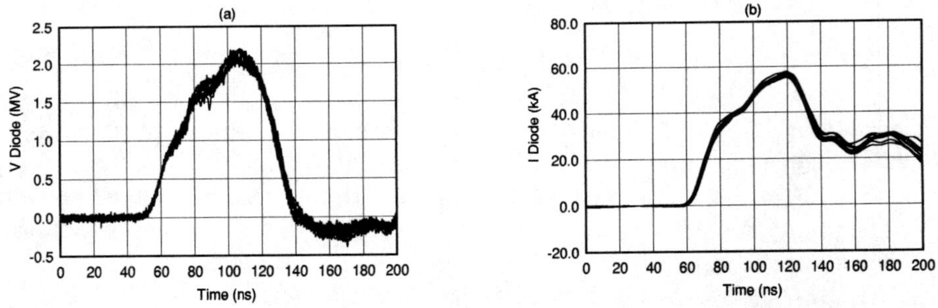

FIGURE 2. Ten shot overlay of diode (a) voltage and (b) current using a 0.75 mm diameter rod.

Diode behavior for this case is excellent with peak values repeatable within +/- 6%.

The contrast between performance of the 0.50 and 0.75 mm rods is emphasized in Fig. 3 in a side by side comparison of the diode impedance overlay.

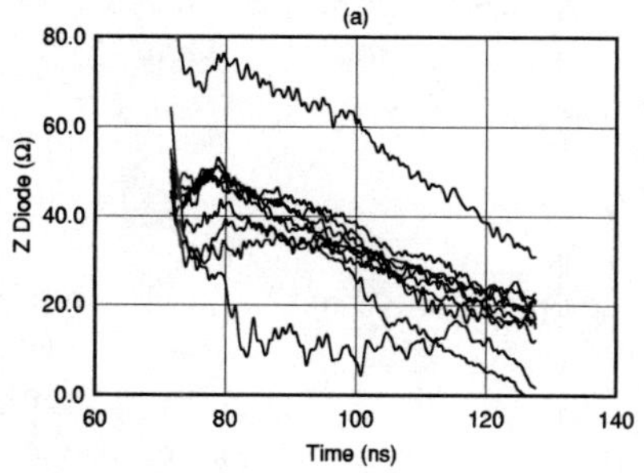

FIGURE 3. Ten shot diode impedance overlay. (a) 0.50 mm diameter rod, (b) 0.75 mm diameter rod.

RADIATION MEASUREMENTS

Radiation was measured using lithium fluoride thermoluminescent detectors (TLDs) and PIN diodes. In all shots there was a 0.75"x12"x12" aluminum plate placed 32 cm downstream of the cathode plate. Combined with shielding of the diode vacuum wall there is 1" aluminum equivalent shielding for all radiation detectors downstream of the diode. Both the TLDs and PIN diode are located 122 cm downstream of the cathode plate and 10 degrees off axis. The TLD's are locally shielded by 3 mm aluminum, and the PIN diode is locally shielded by 25.4 mm steel. Results of TLD and PIN diode measurements are in Fig. 4.

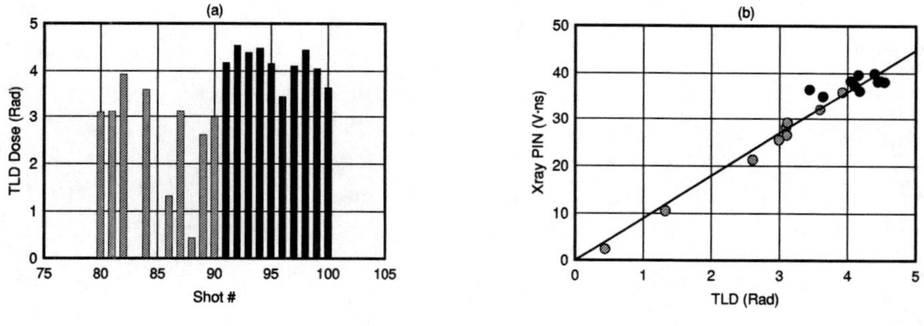

FIGURE 4. (a) Shot histogram of TLD measurements. (b) Integrated PIN diode signal versus TLD Dose. In both plots: grey data = 0.5 mm rod, and black data = 0.75 mm rod. The TLD dose has been scaled for a 1 m source distance.

Figure 4(a) gives the measured TLD Dose for shot #'s 80-100. The dose has been scaled for a 1 m source distance. The grey data represents 0.5 mm diameter rod shots, and the black data is for 0.75 mm diameter rods. The trend of erratic 0.50 mm rod data versus the significantly more uniform 0.75 mm rod data observed from comparison of the diode waveforms is repeated in the dose histogram.

Figure 4(b) shows a linear correlation between the integrated PIN diode signal and the measured TLD dose. Again the grey data represents 0.5 mm diameter rod shots, and the black data is for 0.75 mm diameter rods. The slope of the curve is 9.0 (V·ns/rad) which is the linear calibration constant for the PIN diode.

SUMMARY

Erratic performance of the rod pinch diode has been previously linked to a large cathode/anode ratio (r_C/r_A), where for the case $r_C/r_A > 16$ anomalous low diode impedance has been observed[1]. A large aspect ratio may cause a significant number of electrons to miss the anode rod. This results in enhanced ion current and subsequent backward electron trajectories which can yield low diode impedance. Note that for all of the 0.5 mm shots $r_C/r_A = 16$. The 0.75 mm shots used several different anode apertures corresponding to $11.3 < r_C/r_A < 14.7$.

The 0.75 mm tungsten rod shots consistently gave endpoint energies in the range 2.00-2.25 MeV with dose > 3 rad. Additional tests will be performed to increase the sample size and further confirm the consistency requirement. The endpoint energy was slightly low in some of 0.75 mm shots. All of these tests used a 40 kV Marx charging voltage which may comfortably be increased ~10 % if it is determined the present energy results are insufficient. The dose results easily exceed the stated requirement. From preliminary Cygnus performance in three of the requirement categories we conclude it is optimistic these requirements will be satisfied. The remaining source requirement, spot size, has been measured and will be analyzed shortly.

ACKNOWLEDGMENTS

The authors gratefully acknowledge insight and suggestions offered by Gerald Cooperstein, Robert Commisso, Frank Young (Naval Research Laboratory) Dean Rovang (Sandia National Laboratory), and Bryan Oliver (Mission Research Corporation). This work was supported under auspices of the U.S. Department of Energy.

REFERENCES

1. G. Cooperstein et al, Theoretical Modeling and Experimental Characterization of a Rod-Pinch Diode, Phys. Plasmas **8**, 4618 (2001).
2. D. Weidenheimer et al, Design of a Driver for the Cygnus X-ray Source, Proceedings of the 13th IEEE International Pulsed Power Conference, June 17-22, 2001.

Status of the Dual Axis Radiographic Hydrodynamics Test (DARHT) Facility[1]

Michael J. Burns[c], George J. Caporaso[b], Bruce E. Carlsten[c], Yu-Jiuan Chen[b], Ken P. Chow[a], Edward G. Cook[b], Harold A. Davis[c], Carl A. Ekdahl[c], William M. Fawley[a], Clifford M. Fortgang[c], Thomas P. Hughes[d], B.R. Trent McCuistian[c], Kurt E. Nielsen[c], Henry L. Rutkowski[a], Steve Sampayan[b], Will L. Waldron[a], James A. Watson[b], Glenn A. Westenskow[b], Simon S. Yu[a]

[a]*E.O. Lawrence Berkeley National Laboratory, 1 Cyclotron Rd., Berkeley, CA 94720 USA*
[b]*Lawrence Livermore National Laboratory, 7000 East Ave., Livermore, CA 94550 USA*
[c]*Los Alamos National Laboratory, P.O Box 1663, Los Alamos, NM 87545 USA*
[d]*Mission Research Corporation, 5001 Indian School Rd., Albuquerque, NM 87110 USA*

Abstract. The Dual-Axis Radiographic Hydrodynamics Test (DARHT) facility will employ two perpendicular electron Linear Induction Accelerators to produce intense, bremsstrahlung x-ray pulses for flash radiography. We intend to produce measurements containing three-dimensional information with sub-millimeter spatial resolution of the interior features of very dense, explosively-driven objects. The facility will be completed in two phases with the first phase having become operational in July 1999 utilizing a single-pulse, 20-MeV, 2-kA, 60-ns accelerator, a high-resolution electro-optical x-ray imaging system, and other hydrodynamics testing systems. We will briefly describe this machine. The first electron beams will be generated in the second phase of DARHT this year. The second DARHT accelerator consists of a 18.4-MeV, 2-kA, 2-microsecond pulse-width accelerator. Four short electron micropulses of variable pulse-width and spacing will be chopped out of the original, long accelerator pulse for producing time-resolved x-ray images. The second phase also features an extended, high-resolution electro-optical x-ray system with a framing speed of about 2-MHz. We will discuss this accelerator by summarizing the overall design of the long-pulse injector and accelerator. We will also discuss the fast kicker used to separate the long-pulse beam into short bursts suitable for radiography.

INTRODUCTION

Non-nuclear radiographic hydrodynamic tests are key to the maintenance of a smaller, aging U.S. nuclear weapons stockpile without underground nuclear testing. These tests utilize very powerful x-ray sources to radiograph a full-scale, non-nuclear mock-up of a nuclear weapon primary during the late stages of its operation. DARHT, the next-generation U.S. radiographic hydrodynamic testing facility, will make

[1]Work supported by the U.S. Department of Energy under contracts W-7405-Eng-36 (LANL), DE-AC03-76SF00098 (LBNL), and W-7405-Eng-48 (LLNL).

CP650, *Beams 2002: 14th International Conference on High-Power Particle Beams*, edited by T. A. Mehlhorn and M. A. Sweeney

possible images with sub-millimeter resolution, from two views (the minimum for large-scale 3D information), and with multiple views along one axis. A photograph of the DARHT facility is shown in Fig. 1

Figure 1 Aerial view of the DARHT facility. The Phase 1 accelerator is within the portion of the building running from top to bottom in the photograph. The long-pulse Phase 2 accelerator resides within the perpendicular wing of the building

In Section 2 of this paper we will very briefly describe the conventional, single-pulse, LIA that makes-up the operational "first axis" (or Phase 1) of DARHT. In Section 3 we will describe the long-pulse LIA that makes-up the "second axis" (Phase 2).

DARHT PHASE 1

The DARHT accelerators were optimized for dense-object flash radiography and feature short micropulse length (to limit motion blur) of <~100-ns, moderate energy (to limit x-ray scatter) of ~ 10-20-MeV, high beam current (for x-ray intensity), and good beam quality (to focus to a small x-ray spot size, a significant factor for image resolution). The parameters of both accelerators are shown in Table 1.

Table 1: DARHT system parameters

Parameter	First Axis	Second Axis
Spot size (mm)	2.0 @ 50% MTF	$\leq$2.1
Dose (R@1m)	500	variable
Pulse width (ns)	60	variable
DQE (%)	$\geq$30	$\geq$40
F_c (cycles/mm)	$\geq$0.3 (-3dB)	$\geq$0.3 (-3dB)
Pixel Size (mm)	0.9 x 0.9	1 x 1
1st conjugate (m)	0.8 to 1.5	1.1
Magnification	$\leq$4	$\leq$4
Accelerator Parameters		
Energy (MeV)	19.8	18.4
Current (kA)	2	2
Pulse width (ns)	60	2000
Energy spread	< ± 1%	< ± 1%
Norm. emittance (4-rms, πcm-rad)	0.15	0.15
Grad. (MeV/m)	0.47	0.35

The Phase 1 injector [1] consists of a 3.0-μF/120-kV capacitor bank switched through the primary of a 1:15 Stangenes iron-core autotransformer by an air-blown spark gap that pulse-charges a glycol Blumlein. The 6-ft. diam. Blumlein (7.65-Ω inner line, 7.3-Ω outer line, pulse-charged to 1.5-MV in 4.6-μs) discharges via 4 laser-triggered spark gaps (0.7-ns 1-σ jitter) into a series of three transmission lines with impedance changes to step-up the voltage to 4-MV at the diode. The vacuum diode consists of a 170-Ω liquid resistor load in parallel with the 181-mm A-K gap and has a 10-90% risetime of 20-ns. A flat, 64-mm diameter velvet emitter is used to produce 2-kA at 3.75-MeV.

The accelerator consists of 64, 0.25-MV induction cells [2]. Each has a 148.2-mm-diam. bore, a 19.1-mm accelerating gap, 11 oil-insulated TDK PE16B nickel-iron-zinc ferrite toroids (237-mm ID, 503-mm OD, 25.4-mm thick), a cross-linked polystyrene insulator (ε=2.5), a quadrufilar-wound solenoid magnet with homogenizer rings, two cosine-wound dipole trim magnets, and a cosine-wound quadrupole magnet (to compensate for the weak quadrupole field introduced on-axis by the pulse-power feed to the cells).

Thirty-two water-insulated, 11-Ω, 0.25-MV Blumleins with coaxial midplane-triggered switches operating in SF_6 drive two cells each through four DS 2158 44-Ω cables. Each Blumlein charging unit has two 1.4-μF/40-kV capacitors switched via two parallel EEV CS-1722 thyratrons into a 1:11 step-up transformer to pulse-charge the line in 5-μs. An independent trigger is supplied for each Blumlein consisting of two 30-nF/70-kV capacitors switched by a EEV CX-1725 thyratron into a 1:4 step-up transformer to drive a magnetic pulse-compressor to supply a 200-kV/10-ns risetime pulse to the trigger cable. Typical system jitter is <1-ns (1σ) for any consecutive 100 shots.

DARHT PHASE 2

The objectives of the second phase of DARHT are to provide a second perpendicular view and to generate four x-ray images over a span of two-microseconds. To do this, a two-microsecond pulse-width LIA has been constructed. Four micropulses of arbitrary pulse width and spacing are chopped from that beam using a fast Kicker driven by a solid-state modulator.

The Phase 2 injector consists of a Marx generator feeding a vertical insulating column that drives a flat, 165-mm diam. dispenser-cathode with polished stainless steel electrodes. The Marx consists of 88 type E PFN stages driving a matched load at 3.2-MV with a nominal 500-ns rise, 2-μs flat pulse. A triggered crowbar switch is used to shorten the pulse tail. The insulator is 444.5-cm tall and oil-filled. The peak electric field stress on the cathode shroud is 120-kV/cm with 2-kA beam current. The cathode operates at 1100 C and requires about 3 kW of heater power to produce a fully space-charge limited emission at a current density of 9.4 A/cm^2 with a spatial variation of <5% [3]. We immediately accelerate the beam out of the injector to 4.5-MeV through eight induction cells with an accelerating potential of 0.17MV and a bore of 356-mm. We then eliminate most of the 500-nms beam risetime in the Beam Clean-Up Zone (BCUZ) that consists of a section of beam-pipe about 3-m long, three solenoids, and two apertures, each 152-mm in diameter.

After the BCUZ, the remaining 70 induction cells have a 254-mm diam. pipe and supply 197-kV. All of the oil-insulated cells use 4 individual cores of Allied Signal 2605SC Metglas with either 0.48 V-sec core capacity or 0.43 V-sec in the large bore in injector cells. A 200-ns rise-time leads to a 2.02-μs flattop.

Each induction cell has an individual cell-driver which contains four, 7-section E-network PFNs in a Marx configuration. Each driver has 20-Ω impedance and will deliver a 2.4-μs flattop into a resistive load of 5-Ω for a total drive current of 10-kA at 200kV. To compensate for the non-linear magnetization current of the Metglas cores, the PFN impedance varies from front to back and can be reduced 20% to tune for a ±0.5% flattop.

The principal element of the beam transport section is the fast deflector, or kicker system [4], used to generate four micropulses from the primary accelerator beam. There are four equal size electrodes enclosed within a vacuum housing that has a DC bias magnetic dipole wound over the enclosure. A subsequent drift space of several meters allows a substantial relative deflection to develop between the output beam positions of the kicker and drift. The bias dipole will deflect the beam off-axis into a large-bore, weak quadrupole magnet septum beam pipe. This magnet deflects the off-axis beam into a strong dipole magnet that transports the beam into a dump. When an x-ray pulse is desired the kicker electrodes are pulsed and overcome the bias field allowing a short segment of beam to travel down the axis through the null field region of the quadrupole magnet and on to the x-ray converter target. The Kicker modulator is a solid-state inductive adder with a nominal output voltage of ±18-kV. Corrections for beam/Kicker interactions and the details of the power feed to the Kicker requires the ability to place fine modulation onto each Kicker pulse. In addition, micropulse width and spacing must be adjustable from the control room.

SUMMARY

The first of two DARHT LIAs began operating in 1999 and produces 2-kA at 19.8-MeV in a 60-ns pulse. The second machine is a 2-kA, 18.4-MeV, 2-microsecond pulse width accelerator. Four micropulses of arbitrary pulse width and spacing are chopped out of the beam using a fast Kicker driven by a solid-state modulator. The first beam from the Phase 2 injector will soon be generated and we plan for the accelerator to be completed by December 2002. After 18-months of commissioning we plan on beginning full operations by June 2004.

REFERENCES

1. Downing, J.N. et al, "Pulsed Power Systems for the DARHT Accelerators", *IEEE 1991 Part. Accel. Conf.,* San Francisco, CA, 1991
2. Burns, M.J. et al, "Cell Design for the DARHT Linear Induction Accelerators", ", *IEEE 1991 Part. Accel. Conf.,* San Francisco, CA
3. D. Simmons and C. Fortgang, "Using Multispectral Imaging to Measure Temperature Profiles and Emissivity of a Large Thermionic Dispenser Cathode", submitted to *Review of Scientific Instruments*
4. Chen, Y.J.(Judy), et al, "Precision Fast Kickers For Kiloampere Electron Beams", *IEEE 1999 Part. Accel. Conf.,* New York, NY, 1999

A Composite Target Concept for Multi-Pulse Radiography

James McCarrick

Lawrence Livermore National Laboratory, PO Box 808, L-645, Livermore, CA 94551 USA

Abstract. Multi-pulse radiographic accelerators, such as the DARHT-II facility under construction at Los Alamos National Laboratory, generate X-rays by striking a solid, high-Z "bremsstrahlung converter target" with a tightly focused, high-current, relativistic electron beam. In the process, the converter target is heated to a plasma state and significant thermal expansion can occur during the interval between pulses, depleting the beam focus region of target material and reducing the X-ray output. Beam scatter and various other technical constraints prevent using thicker targets, moving the electron beam to a fresh target region, or moving fresh target material into a fixed beam location. The remaining possibility is to optimize the choice of target material to maximize the effective heat capacity while maintaining dose and minimizing scatter. A kinematic model of a "composite target" of high-Z material in a low-Z matrix is presented here as a candidate for such an optimized target. The thermal destruction of the target cannot be avoided but can be slowed sufficiently to maintain the necessary output dose. Methods of manufacturing such targets, and their radiographic performance, are being tested at the ETA-II accelerator facility at Lawrence Livermore National Laboratory.

INTRODUCTION

Radiographic accelerators are used as transmissive X-ray diagnostics in high density, fast timescale hydrodynamic experiments. X-ray production is via bremsstrahlung from a high-current, relativistic electron beam striking a solid, high atomic number ("high Z") "converter target." The converter target on most existing single-pulse machines is simply a piece of refractory metal such as tungsten or tantalum. Unfortunately, the introduction of multi-pulse accelerator facilities such as DARHT-II at Los Alamos National Laboratory [1] precludes such a simple target system. A single beam pulse deposits enough energy into the converter material to raise its temperature to order 1 electron volt (~11600 K), into a plasma state. The resulting thermal expansion is slow on the timescale of a single accelerator pulse (~50 ns) and unimportant in the operation of a single pulse machine. However, the expansion is significant relative to the interval between pulses (~500 ns) on a multi-pulse machine. Radial expansion lowers the integrated line density of target material in the electron beam focal region, lowering the X-ray output (dose).

Without any additional constraints, there are numerous ways to maintain dose. The most obvious is a thicker target, so that enough material remains even after several pulses. More subtle is to use a dynamic target, replenishing material in the focal region between pulses; or to move the electron beam to a fresh location on a stationary target.

CP650, *Beams 2002: 14th International Conference on High-Power Particle Beams*,
edited by T. A. Mehlhorn and M. A. Sweeney

However, a thicker target increases scattering of the electron beam and resulting X-ray spot size, lowering the resolution of the system. A moving target requires fast conversion of a significant amount of stored energy to achieve the required replenishing rate. Moving the beam has its own technical challenges and also changes the radiographic axis of the system. The remaining choice is optimize the target material itself to absorb as much energy as possible without expansion.

BASIC REQUIREMENTS

There are two basic constraints on the choice of target material. The most obvious is that there must be enough high-Z material present to generate sufficient X-ray dose. The second is that the scattering of the electron beam in the target must be low enough that the system resolution is maintained. We wish to add to these some measure of "thermal strength." Let us use a highly simplified material model that lumps heat capacity, all energies associated with phase changes, and perhaps ionization energies into a single quantity we will call "the bonding energy," E_B. We will define it as the amount of energy per atom to go from fully bound in a solid lattice to completely unbound, in a material at constant volume (heating takes place much faster than any expansion). E_B can be estimated by integrating over experimentally determined equations of state, and it does not vary dramatically for different materials on a per-atom basis. Other studies looking to optimize materials for beam interactions [2,3] did not find any naturally occurring compounds that had major advantages.

The idea we propose is to "custom build" a material by creating an artificial ratio of high- to low-Z material. The high-Z material generates X-rays and, as a side effect, absorbs energy from the electron beam. The low-Z material serves as a "reservoir" of bonds to be broken before the material can undergo thermal expansion, without contributing much to energy deposition or beam scatter (or, unfortunately, dose). One needs to transfer the energy from the high-Z material to the low-Z as fast as possible. Radiative transfer is not important at these timescales and temperatures. For conductive heat transfer to occur, we want the diffusion time of heat across the high-Z material to be on the same timescale as a beam pulse. For a material such as tungsten, the diffusion length $d \sim (\kappa\tau/\rho C_p)^{1/2} \sim 2$ microns, where κ is the thermal conductivity (~170 W/m-K), τ is the pulse length (~50 ns), ρ is the density (19300 kg/m^3), and Cp the specific heat (~130 J/kg-K). This invokes an analogy to composite structural materials, where in this case we have an "aggregate" of small high-Z particles embedded in a low-Z "matrix". In principle, the task is much simpler than the designed of advanced structural composites: there are no constraints other than minimal self-support on mechanical strength.

KINEMATIC MODEL

Consider a simple energy-balance model that states the energy available for thermal expansion, E_{th}, is the total energy deposited in the material by the beam, less the total bonding energy. Given the proper material scale sizes, we will assume the different types of atoms in the composite are in local thermal equilibrium, and work in terms of energy per atom. Consider a composite of two materials, made of amounts equivalent to lengths L_H and L_L of pure high- and low-Z, respectively. For each material (with proper H or L subscript), denote the density by ρ, mass per atom m, bonding energy per atom E_B, and energy deposition traditionally denoted as dE/dx in units of energy per unit mass per unit area of material, per beam electron. Define the following dimensionless quantities: $\lambda = \rho_L L_L/\rho_H L_H$, $M = m_H/m_L$, $f_D = (dE/dx)_L/(dE/dx)_H$ and $f_B = E_{B,L}/E_{B,H}$. If there were only a solid slab of pure high-Z material, the energy deposited per atom would be $E_{D,o} = (m_H I\tau\, dE/dx)/(Aq)$ and the energy available $E_o = E_{D,o} - E_{B,H}$, given an electron beam of current I, pulse length τ, impinging on an area A; q is the electron charge. Normalizing to E_o and defining a last dimensionless parameter $\varepsilon = E_{B,H}/E_{D,o}$,

$$\frac{E_{th}}{E_o} = \frac{(f_D\lambda + 1) - \varepsilon(f_B M\lambda + 1)}{(1-\varepsilon)(1+M\lambda)}. \tag{1}$$

This expression reduces to 1 in the case of λ=0, no low-Z matrix; it can go negative, which in this simple two-state model simply means not enough energy has been added to break the bonds and allow thermal expansion. Consider a specific case of tantalum and carbon, M=15, and beam parameters corresponding to $E_{D,o} \sim 12.5$. E_B is estimated using integration of state equations and gives $f_D \sim 1$ (strictly speaking it is a weak function of λ), $\varepsilon \sim 0.2$, $f_B \sim 0.6$. λ controls the ratio of the materials, which fixes the design because the amount of Ta is determined by the desired dose. A curve of E_{th}/E_o is shown in figure 1.

HIGH-Z MATERIAL VELOCITY

The energy curve looks highly favorable but is not the complete picture. The average thermal energy of the material can be been lowered considerably, but the average mass goes down as well and this leads to a higher average asymptotic expansion velocity. The question is, at what velocity does the high-Z (X-ray generating) material move? Let us look at two extreme cases. The first is the "ideal" case of no collisions between species, so that the high-Z expands at its own thermal speed $v_H \sim (E_{th}/m_H)^{1/2}$ based on the equilibrium energy. Unfortunately, for such a dense vapor or plasma state, there will be extensive collisions between the two materials. Consider the extreme of a single equilibrium velocity due to the collisional drag; the RMS thermal speed v_{RMS} is the value that conserves total energy, and its value is high since it is weighted by the large number of low-Z atoms. Normalizing both of these quantities to the no-matrix case $v_o = (E_o/m_H)^{1/2}$ gives the following expressions, graphed for the Ta/C case in figure 1.

$$\frac{v_H}{v_o} = \sqrt{\frac{E_{th}}{E_o}}, \quad \frac{v_{RMS}}{v_o} = \sqrt{\frac{E_{th}}{E_o}\frac{1+M\lambda}{1+\lambda}}. \tag{2}$$

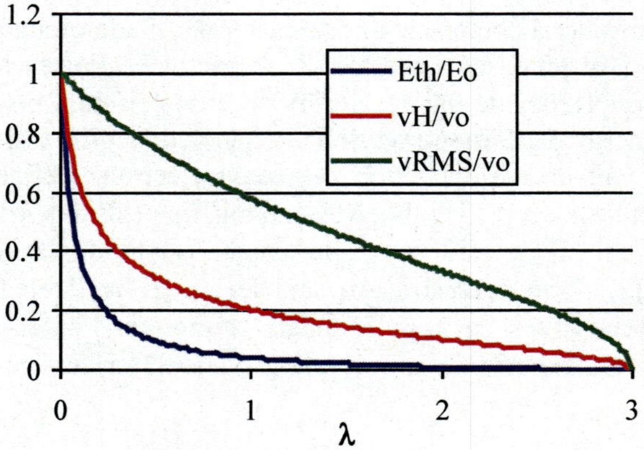

FIGURE 1. Sample curves of normalized energy and velocity for a mixture of Ta and C.

CONCLUSIONS

A composite mixture of high- and low-Z materials reduces the asymptotic expansion velocity of converter target material by providing an "energy reservoir" in the form of a larger number of atomic bonds to be broken, while preserving dose relative to a solid high-Z target. Preliminary hydrodynamic modeling with the LASNEX code shows favorable agreement with the simple kinematic model given above [4]. Early experiments using visible-light imagine of expanding target plasmas also show favorable results, but have not clearly resolved the question of the expansion velocity of the high-Z material. A number of material mixtures and manufacturing processes for achieving the requisite aggregate scale size will be tested at the ETA-II accelerator facility at Lawrence Livermore National Laboratory.

ACKNOWLEDGMENTS

This work was performed under the auspices of the U. S. Department of Energy by the University of California, Lawrence Livermore National Laboratory under Contract No. W-7405-Eng-48.

REFERENCES

1. M. J. Burns, et al., "DARHT Accelerators Update and Plans for Initial Operation," *Proceedings of the 1999 Particle Accelerator Conference,* New York, NY, 1999, pp. 617-621.
2. E. J. Lauer, "Materials that Resist Melting in Electron Beams," Beam Research Program Memo 89-13, Lawrence Livermore National Laboratory, Livermore, CA, 1989.
3. C. M. Snell, T. J. T. Kwan, "Photon Output for DARHT Beam Incident on Ta, Ta-Nb-V,and TaSi2 Target Foils," Los Alamos National Laboratory Research Note XTM-RN(U)99-004, Los Alamos National Laboratory, Los Alamos, NM, 1999.
4. D. D.-M. Ho, et al., "Target Studies for DARHT-II Multi-Pulse e-Beam Radiography," *Proceedings of the 29th IEEE Conference on Plasma Science,* Banff, Alberta, Canada, 2002

Downstream Transport System for the Second Axis of the Dual-Axis Radiographic Hydrodynamic Test Facility

Yu-Jiuan Chen, Lou Bertolini, George J. Caporaso, Darwin D.-M. Ho, James F. McCarrick, Arthur C. Paul, Philip A. Pincosy, Brian R. Poole, Li-Fang Wang and Glen A. Westenskow

Lawrence Livermore National Laboratory, Livermore, CA 94550, USA

Abstract. This paper presents physics design of the DARHT-II downstream system, which consists of a diagnostic beam stop, a novel, fast, high-precision kicker system and the x-ray converter target assembly. The beamline configuration and its beam parameter acceptance, the transverse resistive wall instability modeling, the ion hose instability in the presence of the background gas, and the simulations of beam spill are discussed. We also present the target converter assembly's configuration, and the simulated x-ray spot sizes and doses based on the radiation hydrodynamics code LASNEX and the Monte Carlo radiation transport code MCNP.

BEAMLINE CONFIGURATION AND ACCEPTANCE

The second-axis of the Dual Axis Radiographic Hydrodynamic Test facility (DARHT-II) will perform multiple-pulse x-ray flash radiography [1]. Its downstream system [2] consists of a diagnostic beam stop, a novel, fast, high-precision kicker system [3] and the x-ray converter target assembly [4, 5]. The kicker is used to select multiple short current pulses out of a 2-kA, 18.4-MeV, 2-μs beam pulse provided by the accelerator. These selected beam pulses will be delivered to a x-ray converter target to produce high quality x-ray pulses [6], and the remaining electron beam will be delivered to a beam dump. When the diagnostic beam stop is inline, the beamline configuration provides a capability for diagnosing emittance of the beam exiting the accelerator. The nominal beam envelope and beamline configuration are shown in Fig. 1, and the acceptances are shown in Fig. 2.

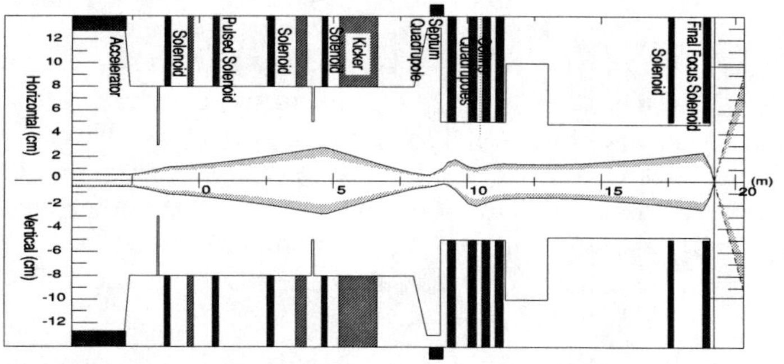

FIGURE 1. The DARHT-II transport line from the accelerator exit to the target

CP650, *Beams 2002: 14th International Conference on High-Power Particle Beams*, edited by T. A. Mehlhorn and M. A. Sweeney
2002 American Institute of Physics 0-7354-0107-1

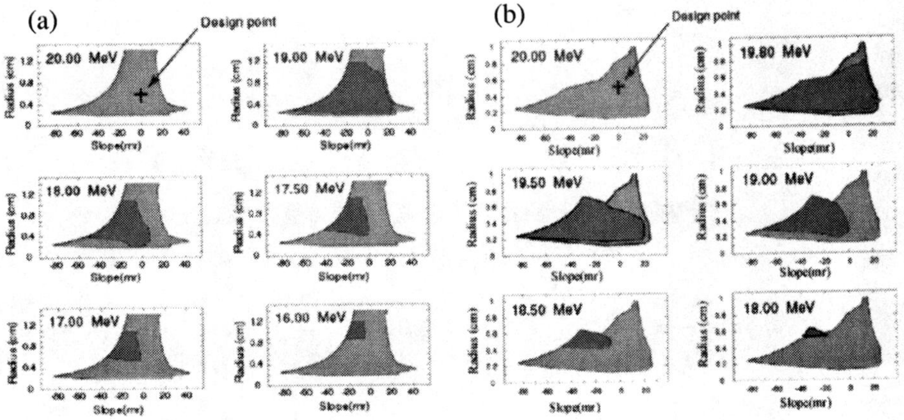

FIGURE 2. The matching acceptance for beam transporting from the accelerator exit to (a) to the quadrupole septum exit and (b) the main dump.

INSTABILITIES

The beam in the transport line upstream of the quadrupole septum and in the main beam dump line is 2 μs long. To minimize the transverse resistive wall instability [7] in this mainly drift space, all transport components have large apertures, and most (70%) are made out of aluminum. Only short beam pulses will be delivered to the converter target. Since the time-averaged current over 2 μs is low, the instability growth is insignificant after the quadrupole septum (see Fig. 3b). The estimated amplification of the instability at the end of the 2 μs upstream of the quadrupole septum is about 2 as shown in Fig. 3a.

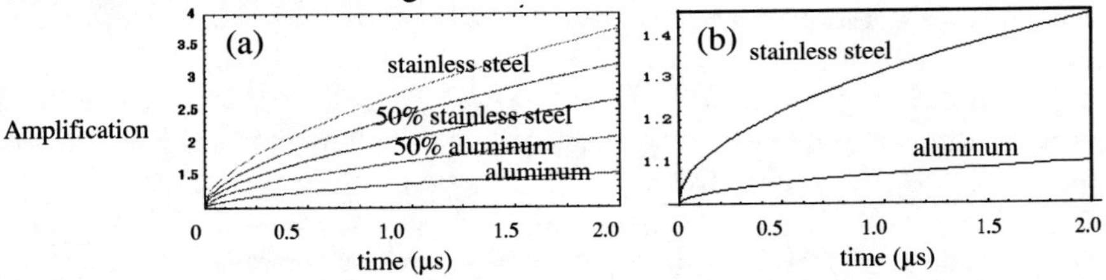

FIGURE 3. Amplification of the transverse resistive wall instability after (a) traveling 12 meters in the DARHT-II beamline and (b) in the target line downstream of the quadrupole septum.

Beam head loss upstream of the kicker and beam loss at the septum will deposit charges on the wall with the charge density in the order of nC/cm^2 and μC/cm^2, respectively. These losses may lead to beam stimulated desorption. To minimize the focusing effects due to beam-induced ionization of the background gas, the average vacuum in the system is 1.5 x 10^{-7} torr. The beam envelope in the long pulse section, i.e., upstream of the septum, can be approximately described with a function $a(1+\mu\sin(\pi z/L))$ and $\mu \cong 5$. Since the ion channel's oscillation frequency is determined by the beam's space charge forces, the rapidly changing DARHT-II envelope provides a strong detuning mechanism [8] to suppress the ion-hose instability as indicated by Fig. 4. Note that the maximum μ value in Fig. 4 is only 2. The calculation was done for beam transport in a 12-m drift space with a preformed ion channel with a Lorentzian ion frequency spread. However, there is no betatron frequency spread for the electrons.

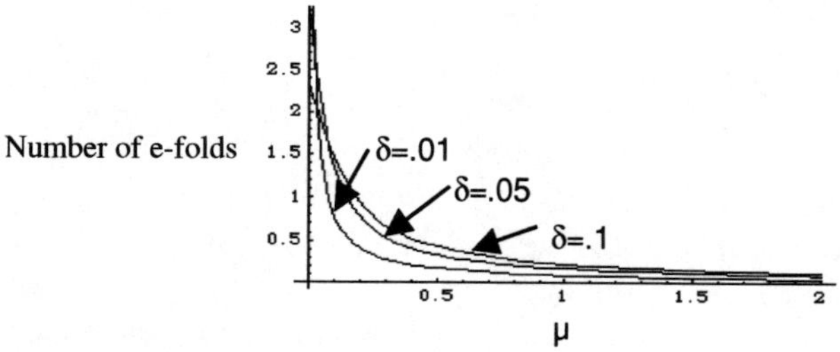

FIGURE 4. The number of e-folding growth for a tickler excitation of the ion-hose instability varies with the envelope variation parameter μ and the fractional ion frequency spread δ.

TARGET

The DARHT-II x-ray converter target is designed to minimize the hydrodynamic expansion of the converter material so that there is enough material to generate the required X-ray dose for all four pulses. Our LASNEX modeling results of target plasma hydro-expansion due to energy deposition of electron beams are shown in Fig. 5. Since the longest plasma channel is only about 8 cm (see Fig. 5), the impact of the beam-plasma interactions on the x-ray spot size and dose is small. The simulated x-ray spot sizes and doses by using the Monte Carlo radiation transport code MCNP, shown in Fig. 6, indicate that the target can meet its requirements.

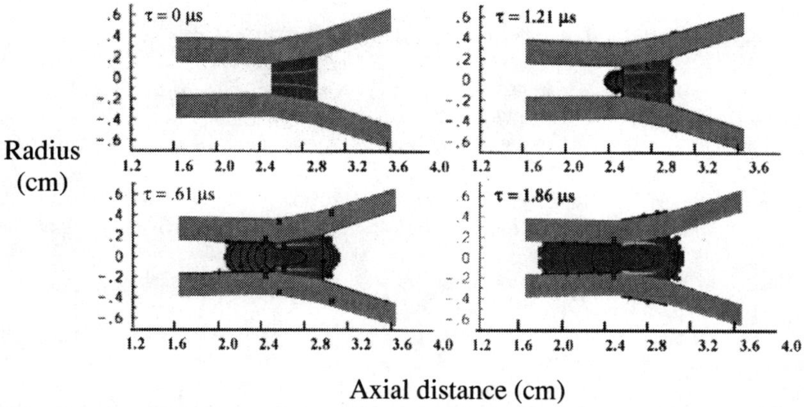

FIGURE 5. The DARHT-II Ta target at the beginning of each pulse. The mass density at 0 μs is 3.95 g/c.c, and the lowest density contour value plotted is 10^{-10} g/c.c.. All beam pulses' spot sizes on the front target surface are 1.08 mm FWHM (Gaussian).

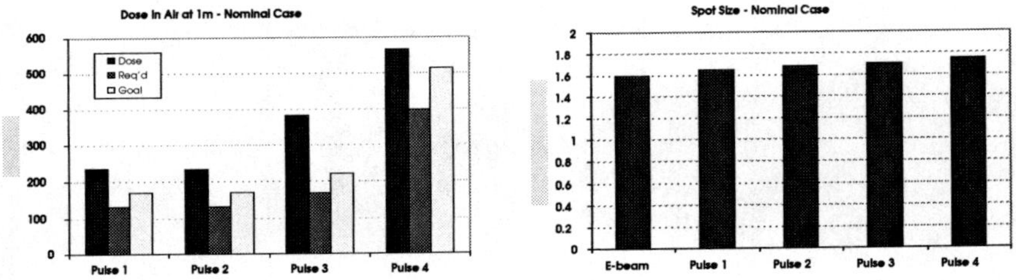

FIGURE 6. The DARHT-II forward x-ray doses and spot sizes.

SUMMARY

We have discussed some physics design issues of the DARHT-II downstream system, which consists of the transport beamline and the x-ray converter target. Since the DARHT-II accelerator will deliver a beam with a long beam head, the downstream transport line is designed to have a large beam acceptance. The beam head loss on the wall should be reasonably small. We have shown that the transverse resistive wall instability and the ion hose instability will not threaten the quality of the 2-kA and 2-μs beam in the long drift sections. The LASNEX and MCNP modeling results, presented in the paper, for the baseline target indicate that there are sufficient margins in the DARHT-II baseline target design.

ACKNOWLEDGMENTS

This work was performed under the auspices of the U.S. Department of Energy by University of California Lawrence Livermore National Laboratory under contract No. W-7405-Eng-48.

REFERENCES

1. M. J. Burns, et. al., "Status of the Dual Axis Radiographic Hydrodynamics Test (DARHT) Facility", *Proc. of BEAMS 2002,* Albuquerque, NM, June 24-28, 2002.
2. G. A. Westenskow, et. al., "The DARHT-II Downstream Transport Beamline", *Proc. of the 2001 IEEE PAC*, Chicago. IL, June, 2001, p.3487.
3. J. A. Watson, et. al., "A Solid-State Modulator for High Speed Kickers", *Proc. of the 2001 IEEE PAC*, Chicago. IL, June, 2001, p. 3738.
4. Y.-J. Chen, et. al., "Final Focusing System for the Second Axis of the Dual-Axis Radiographic Hydrodynamic Test Facility", *Proc. of the 14th High-Power Particle Beams,* Albuquerque, NM, June 24-28, 2002.
5. D. D.-M. Ho, Et. al., "Hydrodynamic Modelling of a Multi-Pulse X-Ray Converter Target for DARHT-II", LLNL Report. UCRL-JC-1442, July 26, 2001.
6. B. R. Poole and Y.-J. Chen, "Particle Simulations of DARHT-II Transport System", *Proc. of the 2001 IEEE PAC*, Chicago. IL, June, 2001, p. 3299.
7. G. J. Caporaso, et. al., *Part. Accel.*, **11**, 71-79 (1980).
8. G. J. Caporaso and J. F. McCarrick, "Ion Hose Instability in Long Pulse Induction Accelerator", *Proc. of the XX International LINAC Conference*, Monterey, CA, Aug. 21-25, 2000, p. 500.

Final Focusing System for the Second Axis of the Dual-Axis Radiographic Hydrodynamic Test Facility

Yu-Jiuan Chen, James F. McCarrick, Arthur C. Paul, Glen A. Westenskow

Lawrence Livermore National Laboratory, Livermore, CA 94550, USA

Abstract. The DARHT-II final focusing system consists of a solenoid and a foil, which is used to confine backstreaming ions. The separation between the converter target and the foil needs to be small to minimize the ion focusing effects. The beam spot size on the foil has to be large enough to ensure survivability of the foil while it is being struck by four high current pulses over 2 microsecond period. We have investigated several final focusing lens and focusing schemes. The simulation results of the beam spot size on the target are presented.

INTRODUCTION

The second-axis of the Dual Axis Radiographic Hydrodynamic Test facility (DARHT-II) [1] will deliver multiple electron pulses to a converter target to produce x-ray pulses (with spot sizes ≤ 2.1 mm) for flash radiography. Several effects, such as aberrations of the focusing system, target plasma created by preceding pulses, and the backstreaming ion effects [2], may impact the spot size. We have designed the DARHT-II x-ray converter target to minimize expansion of target plasma into the incoming beam's path [3,4]. Thus, the spot size changes from pulse to pulse due to the charge neutralization effects of the target plasma are small. However, the time varying focusing forces of backstreaming ions pulled from the desorbed gas at the target surface or from a pre-existing target plasma plume by the electron beam's strong electric field could potentially and drastically change the spot size.

This paper discusses physics design of the DARHT-II final focusing system, which consists of a solenoid and a foil barrier. Using a foil to confine backstreaming ions within a small channel so that ions' focusing effects are tolerable was first suggested by Hughes [5]. The beam spot size on the foil has to be large to ensure survivability of the foil while it is being struck by four high current pulses over 2 microsecond period [3]. In Sec. 2, We discuss the effects of the foil-target separation and the focusing schemes on the beam spot size on the foil and the beam divergence at the target. We discuss the effects of focusing lens' focal length on the beam spot sizes on the foil and at the target, and their corresponding beam parameter acceptances in Sec.3. Summary is given in Sec. 4.

FOIL BARRIER AND FOCUSING SCHEME

To determine the foil-target separation in the DARHT-II final focusing system, we have performed particle simulations and envelope calculations. The DARHT-II system

CP650, *Beams 2002: 14th International Conference on High-Power Particle Beams,*
edited by T. A. Mehlhorn and M. A. Sweeney
2002 American Institute of Physics 0-7354-0107-1

uses a modified DARHT-I final focus lens. For the results presented in this section, the nominal focal length is 32 cm. The DARHT-II beam near the x-ray converter is 18.4 MeV, 2 kA with a 1-mm radius (0.7-mm r.m.s.). Both the foil and the target are grounded conducting plates in the simulations. Ions are emitted from the surface by space-charge-limited emission as soon as beam electrons strike the target surface. The simulations indicate that the charge neutralization fraction (f_i) of those ions, trapped by the foil, near the end of DARHT-II 4th pulse is a function of the distance normalized to the foil-target separation (z/L). To allow a fast parameter study of various focusing schemes, envelope equation was solved by using a fit for the simulated charge neutralization fraction. Foil focusing effects and lens' aberrations are not modeled by the envelope equation. The spherical aberration and chromatic aberration of the solenoid lens are irrelevant to this section's discussion. The foil focusing effects are insignificant due to the small beam size near the foil. Our particle simulations indicate that the spot size changes due to the foil focusing effects are only a few percent.

FIGURE 1. R. m. s. beam envelopes in the foil-barrier traps with various foil-target separations.

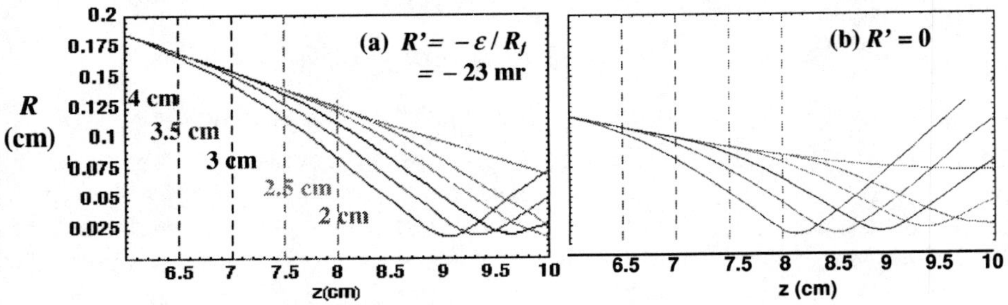

The scaling law for the ion channel's disruption length L_D [6] is given as $L_D \cong (\pi\gamma\beta^2 I_o/f_I I)^{1/2}$, where I and I_o is the beam current and Alfven current, respectively. This scaling law indicates that the suitable foil-target separation depends on the overall beam envelope inside the ion trap region, and that the final spot size is sensitive to the beam radius and ions' neutralization fraction. For a single pulse machine, the foil can be placed closely to the target to minimize the spot size sensitivity. However, the spot size on a closely placed foil will be too small to be practical for a multi-pulse machine since the foil may not survive the impact of multiple beam pulses. We have investigated several tuning schemes with various foil locations to achieve a larger spot size on the foil barrier without sacrificing the nominal spot size on the target. Figure 1 shows the r. m. s. beam envelopes (curves) with a foil (vertical dashes) placed at 4, 3.5, 3, 2.5, 2 or 0 cm in front of the target surface. The beam is focused beyond the target front surface in Fig. 1a and on the front surface in Fig. 1b. The focusing scheme used for Fig. 1a is better for the multi-pulses machine since the foil can placed further upstream from the target, and hence, the beam size on the foil is larger. The beam divergence at the x-ray converter corresponding to the envelopes in Fig. 1b is shown as a function of the foil-target separation in Fig. 2. The envelope slope (R'), thermal spread (θ_{th}) and the effective divergence are plotted. Since the effective divergence for various foil-target separations remains small compared with the scattering angle accumulated in a 1-mm

Ta, the effects of a convergent beam hitting the x-ray converter on the forward x-ray dose should not be an issue.

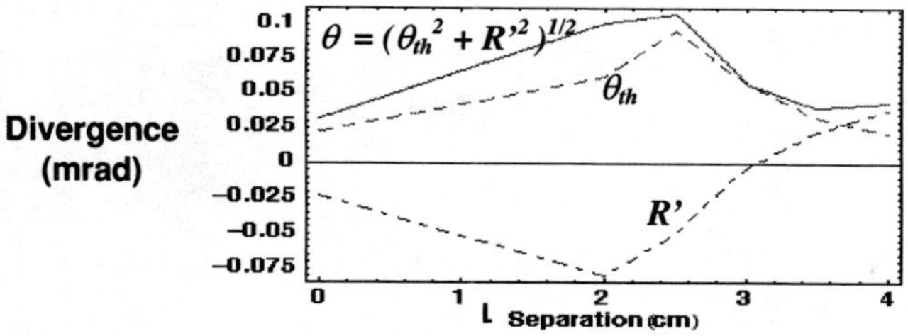

FIGURE 2. The beam divergence at the x-ray converter when an underfocusing scheme is used.

FOCUSING LENS AND FOCAL LENGTH

According to the thin lens approximation, the focal spot size of an emittance dominated beam focused by a solenoid lens with a spherical aberration coefficient C_s is given as

$$R_f^2 \cong \left(\frac{\varepsilon f}{R_i}\right)^2 + \left(2\frac{\delta\gamma}{\gamma}R_i\right)^2 + \left(\frac{C_s R_i^3}{4}\right)^2 + \left(R_i' f\right)^2 \qquad (1)$$

where ε is the beam emittance, R_I and R_i' are the beam envelope and slope entering the final focusing lens. Assuming the coefficients of spherical aberration being the same for two lenses, Eq. 1 suggests that the focal system with a shorter focal length lens can accommodate a wider range of beam parameters. Furthermore, this system can also provide a larger beam spot on the foil. To investigate the benefits, we have studied two focusing systems with particle simulations and ray-tracing simulations. The baseline case consists of the DARHT-I lens with a 29-cm focal length and a foil with a 4-cm separation. We designed a magnet with a similar spherical aberration for the second case, and its focal length is 20 cm. Since Eq. 1 does not applied to an ion channel, we did not include backstreaming ions in the simulations. While foil focusing forces were included in both types of simulations, foil scattering was included only in the particle simulations, which indicate that foil scattering of a 5-mil graphite would have very little effects on the beam envelope. A beam with a ±0.5% energy variation is lunched off at a waist upstream of the lens, and focused on the target plan located at either 29 cm downstream from the center of the baseline lens or 20 cm downstream from the center of the new lens. The beam spot sizes on the target and on the foils for the 29-cm system and the 20-cm system are shown in Figs. 3a and 3b, respectively. The final spot sizes in both systems are very similar as shown by the lowest curves in the graphs. For the DARHT-II pulse format, the FWHM spot size on a graphite foil should be larger than 2.5 mm [3]. Figure 3 indicates that both focusing system can accommodate a wide range of entrance beam radius. However, the spot sizes on the foil are much lager in the 20-cm system, and hence, we can shorten the foil-target separation to minimize the backstreaming ion effects further.

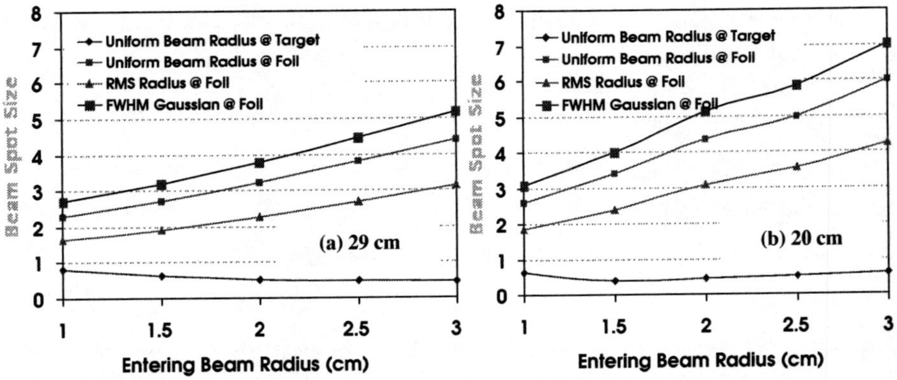

FIGURE 3. The beam spot sizes on the target and on the foil in two final focusing systems.

SUMMARY

The integrated DARHT-II final focusing system consists of a solenoid lens and a foil barrier. We have investigated the effects of the focusing schemes on the foil-target separation requirement and the effects of the focal length on the beam spot size on the foil. We have found that the final focusing system using a DARHT-I final focus lens can provide an adequate performance. However, the mechanisms of backstreaming ion emission from either a solid target surface or target plasma are still not understood, and the maximum foil-target separation depends on the charge neutralization fraction of the ions trapped by the foil. A focusing system using a shorter focal length lens will provide more flexibility to minimize the backstreaming ions' focusing effects.

ACKNOWLEDGMENTS

The authors would like to thank George Caporaso for valuable discussions. This work was performed under the auspices of the U.S. Department of Energy by University of California Lawrence Livermore National Laboratory under contract No. W-7405-Eng-48.

REFERENCES

1. M. J. Burns, et. al., "Status of the Dual Axis Radiographic Hydrodynamics Test (DARHT) Facility", this Proceedings.
2. D. R. Welch and T. P. Hughes, Laser Part. Beams **16**, 285 (1998).
3. D. D.-M. Ho, et. al., "Hydrodynamic Modeling of a Multi-Pulse X-Ray Converter Target for DARHT-II", LLNL Report No. UCRL-JC-1442, July 26, 2001.
4. Y.-J. Chen, et. al., "Downstream Transport System for the Second Axis of the Dual-Axis Radiographic Hydrodynamic Test Facility", this Proceedings.
5. T. P. Hughes, "Target Calculations for ITS Short-Focus Experiment", MRC/ABQ-N-589, September 1997.
6. Caporaso, G. J. and Chen, Y.-J., *Proc. of the XIX LINAC Conference*, Chicago, IL, 23-28 August, 1998, p. 831 - 833.

Recent Immersed Bz X-ray Diode Experiments

G. M. Cooper, J. McLean, R. Davitt. T. J. Goldsack

Pulsed Power Group, Atomic Weapons Establishment,
Aldermaston, Reading, Berkshire, RG7 4PR, United Kingdom.

Abstract. The immersed Bz diode is being fielded on one of the AWE Superswarf machines which provides a 55ns, 5.5MV, 35kA electron beam. The external magnetic field, up to 25Tesla, is produced by a solenoid which is driven by a 624μF, 22kV capacitor bank. The magnetic field constrains the electron beam to a small diameter at the target which results in a small x-ray source size.

Recent experiments to try and reduce the source size include investigation of shaped field solenoids and the effects of reducing the cathode diameter. The inclusion of a time resolved source size diagnostic has provided more information on the behaviour of the diode.

One of the better Bz shots has produced 72R@1m with a 4.0mm spot. This compares to a standard paraxial diode 80R@1m with a 5.3mm spot and the enhanced vacuum cell paraxial diode 65R@1m with a 4.0mm spot.

Future investigations aimed at reducing the spot size will include providing a better vacuum in the diode and a possible reduction in the pre-pulse on the diode.

INTRODUCTION

The proposed new Hydrodynamics Research Facility (HRF) at Aldermaston is designed to provide five high resolution images of a hydrodynamic experiment. The radiographic sources for this facility will be five, 14MV IVA machines each driving a diode capable of 1000R@1m from a 2mm spot. This paper describes the work being undertaken on the Immersed Bz diode to provide evidence of its suitability for the HRF.

During the past few years work on the immersed Bz diode has been undertaken on a Superswarf machine [1]. This provides a 55ns, 5.3MV, 35kA electron beam which is focussed onto a tantalum bremsstrahlung converter by an externally applied axial magnetic field. The work has been undertaken in collaboration with the Sandia National Laboratories and Mission Research Corporation who have provided theoretical input [2].

IMMERSED Bz DIODE

The pulsed external magnetic field is produced by a solenoid, which is driven by a 627μF, 22kV capacitor bank. The solenoid is constructed from 3mm by 2mm copper wire wound in a number of rows. The rows are potted in epoxy resin and surrounded with epoxy and Aramid fibre (Kevlar) to provide the mechanical strength. The

CP650, *Beams 2002: 14th International Conference on High-Power Particle Beams,*
edited by T. A. Mehlhorn and M. A. Sweeney
2002 American Institute of Physics 0-7354-0107-1

solenoid is then placed around a stainless steel propagation tube, which acts as the vacuum vessel for the diode. The diode consists of a stainless steel cathode positioned at the entrance to the solenoid and a tantalum converter target at the peak field in the solenoid. The magnetic field and brass field shaper shunt the MITL sheath electrons to the wall, while a cold beam is drawn from the cathode and is transported in the axial magnetic field to the tantalum converter target. Three different diameter propagation tubes have been used (Ø60mm, Ø44mm and Ø31mm), the 60mm bore can achieve a peak field of 18T at the target, the 44mm bore 20T and the 31mm 25T.

As the bank energy is limited a more efficient use of this energy is required. To concentrate the field between the Cathode and the converter target two different solenoids have been built. The first has a uniform axial field between the cathode and the anode (Figure 1), the second is based around a cone shaped propagation tube which allows the axial field to build gradually to a peak at the target (Figure 2). A comparison of the field can be seen in figure 3.

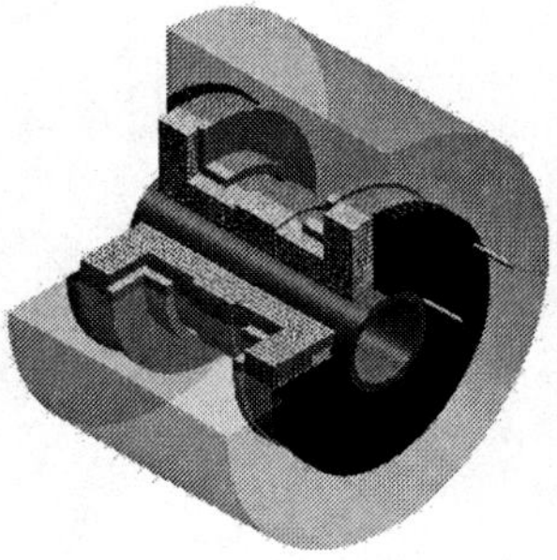

FIGURE 1. Uniform field solenoid.

FIGURE 2. Cone shape solenoid

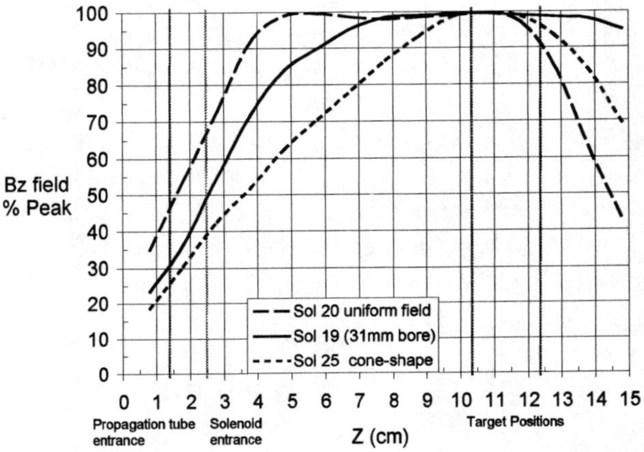

FIGURE 3. B field (% peak) along solenoid axis

RESULTS

Three shots were fired using the uniform field solenoid at 20T, each failed to produce any dose. Then the solenoid was moved back from the propagation tube entrance and the diode produced 24R@1m. The reason for these poor shots was believed to be gas entering the diode at the point where the propagation tube joins the front plate. To eliminate this join a new propagation tube was machined from a solid billet of stainless steel. However, with the new hardware the performance of this solenoid (7R@1m, 4.9mm spot size) is still poor compared to a standard solenoid (50R@1m, 4.2mm spot size). Further investigation into the performance of this solenoid is required.

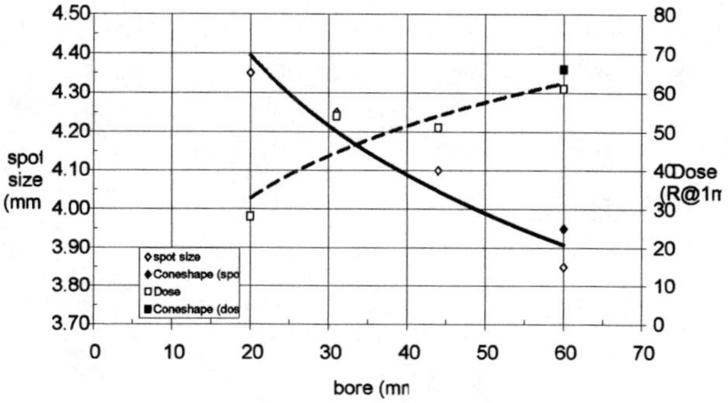

FIGURE 4. Comparison of solenoid bore

The cone shaped solenoid performs as a 60mm bore standard solenoid (Figure 4) but is much more efficient. A series of shots have been fired to look at the effects of the proximity of the propagation tube wall to the cathode (Figure 4). As the cathode to wall distance is increased the spot size is reduced and the dose output increased. This is possibly due to lack of magnetic insulation at the solenoid entrance. With the larger bore solenoids the Anode-Cathode (A-K) gap can be reduced with little reduction in dose. For example the 31mm bore solenoids operate best with the cathode at the entrance to the solenoid (A-K gap of 80mm). The 60mm bore solenoid performed satisfactorily with the cathode pushed in to an A-K gap of 40mm. The cone shape solenoid under performed when a 3mm diameter cathode was used with an A-K of 40mm or less. When the cathode diameter was reduced to 2mm, this loss of performance occurred at an A-K gap of 50mm.

A series of shots were fired using 3 and 2mm diameter cathodes to investigate how the spot size scaled with voltage (Figure 5). They show a spot size dependency on the applied voltage but the diameter of the cathode does not make much difference to the spot size.

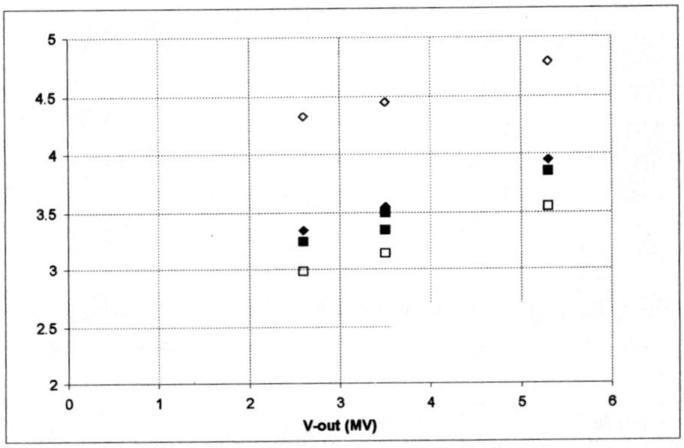

FIGURE 5. The effect of voltage and cathode diameter on the spot size.

Figure 6 shows the results from an array of pin diodes which look across a rolled edge to give a time resolved spot size measurement. From the data it can be seen that the spot size increases as the impedance decreases. There is no evidence of a rapid increase in spot size within the first 10ns, as expected when the ions from the target interact with the electron beam.

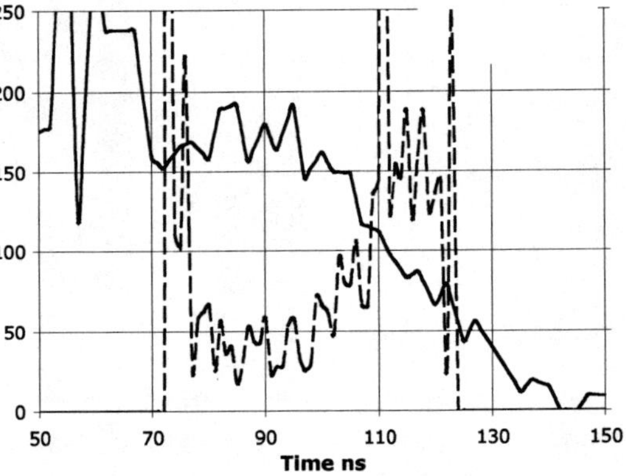

FIGURE 6. Spot size increases as impedance decreases.

REFERENCES

1. Cooper, G.M., McLean, J., Davitt, R., *Proceedings of the 28th IEEE International Conference of Plasma Science and 13th IEEE International Pulsed Power Conference,* (IEEE, New York, 2001), p393.

2. Oliver, B.V., Welch, D.R., Olson, C.L., Rosenthal, S.E., and Rovang, D.C., Theory and Simulation of Electron Beam Dynamics in the AWE Superswarf Magnetically Immersed Diode, *26th IEEE International Conference of Plasma Science.*

Further Work on the Implementation of Rod Pinch Diodes on Negative Polarity Pulse Power Drivers

Jim R. Threadgold, Adrian G. Pearce, Mark A. Sinclair, Ian Crotch

AWE Aldermaston, Reading, Berkshire, RG7 4PR England.

Abstract. Recent work by the AWE Pulsed Power Group has been to implement a rod pinch diode on some of the UK lower voltage (800kV to 4MV) negative polarity pulse power drivers. The rod pinch diode is normally fielded on positive polarity drivers which eliminates stray emission from the cathode emitter to the surrounding support structure. All the UK radiographic machines operate in negative polarity and cannot readily be switched to positive. This has consequences for electron trajectories, potential loss current paths and x-ray photon attenuation in the anode support structure. Although successful in its implementation, the diode performance is partly limited by the x-ray photon attenuation in the anode support structure. This poster reports on the further work undertaken in implementing the existing diode design on higher voltage drivers as well as experiments with an alternative diode designed to overcome attenuation from the anode structure.

INTRODUCTION

The Atomic Weapons Establishment, Aldermaston, UK has a number of Pulsed Power driven flash X-ray machines which are used in the diagnosis of hydrodynamic experiments. Long term plans envisage the upgrading of existing facilities through the construction of a hydrodynamic research facility (HRF) with multi axis radiography. As part of this project, the Pulsed Power Group have been carrying out a research programme aimed at reducing the X-ray spot size. Part of this research program has been to investigate the use of rod pinch diodes on negative polarity pulsed power drivers.

ROD PINCH PROGRAM AT AWE

The rod pinch program at AWE was started approximately two years ago. Experiments were carried out on different diode designs mainly on two flash X-ray machines, MEVEX (at 800kV) and Superswarf (at voltages up to 3.5MV) [1,2]. The experimental results on MEVEX were encouraging achieving 600mR @ 1m with a 1.2mm spot (compared with normal self magnetic pinch operation of 1200mR @ 1m with a 2.4mm spot). Unfortunately results on Superswarf did not scale. This was considered to be mainly due to the design of the Superswarf machines. The Superswarf machines contain an inductive "Herbucter" device within the Blumlien section which increases the output voltage of the machine at the expense of a slower

CP650, *Beams 2002: 14th International Conference on High-Power Particle Beams,*
edited by T. A. Mehlhorn and M. A. Sweeney
2002 American Institute of Physics 0-7354-0107-1

voltage rise time. It is thought that this slower rise time was delaying the formation of the magnetic pinch at the rod tip making the diode less efficient.

RECENT WORK

Experiments have been carried out on two basic diode designs. A re-entrant rod diode as previously used on MEVEX and Superswarf and a novel toroidal cathode (doughnut) diode.

The re-entrant rod design had previously been shown to work at lower voltages and it was decided to continue with this design to push it to the maximum voltage that could be produced on the AWE machines. However as the design is limited by attenuation in the anode support structure, an alternative diode was designed to try to improve this.

MOGUL D

Experiments using the previous re-entrant rod design were taken onto the Mogul D machine. Mogul D is normally operated at voltages up to 7MV with a paraxial diode producing 150R @ 1m in a 4.5mm spot. Mogul D is operated as part of the hydrodynamic firing program and diode research experiments have to be fitted in without compromising its normal operation. Initial experiments were therefore carried out using hardware that could be supported on the standard Mogul D Magnetically Insulated Transmission Line (MITL) and front end. Later experiments were carried out using a new larger diameter, more rigid stalk which could support increased sized diode components.

All of the experiments were carried out using a scaled up version of the angled rod pinch diode design that was first used on MEVEX.

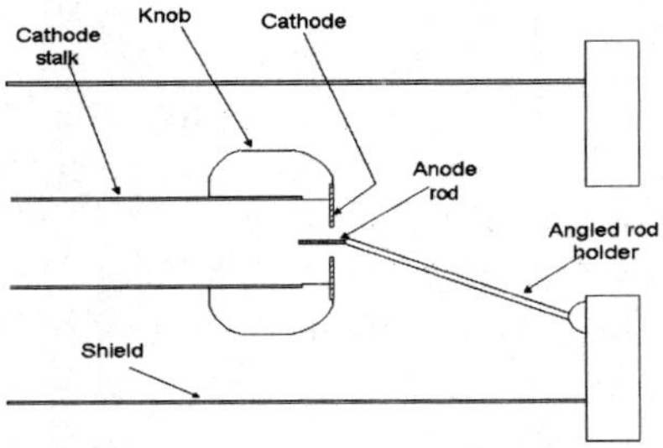

Figure 1. MEVEX Angled Rod Pinch Diode Design

The diode design has a brass field shaper fitted onto the central negative stalk. An annular cathode is fitted into the end of the knob. The majority of the anode rod is made of 3mm diameter aluminium and approaches to just in front of the cathode plane at an angle. A thin tungsten anode rod (1mm diameter) is mounted into the end of the aluminium rod and passes through the cathode aperture perpendicular to the cathode plane.

The initial series of experiments were carried out using a relatively small brass field shaper slid onto a 50mm diameter hollow tube fitted onto the standard stalk. This field shaper proved to be too small and suffered from electron emission on all shots fired (voltages from 2.6MV to 3.6MV). Considerable electron damage was noted at the base of the rod and doses were very low at 2-3 R @ 1m This is consistent with electron flow not coupling to the rod.

A decision was made to implement a larger (100mm diameter) hollow aluminium stalk and increased diameter field shaper which would be able to withstand fields greater than that which Mogul D would produce. Shots with this larger front end showed immediate improvement with doses of up to 11R @ 1m produced from an output voltage of 2.75MV. Although the recorded currents were similar to those from the earlier shots at around 40 – 50kA very little electron damage occurred at the rod base. This suggested that the electrons were now coupling to the rod and pinching at the tip.

Earlier shots have all shown some dose produced at the aluminium/tungesten rod joint. A side on pinhole camera viewing the tip section of the rod have shown that changes in the method of producing the joint have reduced the dose produced there. Line scans through the images show that the dose produced is negligible with respect to that produced at the tip.

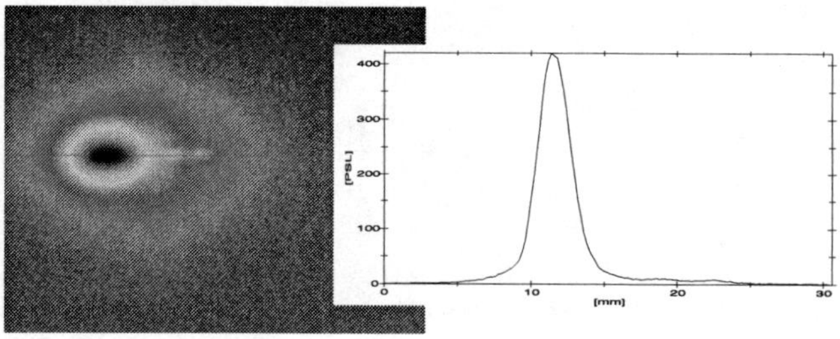

Figure 2. Pinhole Camera Image

TOROIDAL CATHODE DIODE

The toroidal cathode (or doughnut) diode was designed to enable an unrestricted view of the tip of the rod from the radiographic target.

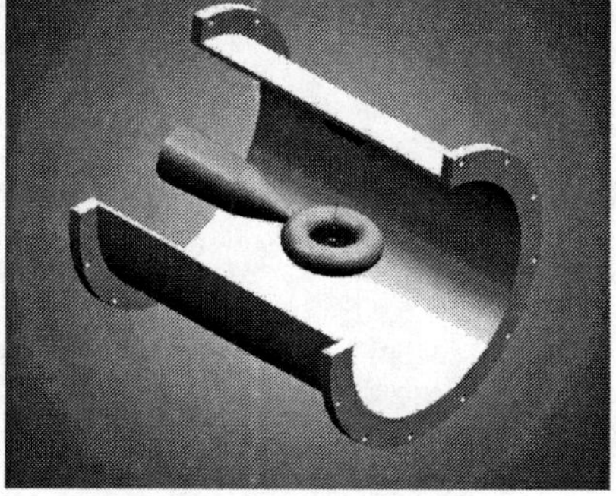

Figure 3. Toroidal Cathode Diode Design

A toroidal cathode is fitted on the central stalk with a removable cathode disc containing an aperture in its centre. The anode rod is mounted perpendicularly from the side wall of the machine vacuum chamber such that it passes through and beyond the cathode aperture. This arrangement is extremely useful for diagnostic purposes as it enables an unrestricted view of both the base and tip of the rod. Also, as the rod is positioned within a cylindrical vessel of equal wall thickness and therefore an equal amount of attenuating material, a polar plot of the dose from the rod can be measured at all angles apart from a small area within the shadow of the toroid and the rod holder at its base. 3D field modelling showed it difficult to make a toroid that could fit within the Mini-B vacuum vessel and hold off full voltage of 2.2MV. A slightly undersize toroid was made and has been tested with Mini-B underrun to approximately 1.1MV. This has proved successful and a full polar plot has been measured of a blunt ended 1mm rod extending 15mm through a cathode aperture of 10mm

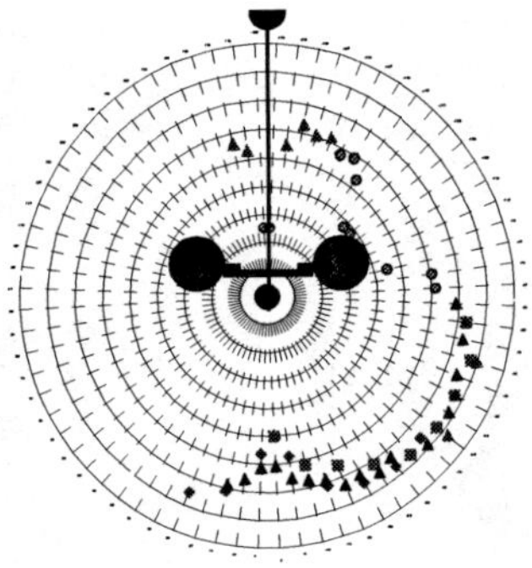

Figure 4. Polar Dose Plot

Points which were affected by the shadow of the toroid and the rod holder have been shown as hollow squares. It can be seen that the dose falls notedly in line with the rod at both the tip and base ends. The plot also shows that the dose recorded in the backwards direction i.e. that towards the base end of the rod, is approximately 15% less than that in the forward direction. Further work is planned to map the polar dose plots of different geometry diode tips such as tapered rods.

References

1. Jim Threadgold, Dave Forster, Adrian Pearce. "Proceedings of UK Pulsed Power 2001."
2. Adrian Pearce, Dave Forster, Jim Threadgold, Graham Cooper, Tim Goldsack, John Mclean. "Proceedings of Pulsed Power Plasma Science 2001"

Optimization of diode performance for deep penetration flash radiographic applications at AWE

John O' Malley

Atomic Weapons Establishment, Aldermaston, Reading, Berkshire, United Kingdom, RG7 4PR

Abstract. AWE is attempting to define x-ray sources for transmission radiography of thick, high atomic number explosively driven hydrodynamic experiments in a new hydrodynamics research facility (HRF). Figures of merit (FOM) with which to compare radiographic imaging systems are presented. These consider the effects of experimental magnifications, detector characteristics and scattered x-ray backgrounds in addition to the x-ray source intensity, spot size and spectrum. The application of these FOM to long-term (10-15 year) source options for the HRF is also presented and a zone of potential source characteristics is defined as suitable for long term HRF goals.

INTRODUCTION

AWE performs experiments to investigate the hydrodynamic behavior of explosively driven metal systems in order to underwrite the UK nuclear deterrent. Explosive/metal assemblies are detonated under remote control within specially designed containment buildings called firing chambers. During the experiment a very brief, intense, collimated flash of high energy x-rays are used to take a snapshot of the configuration of the assembly. From the resulting image, measurements of the density distribution of the components in the device are inferred and compared with codes[1].

In order to meet the requirements of the Comprehensive Test Ban Treaty there has been a re-evaluation of the accuracy of density information required from these experiments and this has led to the proposal for a new hydrodynamics research facility to replace existing firing chambers with a new, multi-axis flash x-ray facility.

The work described in this paper is an attempt to determine the trade off in flash x-ray source intensity (defined as dose at 1m), and spot size to produce an acceptable source to meet the long-term goals (10-15years) of the HRF. Other uncertainty studies, outside of the scope of this report have identified that a radiographic source delivering 1000R of 10MeV bremsstrahlung radiation at 1m with a spot size of 2mm by AWE definitions would be adequate to meet long term HRF goals when used in conjunction with opto-electronic gamma camera detectors. The scoping study therefore has addressed the question of what other possible ranges of spot size, dose and energy of source would yield the equivalent radiographic information as such a source.

CP650, *Beams 2002: 14th International Conference on High-Power Particle Beams*, edited by T. A. Mehlhorn and M. A. Sweeney
2002 American Institute of Physics 0-7354-0107-1

INFORMATION THEORY BASED FIGURES OF MERIT (FOM)

Two FOM have been investigated. The first, simpler FOM is based on a closed form of the Felgett-Linfoot radiographic FOM as used by Watson [2]. This calculates the amount of information in terms of bits per pixel in the object plane contained in an image.

$$FOM \propto (MTFA)\log_2(1+SNR) \tag{1}$$

In (1) the MTFA is the modulation transfer function area and SNR is the signal to noise ratio in the image.

Using Watson's expressions for MTFA and SNR [2], this can be expressed simply for a 10MeV peak Bremsstrahlung energy, flash radiographic system as:

$$FOM_{simple} \propto \frac{\log_2\left(1+\frac{P}{C_1}\sqrt{8x10^6 DTQ_{eff}}\right)}{\sqrt{2\left(\frac{(M-1)S}{M}\right)^2+\left(\frac{1}{Mf_c}\right)^2+2(VW)^2}} \tag{2}$$

In (2) above, P is the pixel size in the object under consideration, C_1 is the source to detector distance (first conjugate), D is the dose at 1m from the source in R, T is the object transmission, Q_{eff} is the detector efficiency, M is the magnification of the image, S is the spot size of the source (AWE definition), f_c is the frequency at which the detector modulation transfer function is 0.5, V is the velocity of motion of the object under consideration and W is the temporal width of the x-ray pulse.

A problem with the simple FOM is that radiographic systems utilising very low signals on detectors are not penalised. In real imaging systems detectors will have some small level of noise which may become large with respect to the signal in small signal conditions. There may also be small amounts of scattered x-ray contribution to the image, the variance of which degrades the image information content.

This FOM has therefore also been extended to consider the degradation in the SNR due to small amounts of detector variance or Poisson statistics associated with any small amounts of scattered x-ray background on the image. The complex FOM is defined in (3) below.

In the complex FOM the SNR is expressed as a sum of terms dues to the Poisson statistics of the signal dose combined with a scatter background and any small detector backgrounds. The scatter background consists of two terms. One term simply adds a scatter signal to the detector for each x-ray machine in addition to the main beam source. This therefore represents glows or crosstalk in a multi-axis facility. The second term represents Compton scattering from protection masses in the main beam. In the complex FOM n represents the number of x-ray machines and G is a scatter glow at the detector per x-ray source.

$$FOM_{comp} \propto \frac{\log_2\left[1+\frac{8x10^6 PTDQ_{eff}}{C_1^2\sqrt{\frac{8x10^6 TDQ_{eff}}{C_1^2}\left(1+\frac{0.2}{((M-1)C_1-0.4)^2}\right)+8x10^6(n-1)GM^2+480M^2}}\right]}{\sqrt{2\left(\frac{(M-1)S}{M}\right)^2+\left(\frac{1}{Mf_c}\right)^2+2(VW)^2}} \tag{3}$$

SCOPING STUDY

Both the simple and complex FOM have been applied to consider the information content associated with radiography of a thick object using opto-electronic gamma camera detector systems (with a Q_{eff} and f_c of 0.3). A scoping study has considered several experimental magnifications (1.3 to 20) at various first conjugate distances (1m and 1.645m) and machine spectra (5MeV, 10MeV and 15MeV.

Figure 1 shows that for 1000R, 2mm spot size machines with gamma camera systems the optimum FOM is 2.5 in the simple case but the information content is reduced by 0.5 bits per pixel in the complex case where scatter and detector variance is also considered.

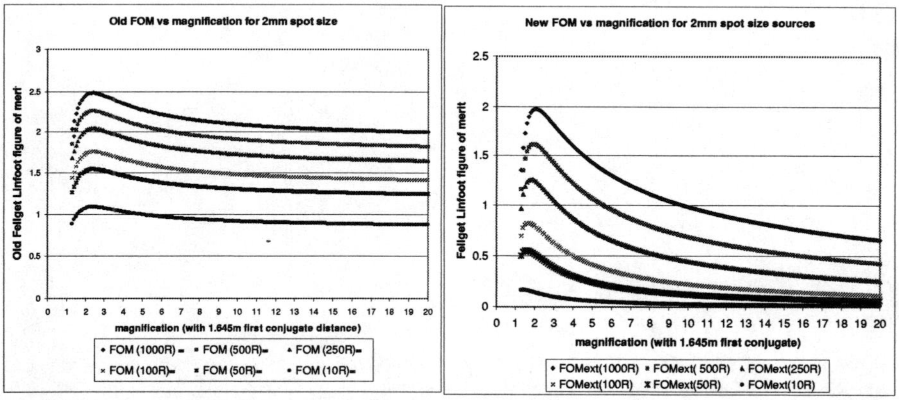

FIGURE 1. New and old FOM for five, 2mm spot size machines with a 10MeV spectrum and first conjugate distance of 1.645m for various doses and magnifications.

Since the difference in the two FOM are due to differences in SNR assumptions these should vary only with the source photon statistics. The difference between the two FOM have been determined at 10MeV peak bremsstrahlung source energy, for source doses ranging from 1R to 10,000R at the optimum radiographic magnification deduced from the new FOM with 2mm source sizes. This correction has been used to establish what FOM should be required by the simple method (which ignores both scatter and detector variance) assuming 2 bits per pixel by the new FOM is adequate.

The scoping study can then be used to calculate a zone of sources for which the information theory based FOM is greater than 2 by the complex method or greater than the appropriate simple FOM figure depending on the source photon statistics. The simple FOM goals are found to vary from 1.6 bits per pixel at low doses to 2.3 bits per pixel at high doses.

SUMMARY OF RESULTS AND CONCLUSIONS

The large number of scoping calculations performed using the two FOM methods can be summarised in a series of charts identifying a zone of sources which may be suitable for long term HRF goals. The charts show zones of source characteristics which appear suitable by either the simple or the complex FOM. Between the two zones it is possible to raise information content on radiographic images by reducing detector backgrounds or scatter backgrounds without changing source characteristics. The FOM study is being used at AWE to drive diode research to mitigate the risk of achieving cost effective, high quality flash radiographic sources. The study shows that smaller spot size sources may be considered with lower intensities than 1000R for deep penetration problems. Work on pinch diodes including self-magnetic pinch, rod pinch and plasma filled rod pinch diodes are therefore being considered in parallel with paraxial and immersed Bz diode concepts for long term HRF goals. In addition, spot collimation can be considered as this may modify the output dose or spot size of a source to a new configuration closer to the required zone.

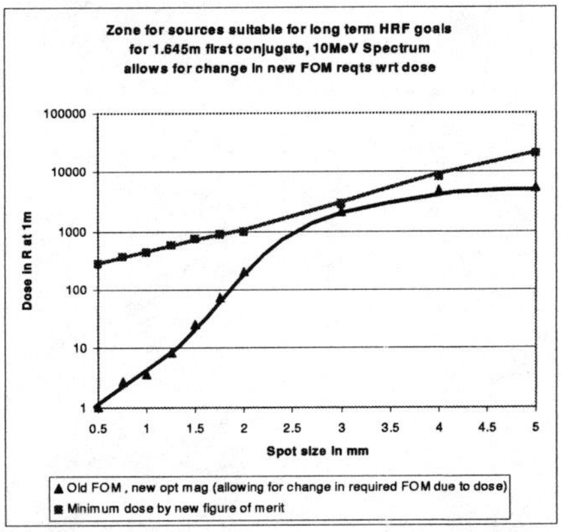

FIGURE 2. New and old FOM for five, 2mm spot size machines with a 10MeV spectrum and first conjugate distance of 1.645m for various doses.

REFERENCES

1. Quillin S, Crotch I, McAlpin S, O'Malley J, "The Use of MCNP in Flash Radiographic Applications at AWE," in Proceedings of MC2000, edited by A. Kling et al, Springer, New York, 2001, pp 357-368
2. Watson S, "A Closed-Form, Fellget-Linfoot Radiographic Figure of Merit," DARHT Technical note No 94, June 1998

Experiments on PIM in Support of the Development of IVA Technology for Radiography at AWE

Stephen G. Clough,[a] Kenneth J.Thomas,[a] Mark C.Williamson,[a] Martin J.Phillips,[a] Ian D Smith,[b] Vernon L Bailey,[b] Hiroshi J Kishi,[b] John E. Maenchen,[c] David L. Johnson[c]

[a] *Pulsed Power Group, Building H36, AWE, Aldermaston, Reading RG7 4PR, U.K*
[b] *Titan Pulsed Sciences Division, 2700 Merced Street, San Leandro, CA 94577*
[c] *Sandia National Laboratories, P.O.Box 5800, Albuquerque, NM 87185*

Abstract. The PIM machine has been designed and constructed at AWE as part of a program to investigate IVA technology for radiographic applications. PIM, as originally constructed, was a prospective single module of a 14 MV, 100 kA, ten module machine. The design of such a machine is a primary goal of the program as several are required to provide multi-axis radiography in a new Hydrodynamics Research Facility (HRF). Another goal is to design lower voltage machines (ranging from 1 to 5 MV) utilizing PIM style components. The original PIM machine consisted of a single inductive cavity pulsed by a 10 ohm water dielectric Blumlein pulse forming line (PFL) charged by a Marx generator. These components successfully achieved their design voltages and data on the prepulse was obtained showing it to be worse than expected. This information provided a basis for design work on the 14 MV HRF IVA, carried out by Titan-PSD, resulting in a proposal for a prepulse switch, a prototype of which should be installed on PIM by the end of this year. The original single, coaxial switch used to initiate the Blumlein has been replaced by a prototype laser triggered switching arrangement, also designed by Titan-PSD, which it was desired to test prior to its eventual use in the HRF. Despite problems with the laser, which will necessitate further experiments, it was determined that laser triggering with low jitter was occurring. A split oil co-ax feed has now been used to install a second cavity, in parallel with the first, on the PIM Blumlein. This two cavity configuration provides a prototype for future radiographic machines operating at up to 3 MV and a test facility for diode research.

THE PIM MACHINE

PIM (Prototype IVA Module) (Fig. 1), as originally constructed [1], was a prospective single module of a 14 MV, 100 kA, ten module IVA (Inductive Voltage Adder) [2] machine. The design of such a machine is a primary goal of the AWE program as several are required to provide multi-axis radiography in a new Hydrodynamics Research Facility (HRF). The single inductive cavity is pulsed by a 10 ohm water dielectric Blumlein pulse forming line. This is charged by a Marx generator based on the Sandia hanging Marx design using 32 1.35uF 100kV DC capacitors. The Blumlein acts as a fast pulse forming section initiated by a switch. The switch initially installed was a co- axial two stage Rimfire[3] and Trigatron spark gap. The inductive

CP650, *Beams 2002: 14th International Conference on High-Power Particle Beams,*
edited by T. A. Mehlhorn and M. A. Sweeney
2002 American Institute of Physics 0-7354-0107-1

FIGURE 1. The original PIM machine.

cavity incorporates ferromagnetic cores formed from Metglas alloy tape and the fast pulse appears across a perspex (lucite) insulated accelerating gap. The pulse travels along a short vacuum transmission line to a ball/plane e-beam diode test load.

The Blumlein PFL and its Rimfire/Trigatron switch were tested using a copper sulphate load. A peak output voltage of 1.7MV was achieved from the PFL. The Inductive cavity, when added, successfully achieved a peak voltage of 1.4MV. The output voltage waveforms are shown in Fig. 2 for three levels of Marx DC charge. The prepulse, as seen in these waveforms, was worse than expected from simple modelling based on results with the smaller LINX [4] machine previously built at AWE, being ~ 8 % of peak voltage at the highest voltage level. This prepulse data provided a basis for modelling and design work for the 14 MV HRF IVA, carried out by Titan-PSD. As a result of that work a prepulse switch for the 14 MV IVA may be tested on PIM this year which should reduce the prepulse to less than 0.5% of peak.

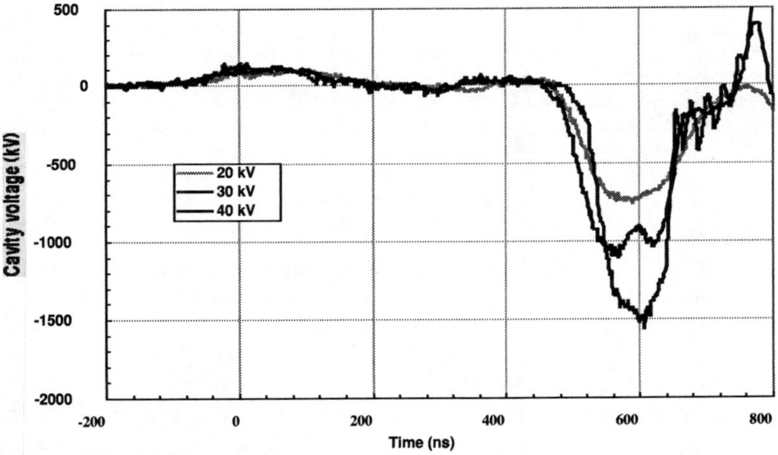

FIGURE 2. PIM cavity output voltage waveforms for 3 levels of Marx DC charge.

LASER TRIGGERING

A laser triggered switching arrangement, also designed by Titan-PSD for the HRF IVA and based on a system used in DARHT [5] and AIRIX [6], was tested using the PIM Blumlein. This replaced the coaxial rimfire/trigatron switch originally installed. Fig. 3 shows the laser triggering system during installation and when installed on the PIM Blumlein. There are two switches, compared to the one previously used, and the beneficial effect of the resulting lower inductance on pulse rise-time was demonstrated. Problems with the laser used limited the data that could be obtained initially. However, it was determined that laser triggering with low jitter was occurring. Laser triggering experiments will resume shortly, now that those problems have been resolved.

TWO CAVITY VERSION OF PIM

A second cavity has now been installed on PIM driven, in parallel with the first, by the one Blumlein. This necessitated the design and construction of an oil insulated coaxial transmission line to connect the Blumlein to the two cavities. It is hoped that this two cavity configuration will provide a prototype for future radiographic machines operating at up to 3 MV. It will also provide a test facility for diode research on an IVA machine at this voltage and experiments are now proceeding to test the pinch and paraxial diodes, currently in use at AWE, to determine whether their operation on an IVA machine causes any problems. It is also intended to test the rod pinch [7] and Plasma Filled Rod Pinch (PFRP) [8] diodes developed by NRL as these are proven to provide the ~ 1 mm spot sizes desired for high resolution flash radiography. Further up-rates of PIM may lead to machines with a total of 3 or 4 cavities to allow a suite of X-ray machines to be installed in the HRF to provide a radiographic capability over a wide range of voltages, as in AWE's current facilities.

FIGURE 3. The laser triggering system during assembly and fully installed on the PIM Blumlein.

CONCLUSIONS AND FUTURE PROGRAM

The PIM machine has been built as part of a continuing program to develop AWE's experience of IVA technology. It is now providing a facility for prototyping aspects of the design of the 14 MV, 10 module IVA machines that it is intended to install in AWE's new Hydrodynamics Research Facility. Experiments on the laser triggering system are to be resumed soon and at the end of this year it is hoped to test a prepulse reduction system. It is also being used to conduct some diode experiments to research IVA specific issues in the operation of the focused e-beam diodes used for flash radiography. Another goal of the program is to develop lower voltage machines to allow a full range of radiographic experiments to be carried out in the new facility. These may utilize PIM or HRF IVA components with one or two cavities as already constructed or a total of three or four cavities in configurations that may be realized in future upgrades of PIM. These possible future machines may use either a rod pinch diode or the Plasma Filled Rod Pinch (PFRP) developed by NRL to generate spot sizes of the order of 1 mm and these diodes will be tested on PIM this year.

ACKNOWLEDGEMENTS

This work was funded by the UK Ministry of Defense. Thanks are also due to many members of the Pulsed Power Group at AWE and people at Titan-PSD and SNLA not mentioned in the list of authors for their assistance. The UK authors owe a great deal to Dave Forster who instigated this program before his retirement from AWE.

REFERENCES

1. Thomas, K.J., Williamson, M.C., Phillips, M.J., and Clough, S.G., *Proc. of International Conf. on Pulsed Power and Plasma Physics*, 306-309 (2001).
2. Ramirez, J.J, et al, *Proc. of 5th IEEE Pulsed Power Conf.*, 143 (1985).
3. Humphreys, D.R., et al, *Proc. of 5th IEEE Pulsed Power Conf.*, 262 (1985).
4. Thomas, K.J., and Forster, D.W., *Proc. IEE (UK) Pulsed Power '96*, 2/1-2/3 (1996).
5. Downing, J.N., et al, *Proc. 8th IEEE Pulsed Power Conf.*, 949-952 (1991).
6. Voisin, L., et al, *Proc. of 11th IEEE International. Pulsed Power Conf.*, 1315-1319 (1997).
7. Mahaffey, R., et al, *Appl. Phys. Lett.*, **33**, 795 (1978).
8. Weber, B.V., et al, *Proc. of International Conf. on Pulsed Power and Plasma Physics*, 462-465 (2001).

Real-Time Measurement of the Focal-Spot Intensity-Distribution for Megavolt Flash X-ray Machines

Gideon Barnea, Meir Markovits and Itamar Navon

Rafael, P.O. Box 2250, Haifa, 31021, Israel

Abstract. A new method for measuring the 2D intensity-distribution of the focal-spot of megavolt flash X-ray machines is demonstrated. A position-sensitive plane detector views the radiation source through an X-ray-opaque large square aperture. The square aperture was introduced in a previous publication [1] and it was shown that in this case, the focal-spot intensity-distribution is given by the second order mixed Cartesian partial derivative of the penumbral image intensity distribution. This technique is demonstrated by measuring the intensity distribution of the focal-spot of a pulsed X-ray machine (1 MV, 40 kA, 50 ns). The results are compared with the physical damage visible on the metal target.

INTRODUCTION

The size and shape of a radiation source is of interest in numerous applications. In radiography (both medical and industrial), the focal spot size determines the degree of blurring due to magnification. In radiotherapy, the shape and size of the focal spot in a medical accelerator is needed to determine primary beam penumbra. In inertial confinement fusion (ICF) much can de learned on the size and shape of the compressed core by imaging the neutron radiation emanating from it. Other applications may be found in nuclear medicine and x-ray astronomy.

Numerous techniques for source characterization are described in the literature [2]. However, the case of highly penetrating radiation and short pulse duration imposes special difficulties. First, the aperture substrate must be several mean-free-paths thick to provide adequate image contrast. This precludes the application of ordinary pinhole methods as well as various multiplex methods (Fresnel zone plate, pinhole arrays, etc.) [2]. Secondly, camera movement is unviable, preventing application of the "moving slit" method [3] and methods based on computerized tomography [4,5]. Only a few techniques remain to be considered: The "multislit" method [6] is only capable to generate a set of parallel line-integrals of the source intensity distribution. The "rollbar" methods (single or double) [7], only provide line-integrals parallel to one (or 2 mutually perpendicular) directions. Only two earlier methods are capable of providing full 2D source images: the "tapered-pinhole" method [8] and a method based on Fourier-transform-deconvolution analysis [9-12]. The 1st is low in photon collection

CP650, *Beams 2002: 14th International Conference on High-Power Particle Beams,*
edited by T. A. Mehlhorn and M. A. Sweeney

efficiency and hard to manufacture. The 2nd is known to generate undesirable artifacts, is sensitive to small geometrical uncertainties (exact aperture size, exact camera distances, etc.), and is affected by the 3-dimentionality of real objects [11,12].

THEORY

An isotropic 2-D source (intensity distribution: S(x',y')) is viewed by a flat linear-response position-sensitive detector (image intensity: D(x,y)), through an aperture (transmittance: A(x",y")). The source, aperture and detector planes are parallel to each other. L_1 is the distance between the source plane and the aperture plane and L_2 is the distance between the aperture plane and the detector plane. θ is the angle between the normal to the planes and the line of sight from a source point (x',y') to a detector point (x,y). The aperture substrate is assumed to be perfectly opaque to the radiation. In this case, the image intensity D(x,y) is given by the following equation [2].

$$D(x,y) = k\int dx'\int dy'\cos^3(\theta)A(x'',y'')S(x',y') \tag{1}$$

where $k = \Phi/(L_1+L_2)^2$ is a constant, and Φ is the energy fluence at unit distance. x'=x"(1+1/m)-x/m and y'=y"(1+1/m)-y/m, where $m=L_2/L_1$ is the magnification. cos(θ) is a complicated function of (x,y,x',y'). Fortunately, in our case (L_1+L_2) is large compared to the source dimensions or the image size. In this case, cos(θ) ≅1. Eq. 1 (with cos(θ)=1) is the basis of the Fourier-transform-deconvolution method [9-12] discussed above.

A much simpler equation is obtained if one considers a large square aperture [1]. If the source is viewed through a corner (x_c,y_c) of the aperture, eq. 1 reduces to:

$$D(x,y) = k\int_{x_0}^{x_1} dx'\int_{y_0}^{y_1} dy' S(x',y') \tag{2}$$

where (x_0,y_0) is the projection of (x_c,y_c) onto the source plane as viewed from the detector point (x,y) (eq. 3). (x_0,y_1) and (x_1,y_0) are points exterior to the source.

$$x_0 = x_c(1+1/m) - x/m \quad ; \quad y_0 = y_c(1+1/m) - y/m \tag{3}$$

Obviously,
$$S(x_0,y_1) = S(x_1,y_0) = S(x_1,y_1) = 0 \tag{4}$$

and, $S(x_0,y_0)$ is simply given by eq. (5) (the constants are irrelevant).

$$S(x_0,y_0) \propto \left|\frac{\partial^2 D(x,y)}{\partial x \partial y}\right|_{x,y\ \mathit{derived\ from\ eq.(3)}} \tag{5}$$

Contrary to the Fourier-transform-deconvolution method, the source intensity distribution given by eq. 5 is insensitive to geometrical uncertainties (aperture size, source to aperture distance), apart from scaling which depends linearly on the magnification (Eq. 3). The three-dimensionality of the object does not affect the result. In this case, eq. 5 merely gives the line integral of the (self-attenuated) source intensity along the direction defined by the detector point (x,y) and the aperture corner (x_c,y_c).

Each of the four corners provides a complete image of the source. Ideally, the four images are identical. In reality, some source halo may exist that exceed the aperture limits and eq. 4 may not be strictly observed, causing small variations between source images obtained from different aperture corners. The source's three-dimensionality also contributes to these variations. In practice, the corner which gives the highest value for the source's peak intensity ("best corner") is chosen. This minimizes the inaccuracy. The aperture is self-shielding, therefore, scatter is negligible. Moreover, the differentiation process greatly eliminates the effect of any residual slowly-varying scatter fluence.

A drawback of our method is that differentiation is involved, thus inherently emphasizing noise. Therefore, some kind of noise reduction technique should be applied on the raw data, before differentiating. Numerous examples may be found in the literature. However, a specially designed filter is preferable: Around any given image point (i,j), one looks for the best-fit two-dimensional polynom D*(x,y) to the (2n+1)x(2n+1) nearest-neighbor data points (n is a chosen parameter):

$$D^*(x,y) = \sum_{p,q} a_{pq}(i,j)\cdot(x-x_i)^p(y-y_j)^q \quad ; (0 \le p+q \le m) \qquad (6)$$

Only the a_{11}(i,j) coefficients are needed Obviously m must be ≥ 2. It turns out that the least-square solution for a_{11}(i,j) with m=2 or m=3 is given by:

$$a_{11}(i,j) = \sum_{k,l=-n}^{n} k\cdot l\cdot D(i-k,j-l) \qquad (7)$$

where D(i,j) ≡ D(x_i,y_j). Finally, the source's image matrix is given by: $S \propto |a_{11}|$.

EXPERIMENTAL SETUP AND RESULTS

The present technique was applied to measure the focal-spot intensity distribution of a pulsed X-ray machine (1 MV, 40 kA, 50 ns). The aperture's substrate was made of a tungsten alloy, the aperture's width was 12 mm and its thickness was 100 mm. The aperture was imaged onto a storage phosphor screen. The distance from the source's plane to the aperture's front-end (L_1) was 15 cm and the distance from the aperture's front-end to the screen (L_2) was 90 cm. The magnification, m, is 6 and detector blur is negligible. The screen was read using a 16-bit laser-scanner with a pixel size of 0.2 mm. The data was filtered with a 10x10 pixel-averaging window and resized accordingly, corresponding to a resolution of 1/3 mm at the source plane.

In most cases an oval or banana-shaped source shape was observed. A strong correlation was found between the decoded source shape and the physical damage visible on the metal target, demonstrating the effectiveness of the present method. This correlation exists even when, in very rare occasions, the electron beam splits into two. While detrimental to radiography, this event is superb for a good demonstration of the

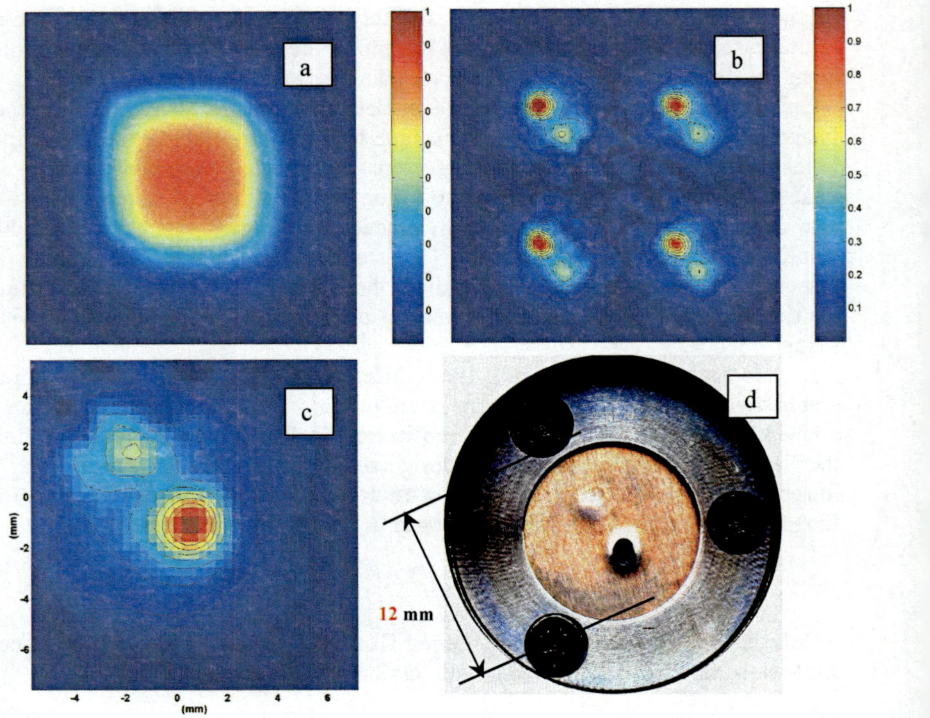

FIGURE 1. (a): Detector's raw data (image of D). (b): Full four-corner decoding (image of $|a_{11}|$, eq. 7 with n=2), detector plane. (c):"Best corner" decoding, source image matrix S, source plane, contours are at 0.1, 0.2, .. , 0.9 relative intensity levels. (d): Physical damage visible on the metal target (rear view). Note the mound at 11 o'clock. The two intensity peaks are 3.7 ± 0.5 mm apart.

present technique, as shown in figure 1.

REFERENCES

1. Barnea G., *Rev. Sci. Instrum.* **65**, 1949-1953 (1994).
2. Barrett H.H. and Swindell W. *Radiological Imaging* (Academic, New York, 1981).
3. Lowenthal E., Loewinger E., Bar-Avraham E. and Barnea G. *Med. Phys*, **19**, 687-690 (1992).
4. Gillman G. and Macleod I. *Comp. Graphics Image Process.* **11**, 227-241 (1979).
5. Jaffray D. A., Battista J. J., Fenster A. and Munro P. *Med. Phys*, **20**, 1417-1427 (1993).
6. Lutz W. R., Maleki N. and Bjarngard B. E., *Med. Phys*, **15**, 614-617 (1988).
7. Richardson R. A. and Houck T. L., *Proceedings of the LINAC98 Conference*, Chicago, IL. 908-910 (1998), Richardson R. A., Sampayan S. and Weir J., *Proceedings of the LINAC2000 Conference*, Monterey, CA, **1**, 161-163, (2000).
8. Crist C.E. et. al *Proceedings of the LINAC98 Conference*, Chicago, IL. 195-197 (1998),
9. Nugent K. A. and Luther-Davies B., *J.Appl.Phys.*, **58**, 2508-2515 (1985).
10. Watson S. A., *Proceedings of the 1993 Particle Accelerator Conference*, **3**, 2447-2409 (1993).
11. Nugent K. A. and Luther-Davies B., *Appl. Opt.*, **25**, 1008-1014, (1986).
12. Ress D., Lerche R. A. Ellis R. J. Lane S. M. and Nugent K. A., *Science*, **241**, 956-958 (1988).

2 MeV Average γ-ray Energy, 6 ns Duration Bremsstrahlung Radiation from the Conical Anode Diode

B.A. Kablambaev, N.A. Ratakhin, V.K. Petin

High Current Electronics Institute, 4 Academichesky Ave., Tomsk, 634055, Russia

Abstract. This paper presents the results of experimental research on generation of short γ-ray pulses. The experiments have been performed using the MIG facility. This facility is a high-voltage generator with a water-filled transfer capacitor, which is based on a line pulse transformer (240 kJ) [1]. The water-filled transfer capacitor is discharged compression transfer capacitor and into water transmission line through the output switch, and thus causes the current to flow through the POS. The use of a conical Al-Ta compound anode in the diode with a pinched electron flow has made it possible to decrease the γ-ray duration and to increase the average γ-ray energy by decreasing the X-ray generation efficiency later in the voltage pulse. We have managed to obtain the following γ-ray parameters: an average γ-ray energy of 2 MeV, a FWHM of 6÷6.5 ns, a pulse risetime of 2÷2.5 ns, and a doze of $1.5{\cdot}10^{12}$ rad/s over an area of more than 15 cm^2.

This paper presents the results of experimental research on generation of short γ-ray pulses. The experiments have been performed using the MIG facility [1]. This facility is a high-voltage transformer with two water-filled transfer capacitors, water transmission line, vacuum chamber with POS and electron diode. The inductive energy storage consists of coaxial electrodes, which are placed in vacuum chamber. The current is switched by plasma bridge at the phase of energy accumulation. This bridge is formed by 96 plasma guns. Plasma guns have simultaneous triggering with start of linear transformer. The quantity of injected plasma was regulated by current value in these plasma guns. The experiments have shown that the best stability and highest diode voltage was observed at the opening current about 0,7÷0,8 of the short-circuit one. In our experiments next characteristics were recorded: voltage near isolator was measured by capacitive divider, currents and their derivatives downstream and upstream were measured by Rogovsky coils and differentiating magnetic coils, γ -ray impulse was registered by pin-diode, dose after anode was measured by dosimeters IS-7, spectral distribution of radiation was registered by multi-channel spectrometers. Diode voltage was estimated through the equation :

$$U_{\text{diode}} = U_{\text{insul}} - \frac{d(L_1 I_1)}{dt} - \frac{d(L_2 I_2)}{dt}$$

Indices 1 and 2 are to the sensors, which are situated upstream and downstream of POS respectively.

In earlier POS experiments on generation γ-ray in pinch-diode with flat anode and annular cathode we registered the ≈1,2 MeV average γ-ray energy /2/. The analysis of

CP650, *Beams 2002: 14th International Conference on High-Power Particle Beams*, edited by T. A. Mehlhorn and M. A. Sweeney

these and others our experiments showed that under existing conditions peak voltage is at starting phase of pinch. That's why γ-ray source represents a ring which is nearly equal-sized with cathode and has polar pattern elongated along the axis. Evidently, it is concerned with "normal" entrance in convertor of high-energy electrons. At later phase, when beam pinching the radiation source becomes more isotropic and has "soft" spectrum, because of diode voltage decrease and wide electron distribution entrance in convertor. Over a half of γ-ray energy emanates in the first phase.

The purpose of this experiments was confirmation our conjectures and utilization them for increase average γ-ray energy radiated when braking relativist electrons in convertor.

We carried out a set of special experiments to determine spectral and temporal characteristic of *γ* -radiation generated by different zones of convertor in planar diode with strong electron beam pinch. For that a special lead block was closely placed at the back surface of convertor. It consists of 13 multi-channel spectrometers on basis of copper filters with different thickness. There is a diagrammatic representation of the block (fig. 1). There are 13 channels in lead cylinder *1*: one along centroidal axis, the others at three different radiuses for observation different zones of convertor *2*. In the channels there are LiF-detectors *3* separated by cooper filters *4* with different thickness. In some channels pin-diode were set for registration the duration of γ-ray impulse. The duration of impulse was recorded by pin-diode with permanent sensitivity for all spectral diapason.

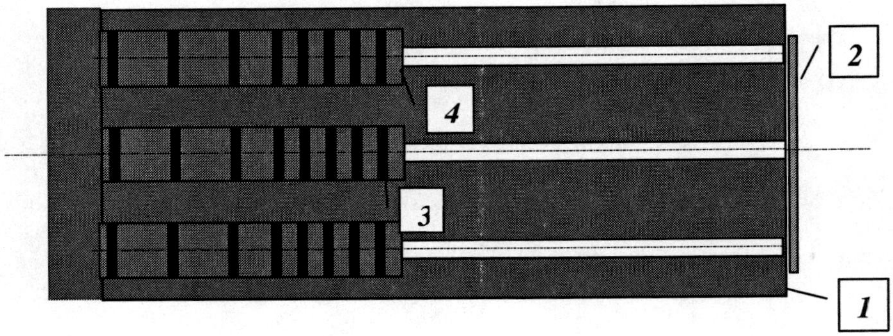

FIGURE 1.

These measurements showed, that in the pinch-diode with flat anode and circula cathode the duration of γ-ray radiation from separate parts of convertor is 1,5÷2 times shorter than the duration from whole convertor (a typical oscillogram of pin-diode signal is represented in fig.2.). The average γ-ray energy of radiation from different zones of convertor differs 2÷3 times. Dose measurements show, that about a half of full γ-ray energy generates by outer circular zone of convertor.

As a result of the experiments diode configuration was found. which chooses high-energy part of radiation and directs it in the adjusted area. This configuration is represented in fig.3: hemispherical cathode with the radius 30 mm and conical emission surface 7 mm wide on inside and conical Ta-Al compound anode. The circular surface of cathode was covered with fleecy material *1* for equal emission and minimizing its delay.

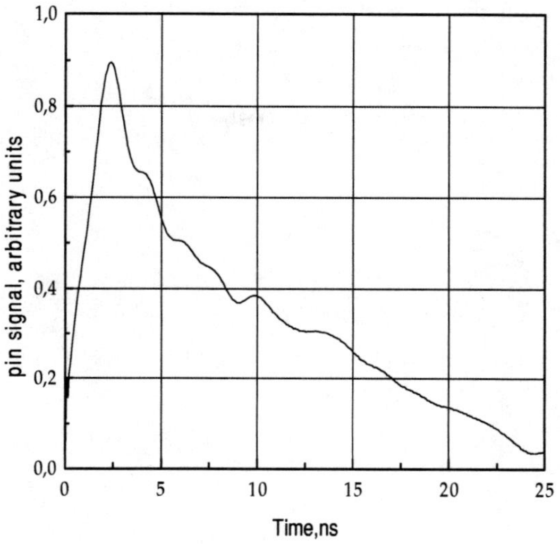

FIGURE 2.

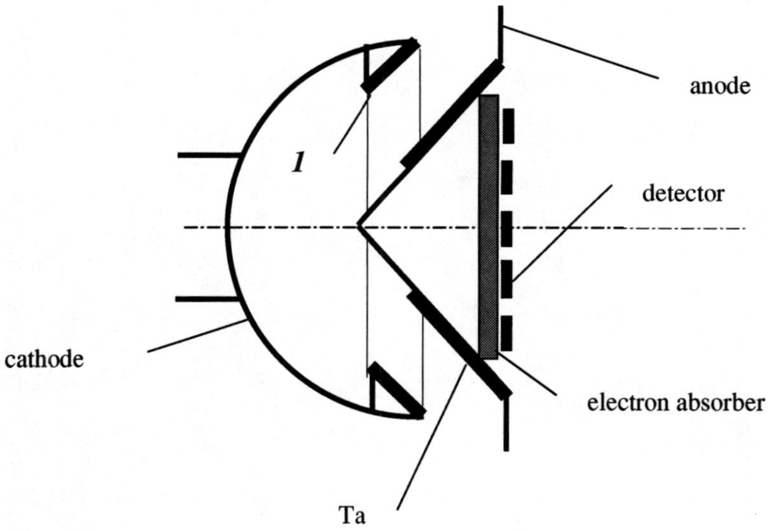

FIGURE 3.

Spark gap varied within the limits of 9÷11 mm, it allows to realize strong electron beam pinch mode.

Using the conical Al-Ta compound anode in the pinch-diode allowed to increase γ-ray average energy and decrease the duration of γ-radiation. It became possible due to the concentration of hard γ-radiation in the adjusted area in its early phase when electrons attack convertor normally and there is narrow polar pattern of γ-radiation. At the same time we decreased the efficiency of generation “soft” γ-radiation at its later pinch phase. It’s the result of deviation of electron beam to the conical anode peak, where light materials with low emissivity are used. We have managed to obtain the following γ-ray parameters: an average γ-ray energy of 1,5÷2 MeV, a FWHM of 6÷6.5 ns, a pulse risetime of 2÷2.5 ns, and a doze of $1.5 \cdot 10^{12}$ rad/s over an area of more than 15 cm^2.

References

1. Luchinskii A.V., Ratakhin N.A., Fedushchak V.F., Shepelev A.N. Multipurpose transformer-type pulse generator. Izvestiya Vysshikh Uchebnykh Zavedenii Fizika, vol.40, Dec. 1997, pp. 67-75.
2. B. Kablambaev, N. Ratakhin, S. Shlyakhtun. Generation of 10^{13} r/s dose power gamma radiation on 240 kJ MIG facility with POS. 13th International conference on high-power particle beems-2000, Nagaoka, june 25-30, 2000, pp.1063-1066.

A Characterization of the Radiation from a Rod-Pinch Diode*

Stephen B. Swanekamp,[b] Raymond J. Allen,[a] David D. Hinshelwood,[a] David Mosher,[a] and Joseph W. Schumer[a]

[a]*Naval Research Laboratory, Plasma Physics Division, Washington, DC USA*
[b]*Titan/Jaycor, Mclean, VA*

Abstract. Coupled PIC-Monte-Carlo simulations of the electron-flow and radiation production in a rod-pinch diode show that multiple scatterings in the rod produce incident electron energies that ranging from zero to slightly higher than the applied voltage. It is speculated that those electrons that gain energy do so by remaining in phase with a rapidly varying electric field near the tip of the rod. The simulations also show that multiple passes in the rod produce a wide spread in incident electron angles. For diode voltages of V=2 MV, the angular distribution of electrons incident on the rod is broad and peaked near 90° to the axis of the rod with a larger fraction of electrons striking the rod at angles less than 90°. The electron angular distribution for V=4 MV is narrower and peaked at 105° with a larger fraction of electrons incident on the rod with angles greater than 90°. The photon distributions are peaked along the direction of the high-energy electrons. For V=2 MV the dose filtered through 2¼-cm thick Plexiglas is peaked at 90° and is 1.8 times higher than the forward-directed [0°] dose. For V=4 MV the dose filtered through 2¼-cm thick Plexiglas is peaked at 120° and is 2.3 times higher than the forward-directed dose. Similar angular variation of the dose has been observed on the 4-MV Asterix accelerator [2] and on 1-2 MV accelerators at the Atomic Weapons Establishment [8].

INTRODUCTION

The rod-pinch diode is being developed as a bright, intense source of x-rays for radiography.[1] The rod-pinch diode has been successfully operated at voltages of up to 4 MV with peak currents of 135 kA producing doses of 16-20 rad(Si) at 1 meter with a source diameters of 1-2 mm.[2] The rod pinch diode consists of a small diameter (~1 mm), tapered, high-atomic-number anode rod that extends through a hole in a thin annular cathode. Electrons emitted from the cathode are accelerated across the anode-cathode gap by an applied voltage and are simultaneously bent by their own self-magnetic field. Electron deposition is concentrated at the tip of the rod producing an intense, bright source of bremsstrahlung x-rays.[3,4] The charged particle flow in the diode

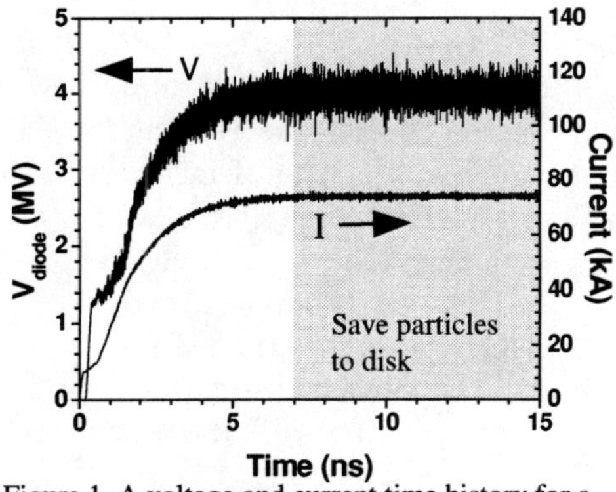

Figure 1. A voltage and current time history for a 4 MV simulation.

CP650, *Beams 2002: 14th International Conference on High-Power Particle Beams,*
edited by T. A. Mehlhorn and M. A. Sweeney

and the subsequent production of x-rays is a complicated process requiring numerical simulation.

In this paper, we present results from 2½ D electromagnetic LSP [5] particle-in-cell simulations (PIC) of the diode in which electron scattering and energy loss in the anode rod are treated by ITS[6] Monte-Carlo (MC) algorithms. The simulation geometry consists of a 1 mm diameter tungsten anode rod that extends through a 16-mm diameter, 3-mm thick hole in the cathode. The rod extends 16 mm beyond the end of the cathode hole and is tapered to a point over the last 10 mm. The technique used here is similar to one used previously to determine the forward-directed dose from a rod-pinch diode.[5] Here, emphasis is placed on how the electron distributions in energy and angle striking the surface of the rod relate to the angular dependence of the photon flux emerging from the rod and the dose at 1 meter.

The basic steps of the algorithm are as follows. First, the PIC-MC code is run to a quasi-steady state for a particular voltage and information about the electrons that strike the surface of the rod are stored for post-processing electron/photon transport. A sample time history for a 4 MV simulation is shown in Fig. 1 and information is stored to disk during the shaded times while the voltage is approximately constant. These electrons are then used to calculate the energy and angular distribution of photons leaving the rod. Dose is then calculated by convolving the photon distribution with the detector response function and is given by

$$\frac{Dose(\theta)}{Q_{inc}} = \int_0^\infty \rho(h\nu,\theta)\frac{d^2N}{R^2 d\Omega d(h\nu)} d(h\nu) \ , \qquad (1)$$

where $\rho(h\nu,\theta)$ is the detector response in rads-cm^2/photon obtained from a mono-energetic flux of photons, $d^2N/d\Omega d(h\nu)$ is the photon number spectrum per solid angle per incident electron charge calculated from ITS, and R is the distance between the source and the detector. In general, the θ dependence of the detector response is due to attenuation from the vacuum vessel and any other filters between the source and the detector. When the detector is properly equilibrated, the detector response can be calculated from tabulated attenuation and absorption coefficients.[7]

For comparison with experiments, it is useful to renormalize Eq. (1) to the total instantaneous charge in the diode. This is done in two steps. First, Eq. (1) is multiplied by the average number of passes of an electron through the rod defined by $N_{passes} \equiv Q_{inc}/Q_e$, where Q_e is the absorbed electron charge. The average number of passes can be calculated from the PIC simulations as the ratio of incident electron current to the absorbed electron current. The simulations show that electrons make about 3 passes through the rod at 2 MV increasing linearly to 5 passes at 10 MV. The next step is to multiply Eq. (1) by $Q_e/Q_T = 1 - Q_i/Q_T$, where $Q_T = Q_e + Q_i$ is the total ion and electron charge in the diode. The ratio Q_i/Q_T can be determined from the ion current fraction obtained from PIC simulations. The simulations show that the ion current fraction is a weak function of voltage increasing nearly linearly from 0.3 at 2 MV to 0.45 at 10 MV.

The electron distributions on the surface of the tungsten rod for V = 2 MV and V = 4 MV are shown in Fig. 2. Peaks in the distributions occur near the endpoint voltage and near 90° [0° is along the axis toward the tip of the rod] corresponding to the unscattered electrons. The remainder of the distribution arises from multiple scattering processes in the tungsten rod as the electrons make repeated passes. The simulations also show that the width of the electron angular distribution on the rod is narrower for the 4 MV case than the 2 MV case. In addition, the 4 MV case shows a peak in the distribution near 110° with a larger fraction of electrons striking the rod at angles greater than 90°. In contrast the 2 MV case shows a peak in the distribution near 90° with a larger fraction of electrons incident on the rod at angles less than 90°. The simulations also predict that some of the electrons gain energy above the endpoint energy. One possible explanation is that a small fraction of the electrons gain energy as they oscillate in phase with a rapidly fluctuating electric field near the tip of the rod. This rapidly fluctuating field is responsible for the high frequency voltage oscillation seen in Fig. 1. Another possible reason that some particles gain energy is a numerical grid heating instability that can occur near the rod tip where the electron density exceeds 10^{15} cm^{-3}. However, simulations with better spatial resolution showed essentially identical energy distributions.

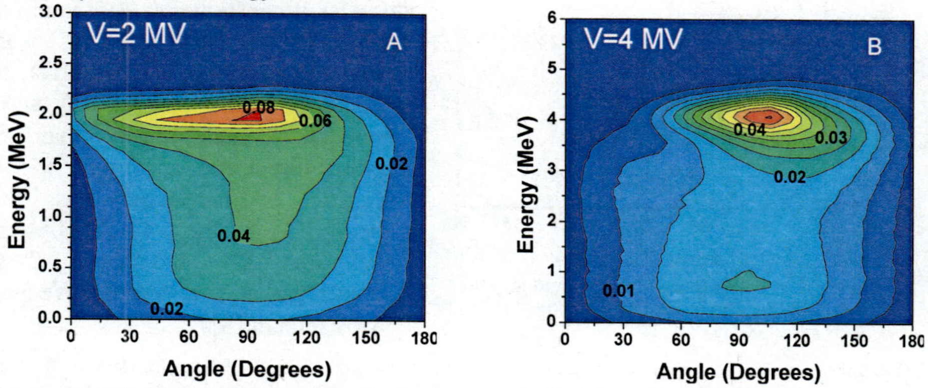

Figure 2. The doubly differential distribution in energy and angle of electrons incident on the surface of the tungsten rod for: A) V=2 MV, B) V=4 MV.

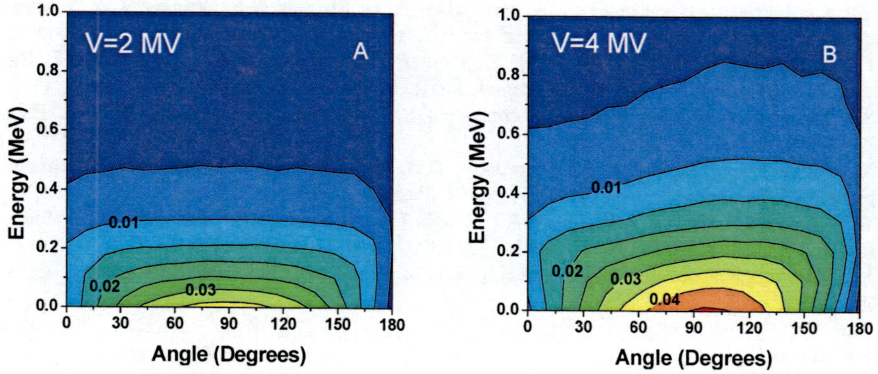

Figure 3. The doubly differential distribution in energy and angle of photons emitted from the tungsten anode rod for: A) V=2 MV, B) V=4 MV.

The electron distribution on the surface of the rod can be used as input to ITS to predict the radiation emitted from the rod. The doubly differential distribution [$d^2N/d\Omega d(h\nu)$] in energy and angle for photons emitted from the tungsten rod is shown in Fig. 3. This figure shows that, for V = 2 MV, the photons are peaked at 90° with a symmetric distribution about 90°. For V = 4 MV the distribution is peaked > 90° with more photons at angles greater than 90° than at angle less than 90°. These distributions are consistent with the electron distributions on the surface of the rod shown in Fig. 2.

The photon distributions given in Fig. 3 can be used in Eq. 1 to generate angular dose maps. A 2¼-cm thick Plexiglas dome is used as a filter so that only photons above 50 keV contribute to the dose. The detector was a thermo luminescent dosimeter consisting of a 1×1×6-mm parallelepiped of CaF_2 surrounded by a 2.3-cm thick aluminum cylinder as an equilibrator. The normalized angular dose maps for V=2 MV and V=4 MV are shown in Fig. 4. For V=2 MV, the dose 1.8 times higher at 90° than in the forward direction and is nearly symmetric about 90°. In contrast, the 4 MV dose peaks at about 120° and is about 2.3 times higher than the forward directed dose. Similar angular variation of the dose has been observed on the 4-MV Asterix accelerator [2] and on 1-2 MV accelerators at the Atomic Weapons Establishment [8].

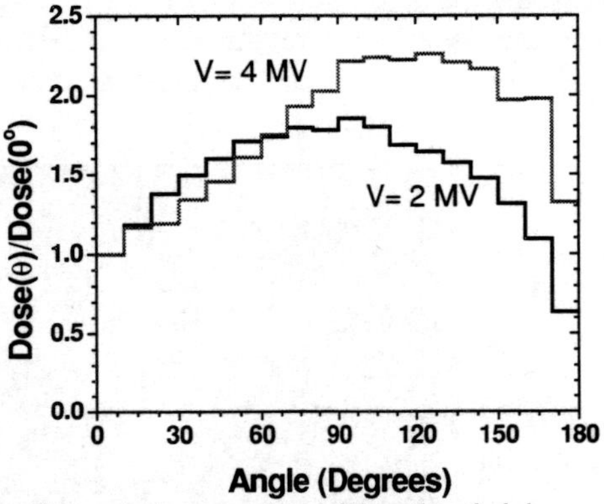

Figure 4. A comparison of the azimuthal dose profile for V=2 MV and V=4 MV.

*The authors benefited from discussions with Dr. D.V. Rose and Dr. D.R. Welch of Mission Research Corporation. This work was supported by the US Department of Energy through Sandia, Los Alamos, and Lawrence Livermore National Laboratories.

1. R.J. Commisso, G. Cooperstein, D.D. Hinshelwood, D. Mosher, P.F. Ottinger, J.W. Schumer, S.J. Stephanakis, S.B. Swanekamp, B.V. Weber, and F.C. Young, IEEE Trans. Plasma Sci. **30**, 338 (2002).
2. F.C. Young, R.J. Commisso, R.J. Allen, D. Mosher, S.B. Swanekamp, G. Cooperstein, F. Bayol, P. Charre, A. Garrigues, C. Gonzales, F. Pompier, and R. Vezinet, submitted to Phys. Plasmas.
3. S.B. Swanekamp, R.J. Commisso, G. Cooperstein, P.F. Ottinger, and J.W. Schumer, Phys. Plasmas **7**, 5214 (2000).
4. G. Cooperstein, J.R. Boller, R.J Commisso, D.D. Hinshelwood, D. Mosher, P.F. Ottinger, J.W. Schumer, S.J. Stephanakis, and B.V. Weber, Phys. Plasmas **8**, 4618 (2001).
5. D.V. Rose, D.R. Welch, B.V. Oliver, R.E. Clark, D.L. Johnson, J.E. Maenchen, P.R. Menge, C.L. Olson, and D.C. Rovang, J. Appl. Phys. **91**, 3328 (2002).
6. J.A. Halbleib, R.P. Kensek, G.D. Valdez, S.M. Seltzer, and M.J. Bergerm IEEE Trans. Nucl. Sci. **41**, 1025 (1992).
7. See http://physics.nist.gov/PhysRefData/XrayMassCoef/tab3.html for tabulated attenuation and absorption coefficients.
8. J. Threadgold, et al. these proceedings.

Characterization of the Rod-Pinch Diode at 2 to 4 Mv as a High-Resolution Source for Flash Radiography*

R.J. Commisso,[a] F.C. Young,[b] F. Bayol,[c] R.J. Allen,[a] J.R. Boller,[b] P. Charre,[c] G. Cooperstein,[a] A. Garrigues,[c] C. Gonzales,[c] D. Mosher,[a] F. Pompier,[c] S.B. Swanekamp,[b] and R. Vezinet[c]

[a]*Plasma Physics Division, Naval Research Laboratory, Washington, DC 20375*
[b]*Titan/Jaycor, McLean, VA 22102*
[c]*Centre d' Etudes de Gramat, Gramat, France*

Abstract. The ASTERIX generator is used to evaluatate the rod-pinch electron-beam diode as an intense source of x-rays for high-resolution, pulsed (30- to 40-ns FWHM) radiography at peak diode voltages of voltages of 2.4 to 4.4 MV and peak diode currents of 55 to 135 kA. At 4 MV, tungsten anode rods of 1-mm or 2-mm diameter produce on-axis doses at 1 meter of 16 rad(Si) or 20 rad(Si), respectively. The on-axis source diameter based on the full-width at half-maximum (FWHM) of the line-spread-function (LSF) is 0.9 ± 0.1 mm for a 1-mm diameter rod and 1.4 ± 0.1 mm for a 2-mm diam rod, independent of voltage. The LANL source diameter is nearly twice the FWHM. The measured rod-pinch current is reproduced with a diode model that includes ions and accounts for anode and cathode plasma expansion. A composite diode with a large diameter carbon-rod anode followed by a smaller-diameter tungsten-tip converter shows promise for applications where a small central source feature is desired.

The rod-pinch diode [1,2] provides an intense small-diameter bremsstrahlung source for pulsed radiography. Operation of this diode at 1 to 2.3 MV has been investigated [3-5] and particle-in-cell simulations have been carried out to characterize the charged-particle flow in the diode [2,6,7]. A model of the diode operation based on space-charge-limited current at low voltage and self-magnetically-limited critical current at high voltage has been applied to experiments at 1 to 2 MV with peak currents up to 70 kA [2]. This report summarizes results from the first experiments to extend rod-pinch-diode operation up to 4 MV with peak currents up to 135 kA [8,9]. For this work, the ASTERIX generator, located at Centre d' Etudes de Gramat in Gramat, France was modified to operate in positive polarity [10]. Marx charging voltages between 45 and 75 kV are used.

The experimental setup is illustrated in Fig. 1. Detailed descriptions of the diode hardware and the electrical measurements are found in Ref. [10]. A 6.4-mm diam carbon rod extends either 47 or 24 cm from the anode stalk, depending on the Marx charge, where it connects to a smaller-diameter tungsten-rod anode. The tungsten anode projects through the annular carbon cathode and extends 16 mm beyond the cathode with the last 10 mm tapered to a point. Tungsten anodes of 0.5-, 1-, and 2-mm diameter are used, and the ratio of the cathode-to-anode radius, R_C/R_A, ranges from 5.5

CP650, *Beams 2002: 14th International Conference on High-Power Particle Beams*, edited by T. A. Mehlhorn and M. A. Sweeney

to 20. A composite diode was also tested. For this diode, the 3.2-mm diam carbon rod extends 18 mm beyond the cathode with the last 8 mm tapered, and transitions to a 0.5-mm diam, 6-mm long, tapered tungsten tip. The dose on axis (forward direction) is measured with radiophotoluminescent detectors that are cross calibrated against CaF_2 thermoluminescent dosimeters. The on-axis source diameter is determined from analysis of the photographic image of a 1-m radius, tungsten rolled edge [3]. The variation in film density transverse to the rolled edge gives the edge-spread-function (ESF), while the derivative of the ESF is the line-spread-function (LSF).

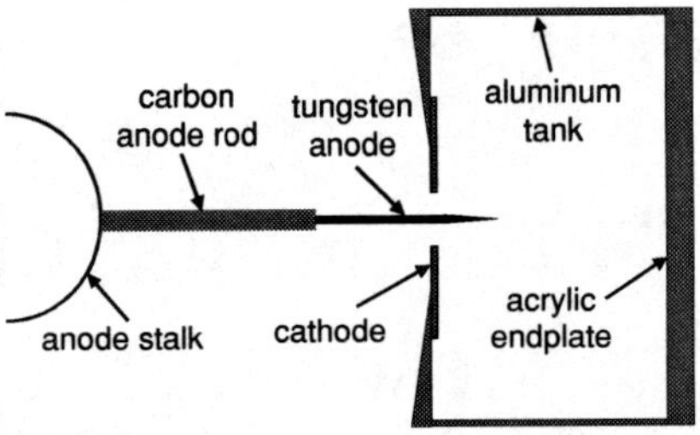

FIGURE 1. ASTERIX rod-pinch geometry.

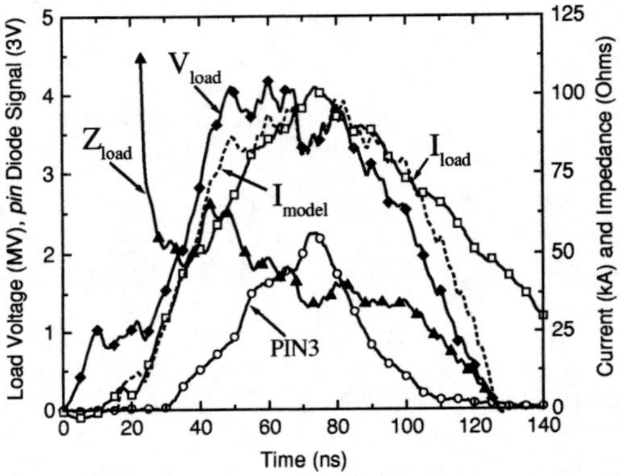

FIGURE 2. Electrical waveforms (V_{load}, I_{load}, and Z_{load}), x-ray signal (PIN3), and model current (I_{model}) for shot 6400.

Diode electrical waveforms for shot 6400 (2-mm diam rod with $R_C/R_A = 7.5$) are shown in Fig. 2. The load voltage, V_{load}, total current (electron + ion), I_{load}, and impedance, Z_{load} (= V_{load}/I_{load}), are plotted as functions of time. Also shown is the radiation signal from a *pin* diode detector located at 10° relative to the diode axis. ASTERIX couples efficiently to this diode, producing a 34-ns (FWHM) radiation pulse and a dose at 1 m of 20 rad(Si). The electrical behavior of the diode is analyzed with the rod-pinch-diode model described in Ref. [2]. This model generates a current waveform, I_{model}, from the measured diode voltage by combining the space-charge-limited current and the critical current. It includes formation times and expansion velocities to account phenomenologically for anode- and cathode-plasma expansion, as well as the ion formation time and the scale factor, α, in the critical-current formula, which is a function of R_C/R_A in the presence of ions. The result of this analysis for shot 6400 with $\alpha = 2.0$ and an anode plasma velocity of 1.0 cm/μs is given in Fig. 2 (dashed line). For shots with acceptable impedance [8,9], the model current reproduces the magnitude and general shape of the measured current using values of the model parameters consistent with previous analyses at lower voltage [2]. Anode expansion velocities are 0.7 to 1.0 cm/μs for 2-mm diam rods, 1.0 to 1.5 cm/μs for 1-mm diam rods, and 2.0 cm/μs for a 0.5-mm diam rod. Increased energy deposition per unit volume in the smaller diameter rods may account for the larger expansion velocities. A more detailed discussion and analysis of the diode electrical behavior can be found in Refs. [9,10].

The dose/charge in the forward direction is plotted as a function of the characteristic load voltage (8-ns average of that parameter about the maximum of this radiation signal [8]) in Fig. 3 for shots with acceptable impedance [8,9]. The charge, Q, is determined by integrating I_{load} until the time after peak radiation when the *pin* diode signal has decreased to 5% of its maximum value. Results from Gamble II [3] for 1-mm-diam anodes are included in Fig. 3. Doses as large as 20 rad(Si) for a 2-mm-diam anode and 16 rad(Si) for a 1-mm-diam anode are achieved at the highest voltages on ASTERIX, compared with 2.5 rad(Si) for the lower voltage on Gamble II. Power-law fits to the 2-mm-diam (ASTERIX only) and 1-mm-diam (ASTERIX + Gamble II) data suggest that the dose/Q scales with V_{load} to the 1.3 to 1.6 power. At higher voltages, the 2-mm-diam anode is a more efficient radiation converter than the 1-mm-diam anode. The smaller dose/Q for the 1-mm-diam anode at higher voltage may be a result of the rod becoming sub-range to electrons, especially in the taper.

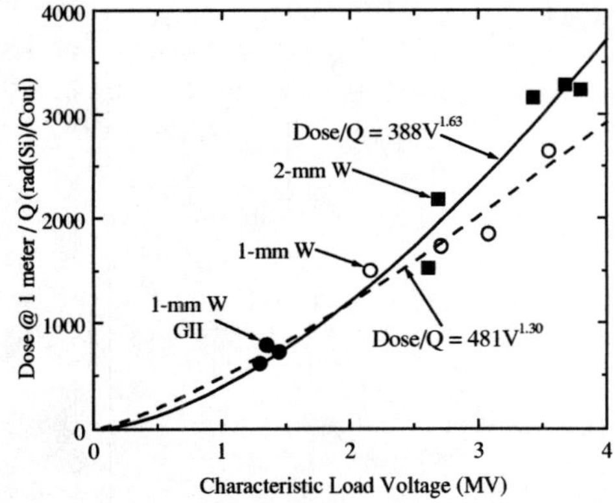

FIGURE 3. Normalized dose as a function of with the characteristic voltage (Fig. from Ref. [8]).

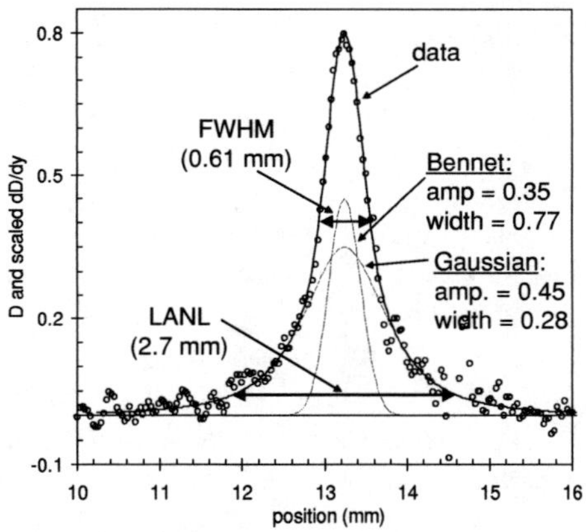

FIGURE 4. Fit to measured LSF for composite rod pinch using Bennet and Gaussian distributions.

One measure of the source diameter is the FWHM of the LSF. The central feature of the radial distribution dominates this measure. The average value of the FWHM source diameter is 0.9 ± 0.1 mm for a 1-mm-diam rod (averaged over seven acceptable-impedance shots) and 1.4 ± 0.1 mm for a 2-mm-diam rod (averaged over five acceptable-impedance shots). A more common measure of source diameter is determined from the wave number at the 50% point of the modulation transfer function (Fourier transform of the LSF), the so-called Los Alamos definition. This measure incorporates the wings of the source distribution and results in a larger diameter. Average values of the LANL source diameter are nearly twice the FWHM for 1-mm and 2-mm-diam rods. Both measures of source diameter increase with anode diameter, but somewhat less than one-to-one, similar to the behavior observed on

Gamble II [3]. For fixed anode diameter, the source diameter is independent of characteristic voltage from 2.2 to 3.8 MV.

The composite diode is designed to avoid the sensitivity of impedance decay to AK gap ($R_C - R_A$) for small (≤1 mm) diameter anodes. A larger diameter, low-Z anode extends through the cathode while keeping R_C/R_A ~ constant. Rapid and efficient e-beam propagation along this anode from the cathode to a smaller diameter, higher-Z tip is relied on to maintain a small source size and high dose. With larger R_A, the specific energy deposition on the anode surface and the fractional change in R_A from plasma expansion are reduced. The measured electrical behavior of the composite diode with a 0.5-mm diam tip is acceptable. The on-axis dose at 1 m is 9 rad(Si) (shot 6406), 5.4x larger than the average dose measured for a standard, 0.5-mm-diam-anode, rod-pinch that suffers from rapid impedance decay. However, this dose is 1.8x smaller than the dose from a standard, 1-mm-diam-anode rod-pinch, suggesting a non-optimum diode geometry and/or reduced efficiency with smaller diameter converters. The LSF for a composite diode (shot 6406) is shown in Fig. 4 (circles). The FWHM of the LSF is 0.61 mm, close to the tungsten-tip diameter (0.5 mm) and nearly 5x smaller that the LANL source size of 2.7 mm, which is close to the carbon-anode diameter (3.2 mm). A good fit to the LSF is obtained by summing a Bennet distribution and a Gaussian distribution of higher amplitude and smaller width (see Fig. 4). The small central feature of the LSF can have important applications to high-resolution radiography.

The largest figure of merit (dose @ 1 m/[LANL source diameter]^2) is 6.2 rad(Si)/mm^2, achieved with a 1-mm-diam tapered rod and a characteristic voltage of 3.6 MV. The results of this investigation indicate that the rod-pinch can be a very useful source for pulsed radiography at voltages up to 4 MV.. Additional experiments are in progress on ASTERIX to evaluate the rod-pinch at 6 MV.

The authors thank Dr. J. F. Leon for suggesting this collaboration. They also thank him and Drs. J. Maenchen (SNL) and R. D. Fulton (LANL) for supporting this work.

*Work supported by CEG, LANL, LLNL, and SNL.

REFERENCES

1. R.A. Mahaffey, et al., Appl. Phys. Lett. **33**, 795 (1978).
2. G. Cooperstein, et al., Phys. Plasmas **8**, 4618 (2001).
3. R.J. Commisso, et al., IEEE Trans. Plasma Sci. **30**, 338 (2002).
4. P.R. Menge, et al., *Proc. 2001 Pulsed Power Plasma Science Conference*, Las Vegas, NV (2001), IEEE Cat. No. 01CH37251, p. 454.
5. P.R. Menge, et al., *Conference Record, 27th International Conference on Plasma Science*, New Orleans, LA (2000), IEEE Cat. No. 00CH37087, p. 232.
6. S.B. Swanekamp, et al., Phys. Plasmas **7**, 5214 (2000).
7. D.V. Rose, et al., J. Appl. Phys. **91**, 3328 (2002).
8. F. Bayol, et al., Proc. *2001 Pulsed Power Plasma Science Conference*, Las Vegas, NV (2001), IEEE Cat. No. 01CH37251, p. 450.
9. F.C. Young, et al, accepted for pub, Phys. Plasmas.
10. R.J. Allen, et al. *Proc. Pulsed Power Plasma Science Conference*, Las Vegas, NV (2001), IEEE Cat. No. 01CH37251, p. 446.

Non-Laminar Flow Model for the Impedance of a Rod-Pinch Diode

Paul F. Ottinger,[a] Joseph W. Schumer,[a] Sean D. Strasburg,[a,1] Stephen B. Swanekamp,[b] and Bryan V. Oliver[c]

[a]*Plasma Physics Division, Naval Research Laboratory, Washington, DC 20375*
[b]*JAYCOR, McLean, VA 22102*
[c]*Mission Research Corporation, Albuquerque, NM 87110*

Abstract. A previous laminar flow model for the rod-pinch diode is extended to include a transverse pressure term to study the effects of non-laminar flow. The non-laminar nature of the flow has a significant impact on the diode impedance. Results show that the introduction of the transverse pressure decreases the diode impedance predicted by the model bringing it into better agreement with experimental data.

The rod-pinch diode[1,2] is a cylindrical, pinched-beam diode that is being developed as a radiography source[3]. The diode consists of a small-diameter anode rod of radius r_a that extends a distance l through and beyond the hole of an annular cathode of radius r_c. The diode has been operated at 1 to 2 MV with an impedance of 20 to 50 Ω, a FWHM pulse width of 20 to 50 ns, and an anode radius as small as 0.25 mm[2,4]. The diode is designed to run at critical current so that electrons emitted from the cathode flow axially along the anode rod in a sheath extending from r_c to r_s and pinch radially onto the rod at its tip. Typically, ion emission from the anode is required for propagation of the pinch along the rod. In order to assure that a given diode will be properly designed to run at critical current requires a detailed knowledge of the diode impedance characteristics. A laminar flow model[5] has been developed recently to describe the rod-pinch diode. Although this model provides considerable insight into diode operation, PIC simulations show that the electron flow is not laminar (i.e., their orbits have large radial excursions as they flow axially along the rod)[2,6,7]. Here, the earlier model is extended to include transverse pressure in the electron flow in order to study its effect on the diode impedance.

In deriving the equations, quasi-equilibrium (i.e., $\partial/\partial t = 0$) and azimuthal symmetry (i.e., $\partial/\partial\theta = 0$) are assumed, and axial variations are assumed to be negligible (i.e., $\partial/\partial z << \partial/\partial r$). The equations for the fields and electron fluid are

[1] National Research Council Research Associate at NRL.

CP650, *Beams 2002: 14th International Conference on High-Power Particle Beams*, edited by T. A. Mehlhorn and M. A. Sweeney

$$\nabla \cdot \mathbf{E} = -\frac{1}{r}\frac{\partial}{\partial r}\left(r\frac{\partial \phi}{\partial r} \right) = 4\pi e(Zn_i - n_e) \quad , \tag{1}$$

$$(\nabla \times \mathbf{B})_z = \frac{1}{r}\frac{\partial}{\partial r}(rB) = -4\pi e n_e v_e / c \quad , \tag{2}$$

$$0 = -en_e\left(-\frac{\partial \phi}{\partial r} - \frac{v_e B}{c} \right) - \frac{1}{r}\frac{\partial}{\partial r}(rP_{rr}) + \frac{P_{\theta\theta}}{r} \quad , \tag{3}$$

$$\gamma = 1 + \frac{e\phi}{mc^2} = \left(1 - \frac{v_e^2}{c^2} \right)^{-1/2} \quad , \tag{4}$$

where **E** and **B** are the electric and magnetic fields, ϕ is the potential, n_i and n_e are the ion and electron densities, e and m are the electron charge and mass, v_e is the axial electron fluid velocity, γ is the relativistic factor for the electron fluid, Z is the ion charge state, and P_{rr} and $P_{\theta\theta}$ are the radial and azimuthal diagonal terms from the electron pressure tensor. In writing Eq. (4), it is assumed that the energy associated with the beam pressure is small compared to the energy associated with the flow. The boundary conditions on the potential are $\phi(r_a) = \phi_a$, $\phi(r_c) = 0$, and $\phi'(r_a) = \phi'(r_c) = 0$, where space charge limited emission is assumed at both the anode and cathode and $'$ denotes differentiation with respect to r.

In order to close this set of equations, expressions for the ion density and electron pressure are required. Ions emitted from the anode are assumed to flow radially outward. Consequently, charge continuity and energy conservation provide the needed ion equations

$$eZn_i v_i = \frac{r_c j_{ic}}{r} \quad , \tag{5}$$

$$\frac{Mv_i^2}{2} = eZ(\phi_a - \phi) \quad , \tag{6}$$

where v_i is the radial ion velocity, M is the ion mass, $j_{ic} = eZn_i v_{ic}$ is the ion current density at the cathode, and $v_{ic} = (2Ze\phi_a/M)^{1/2}$ is the ion velocity at the cathode. For simplicity, it will be assumed that $P_{\theta\theta}$ can be neglected. The analysis could easily be generalized to include $P_{\theta\theta}$ by writing $P_\perp = \partial(rP_{rr})/\partial r + P_{\theta\theta}$ and following the same procedure as outlined below. The electron pressure is assumed here to have the form

$$P_{rr} = \begin{cases} \dfrac{mc^2 n_0}{r/r_c}\left(1-\dfrac{r}{r_c}\right)^2\left[p_2 + p_3\left(1-\dfrac{r}{r_c}\right) + p_4\left(1-\dfrac{r}{r_c}\right)^2\right] & \text{for } r_c \ge r \ge r_s \\ 0 & \text{for } r_s \ge r \ge r_a \end{cases} \quad , \tag{7}$$

where n_0 is defined below, and $p_4 = -p_3/x_s - p_2/x_s^2$ with $x_s = 1 - r_s/r_c$, so that $P_{rr}(r_c) = P_{rr}(r_s) = 0$ and that a Taylor expansion of the solution is easily obtained near the cathode. Taylor series expansions of the various functions are used for numerical convenience in starting the integration of the equations developed below at r_c and r_a. Note that p_2 must be much less than $\langle n_e v_e^2\rangle/n_0 c^2$ (and $p_3 \sim p_2$) so that the perpendicular energy associated with the pressure is small compared to the kinetic energy associated with the flow as assumed in Eq. (4). Again, the analysis can be generalized by using the same procedure outlined below with other forms for P_{rr}.

The equations are normalized using $\Phi = e\phi/mc^2$, $\gamma = 1 + \Phi$, $R = \omega_{pe}r/c$, $N_{e,i} = n_{e,i}/n_0$, $V_{e,i} = v_{e,i}/c$, and $b = B(R)\omega_{ce}/B_c\omega_{pe}$, where $\omega_{pe} = (4\pi e^2 n_0/m)^{1/2}$ is the electron plasma frequency, $\omega_{ce} = eB_c/mc$ is the electron cyclotron frequency, B_c is the magnetic field at the cathode, and n_0 is chosen so that $R_c = \omega_{pe}r_c/c = 1$, $R = r/r_c$, $R_a = r_a/r_c$,and $R_s = r_s/r_c$.

With the inclusion of the electron pressure term, the equation for γ becomes

$$\frac{1}{R}\frac{d}{dR}\left(R\frac{d\gamma}{dR}\right)=\begin{cases}\dfrac{\gamma}{\gamma^2-1}\left(\dfrac{d\gamma}{dR}\right)^2+\dfrac{(\gamma^2-1)K}{R(\gamma_a-\gamma)^{1/2}}+G & \text{for } R_c \ge R \ge R_s \\ -\dfrac{K}{R(\gamma_a-\gamma)^{1/2}} & \text{for } R_s \ge R \ge R_a\end{cases}, \quad (8)$$

where $K = (M/2mZ)^{1/2}R_cJ_{ic}$ is a constant, $J_{ic} = j_{ic}/en_0c$ is the normalized ion current density at the cathode, $\gamma_a = 1 + \Phi_a$,

$$G(R)=\frac{\gamma(\gamma^2-1)^{1/2}}{R}\frac{d}{dR}\left[\frac{\gamma}{N_e(\gamma^2-1)^{1/2}}\frac{d}{dR}(RP)\right], \quad (9)$$

$$N_e(R)=\frac{\gamma^2}{R(\gamma^2-1)}\frac{d\gamma}{dR}-\frac{\gamma}{(\gamma^2-1)^2}\left(\frac{d\gamma}{dR}\right)^2+\frac{\gamma^2}{(\gamma^2-1)}\frac{d^2\gamma}{dR^2}-\frac{G}{(\gamma^2-1)}, \quad (10)$$

and $P(R) = P_{rr}(R)/(mc^2n_0)$ is the normalized electron pressure. Aside from the additional pressure term, Eq. (8) is the same equation as derived previously in Ref. 5.

The result is a two-point boundary value problem which can be solved using a shooting technique[8]. The shooting technique finds values for the eigenvalues K and R_s by matching γ and γ' at R_s from solutions of Eq. (8) integrated from the anode and cathode toward R_s. Taylor series expansions are used for numerical convenience to start the solutions at the anode and cathode. Iteration of the solution is also required to obtain a solution that is consistent with the desired pressure profile. First, Eq. (8) is solved for $\gamma_0(R)$ with $P = 0$ (and, therefore, $G_0 = 0$) using the shooting technique. Using $\gamma = \gamma_0(R)$ and $G = G_0 = 0$, $N_{e0}(R)$ is obtained from Eq. (10). These solutions [i.e., $\gamma_0(R)$ and $N_{e0}(R)$] are substituted back into Eq. (9) to generate $G_1(R)$ for a given P(R) (here Eq. (7) is used for P(R)). Using the shooting technique again, Eq. (8) is resolved with $G = G_1(R)$ to find $\gamma_1(R)$. Then $N_{e1}(R)$ is found from Eq. (10) using $\gamma = \gamma_1(R)$ and $G = G_1(R)$. Finally, $\gamma_1(R)$ and $N_{e1}(R)$ are used in Eq. (9) to generate $G_2(R)$. The process is repeated until the solutions converge.

Solutions have been found for various values of p_2. The ion current and sheath position are little changed as p_2 increases, but the total current does increase so that the magnetic pressure can balance the increasing electron pressure. Results are shown in Figs. 1 and 2 for $p_2 = 0.6$ which provide a reasonable match to results from PIC simulations performed with MAGIC[2] and LSP[5,7]. The rod-pinch diode has been shown to operate in critical current flow[2] so that the diode current $I = \alpha I_{crit}$, where $I_{crit}(kA) = 8.5(\gamma_a^2 - 1)^{1/2}/\ln(r_c/r_a)$. The factor α is plotted in Fig. 1 as a function of r_c/r_a for various cases. Results from the laminar flow model ($p_2 = 0$) fall significantly below the PIC simulation results, however, results from the non-laminar flow model developed here are in reasonable agreement with the PIC simulation results when $p_2 = 0.6$. The diode impedance, ϕ_a/I, is plotted as function of the diode gap $D = r_c - r_a$ in Fig. 2. Reasonable agreement with data taken on SABRE[5] is obtained when $p_2 = 0.6$ is used in the non-laminar flow model. Similar agreement is found for data taken on Gamble II[2].

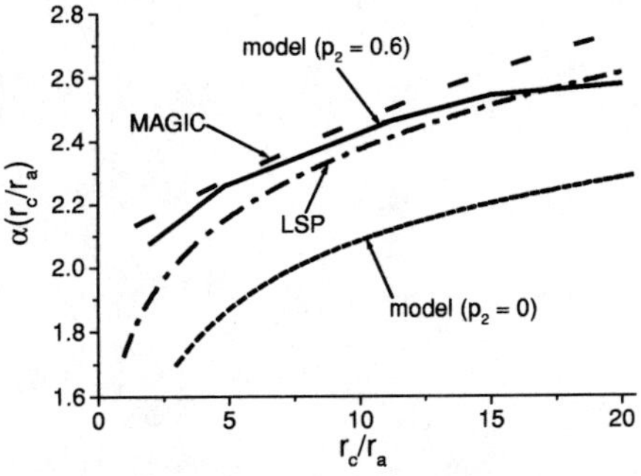

FIGURE 1. Plot of α versus

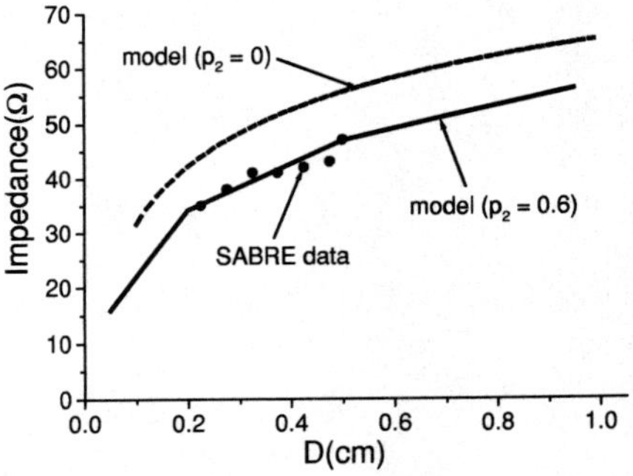

FIGURE 2. Plot of diode impedance versus

These results show that the introduction of the transverse pressure decreases the diode impedance predicted by the model bringing it into agreement with experimental data. For the cases studied, $\langle n_e v_e^2 \rangle / n_0 c^2 \sim 5$ at the radial location where the pressure is maximum, so that the condition on p_2 stated above becomes $p_2 \ll 5$. Thus, the assumption used to write Eq. (4) that the energy associated with the beam pressure is small compared to the energy associated with the flow is satisfied with p_2 as large as 0.6 [implying $v_\perp/v_e \sim (0.6/5)^{1/2} \sim 0.35$]. Future work will investigate a self-consistent, relativistic Vlasov approach that will eliminate the need for this assumption.

ACKNOWLEDGEMENTS

Work supported by the US DOE through SNL, LANL, and LLNL.

REFERENCES

1. R.A. Mahaffey, J. Golden, S.A. Goldstein, and G. Cooperstein, *Appl. Phys. Lett.* **33**, 795-797 (1978).
2. G. Cooperstein, et al., *Phys. Plasmas* **8**, 4618-4636 (2001).
3. R.J. Commisso, et al., *IEEE Trans. Plasma Sci.* **30**, 338-351 (2002).
4. P.R. Menge, et al., *Proceedings of the 13th International Pulsed Power Conference*, (Las Vegas, NV, June 2001), pp. 454-457.
5. B.V. Oliver, T.C. Genoni, D.V. Rose, and D.R. Welch, *Proceedings of the 13th International Pulsed Power Conference*, (Las Vegas, NV, June 2001), pp. 458-461.
6. S.B. Swanekamp, et al., *Phys. Plasmas* **7**, 5214-5222 (2000).
7. D.V. Rose, et al., 2000 IEEE International Conference on Plasma Science, (New Orleans, LA, June 2000), p. 232.
8. S.M. Roberts and J.S. Shipman, *Two-Point Boundary Value Problems: Shooting Methods* (American Elsevier, New York, 1972).

The Plasma-Filled Rod-Pinch Diode: A New Technique to Concentrate MeV Electron Beams to Ultra-High Power and Energy Densities

B. V. Weber,[a] G. Cooperstein,[a] D. D. Hinshelwood,[a] D. Mosher,[a] J. W. Schumer,[a] S. J. Stephanakis,[a] S. B. Strasburg,[b] S. B. Swanekamp,[c] F. C. Young[c]

[a]*Plasma Physics Division, Naval Research Laboratory, Washington, DC 20375*
[b]*NRL National Research Council Associate*
[c]*Titan/Jaycor, McLean, VA, 22102*

Abstract. The plasma-filled rod-pinch diode produces an MeV electron beam with ultra-high power density (~30 TW/cm^2) and energy density (600 MJ/cm^3) at the tip of a tapered tungsten rod. Current flows through an injected plasma between the grounded cathode and the high-voltage anode rod. Once J×B forces translate the current to the tip, an e-beam is drawn across a vacuum gap that opens there. Intense e-beam deposition on the anode tip creates an ultra-bright x-ray source. The x-ray source distribution has a very narrow central part (0.4 mm FWHM) with large-scale (few mm) wings, attributed to expanding rod material. Despite the wings, resolution target data show the ability to resolve (with contrast ratio 1.2) 2.5 line-pairs per mm at 5× magnification, the same as would be obtained from a 0.42-mm diam, uniformly emitting disk source.

The plasma-filled rod-pinch (PFRP) diode [1] consists of a small diameter (1 mm or less) rod anode that extends through a circular hole in a planar cathode as shown in Fig. 1. The rod is connected to the high-voltage center conductor of the Gamble II generator (1.5 MV, 2Ω, 60 ns). A plasma is injected on the downstream side of the cathode using six coaxial plasma guns, about 1 μs prior to the start of the generator current. The plasma makes a short-circuit connection between the cathode and the rod, so that current flows with negligible voltage, as shown in Fig. 2 for shot 7956. After a low-impedance phase that lasts for 50 ns, the voltage increases to 1 MV and an electron beam forms at the rod tip. The current reaches 600 kA and the total coupled energy is 33 kJ. The x-ray pulse width is 60 ns. This shot produced an x-ray dose of 20 rad at 1 m through a 3-mm aluminum filter.

The x-ray emission is primarily from the last 2 mm of the tapered rod as indicated by the time-integrated pinhole camera image in Fig. 2a. The e-

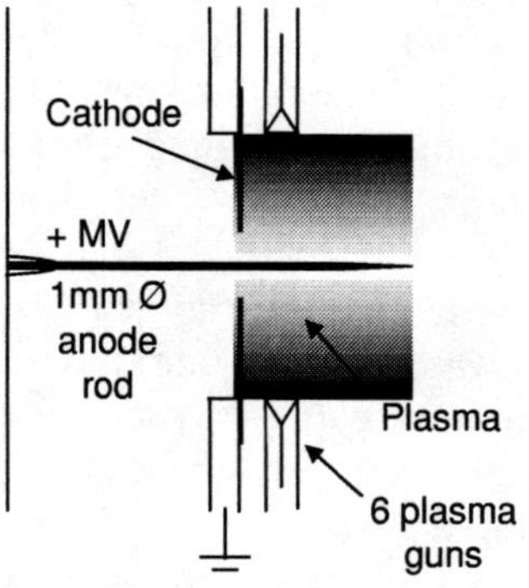

FIGURE 1. Plasma-filled rod-pinch-diode setup.

CP650, *Beams 2002: 14th International Conference on High-Power Particle Beams,*
edited by T. A. Mehlhorn and M. A. Sweeney

beam power and energy densities are estimated using a Cyltran [2] analysis assuming radially-incident electrons along the last 2 mm of the conical tungsten tip. The electron energy distribution is derived from the measured current and voltage waveforms. This analysis indicates 13 kJ of e-beam energy is required to produce the measured dose. If this energy is absorbed in the conical tip, the energy density is 650 MJ/cm^3 and the average power density is 32 TW/cm^2. Expansion during the pulse (suggested in Fig. 2a) will reduce these values, but the potential for creating high energy-density plasma conditions is clear.

A simple picture for the operation of the PFRP diode is shown in Fig. 3. The current sweeps up the plasma as it moves toward the rod tip. When the current moves past the tip, a vacuum gap forms, perhaps by an erosion mechanism, in the rarefied plasma left behind. An intense e-beam (indicated by arrows in Fig. 3) forms at the tip and the strong magnetic field (several MG) along the rod restricts the beam to the vicinity of the tip.

The interferogram [3] in Fig. 4 was taken during the low-impedance phase, prior to the start of x-ray emission. The high-density current sheath is similar to the picture on the left in Fig. 3. Similar phenomena are observed in MHD

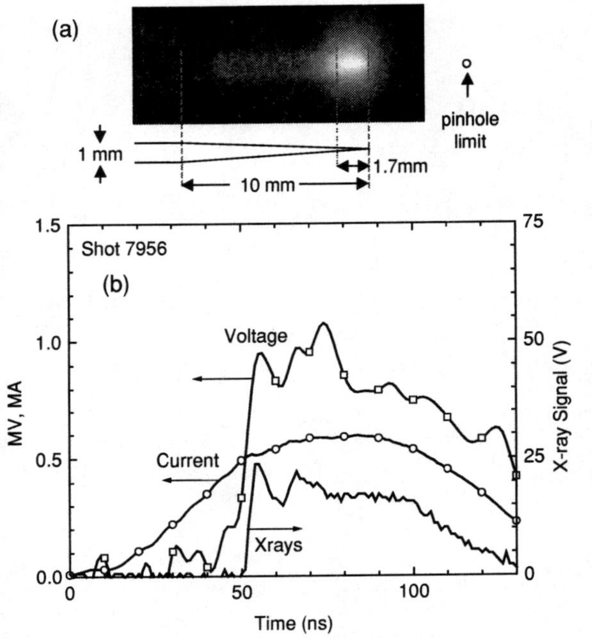

FIGURE 2. (a) X-ray pinhole camera image and (b) data from shot 7956 on Gamble II.

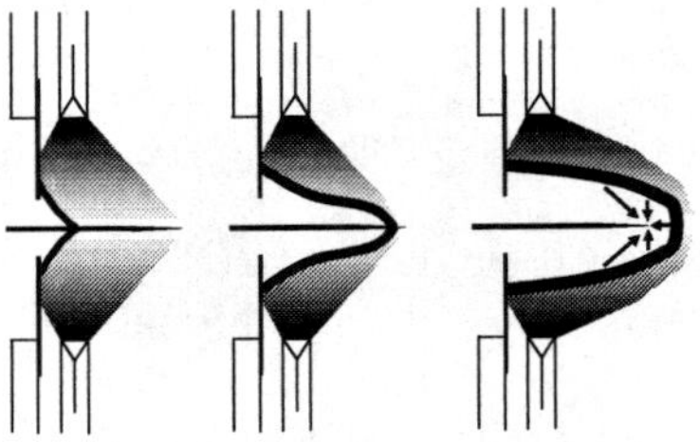

FIGURE 3. Simple picture of PFRP operation. The dark line represents the current sheath and the shaded area represents plasma.

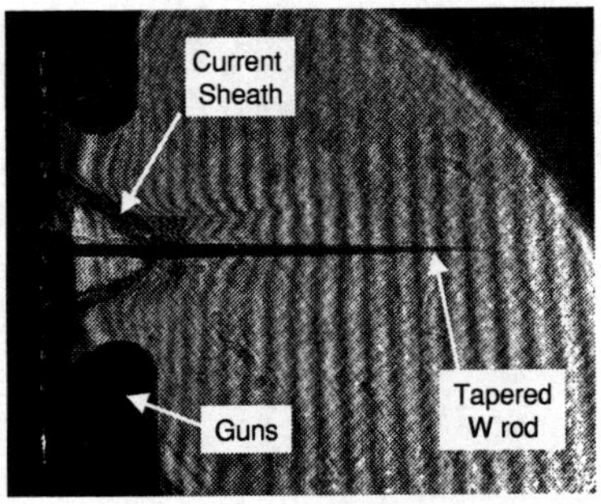

FIGURE 4. Interferogram during the low-impedance phase showing current sheath propagating toward the tip.

modeling.[4] Further measurements and modeling will be pursued to investigate PFRP operation through the rest of the low-impedance and high-impedance phases. If this picture is confirmed, it shows an interesting difference between plasma-filled and vacuum rod-pinch diodes. For the vacuum rod-pinch diode, the electron beam starts at the cathode and propagates toward the tip.[5] For the PFRP diode, the beam starts at the tip, potentially reducing x-ray emission from the rest of the rod and resulting in a smaller source diameter.

The shape of the x-ray source is measured using the rolled-edge technique shown in Fig. 5. The exposure on the film in optical-density (OD) units, projected to the source is the edge-spread function and is plotted against the left axis in Fig. 5. The derivative is the line-spread function (LS) and is plotted against the right axis. The LS has a narrow (0.4-mm FWHM) central feature, and low-amplitude "wings" that extend about 1 mm on either side. The central feature is what would be produced by a solid uniformly-radiating cone of 0.5-mm base diameter (indicated by the solid rod tip in Fig. 5). The wings could be from x-rays produced by beam interactions in expanding tip material, indicated by gray shading in Fig. 5. Two other possibilities for the wings, radiation from electrons striking the rod holder and scattered radiation have been ruled out. Radiation from the holder is minimized by using a carbon holder. Lead bricks are used to block radiation from the holder to measure the LS with the rolled edge moved 40° off the rod axis. This LS shows the same wings, demonstrating that radiation from the holder is not significant. Gamble II shots with low-current vacuum (no plasma) rod-pinch diodes [6] have a similar scattering environment, but deposit an order-of-magnitude less energy at the tip. The LS for these shots do not have large wings although scattering would be a similar proportion of the LS.

For radiographic applications, x-ray sources are often characterized by a source size derived from the LS. One example is the "LANL" source size defined as the diameter of a uniformly emitting circular disk that has the same 50% point on the modulation transfer function (MTF) of the LS. For the LS in Fig. 5, the LANL source size is 1.7 mm, much greater than the 0.4-mm FWHM.

The resolution capability of the PFRP x-ray source is much better than might be expected based on the LANL source size, due to the narrow central feature in the LS. This is demonstrated using a resolution bar target consisting of sets of three slots machined in 3.5-mm thick tungsten. The slot width is equal to the thickness of the bar

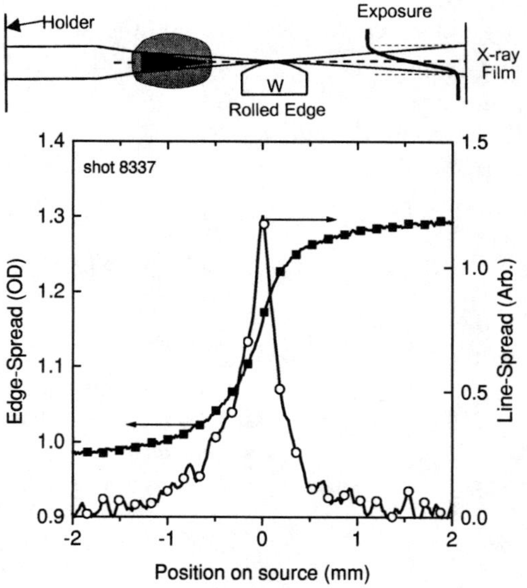

FIGURE 5. Rolled edge measurement of x-ray source distribution (above) and source distributions (below).

between the slots. The smallest slots (0.2 mm) correspond to 2.5 line-pairs per mm (lp/mm). The resolution target was aligned with the rod axis and imaged at 5× magnification to produce the film image shown in Fig. 6. Line-outs of the measured film density are also shown. The slots at 2.5 lp/mm are clearly resolved. Measurements with vacuum rod-pinch sources comparably resolve only the slots at 1.5 lp/mm.

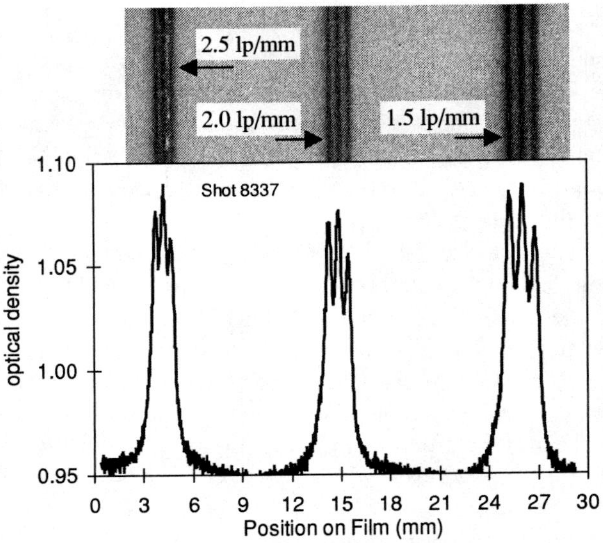

FIGURE 6. Film image (above) and line-out data (below) for a resolution target with 5 × magnification.

The contrast (peak-to-valley ratio) is about 1.2 for the 2.5 lp/mm slots. This contrast is the same as would be obtained from a 0.42-mm diam uniformly emitting disk. For a larger diameter source, the three lines merge into two or one. The contrast increases as the slot width increases, but not as much as would occur for a 0.42-mm diam disk, an effect caused by the large wings in the LS. For radiographic applications where this 1.2 contrast is acceptable, the PFRP may be the brightest MV source available: 20 rad @ 1m through 3 mm aluminum (or 10 rad through 2.5 cm aluminum), and a source size equivalent to a disk with less than 0.5 mm diameter (but with wings).

Other applications for the PFRP include high energy-density plasma production and warm x-ray production from reflexing bremsstrahlung converters. These applications and the basic science of PFRP operation will be the subject of future research.

REFERENCES

Work supported by ONR.

1. B. V. Weber, *et al.*, in Proc. PPPS-2001, p. 462.
2. ORNL Report CCC-467, ITS 3.0, November, 1999.
3. Holographic interferometer designed and constructed by HY-Tech Research Corp., Radford, VA.
4 S. Strasburg, *et al.*, these Proceedings.
5. G. Cooperstein, *et al.*, *Phys. Plasmas* **8**, 4618 (2001).
6. R.J. Commisso, *et al.*, *IEEE Trans. Plasma Sci.* **30**, 338 (2002).

Rod-Pinch Diode Currents in the Magnetically Limited Regime

Victor V. Ryzhov[a], Svyatoslav Ya. Belomyttsev [a], Alexander V. Kirikov [a], Igor Yu.Turchanovskii [a] and Vladimir P. Tarakanov [b]

[a]Institute of High Current Electronics, 4 Akademicheskii Ave., Tomsk, 634055, Russia
[b]High Energy Density Research Investigation Center, Moscow, Russia

Abstract. In this work, relations for the Rod-Pinch Diode (RPD) currents, which should hold true for the magnetically limited (ML) regime, have been derived based on the general criterion of magnetic insulation of an electron beam in a cylindrical diode. It has been shown that to appreciate the physics of the processes in question the external currents, which play a leading part in the ML regime, should be distinguished. Simple formulae have been derived, which, given the relation between the currents, make possible prediction of the state of the electron beam in the diode and calculations of the scaling factor α for the critical current.

CONDITIONS OF MAGNETIC INSULATION FOR RPD

In [1], critical conditions have been found for magnetic insulation of an electron beam in a cylindrical diode. According to this work, the electron beam I_b drifts out of the gap, when

$$I_b + 2I_1 = 2I_C \quad (1)$$

Here, I_1 is the external current, and I_C is the critical current in the one-electron approximation. In the case where the beam is pinched, reaching the end of the anode (the steady-state regime), the condition

$$I_b + 2I_1 \geq 2I_C \quad (2)$$

should be fulfilled. The beam goes out of the pinched state if

$$I_b + 2I_1 \leq 2I_C \quad (3)$$

For every diode geometry, the application of conditions (1-3) requires the proper division of the diode current I_D into the external current and the current in the gap. For any RPD, the gap currents are the currents of electrons and

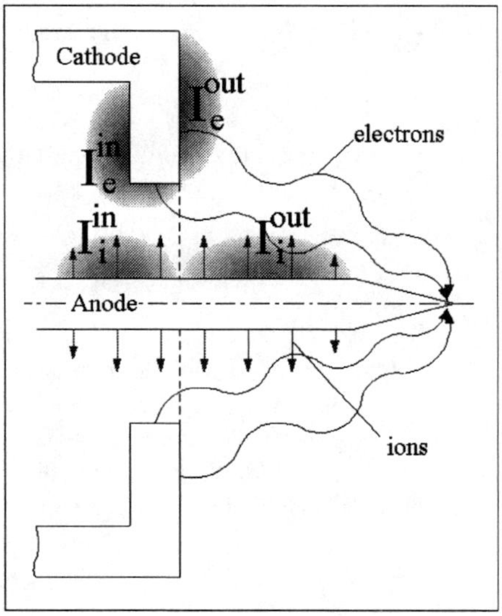

FIGURE 1. Schematic representation of the rod-pinch diode in the ML phase and division of the currents into the gap currents (in) and the external currents (out) relative to the plane of the lateral surface of the cathode shown by the dotted line.

[a] e-mail: ryzhov@to.hcei.tsc.ru

CP650, *Beams 2002: 14th International Conference on High-Power Particle Beams*,
edited by T. A. Mehlhorn and M. A. Sweeney

ions to the left of the right boundary of the cathode face. All current of electrons and ions emitted to the right of this boundary are external to the electron beam drifting out of the gap (Fig.1).

For a RPD in ML regime, conditions (1-3) therefore take the form

$$I_e^{in} + 2(I_e^{out} + I_i^{out}) = 2I_C, \tag{1RPD}$$

$$I_e^{in} + 2(I_e^{out} + I_i^{out}) > 2I_C, \tag{2RPD}$$

$$I_e^{in} + 2(I_e^{out} + I_i^{out}) \le 2I_C, \tag{3RPD}$$

SCALING FACTOR α FOR RPD

In the works performed at NRL [2,3], an empirical criterion is introduced, being taken as the condition of the RPD switching to the regime of magnetic insulation. According to this criterion, the RPD switches to the ML mode, once the diode current approaches the critical current I_{CRIT}, which is associated with the critical current in the one-electron approximation, I_C:

$$I_D = I_{CRIT} = \alpha I_C, \tag{4}$$

where α "is an empirical scaling factor for the critical current".

Let us consider the problem with respect to the values of α which can be taken on during the magnetic insulation of the beam in a RPD. Assume that when the beam leaves the gap conditions (1RPD) and the condition of magnetic insulation NRL (4) are fulfilled. By solving them simultaneously, for α we obtain

$$\alpha = (2 - (I_e^{out} + I_i^{out})/I_C), \quad \alpha \le 2. \tag{5}$$

At the instance the electron beam goes out of the gap, the external ion current is still low (the beam has not yet a chance to heat the anode and to ionize the desorbed gas). Therefore, $I_i^{out} << I_e^{out}$, and formula (5) takes the simple form

$$\alpha = (2 - I_e^{out}/I_C), \quad \alpha \le 2. \tag{6}$$

Whence it follows that at the instant the beam moves out of the gap (the initial moment of magnetic insulation) the factor α is smaller than two. The equality $\alpha = 2$ is found only in the case of self-magnetic insulation of the beam, when the influence of the electron flow from the lateral surface of the cathode and the ion current can be ignored (in the case where their sum is much smaller than the critical current).

For the stage where the beam is pinched, from the simultaneous solution of (2RPD) and (4) we derive

$$\alpha > (2 - (I_e^{out} + I_i^{out})/I_C) \tag{7}$$

For a steady-state pinch, the values of the currents $I_i^{out} >> I_e^{out}$ and one can get the simple formula

$$\alpha > (2 - I_i^{out}/I_C), \tag{8}$$

The current I_i^{out} can be rather high and therefore α for a steady-state pinch can be smaller than two as well as larger than this value.

In [2], experimental and theoretical studies have been conducted on the steady-state stage of the ML regime and the values of α taken on in this regime have been analyzed. The results of simulation have shown that for these regimes $\alpha_{Theory} > 2$, whereas from experimental data it follows that $\alpha_{EXP} < 2$. Relation (8) allows for both cases, but the discrepancy between the values of α means that the values of the currents I_e^{out} and I_i^{out} determined in simulation and, consequently, the conditions of modeling do not correspond to the experiment.

CALCULATIONS OF ML CHARACTERISTICS FOR RPD

To check the validity of relations (1-3RPD), the RPD operation in the modes studied experimentally on the Gamble II setup [2] has been simulated using the KARAT code [4]. Analysis of the calculation data and oscillograms of the RPD currents in the Gamble II setup has revealed that in the ML regime four characteristic moments can be distinguished. The particle flow distribution for these moments are depicted in FIG. 2.

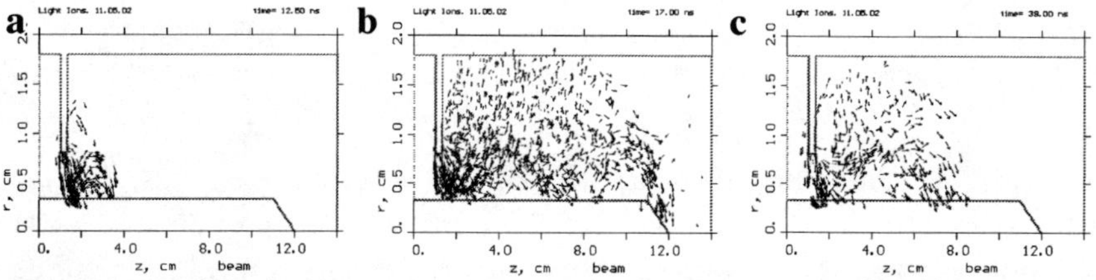

FIGURE 2. Electron pulse plot obtained from KARAT simulation for Gambel II rode –pinch diode short # 7402. a) $t = t_{Start}$, b) $t = t_{SS}$, and $t = t_{End}$ c) $t > t_{End}$

The start of operation in the ML regime, $t = t_{Start}$

The moment at which the ML regime becomes operative is easily determined from oscillograms of the electron current show in FIG. 3 (a characteristic peak of the current $t = t_{Start}$). At this moment, conditions (1RPD) should be fulfilled. The oscillograms and distribution of the diode current indicate that the ion current at this moment is still small, and the role of the external current is played by the electron current from the lateral surface. In this case, from (1RPD) we can obtain the approximated expression $I_e^{in} + 2I_e^{out} \approx 2I_C$. In so doing, to estimate the value of α it is possible to use formula (6) according to which $\alpha = 1.22$ in the case in question.

Transportation of the electron beam along the anode rod, $\Delta t_{tr} = (t_{SS}-t_{Start}))$

A detailed description of this stage can be found elsewhere [2]. At the moment in time t_{SS} condition (2RPD) is fulfilled in the form $I_e^{in} + 2I_i^{out} \geq 2I_C$; $\Delta t_{tr} \approx 10ns$

To estimate the value of α, one can use formula (8) according to which $\alpha > 1.5$. It agrees closely with $\alpha_{Exper} = 1.6$.

Steady-State Pinch, $\Delta t_{Pinch} = (t_{End} - t_{SS})$

The latter condition remains valid up to the moment in time t = t_{End} at which the curves of the time dependence of the magnetic cutoff current $I_S = 2I_C$ и $(I_e^{in} + 2I_i^{out})$ intersect. Calculations show that in the steady-state stage of the pinch, which takes the interval from $t = t_{SS}$ to $t = t_{end}$, the diode currents satisfy the condition $I_e^{in} + 2I_i^{out} > I_S$.

In the case under consideration, the pinch lifetime $\Delta t_{Pinch} = t_{END} - t_{SS} = 50ns$. Calculations have shown that during the ML regime the value of the coefficient α varies with time in the range $1.26 \le \alpha \le 1.5$.

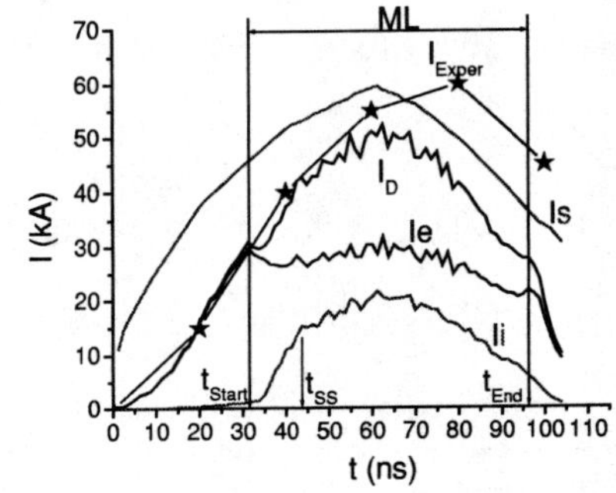

FIGURE 3. Time dependence of the electron I_e, ion I_i, and total diode current $I_D = I_e + I_i$ calculated by the KARAT code for Gamble II rode –pinch diode short # 7402. I_{exp} -experimental value of total diode current $I_S = 2I_C$.

Return to the gap, $t > t_{END}$

Because the electron current remains invariant and the ion current decreases, a decrease of the latter to a value $I_i^{out} = I_i(t_{SS})$ leads to violation of the condition of magnetic self-insulation of the beam, (2RPD), and when condition (3RPD) is fulfilled, the beam starts going back toward the gap.

REFERENCES

1. Belomyttsev S. Ya., Ryzhov V.V., *Tech. Phys.Lett.*, v.**27**, No 7, 608-610 (2001).
2. G. Cooperstein, J.R. Boller, R.J. Commisso, et. al., Physics of Plasmas, vol. 8, pp. 4618-4636, (2001)
3. S.B. Swanekamp, R.J. Commisso, G. Cooperstein, P.F. Ottinger, W. Schumer, Physics of Plasmas, vol. 7, №12, pp. 5214-5222, (2000)
4. Tarakanov V.P. "User's Manual for Code KARAT". Springfield: BRA, 1992.

An Electron-Beam-Heating Model for the Gamble II Rod Pinch

David Mosher[a], Joseph Schumer[a], David Hinshelwood[a], Bruce Weber[a], Stavros Stephanakis[a], Stephen Swanekamp[b], and Frank Young[b]

[a]*Plasma Physics Division, Naval Research Laboratory, Washington, DC 20375*
[b]*Titan/Jaycor, McLean, VA 22102*

Abstract. The rod-pinch diode concentrates electron deposition onto the tip of a high-atomic-number, mm-dia. anode rod to create an ultra-bright x-ray source for multi-MV radiography. Here, a technique is presented whereby line-spread functions acquired on-axis and at 90° to the rod are used to determine the electron-deposition distribution. Results show that the smaller measured on-axis spot size for heated rods on Gamble II is due to pinching closer to the tapered tip. For a diode power of 6×10^{10} W, peak electron heating of 1×10^{14} W/cm^3 is calculated. MHD calculations of the e-beam-heated rod response agree with Schlieren measurements of plasma expansion.

The rod-pinch diode [1] concentrates electron-beam deposition onto the tip of a high-atomic-number, mm-dia. anode rod in order to create a very bright x-ray source for radiography applications in the 1- to 4-MeV endpoint-energy regime [2, 3]. This intense beam heating produces expanding plasmas that can, in some cases, increase the diameter of the radiographic source and degrade electrical coupling. Schlieren images of such plasmas during and after heating [4] have been compared to predictions of preliminary 2D MHD simulations that incorporate a simple estimate of electron volumetric heating $P_e(r,z)$ based on electrical measurements and pinhole photographs. Measurements and computations show good general agreement, but disagree in the detailed shape of the expansion envelope. For load configurations with small anode-cathode gaps, tapered rods can provide spot sizes comparable to that of smaller-diameter blunt-end rods with better electrical coupling [2], thereby providing a higher figure of merit: dose/(spot-size)^2. Also, heated tapered tungsten rods provide smaller spot sizes than similar cold rods.[5] The present work was initiated to understand electron deposition in these tapered-rod pinches, to improve electron-deposition modeling for MHD, and to provide a benchmark for PIC/ITS simulations.

Here, line spreads, the derivative of the edge spread ES from rolled-edge images [2], simultaneously acquired on-axis (LS_y) and at 90 degrees to the rod (LS_z), are used to refine $P_e(r,z)$. This distribution is determined by synthesizing LS_y from CYLTRAN computations [6] of the emitted x-ray distribution and shifting the measured LS_z along the rod taper to match the measured LS_y.

The variation of radiation intensity along the rod for cold and hot 1-mm-dia. tapered tungsten rods is given by the plots of $LS_z(z)$ shown in Fig. 1. The plots are arbitrarily shifted to align the rising edges at 18 mm. The precise location of each LS_z on the

CP650, *Beams 2002: 14th International Conference on High-Power Particle Beams*, edited by T. A. Mehlhorn and M. A. Sweeney

taper is determined below. The apex of the taper is roughly at 21 mm, about 17 mm from the cathode. The taper starts at about 13 mm. For z < 11 mm, LS_z has an average value of about 2% of peak.

Incremental line spreads $\delta LS_y(y)$ are computed from CYLTRAN calculations of radiated power P(r) for rods of radius R. CYLTRAN used an incident electron-energy spectrum derived from Gamble II rod-pinch current and voltage traces (peak electron energy about 1.6 MeV), and an incident-electron angular distribution from tapered-rod-pinch PIC/ITS simulations [7]. These incremental line spreads are then calculated from

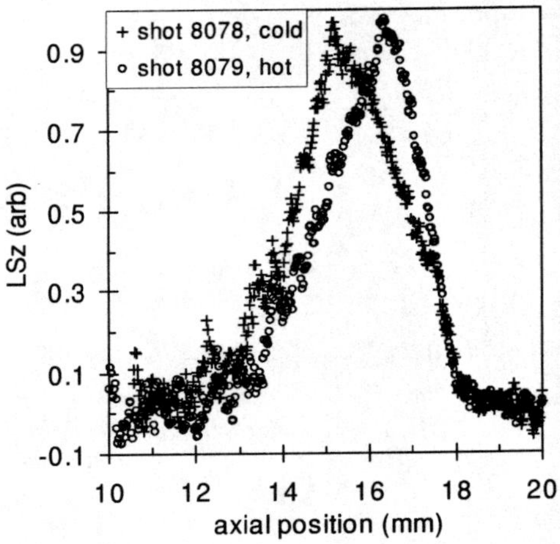

FIGURE 1. Axial line-spreads for cold and hot tapered tungsten rods on Gamble II.

$$\delta LS_y(y,R) = \int_{-X(y)}^{X(y)} P\left(\sqrt{x^2+y^2}\right)dx \;\; ; \;\; X = \sqrt{R^2-y^2} \; . \tag{1}$$

To synthesize the full line spread for the tapered rod, LS_z is located on the rod taper R(z), and $LS_y(y)$ is calculated from

$$LS_y(y) = \int_z dz \cdot \delta LS_y(y, R(z)) \cdot LS_z(z - z_s) \; . \tag{2}$$

Here, z = 0 is the apex of the taper, and z_s is the distance from the apex to where LS_z (smoothed to reduce scatter) rises from zero. The quantity z_s is varied to achieve a best fit to the measured LS_y. Results are shown in Figs 2 and 3 for cold and hot 1-mm-dia. rods. The best-fit middle value of z_s is assumed to represent the correct experimental

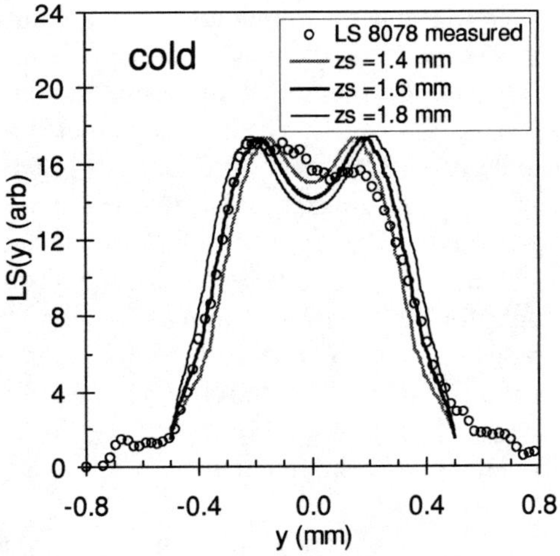

FIGURE 2. Measured cold-rod $LS_y(y)$ and synthesized for several values of z_s

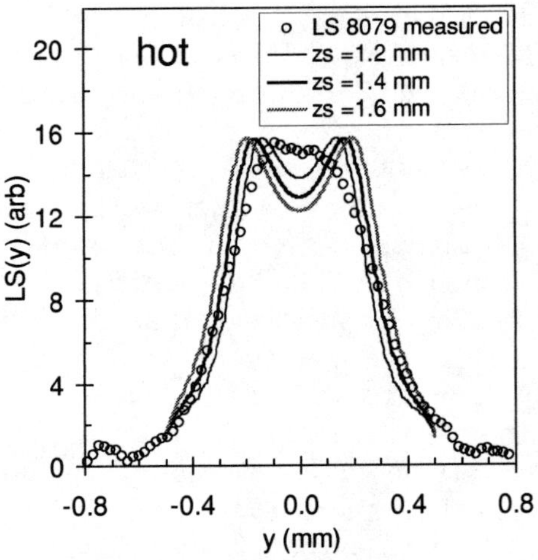

FIGURE 3. Measured hot-rod $LS_y(y)$ and synthesized for several values of z_s

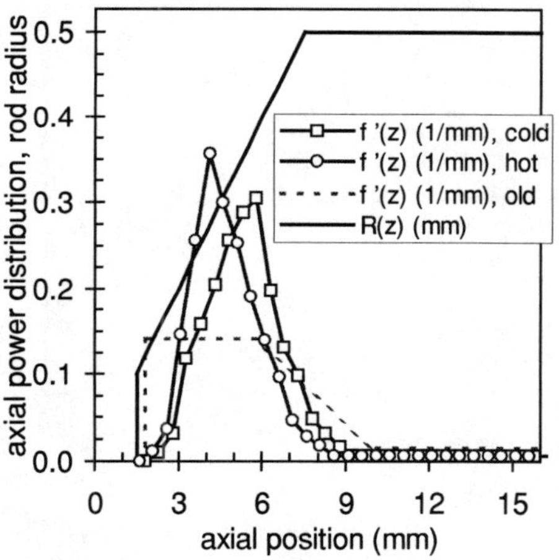

FIGURE 4. LS_z for cold and hot rods, and old MHD model overlaid on the rod taper.

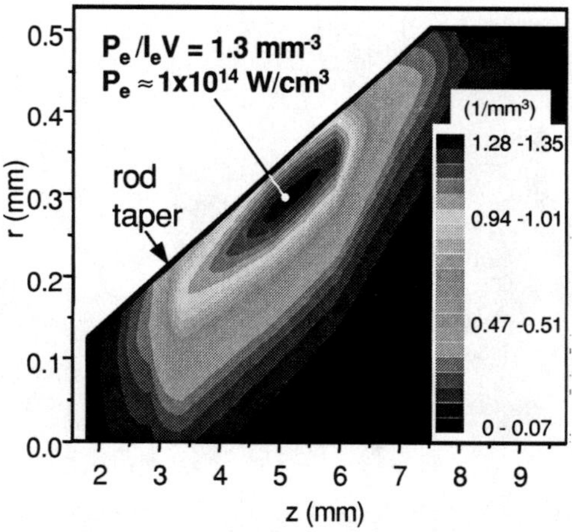

FIGURE 5. Electron-deposition heating for a cold, 1-mm-dia. tapered rod.

location. With these values of z_s, Fig. 4 shows that the heated rod's smaller spot size [5] is explained by peak electron deposition about 1.5 mm closer to the tip. The quantity $f'(z)$ is the smoothed LS_z normalized to its integral over z. The f' for a cold rod was combined with CYLTRAN electron-deposition calculations to construct heating contours, normalized to the incident beam power on the rod (Fig. 5). For Gamble II, beam power is about 6×10^{10} W, so that $P_e \approx 1\times10^{14}$ W/cm^3 is calculated near the surface midway along the rod taper.

The f' labeled "old" in Fig. 4, with assumed uniform heating in a 0.2-mm-thick annulus from the surface of the rod, was used in initial MACH2 MHD computations of rod-plasma expansion. Figure 6 compares measured Schlieren boundaries vs time [4] with those computed from MHD computations ($n_i \approx 3\times10^{17}$ cm^{-3}). Measurements and computations show good general agreement, but disagree in the detailed shape of the expansion envelope. Refinement of the heating model presented here should improve agreement as computed shape errors are likely due to the crude deposition model. Higher-density contours from the MHD results are shown in Fig. 7 at a time close to the end of the radiation pulse. High-density plasma (10% of solid) expands about 0.1

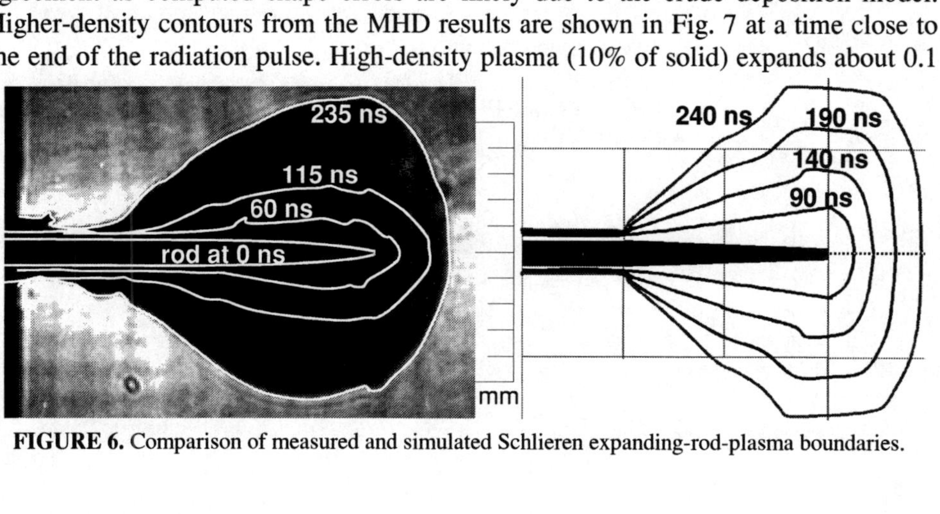

FIGURE 6. Comparison of measured and simulated Schlieren expanding-rod-plasma boundaries.

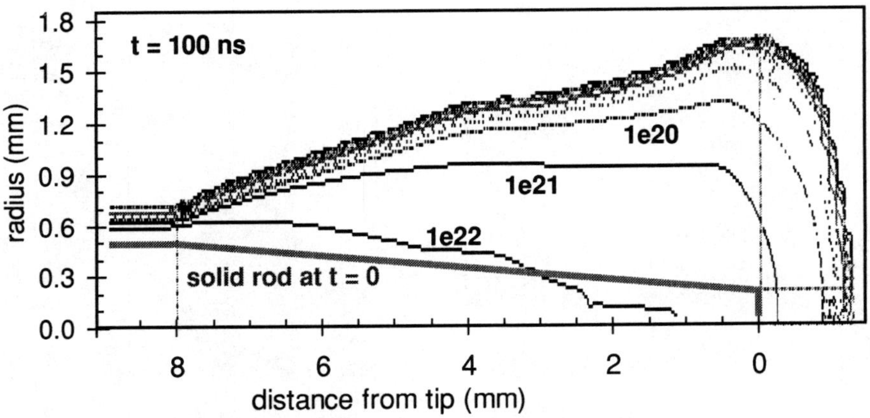

FIGURE 7. High density contours from MHD at t = 90 ns.

mm beyond the nominal 0.5-mm rod radius by the end of the radiation pulse. Hub emission at the base of the rod is likely responsible for extension of the LS_y wings to larger radius.

In summary, line-spread measurements at 0° and 90° to the rod axis have been used to infer 2D r,z electron deposition in cold and heated 1-mm-dia. tungsten rod pinches. Agreement between the simulated and measured LS_y demonstrates that smaller heated-rod spot sizes are due to greater electron deposition closer to the rod tip. It also shows how the end-on spot size can be smaller than the nominal rod diameter (Table 1). Peak electron-heating rates of about 1×10^{14} W/cm^3 are inferred. The new deposition model is expected to refine initial MHD computations of rod-plasma expansion, which already show agreement with Schlieren measurements and can explain a portion of the line spread beyond the rod radius. The inferred $P_e(r,z)$ also provides valuable information for understanding rod-pinch dynamics, comparing to PIC/ITS simulations, optimizing rod-pinch geometry for on- and off-axis radiography, and evaluating other rod-pinch applications such as high-energy-density plasma production.

TABLE 1. Source Sizes in Millimeters

configuration	Δz fwhm	Δy fwhm	Δy LANL
1-mm cold	2.8	0.75	1.1
1-mm hot	2.6	0.61	0.96
0.5-mm blunt	2	0.48	0.89

REFERENCES

Work supported by ONR, LANL, LLNL, and SNL.

1. G. Cooperstein, et al., *Phys Plasmas* **8**, 4618 (2001).
2. R.J. Commisso, et al., *IEEE Trans. Plasma Sci.* **30**, 338 (2002).
3. F.C. Young, et al., submitted to *Phys. Plasmas* and *IEEE Conf. Record Abstracts*, 29th IEEE International Conf. on Plasma Sci., Banff, Alberta, Canada, May 26-30, 2002, p. 248.
4. B.V. Weber, et al., *Bull. Am. Phys. Soc.* **46**, 214 (2001), and *IEEE Conf. Record Abstracts*, 29th IEEE International Conf. on Plasma Sci., Banff, Alberta, Canada, May 26-30, 2002, p. 249.
5. D.D. Hinshelwood, *et al.*, these Proceedings.
6. ORNL Report CCC-467, ITS 3.0, November, 1999.
7. S.B. Swanekamp, ., these Proc., "Characterization of the Radiation from a Rod-Pinch Diode."

Improvement in Rod-Pinch Diode Performance With Anode Heating*

D.D. Hinshelwood, G. Cooperstein, D. Mosher, P.F. Ottinger, J.W. Schumer, S.J. Stephanakis, S.B. Swanekamp[(a)], B.V. Weber, and F.C. Young[(a)]

Naval Research Laboratory, Washington, DC 20375 USA
[(a)]JAYCOR, McLean, VA USA 22102

Abstract. We have studied the performance of rod-pinch diodes with heated anodes. Anode heating is seen to virtually eliminate the proton component of the ion beam, leading to an increase in radiation efficiency. With heated anodes the beam pinches slightly closer to the rod tip which leads to a reduction in spot size when tapered anodes are used. An expected reduction in the rate of impedance collapse is not observed. Heating is easy to implement, and the observed improvements in diode performance merit further study of this approach.

INTRODUCTION

We have studied the use of anode heating as a means of improving the performance of rod-pinch diodes. Anode heating has been shown to eliminate the proton component of the ion beam in standard planar diodes [1]. Removal of the protons is expected to improve diode operation in several ways: Increasing the average ion mass should reduce the ion current and thus improve radiation efficiency. Increasing the anode-plasma atomic mass should reduce the rate of anode-plasma motion and thus

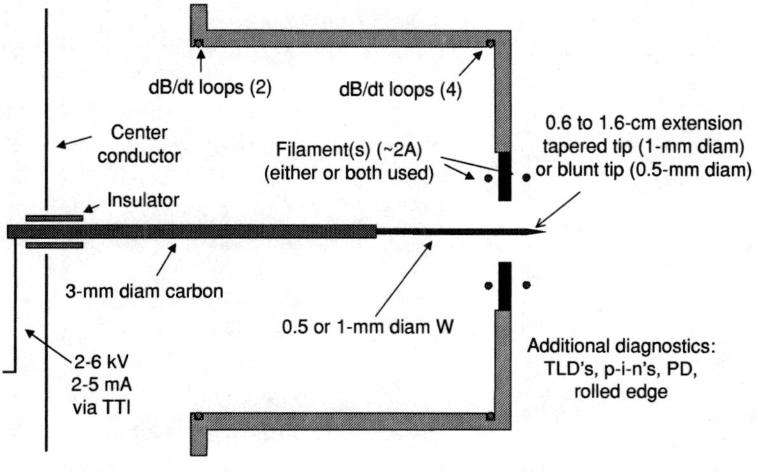

FIGURE 1. Schematic of heated-rod-pinch experimental arrangement

* Work supported by the US DOE through Sandia, Los Alamos, and Livermore National Laboratories

CP650, *Beams 2002: 14th International Conference on High-Power Particle Beams,* edited by T. A. Mehlhorn and M. A. Sweeney

reduce impedance collapse. Finally, reduction in the ion space charge in the diode gap might impede the anomalous impedance collapse that is observed in many experiments involving high diode-aspect-ratios. Reducing this collapse would allow rod-pinch diode operation at high (~60 Ω) impedance.

PIC (MAGIC2d) simulations were carried out beforehand, comparing diodes with protons and with W ions only. As expected, negligible ion current is predicted with W ions alone. The predicted diode impedance varied little with ion mass once the W ions had time to fill the diode gap (the simulations did not model electrode plasma motion). The electron and ion trajectories were only slightly different between the two cases.

The experiments were carried out on Gamble II using the arrangement in Fig 1. The rods, either 0.5 or 1.0-mm diam, were DC-isolated from the anode and biased to several-kV voltages. One or two filaments of 5-mil, thoriated-W wire were located near the diode gap and heated with ~2A of DC current, resulting in a current to the anode of several mA. Based on power balance and visual observation, the rod temperatures were estimated to be about 2000° K.

RESULTS

On many shots the central cathode annulus was graphite. Measurements of activation of this disk, and of nearby carbon targets, indicate that heating reduces the proton flux from the anode by a factor of 10 to 100.

This work comprised several Gamble II pulsing sessions. Data obtained over the course of this study exhibits considerable scatter, which we attribute in part to variations in the machine power pulse and to poor vacuum conditions. Nonetheless, several trends are apparent. Normalized TLD yields for shots under various conditions are shown in Fig. 2. These data illustrate the experimental scatter, but they also show a general increase in radiation efficiency with anode heating. This increase is barely outside the experimental variation but is consistent with the 20% - 50% value that would be expected from elimination of the proton current.

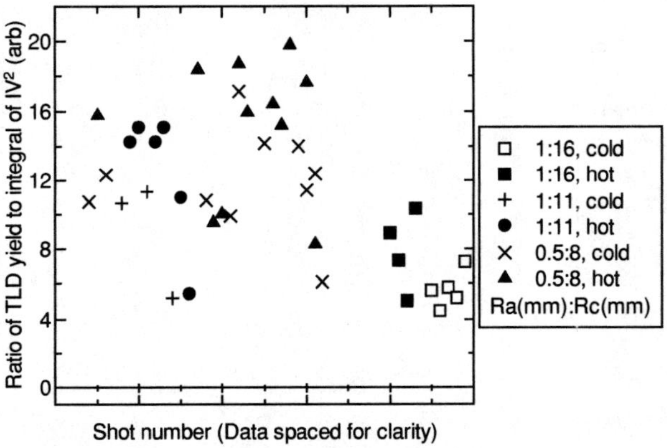

Figure 2. Normalized radiation yield for different geometries.

For many experimental configurations the effect of heating on impedance behavior was inconclusive; for example, with 0.5-mm-diam anodes and 0.8-cm-diam cathodes, a decrease in impedance collapse was observed during one pulsing session but not during another. However, shots with large gaps and large aspect ratios exhibited a consistent effect of heating. Fig. 3 compares the impedance histories of shots with 16-mm cathodes and 1-mm-diam anodes, under different heating conditions. These shots had single filaments, positioned either in front of, or just behind, the cathodes. A marked effect of heating is seen, particularly when the filament is behind the cathode. Similar results were obtained on shots with large (~34X) aspect ratios, where relatively stable impedances exceeding 60 Ω were obtained. Such values have been impossible to obtain on Gamble II without heating.

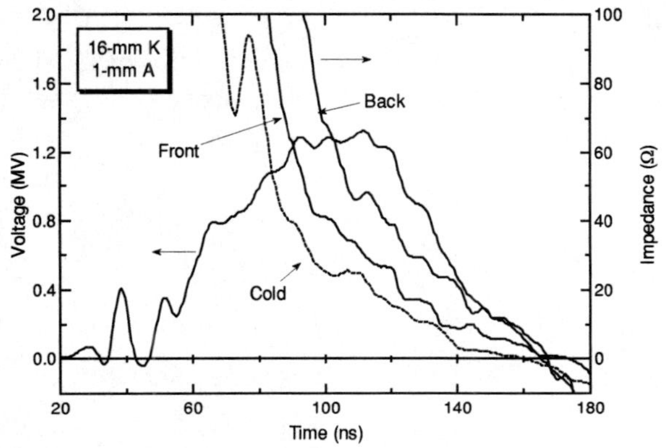

Figure 3. Impedance behavior with a cold anode, an anode heated from the front, and one heated from the rear.

If the main effect of heating on impedance arose from a decreased anode-plasma expansion velocity (our *a priori* assumption), we would expect the greatest change at relatively small gaps, where plasma motion plays a greater role. Instead, we see the greatest effect at large gaps. We attribute this to a reduction in critical current, arising from a reduced ion emission area [2]. When low-mass surface contaminants are removed, anode plasma turn-on requires greater electron energy deposition. Sufficient energy deposition will be less likely under conditions of large gaps and/or high aspect ratios.

We can only speculate as to why a greater reduction in anode plasma velocity is not inferred from the data. The space-charge-limited ion current densities, and therefore the plasma density corresponding to the emission edge, in these experiments are much higher than those in conventional pinched-beam diodes. It may be that at these higher densities the hydrogenic material is entrained in the higher-mass material and moves at the same rate. Alternatively, magnetic pressure may play a larger role in retarding the lighter plasma components.

Three measurements diagnosed the spatial extent of the radiating region. Side-viewing and end-viewing rolled edges were used with film to obtain the time-integrated axial and radial spot sizes. In addition, an array of 18, p-i-n diodes was mounted behind the side-viewing edge on some shots, to obtain time-resolved axial spot-size data. At an anode-extension distance of 0.6 cm, little difference between heated and cold anodes was seen. But at 1.6-cm extension, film measurments show noticeably smaller axial spot sizes for heated anodes. The time-resolved measurements show no significant time variation in spot location or size, but also show reduced spot size with heated anodes.

When tapered rods were used, the decrease in axial spot size led to a corresponding decrease in radial spot size, as seen in Fig. 4. The spot sizes for these data were calculated both by fitting the data to a Gaussian distribution, and by using the LANL spot-size definition. By both metrics the spot size decreases 10% with anode heating.

Cyltran modeling correlates the axial and end-on spot size measurements [3]. The axial measurements define the length of the electron deposition region but are not indexed absolutely to the rod tip location. A series of code runs, with electron deposition over this length but at varying distance to the tip, were performed. For each run, the end-on escaping radiation is collected to predict the end-on spot size. Comparision with experimental data shows good agreement, and indicates that with heated anodes the radiating region moves 0.2 mm closer to the rod tip (1.8 to 1.6 mm).

This behavior is consistent with preliminary MAGIC simulations, where electrons are seen to flow slightly further downstream in the absence of protons. This presumably results from a change in the distribution of positive space charge when protons are replaced by heavy ions.

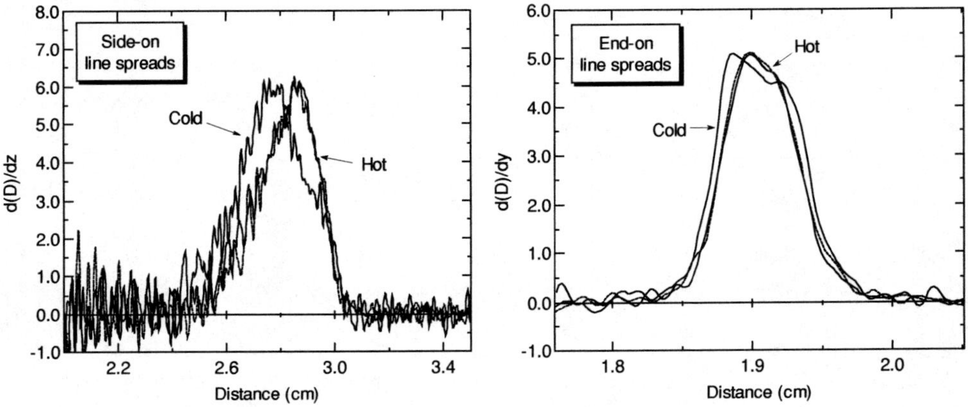

Figure 4. Edge--spread functions derived from rolled-edge data for tapered hot and cold rods.

CONCLUSIONS

Anode heating is seen to provide a small but significant increase in rod-pinch diode performance. Heating essentially eliminates protons from the ion beam, and results in an increased efficiency of radiation production. With tapered rods, heating results in a shorter and smaller-diameter radiation region, leading to a reduced spot size. The effects of heating on impedance behavior are less clear, and gap closure may be less dominated by lighter anode materials than we expected. Heating has, however, allowed operation at impedances previously unobtainable on our machine. Heating is easy to implement as long as access to the anode is possible. The interpretation of these results is hampered by variation in our experimental conditions and we suggest that these experiments should be repeated on more appropriate (better impedance match and vacuum conditions) generators.

REFERENCES

1. B. Weber, et al, these proceedings.
2. G. Cooperstein, et al, these proceedings.
3. D. Mosher, et al, these proceedings.

Anode-target Effects in Electron Beam Driven Radiography

B.V. Oliver, R.E. Clark, T.C. Genoni, T.P. Hughes, D.V. Rose, and D.R. Welch

Mission Research Corporation, Albuquerque, NM 87110

Abstract. Beam/target interactions play an important and sometimes necessary role in electron beam driven x-ray radiography systems. A review of the current theoretical understanding of the problems associated with such interactions is presented. Various proposed methods to mitigate the adverse effects and the areas of future research are also discussed.

A variety of electron beam diodes are being used and developed for the purpose of creating high-brightness flash x-ray radiography sources. In all cases, high energy (multi MeV), high current (multi kA), small spot (multi mm) electron beams are stopped in high atomic number anode-targets (typically Ta or W). Beam stopping in the target creates copious amounts of bremsstrahlung radiation. In addition, beam heating of the target liberates material, either in the form of low density ($\sim 10^{13}$-10^{15} cm^{-3}) ion emission or higher density ($> 10^{15}$ cm^{-3}) plasma. The anode/target material not only influences the operating characteristics (current and voltage) of the diode but can also directly interact with the electron beam. Both can alter the radiation properties (dose and spot-size) and, in some instances, disrupt the beam altogether.

We consider two types of radiographic accelerators: Linear Induction Accelerator (LIA) driven e-beams (e.g. DARHT) and Pulsed-Power Induction Voltage Adder (IVA) systems (e.g. RITS). The beam physics is similar in both cases, however important distinctions do exist. Modern LIAs operate at energies $E_b \sim 20$ MeV and currents $I_b \sim 2$ kA, whereas IVAs operate at ~ 2-10 MeV and 20-50 kA and are thus lower energy, higher current systems. In addition, an LIA e-beam is generated and accelerated upstream of the x-ray converter target. In this case, electric potentials induced on the target are due to the beam space-charge and scale roughly as

$$\phi \sim 30\, I_b\, (1 + 2 \ln(r_w/r_b)), \quad \text{Volts} \tag{1}$$

where I_b is in amps and r_w (r_b) is the wall (beam) radius in cm. Alternatively, in IVA systems, the generation, acceleration, and target interaction all occur in the A-K gap of the diode. There is thus direct coupling of the beam generation to the target/anode. The potential at the target is determined by the applied voltage in the diode.

CP650, *Beams 2002: 14th International Conference on High-Power Particle Beams*, edited by T. A. Mehlhorn and M. A. Sweeney

When material temperatures are raised ~ 400° C lightly bound hydrocarbons can be liberated from the surface[1]. Simple stopping power arguments relate the time rate of change of the material temperature T to the electron beam current density on target via:

$$dT/dt = (dE/dx)\,(J/e\rho C_v) \quad {}^{o}C/s \tag{2}$$

where dE/dx (eV/cm) is the stopping power, J (A/cm^2) the current density, ρ (g/cm^3) the mass density, and C_v (J/g °C) the specific heat. Equation (2) suggests heating rates of roughly ~ 10 J(kA/cm^2) °C/ns and 2.5 J(kA/cm^2) °C/ns for materials like Ta and C, respectively. For the 100-1000 kA/cm^2 current densities considered here (advanced LIAs like DARHT have J~200 kA/cm^2, whereas IVAs are a factor of ten higher J~1 MA/cm^2) the 400° C level for desorption is obtained within 0.1-1 ns. A comparison of the heating rates in Ta for different accelerators is presented in Fig. 1 (courtesy of C. Olson, SNL) where the beam current is plotted vs. spot-size on target (heating rates in carbon are ~ 4 times slower). The desorbed material need only be ionized enough to provide space-charge limited emission of ions from the surface

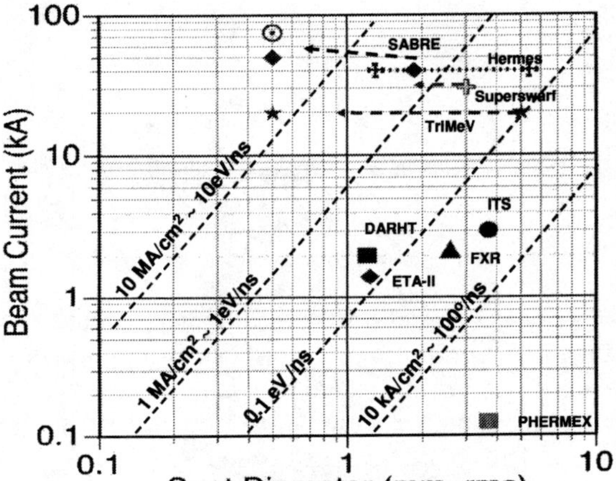

Figure 1. Heating rates in Ta.

$$J_{cl} = 17\,(1/9\pi)(2Zm/M)^{1/2}\,(\gamma-1)^{3/2}/d^2, \quad kA/cm^2 \tag{3}$$

which typically requires less than 5 monolayers of material[2] [γ-1=eϕ/mc^2, Zm/M is the electron to ion mass ratio and d (cm) is the effective A-K gap]. In the case of LIAs the potential ϕ is given by (1), whereas in IVAs it is the diode potential.

FOCAL SWEEPING

In LIA transport near the target, the beam is emittance dominated and the equation for evolution of the beam envelope is

$$\frac{d^2 r_b}{dz^2} = -\frac{2\nu}{\gamma\beta^2}\frac{1}{r_b}\left(1 - f_e - \beta^2(1-f_b)\right) - \left(\frac{\varepsilon}{\gamma\beta}\right)^2 \frac{1}{r_b^3}, \tag{4}$$

where, f_e (f_b) is the charge (current) neutralization fraction, $\nu = eI_b/\beta mc^3 = \gamma I_b/I_A$ is Budker's parameter ($I_A = 17\gamma\beta$ kA), ε (rad-cm) is the beam emittance and $\gamma = (1-\beta^2)^{-1/2}$ is the beam relativistic factor. For beams propagating in vacuum, $f_e = f_b = 0$, and the beam drifts ballistically with a focal length determined by ε. For beam current I_b=2 kA, wall radius r_w=4 cm and r_b=0.1 cm, the potential $\phi \sim 0.17$ MV and the emitted ion current estimated from Eq. (3) (assuming $d=r_b$) is $I_i \sim 12$ A. Thus $f_b = I_i/I_b \sim 0.006$ is negligible. The charge density ratio $f_e = I_i c/I_b(2Ze\phi/M)^{1/2} \approx 0.3(1 + 2\ln(r_w/r_b))^{-1/2} \sim 0.1$ is independent of beam energy and current and is significant enough to influence Eq. (4) and the focal plane[2]. As the ion space-charge moves upstream of the target, f_e becomes a function f_e (z,t) and the interaction length between the beam and back-streaming ions increases. Beam pinching causes a sweeping of the focal plane and changes the spot on target.

In Figure 2, the electron beam and back-streaming ion configuration space is shown from an Lsp[3] simulation of beam focusing on an Al foil target. At a particular time (t=65 ns) protons emitted off the target drift upstream in the beam potential and de-focus the beam. Estimates of the critical length of the ion channel to defocus the beam such that the spot size $r_s > r_b$ yield $l_{crit} = r_b(2I_A/f_e I_b)^{1/2} \sim 10$ cm. The beam temperature is increased with perpendicular energy of order the ion-beam potential ~ $f_e\phi$. This results in a transverse normalized velocity $\beta_\perp \approx (2f_e I_b/I_A)^{1/2} \sim 0.02$. Heated targets or biased foils have been suggested for mitigation of these ions.

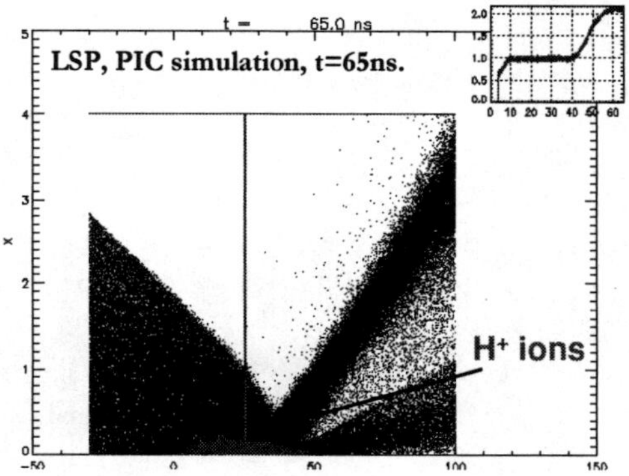

Figure 2. Configuration space of e-beam (20 MeV, 2 kA) and protons from an Al foil.

A similar scenario occurs for beam propagation in over-dense plasma. In the DARHT-2 accelerator, multiple beam pulses of 60 ns duration, separated by 0.5 μs are envisioned. Time-scales are such that fully developed plasmas from early pulses on the target (with plasma density $n_p > 5 \times 10^{15}$ cm^{-3}) can move upstream and influence latter pulses. In this case the plasma fully charge neutralizes and partially current neutralizes the beam such that f_e=1.0 and $1 > f_b(t) > 0$. The plasma return current decays on a time-scale $t_d = 2\pi\sigma r_b^2/c^2 < 10$ ns for plasma conductivity $\sigma = 10^{14}$ s^{-1} and beam radius r_b=0.1 cm. For times $t > t_d$, the return current is nearly zero and $f_b \rightarrow 0$ and the beam focal plane moves upstream causing a sweeping spot[4]. Because the plasma flow velocity is slow, the plasma length l can be assumed constant for the duration of the beam pulse. In this case, the time asymptotic critical plasma length (i.e. $t >> t_d$) is estimated as $l_{crit} = r_b(2I_A/I_b)^{1/2} \sim 3$ cm . Beam emittance growth is marginal since the focusing is nearly linear. Foam targets and tamping foils are suggested for mitigation on DARHT.

Beam propagation in a paraxial diode is similar to the case of propagation in over-dense plasma mentioned above. Here, however, the beam radius r_b and pulse length do not necessarily satisfy $t_p >> t_d$ and the effect of finite current neutralization must be considered. The paraxial diode (see Fig. 3) is a ball shaped cathode where the e-beam is space-charge emitted and focused onto a small (~ 5 cm long) gas filled transport region. It is used extensively for pulsed-power driven flash radiography at AWE, Aldermaston UK and has demonstrated 4.5 mm spot at 5.5 MV and 35 kA[5]. The front foil of the gas cell is the anode and the target is located upstream at the end of the transport region. It is a hybrid between conventional pulsed-power diodes and the LIA system. Like LIAs, the beam is generated and accelerated in vacuum and then transported to target, but beam currents are higher and energies lower. Overall, the beam transport is well described by $\lambda_b/4$ betatron focusing, where the betatron wavelength $\lambda_b=2\pi(I_A/I_{net})^{1/2}$ is determined by the net current I_{net} in the gas transport region for a matched incident beam radius R. The focal length $F=\lambda_b/4$ and can change for time dependent $I_{net}(t)$. Fig. 3 shows the beam configuration space for 35 kA, 5.5 MV beam propagation in 1 torr air at a time corresponding to peak power of the accelerator. The simulation demonstrates the existence of a focal plane consistent with λ_b=20 cm corresponding to I_{net}=20 kA. The spot size $2r_s \cong 2\beta_{\perp 0}F$ = 0.4 cm (where $\beta_{\perp 0}$ is the initial beam transverse velocity), and the beam transverse temperature is related to the compression ratio $\beta_\perp \approx \beta_{\perp 0} R/r_s = R/F = (2R/\pi)(I_{net}/I_A)^{1/2} \sim 0.2$. An important element in obtaining accurate predictions of spot and dose via simulation is the need for detailed physics models of the gas breakdown and plasma conductivity generation. Detailed pulse-shaping and gas composition should be considered for optimization.

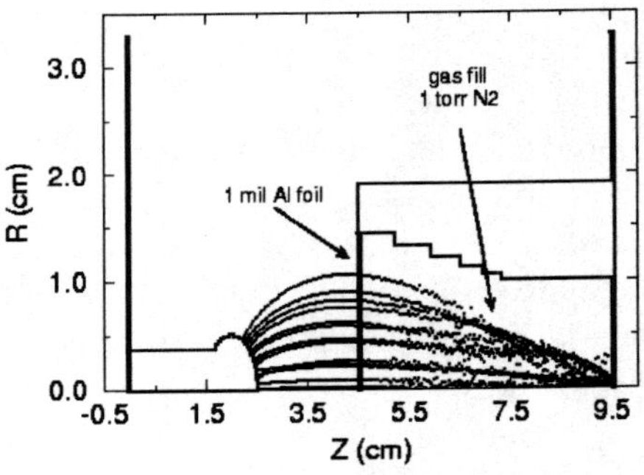

Figure 3. Configuration space of a paraxial e-beam (5.5 MeV, 35 kA) in 1 torr N2.

COLLECTIVE EFFECTS: MAGNETIZED ION-HOSE INSTABILITY AND SELF-INSULATION

Collective oscillations of the e-beam can be driven unstable by the presence of the backstreaming ions. In the ion focused regime, a transverse, 3-D, dipole instability exists known generally as the ion-hose instability[6]. Displacements of the electron beam relative to the backstreaming ion channel result in the generation of a polarization electric field. Finite inertia (of both electron and ion channels) results in out-of-phase oscillations and a convective instability with growth length (distance behind the beam head) determined by the electron collisionless skin-depth in the presence of the beam charge $\lambda_e \approx c/\omega_{pe} = (\gamma mc^2/4\pi e^2 f_e n_b)^{1/2}$ and growth rate (in a frame moving with the beam velocity) given by the ion inertial time in the presence of the e-beam charge[6] $\Gamma \approx \omega_{pi} = (4\pi e^2 n_b/M)^{1/2}$. Recently, the IFR ion-hose instability has been re-visited for cases when a large solenoidal magnetic field is present. In this case, the polarization electric field generated during displacement is perpendicular to the background magnetic field and causes the beam to ExB rotate about the axis, as well as oscillate. The linear analysis of magnetized IFR has been presented previously[7] and demonstrates that in the large field limit ($\Omega_e >> \omega_{pe}$) the linear growth length $\lambda_{ce} \sim (\Omega_e/\omega_{pe})\lambda_e$. For the beams considered here, the time and length scales are < 1 ns and 1 cm for IVA systems and < 10 ns and 10 cm for LIAs, hence nonlinear evolution and saturation are expected. Non-linear analysis based on the spread-mass model[8] suggests saturation of the beam offset amplitude to scale as the inverse of the linear growth-length $1/\lambda_{ce}$ and is thus proportional to $1/B_z$. Roughly speaking, saturation occurs when the beam radial displacement is of order the beam gyro-radius $r_g = \gamma mc^2\beta_\perp/eB_z$, where $\beta_\perp = (2f_eI_b/I_A)^{1/2}$ is again determined by the beam potential near the target (typically $f_e \sim 1$ at the anode of IVA driven systems)[2]. The scaling with inverse field strength implies a method for controlling the e-beam spot by increasing the field strength.

In Figure 4 (a) the beam and ion centroid displacements are shown vs. time for a 2 μs, DAHRT-2 beam in the transport chamber (upstream of the target). Here ion/beam interactions are due to ions created from the breakdown of residual gas in the transport chamber. Agreement is maintained between theory and simulation through the linear phase up to and including the initialization of saturation ($t \approx 475$ ns). Late time asymptotic values of the saturation amplitude are in reasonable agreement for the e-beam offset, although the dynamics of the ion channel are not fully described. The saturation amplitude is not a stable fixed point of the dynamics, hence the need for 3-d simulations to accurately predict the long-time behavior. On the other hand, scaling of the saturation amplitude with $1/B_z$ is observed. In Fig. 4 (b), the spot on target vs. field strength is plotted for an immersed-B diode. Good agreement is obtained between theory, simulation and experiments conducted on the Hermes III and Sabre accelerators at Sandia National Labs (figure courtesy of J. Maenchen, SNL).

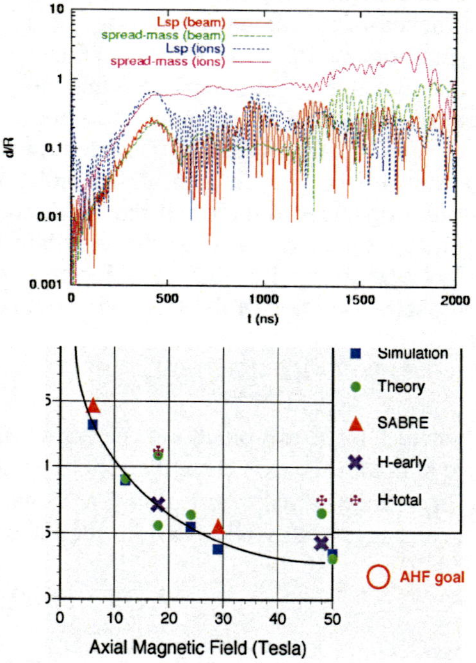

Figure 4. (a) Beam centroid offset normalized to the beam radius, (b) beam spot vs. B

The immersed-B diode[9] is an IVA driven diode where the cathode is immersed in a large solenoidal magnetic field (typically 10-50 Tesla) which guides the beam to the anode/target. Experimental data obtained at SNL and AWE has demonstrated that increased field strengths decrease spot-size. However, above some critical current density on target ~ 100 kA/cm², the diode undergoes impedance collapse and spot growth late in time. It is suspected that charge stripping of ions emitted from the target are responsible for this abnormal diode behavior. As ions cross the diode A-K gap they are stripped and arrive at the cathode with an increased charge. This in turn enhances the space-charge limited electron current from the cathode and the process bootstraps. 3-D Lsp simulations including charge stripping of Carbon emitted from the anode indicate impedance collapse and a significantly larger ion-hose saturation amplitude, resulting in larger spots on target and increased beam temperature. Because protons will not charge strip, a cryogenically frozen layer of D_2 on top of the normal Ta target is proposed on future immersed-B diode experiments at Sandia.

Finally, we briefly discuss the role of ions in the operation of the rod-pinch diode[10]. The rod-pinch is a cylindrical, pinched-beam diode consisting of a small-diameter anode rod of radius r_a that extends a distance l through and beyond the hole of an annular cathode of radius r_c (see Fig. 5). The electron beam self-pinches in a sheath of radius r_s and propagates to the tip of the anode rod. The spot is determined by the rod diameter. It has demonstrated 0.8 mm spot at 1-2 MV operation with 30-50 kA. In this case, ion emission from the anode is necessary for self-insulation of the electron current along the rod. The diode current scales as[11]

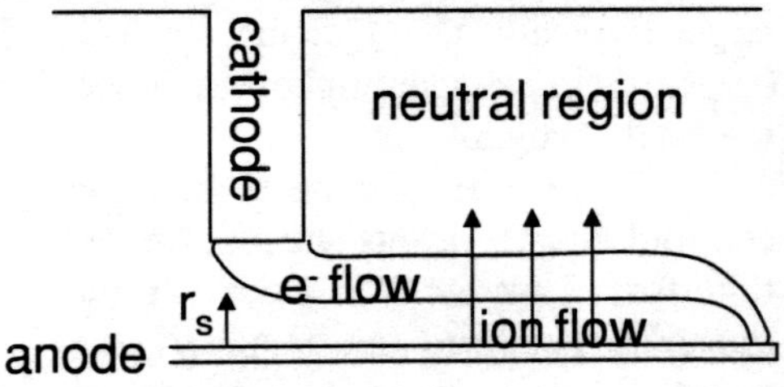

Figure 5. Schematic of a Rod-Pinch diode

$$I = I_e + I_i \cong 17\, r_s\, (4\pi J_i/17)^{1/2}(M(\gamma-1)/2Zm)^{1/4} = 8.5\,(4/3)(\gamma-1)\ \text{kA}, \tag{5}$$

where J_i is the rod-pinch ion current density and is $\cong J_{cl}$ with $d = r_s$. However, in order to maintain electron current efficiency one needs to limit the ion current such that $I_i/I \sim (l/r_c)(Zm(\gamma-1)/M)^{1/2} << 1$. For 2 MV operation and $l/r_c=3$, $I_i/I=0.14$. Heated rods have been suggested by NRL to limit the ion current.

ACKNOWLEDGEMENTS

This work is supported by the US DOE through SNL. We acknowledge discussions with our colleagues at SNL, LANL, LLNL, NRL, and AWE on these topics.

[1] T.W.L. Sanford, et al. *J. Appl. Phys.*, **66**, 10 (1989); R.J. Lipinski, et al. *Phys. Fluids B*, **2**, 2764 (1990)
[2] D.R. Welch and T.P. Hughes, *Laser and Particle Beams*, **16**, 285 (1998)
[3] T.P. Hughes, S.S. Yu and R.E. Clark, *Phys Rev. ST Accel. Beams* **2**, 110401 (1999)
[4] B.V. Oliver, D.R.Welch and T.P. Hughes, "Beam-target Interactions in Single and Multi-pulse Radiography", in MRC report MRC/ABQ-R-1909, pg. 2 1999.
[5] A.Birrell et al. *IEEE Trans. Plasma Sci.* **28**, 1660 (2000)
[6] H.L. Buchanan, *Phys. Fluids* **30**, 1456 (1987)
[7] T.C. Genoni, "Analytic Model of IFR Hose Instability" in MRC report MRC/ABQ-R-1847, p.25, 1997
[8] E.P. Lee, *Phys. Fluids* **21**, 1327 (1978)
[9] M.G. Mazarakis et al. Appl. Phys. Lett **70**, 832 (1997)
[10] G. Cooperstein, et al., *Phys. Plasmas* **8**, 4618-4636 (2001).
[11] B.V. Oliver et al. *Proc. 13th International Pulsed Power Conference*, Las Vegas, NV, **1**, 458 (2001)

ELECTRON BEAMS

The Influence of Anode-Plasma Location on Total Current in Self-Magnetically-Limited Electron-Beam Diodes

G. Cooperstein,[a] P.F. Ottinger,[a] J.W. Schumer,[a] S.B. Swanekamp,[b] R.J. Allen,[a] R.J. Commisso,[a] D.D. Hinshelwood,[a] D. Mosher,[a] S.J. Stephanakis,[a] B.V. Weber,[a] and F.C. Young[b]

[a]*Plasma Physics Division, Naval Research Laboratory, Washington, DC 20375*
[b]*Titan/Jaycor, McLean, VA 22102*

Abstract. The critical current for both planar self-pinched electron-beam diodes and rod-pinch diodes is shown to be affected by the location of anode ion production relative to the cathode location. Specifically, for cathodes that are thin compared with the A-K gap, anode ions produced upstream of the cathode enhance the critical current scaling factor α. On the other hand, when ions are only produced downstream near the point where the electrons first hit the anode at grazing incidence, α approaches the electrons-only value. This new interpretation helps to bring PIC simulation results into better agreement with experimental results for both types of diodes.

It is well known that self-pinched electron-beam diodes, either conventional planar pinched-beam diodes[1] or rod-pinch diodes[2], operate at a self-magnetically-limited total current (ions plus electrons) called the critical current. For a planar pinched-beam diode, illustrated in Fig. 1a, the critical current is given by, $I_{crit} \cong \alpha\, 8500\, (\gamma^2 - 1)^{1/2} (R_C/D)$ where γ is the relativistic factor, R_C is the cathode radius and D is the anode-cathode (A-K) gap spacing. For a rod-pinch diode, illustrated in Fig. 1b, the critical current is, $I_{crit} \cong \alpha\, 8500\, (\gamma^2 - 1)^{1/2} / \ln(R_C/R_A)$ where R_A is the anode radius. The scaling factor α is needed to obtain agreement between PIC simulation results and derived analytical critical current formulas. Without ions, this scale factor α is found to be about 1.4 for both kinds of self-pinched diodes. The presence of anode ions alters the simulation results differently for planar pinched-beam diodes and rod-pinch diodes. For planar pinched-beam diodes with ions, $\alpha \cong 2.1$, independent of R_C/D. For rod-pinch diodes with ions, $\alpha \cong 2.1$ for $R_C/R_A \cong 1$ and $\alpha \cong 2.6$ for $R_C/R_A \cong 16$. This variation with R_C/R_A is related to the enhancement of the electric field near the small-radius anode and its effect on the ion velocities and ion crossing times[2]. For planar pinched-beam diodes with cathodes that are thin compared to the A-K gap spacing, the experimentally accepted value of $\alpha \cong 1.6$ can only be brought into agreement with PIC simulation if anode ions are not produced

CP650, *Beams 2002: 14th International Conference on High-Power Particle Beams,*
edited by T. A. Mehlhorn and M. A. Sweeney

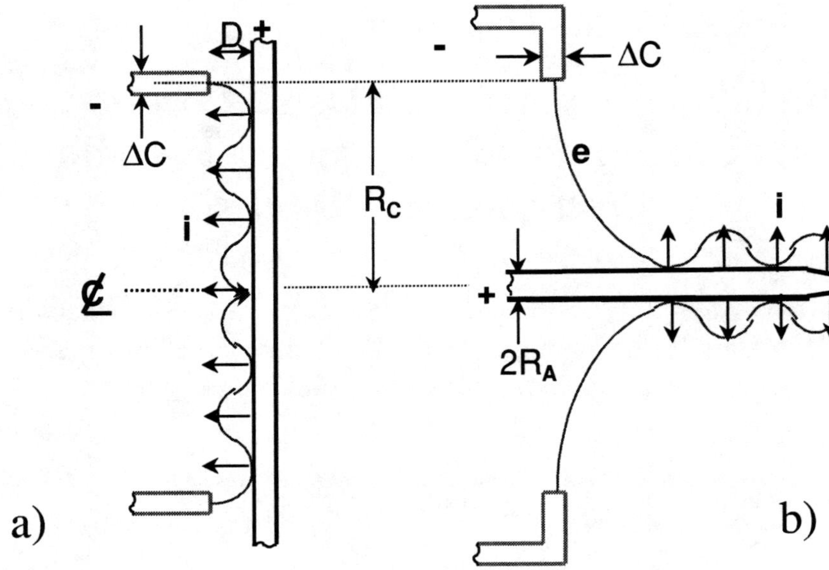

FIGURE 1. a) planar pinched-beam diode, b) rod-pinch diode. R_C, D, and R_A are defined in the text. ΔC is the cathode thickness.

outside the cathode radius ($r > R_C$). In this case, the experimental and theoretical values of α are in reasonable agreement except for small variations in experimentally measured α factors which we relate to the actual position where plasma is formed on the anode. When ions are constrained in the planar-diode simulations to be produced only at smaller radii, e.g. inside the cathode's outer radius, a smaller α is predicted.

Rod-pinch diodes appear to be more strongly affected by the anode-plasma location because these anode ions have a greater effect on α. When anode ions are constrained in the PIC simulations to be produced only downstream of the cathode, α approaches the electrons-only value. We expect that this reduction in α (when ion production is constrained) will manifest itself for both kinds of diodes when the electron beam produced during the space-charge-limited phase is insufficient to produce an anode plasma. This appears to happen for planar pinched-beam diodes with thin cathodes[3] or for rod-pinch diodes with large-diameter low-atomic-number anodes[2]. Low-atomic-number anodes have less electron backscatter leading to reduced anode surface heating. This reduction might also occur with rod-pinch diodes if external anode heating[4] is sufficient to drive low-atomic-number contaminants from the surface. In these cases, ions will be produced only after critical current is approached and an anode plasma is produced by electrons that hit the anode at near-grazing incidence.

For planar pinched-beam diodes, measurements with thin cathodes and high-atomic-number anodes[3] show that the diode current approximately follows the critical current with $\alpha \cong 1.6$ when expanding cathode and anode plasma velocities (~ 2 and 1 cm/μs, respectively) are correctly taken into account, i.e. the A-K gap D is closed at a rate of about 3 cm/μs. When the 2½-dimensional, fully electromagnetic

MAGIC2D PIC code[5] is used to simulate this diode, strong electron-beam pinching is observed, as expected, but $\alpha \cong 2.1$ if anode ions are allowed to be produced from radii larger than the cathode radius. If ions are not present, $\alpha \sim 1.5$ and, as expected[1], electrons do not pinch to a radius smaller than about $R_C - D$. This discrepancy in α goes away if one assumes in the PIC calculations that anode ions are only produced where the electrons first hit the anode at grazing incidence, i.e. at $\sim R_C - D$. Specifically, when a planar pinched-beam diode geometry, shown in Fig. 1a, with $R_C = 6$ cm, $D = 7$ mm and $\Delta C = 3$ mm, is simulated with MAGIC2D, and the simulations are run to equilibrium at 3 MV to insure that the diode is operating in the self-magnetically-limited regime, $\alpha \cong 1.5$ for the electrons-only case, and $\alpha \cong 1.6$, 1.8 and 2.1 for ion emission from 0 to $R_C - D$, R_C, and $R_C + D$, respectively. This agreement between measurements and PIC simulations when ions are emitted only from 0 to $R_C - D$ illustrates how important it is to correctly determine (or control) where ions are produced on the anode.

For rod-pinch diodes, recent data and PIC simulations[2] show a similar discrepancy which is illustrated in Fig. 2. Here, the experimental and PIC-determined

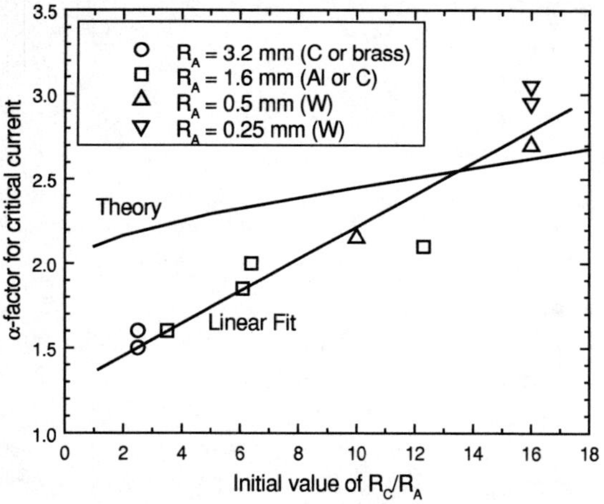

FIGURE 2. Experimental (linear fit to symbols) and PIC-determined (theory) α factors for data in Ref. 2.

α factors are shown for a large range of rod-pinch diode conditions with R_C/R_A varying from 2.5 to 16. There are two points to notice. First, as expected, PIC simulations give the same value of α (2.1) for rod-pinch diodes at small R_C/R_A and planar pinched-beam diodes at large R_C/D. However, at $R_C/R_A = 2.5$, PIC simulations predict $\alpha \cong 2.1$ while the data gives $\alpha \cong 1.5$. This discrepancy was previously noted[2] but was not satisfactory resolved. We now propose that the reason for this discrepancy is related to the fact that rod-pinch diodes with smaller R_C/R_A also had much larger-diameter, and lower-atomic-number anodes than the larger R_C/R_A diodes. Specifically, 6.4-and 3.2-mm diameter carbon and aluminum anodes compared with 0.5-and 1-mm diameter tungsten anodes. We believe that ions are produced both upstream and downstream for the small diameter tungsten anodes during the early non-pinched space-charge limited phase, while ions are produced only downstream of the cathode, at the point where the electrons hit the anode at near grazing incidence, for the larger-diameter, lower-atomic-number anodes.

To confirm this interpretation, the PIC simulations illustrated in Fig. 3 were run for the three cases shown. To guarantee that the electron-only case (not shown) would be

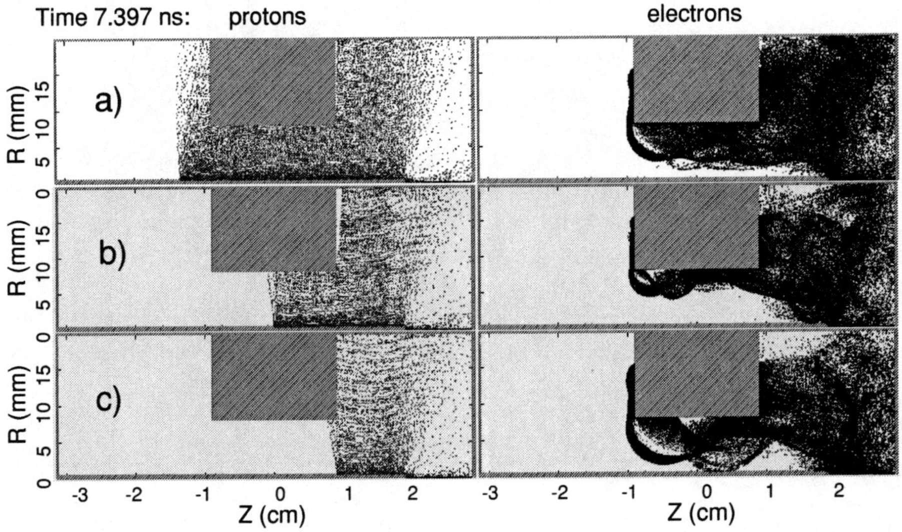

FIGURE 3. Ion (left) and electron (right) positions for $R_A = 0.5$ mm and $R_C/R_A = 16$. Ions are produced from the tip of the anode to a position a) $R_C - R_A$ upstream of the cathode, b) in the center of the cathode, c) flush with the downstream edge of the cathode.

self-magnetically limited, the cathode thickness, ΔC, was chosen as 2 cm and the simulations were run to equilibrium (5 ns) at 3 MV. Note that the electrons pinch to the tip of the rod at 1.8 cm in all cases. Specifically, $\alpha \sim 1.4$ for the electrons-only case, and $\alpha = 2.55$, 2.15 and 1.5 for cases a), b) and c), respectively. This close agreement between simulation and experiment reinforces our interpretation and shows the importance of knowing and controlling exactly where ions are produced along the anode.

ACKNOWLEDGEMENTS

Work supported in part by the US DOE through SNL, LANL, LLNL and the U. S. Office of Naval Research.

REFERENCES

1. A.E. Blaugrund, G. Cooperstein, and S.A. Goldstein, *Phys. Fluids* **20**, 1185 (1977).
2. G. Cooperstein, et al., *Phys. Plasmas* **8**, 4618-4636 (2001).
3. G. Cooperstein, et al., *Proceedings of Beams' 96* (Prague, Czech Republic, June, 1996), p. 1151.
4. D.D. Hinshelwood, et al., this conference.
5. B. Goplen, L. Ludeking, D. Smithe, and G. Warren, Comp. Phys. Comm. **87**, 54 (1995).

Small Compact Pulsed Electron Source for Radiation Technologies

Sergey Korenev

STERIS CORPORATION, STERIS Isomedix Services,
2500 Commerce Drive, Libertyville, IL 60048, USA

Abstract. The small compact pulsed electron source for radiation technologies is considered in the report. The electron source consists of pulsed high voltage Marx generator and vacuum diode with explosive emission cathode. The main parameters of electron source are next: kinetic energy is 100-150 keV, beam current is 5 – 200 A and pulse duration is 100 – 400 nsec. The distribution of absorbed doses in irradiated materials is considered. The physical feasibility of pulsed low energy electron beam for applications is considered.

INTRODUCTION

The development of pulsed electron sources [1,2] has big progress in this techniques and engineering. However, the physical aspects of work for electron sources present scientific interest. The early papers [3,4,5] considered the main principles of design for electron sources for radiation technologies. The main advantages of pulsed electron sources according [4] are high efficiency for dissipation of energy for short time.

This paper is considered compact electron source for radiation technologies with explosive electron emission.

STRUCTURE OF ELECTRON SOURCE

The electron source consists from vacuum diode with vacuum chamber and system, high voltages pulsed Marx generator and beam diagnostics, see Figure 1. The explosive cathode on the basis of carbon fiber materials [6] is used in this vacuum diode. Marx generator with 5 steps allows forming variation voltage 100-150 kV by charged voltage and gaps in the gas sparks. The variation of distance between cathode and anode permits to variation of beam current from 5 to 200 A with pulse duration from 100 to till 400 nsec.

The vacuum chamber manufactures from stainless steel. The pressure of residual gas in vacuum chamber is 10^{-5} Torr.

The special vacuum chamber is used for irradiation of samples.

The diagnostic for electron beam is included the current transformer (Rogowski Coils) and Faraday Cups with low induction current shunts.

CP650, *Beams 2002: 14th International Conference on High-Power Particle Beams,*
edited by T. A. Mehlhorn and M. A. Sweeney

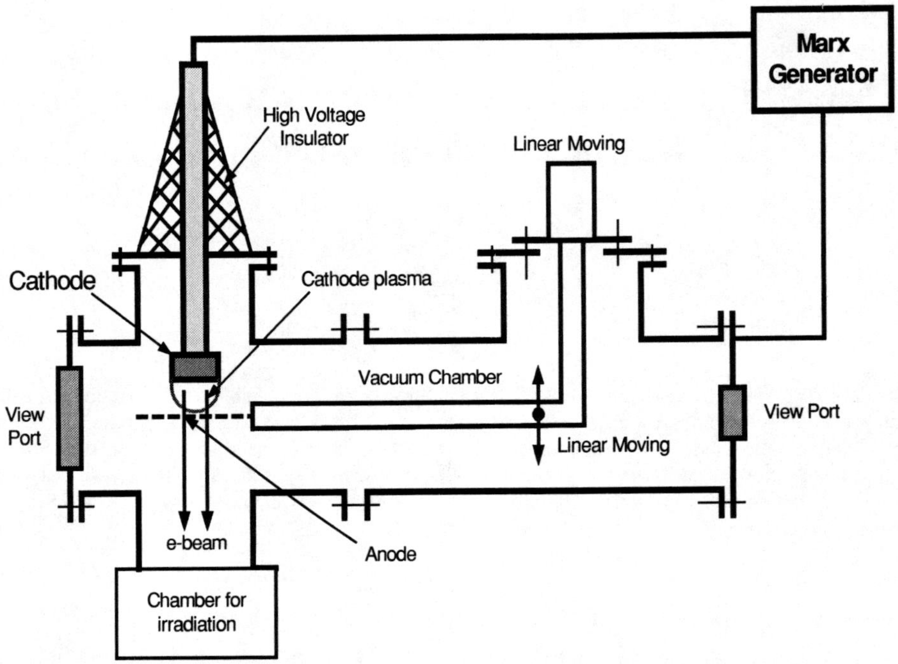

FIGURE 1. The structure of compact electron source

MAIN PARAMETERS AND CHARACTERISTICS

The main parameters of considered electron source are next:

1. Voltage (kinetic energy of electrons)........ 100 – 150 kV (KeV);
2. Beam current...5 – 200 A
3. Beam current pulse duration ....100 – 400 nsec.
4. Repetition...1- 100 Hz.

The typical required distribution of absorbed doses in irradiated sample is given on Figure 2 for one side's irradiation. The value of average absorbed doses D_{ave}on the level of 20% is [7]:

$$D_{ave}=[D_{max}+D_{min}]/2, \qquad (1)$$

where: D_{max} is max absorbed dose, D_{min} is min absorbed doses, see Figure2.

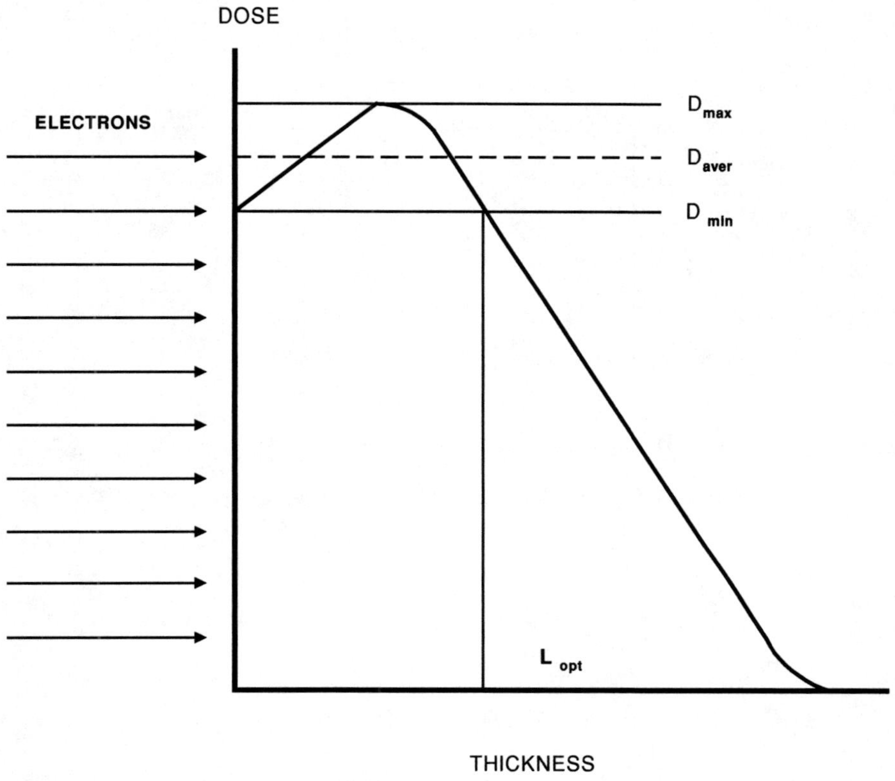

FIGURE 2. Electron beam dose distribution in irradiated sample.

The optimal thickness L_{opt} of an irradiated product is determined by this distribution. The thickness of irradiated product is 0.05 – 0.5 mm for many materials with density 0.1 – 2.0 g/cc and kinetic energy of electrons 100 – 150 keV. The simple calculations of absorbed doses for materials with same thickness shows the level of doses about 50 – 200 kGy/pulse in case of using of pulsed high current accelerators.

The dose distribution for this electron source is presented on Figure 3. This dose distribution shows the high homogeneous of absorbed doses and high homogeneous of current density for electron beam. It is very important parameters of electron source. The carbon fiber cathodes [6] have stable time for forming of cathode plasma and emission characteristics.

According these data we can consider the applications of pulsed electron source for many experiments in condensed matter.

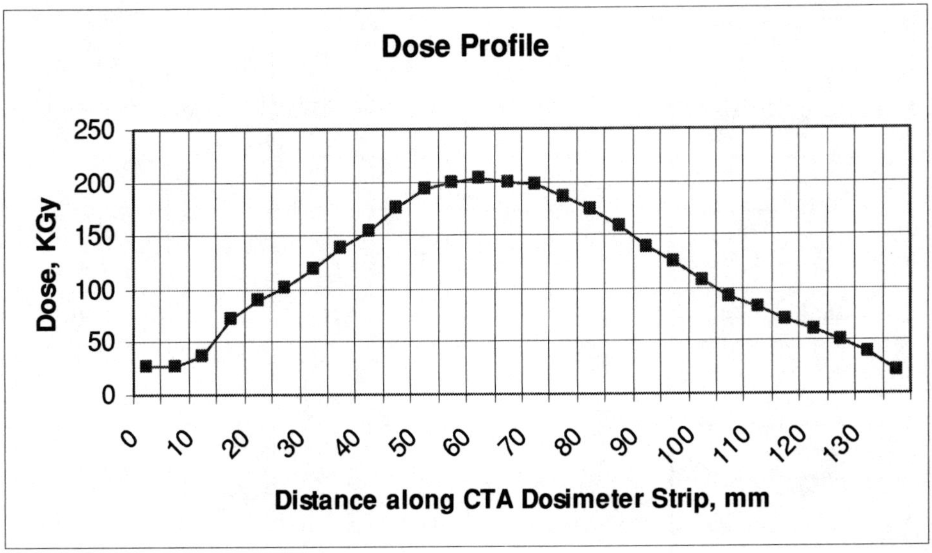

FIGURE 3. Electron beam dose distribution in CTA film dosimeter,

CONCLUSION

As a result of these investigations we can make main conclusion:

1. The pulsed electron source can be used for wide applications for surface treatment of different materials.

REFERENCES

1. Koval'ski G.A. *Emission Electronics.* Moscow Institute of Radio Engineering, Electronics and Automatics, Moscow, 1977.

2. Korenev S.A. *"Electron gun for surface treatment of materials"*, Patent of Russia No 1501829, 1989.

3 Korenev S., Coll B., Perry A. *Surface and Coatings Technology,* **86-87**, 292-301 (1996).

4. Korenev S., Perry A. *Vacuum,* **47**, 1089-1993 (1996).

5. Korenev S.A., Perry A.J., Kalmykov A.V. *Surface & Coatings Technology,* **108-109**, 265-270 (1998).

6. Korenev S.A., Baranov A.M., Kostychenko S.V., Chernenko N.M. *Cathode on the basis of carbon-fibrous plastic material.* Preprint JINR No 13-88-292, Dubna, 1988; *J. Instrum. Exper. Technics* **5**, 194 – 196 (1989).

7 McLaughlin W.L., Boyd A.W., Chadwick K.H., McDonald J.C., Miller A. *Dosimetry for Radiation Processing*. Taylor&Francis, London*New York*Philadelphia, 1989

Propagation of an Intense Electron Beam through a Gap

Michiya Mori, Masaki Kamada, Takashi Nishiguchi, Ritoku Ando and Keiichi Kamada

Department of Physics, Faculty of Science, Kanazawa University, Kanazawa, 920-1192, Japan

Abstract. Experiments and simulations for the propagation of intense relativistic electron beams that pass through a gap are carried out. The radius of the drift tube increases at the gap entrance and it returns to the initial radius after the gap length. Though the injected beam current into the gap is larger than the space charge limiting current at the gap, it passes through the gap without loss when the gap length is short. A criterion is proposed for the gap length through which the beam current passes without loss.

INTRODUCTION

It is well-known that there is a space-charge limiting current when an electron beam propagates through a drift tube immersed in an axial magnetic field. Electrostatic potential in the tube due to the beam's space-charge field limits the beam current propagating through the tube. In the case of a cylindrical beam propagating through a cylindrical tube, the limiting current is calculated by the initial energy (γ_0) of the electrons and the ratio of the beam radius (r_b) to the tube radius (r_t). For an annular beam with an infinite axial length, the limiting current (I_L) is expressed as below

$$I_L = I_A \frac{(\gamma_0^{2/3} - 1)^{3/2}}{2\ln(r_t/r_b)} \qquad (1)$$

where I_A=17 kA is the Alfven current. The limiting current decreases as the radius of the tube increases, when the initial beam energy and the beam radius are not changed.

The above equation is derived for the beam with an infinite axial length. In the experiments, it is not clear how long the axial length is enough for a drift tube to apply the above equation. Propagation characteristics of an intense electron beam through drift tubes with different radii are interesting subjects to study. For example, for autoacceleration using a coaxial cavity with a gap, the gap length is one of the important parameters for the acceleration efficiency. A simple model is utilized to start this problem. Let us think about a drift tube whose radius is suddenly increased from r_1 to r_2 at z (z is the distance from an anode.) and decreased from r_2 to r_1 at $z+d$ (Fig.1). We call the thicker part of the tube a gap with length of d here. An electron beam with current I which exceeds the space-charge limiting current I_L at the gap is injected from z=0. When the gap length d is long enough for the formation of a virtual cathode,

CP650, *Beams 2002: 14th International Conference on High-Power Particle Beams,*
edited by T. A. Mehlhorn and M. A. Sweeney

some of the injected beam electrons are reflected near the entrance of the gap and the current of I_L evaluated by eq.(1) passes through the gap. However, when the length d is not so long, the beam current larger than I_L can pass through the gap. We investigate the relation between the gap length and the propagating current across the gap. In this paper, we report the length of the gap through which the beam passes without loss.

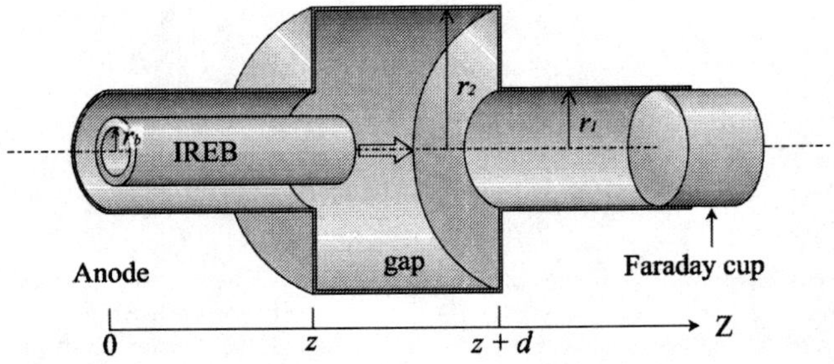

FIGURE 1. An annular intense electron beam is injected through an annular anode at $z=0$. The beam propagates through a drift tube with radius r_1. The radius is increased to r_2 at $z=z$. and decreases to r_1 at $z=z+d$. The thicker radius part is called a gap. The beam that passes through the gap is detected by a Faraday cup.

EXPRIMENTAL RESULTS

An intense relativistic annular electron beam with energy of 500 keV, current of 4~8 kA and duration of 12 ns is utilized in the experiment. The beam propagates through a drift tube with radius of 1.5 cm. A gap with radius of 6 cm is located 20 cm downstream side of the anode. The length of the gap is changed. The beam current is detected at the end of the drift tube with radius of 3 cm by a Faraday cup. In order not to affect the presence of the Faraday cup to the experimental results, the Faraday cup is located 20 cm downstream side of the gap end. Figure 2 shows the detected beam

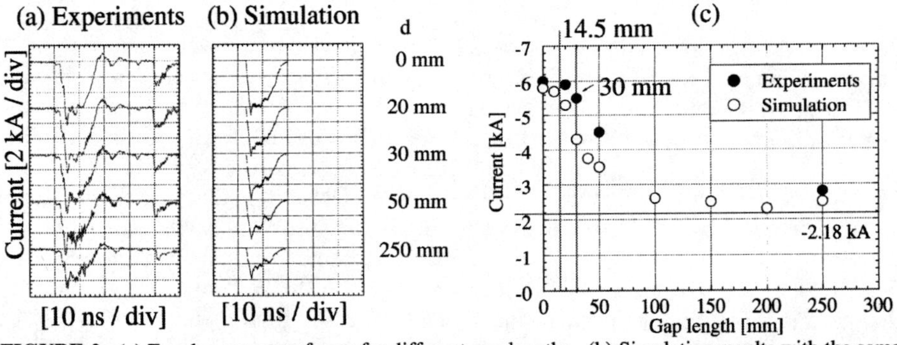

FIGURE 2. (a) Faraday cup waveforms for different gap lengths. (b) Simulation results with the same beam and gap parameters as the experiments. (c) Peak currents are plotted against different gap lengths. The gap lengths through which the beam current passes without loss are indicated by vertical lines.

current for different gap lengths. The space charge limiting current at the gap is evaluated to be 2.18 kA by the eq.(1). Though the beam current of 6 kA is much larger than the limiting current at the gap, it passes through the gap without loss when the gap length is less than ~3 cm. In Fig. 2(c), the maximum beam currents detected by the Faraday cup are plotted against the gap length. The beam current decreases gradually as the gap length increases. When the beam and drift tube radii are changed, the gap length through which the beam passes without loss is also observed experimentally.

SIMULATION RESULTS AND DISCUSSIONS

PIC simulation code KARAT is used. The parameters of the experiment are utilized in the simulation and the simulated results show good agreements with the experiments, as shown in Fig. 2. The simulations show clearly the formation of a virtual cathode when the gap length becomes long, as shown in Fig. 3. The injected beam currents are larger than the limiting current at the gap. When a virtual cathode is formed, a part of the beam current is reflected at the gap and the current passes through the gap decreases. There is a minimum gap length to form a virtual cathode, as shown in Fig. 3. When the gap length is short, the formation of a virtual cathode is incomplete and the injected beam current passes through the gap without loss. The relation between the gap length and the detected beam current at the downstream side of the gap is investigated with different parameters. As the code KARAT simulates well the experimental results, we extend the range of parameters with energy of 400~1750 keV, current of 3~9 kA and gap radius of 40~90 mm in the simulation. The gap length through which the beam passes without loss increases, as the beam energy increases. And that decreases as the beam current increases. That increases as the beam radius increases but that is not affected strongly by the radius of the gap.

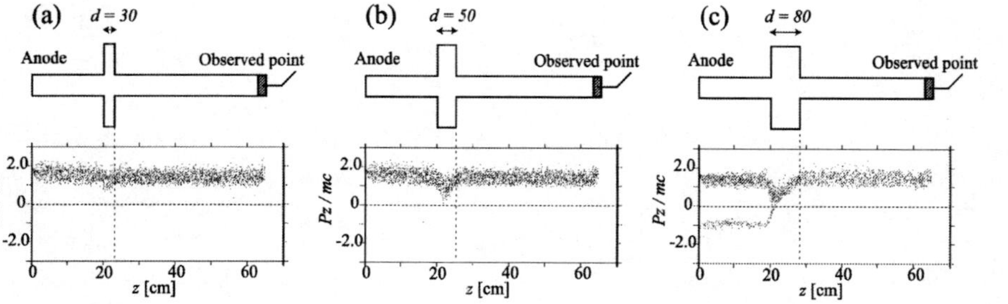

FIGURE 3. Typical simulation results for different gap lengths. The kinetic momentum along axis (Pz) is plotted against z. (a) Gap length = 30 mm. The electron beam passes through the gap without loss. (b) Gap length = 50 mm. For the parameters used in these simulations, this is the maximum length for the beam to pass through without loss. (c) Gap length = 80 mm. A virtual cathode is clearly observed. A part of electrons is reflected.

We looked for a criterion for the gap length through which the beam current passes through without loss. Dependencies of the gap length without loss on the current, energy and radius of the beam indicate the influence of the space-charge limiting current. In Fig. 4, the simulated and experimental gap lengths through which the beam current passes without loss are plotted against calculated values of $2(R - r_b)$ for each data, where R is the calculated maximum tube radius for the beam with current I, energy γ_0 and radius r_b to pass through by eq.(1). From the simulation results, the gap lengths increase with $2(R - r_b)$ linearly. And from the experimental results, they increase also with $2(R - r_b)$. The difference between simulated and experimental results is considered to come from the difference of the voltage and current waveforms. We propose the gap length of $2(R - r_b)$ as a rough criteria that the beam current can pass through the gap without loss. When the gap length is shorter than $2(R - r_b)$, the injected annular beam with current larger than the limiting current at the gap can pass through the gap without loss.

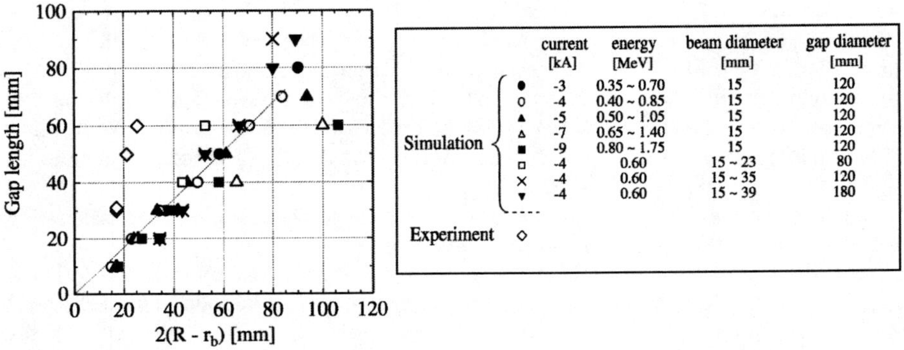

FIGURE 4. The maximum gap lengths through which the beam passes without loss are plotted against $2(R - r_b)$.

CONCLUSIONS

When an intense electron beam is injected into a gap with larger beam current than the space charge limiting current at the gap, there is a minimum gap length to form a virtual cathode. When the gap length is shorter than the minimum gap length to form the virtual cathode, the injected beam current can pass through the gap without loss. A rough criteria about the gap length through which the injected beam current can pass without loss is proposed.

ACKNOWLEDGMENTS

A part of this work is supported by a Grant-in-Aid for Scientific Research from Ministry of Education, Science, Sports and Culture, Japan.

Generation of Trains of Subnanosecond Bunches by Automodulation of IREB Passing through a Series of Coaxial Cavities

Masaki Kamada,[a] Michiya Mori,[a] Yoko Yamada,[a] Ritoku Ando, Naum. S. Ginzburg[b] and Keiichi Kamada[a]

[a] *Department of Physics, Faculty of Science, Kanazawa University, Kanazawa, 920-1192 Japan*
[b] *Institute of Applied Physics, Russian Academy of Sciences, Nizhny Novgorod, Russia*

Abstract. Automodulation of an intense relativistic electron beam was reexamined experimentally to obtain trains of subnanosecond electron bunches. Transmission line theory suggested that the round trip time for the electromagnetic wave in the cavity must be comparable with the rise time of the beam current to obtain high voltage at the gap. Single cavity experiments were carried out using the beam with current rise time of 6 ns. Cavities whose round trip time was less than 6 ns were utilized. The longer cavity resulted in the higher gap voltage and the higher current modulation. For high-frequency current modulation for a beam with slow rise time, we propose a series of cavities with decreasing lengths to improve the current rise time.

INTRODUCTION

An intense relativistic electron beam (IREB) with subnanosecond duration is an interesting subject of study for an application to high power, short pulse millimeter wave generation called superradiance [1]. We reported that an IREB with duration of less than 1 ns was attainable from 10 ns IREB with multi-stage autoacceleration using decreasing length cavities [2]. A single subnanosecond IREB with current of several kA and energy of 1.1 MeV was obtained. On the other hand, trains of subnanosecond IREB pulse should expand the abilities of IREBs. If the high-level current modulation can be obtained by automodulation scheme, trains of subnanosecond electron bunches are available.

Automodulation was intensively studied by M. Friedman, et al. [3]. They reported that electron beams with voltage varying from 0.25 to 1 MeV, current from 1 to 8 kA, and pulse duration from 20 to 300 ns were modulated to a depth of nearly 100 %. However, in our case, only a poorly modulated electron beam was observed with our IREB with pulse duration of 175 ns when four cavities were utilized in accordance with the results reported in [3]. We report here how the relation between the beam current rise time and the round trip time of the cavity for electromagnetic wave affects on the level of current modulation. And a new approach to automodulation for an IREB with slow rise time is proposed.

CP650, *Beams 2002: 14th International Conference on High-Power Particle Beams,*
edited by T. A. Mehlhorn and M. A. Sweeney

EXPERIMENTAL SETUP

Two IREB sources were utilized to generate a short or a long IREB. Pulserad 105A was utilized to generate a short pulse IREB of 500 keV, 4 kA, 12 ns with rise time of 6 ns. Pulserad 220G was utilized to generate a long pulse IREB of 500 keV, 4 kA, 175 ns with rise time of 20 ns. Both IREB sources are produced by Physics International.

Typical waveforms of the diode voltage and the beam current for the short IREB are shown in Fig. 1(a) and (b), respectively. For both IREBs, foilless diode consisted of a carbon cathode with hollow knife-edge of 20 mm diameter and a carbon anode with 29 mm diameter aperture. The anode-cathode spacing in the diode was 5-15 mm. An annular electron beam with diameter of 20 mm and thickness of 1 mm was injected from the diode into a conducting drift tube with inner diameter of 31 mm and following coaxial cavities. The system was immersed in axial magnetic field of 1 T applied by solenoid coils.

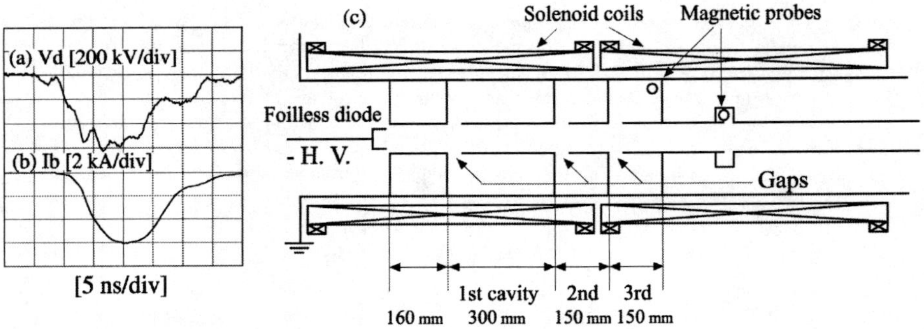

FIGURE 1. Typical waveforms of short IREB diode voltage (a) and diode current (b). Experimental setup of a series of cavities with decreasing length (c).

Two type arrangements of coaxial cavities were utilized in the experiments. A single cavity was utilized to examine how the relation between the rise time of the beam current and the cavity length affected the modulation level. And a series of cavities with decreasing length adjusted to 300-150-150 mm was utilized. In any experiments, the first cavity was connected to the drift tube via gap at z = 16 cm, where z is the distance from the anode. The impedance of any cavity was 76 ohm. The base pressure in the system was maintained below 1 x 10^{-5} Torr.

A Faraday cup was used to measure the beam current. Aluminum foils of various thicknesses were placed in front of the Faraday cup and the transmitted current through aluminum foil was measured to estimate the kinetic energy of beam electrons. Magnetic probes were used to measure the beam current and the current at the end of cavity.

EXPERIMENTAL RESULTS AND DISCUSSIONS

Our goal is to obtain high-level 1 GHz current modulation with a long pulse IREB. Following the experimental results of M. Friedman [3], we employed a series of four cavities with the same length of 75 mm to obtain highly modulated IREB. The round trip time of electromagnetic wave for the cavity is 0.5 ns. The beam decelerated and accelerated with 0.5 ns interval at the gaps was expected to result in the trains of 0.5 ns electron bunches, i.e. 1 GHz current modulation. However, only a weakly modulated electron beam was observed by the long pulse IREB. We considered that the current rise time of 20 ns for the long pulse IREB was too long for high-level modulation in comparison with the round trip time of the cavity of 0.5 ns. The level of current modulation should be related to the induced voltage across the gap. We returned to a single cavity. It was assumed that high-level current fluctuation could be attainable with a series of cavities, if high induced gap voltage was obtained with a single cavity. The induced gap voltage, $V_g(t)$ is expressed as the equation below by transmission line theory.

$$V_g(t) = Z \times (I(t) - 2I(t-T) + \cdots\cdots) \qquad (1),$$

where Z is the impedance of the coaxial cavity, I is the beam current and T is the round trip time for electromagnetic wave in the cavity. The induced gap voltage was calculated for different cavity lengths using a simplified current waveform of the beam. The waveform and calculated gap voltage are shown in Fig.2. The gap voltage increased as the cavity length increased and it was saturated with the cavity length of 900 mm. The round trip time for electromagnetic wave in the cavity with length of 900 mm is 6 ns, which is the same as the rise time of the beam current.

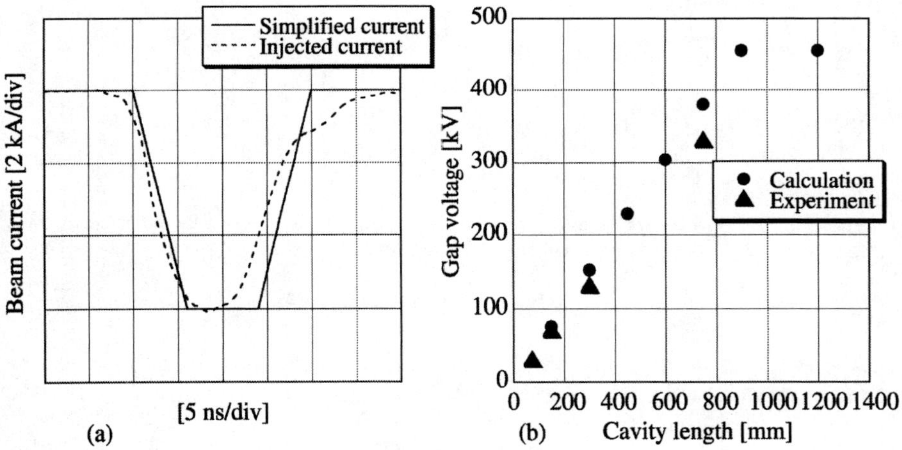

FIGURE 2. (a)Waveform of injected beam current was simplified for calculation. (b) Calculated and experimental results of induced gap voltage for each cavity length. In the experiment, the difference of electron energies between with and without cavity is interpreted to the gap voltage.

A single cavity whose length was changeable was utilized for the short pulse IREB. The cavity length was adjusted to 75, 150, 300, 750 mm. The round trip times for all cavities are shorter than the current rise time of 6 ns. No fluctuation in the beam current was observed for IREB propagation through a smooth drift tube. The experiments with a 75 mm cavity resulted in suppression of the modulation amplitude. The higher level of current modulation was obtained with the longer cavity. When the current modulation was observed, the Fast Fourier transform of the Faraday cup waveforms showed that the fundamental frequency corresponded to the round trip time for electromagnetic wave in each cavity. The kinetic energy of electrons that passed through the gap was measured. The increases of electron energy for each cavity show good correspondence with the calculated gap voltage, as shown in Fig.2. It was observed that the larger energy increase resulted in the larger current fluctuation. We conclude that the round trip time of the cavity must be comparable to the rise time of the beam current for high-level current modulation.

The rise time of the beam current is difficult to improve. Therefore, it is difficult to obtain high-level high-frequency current modulation with an IREB with slow rise time. We propose a series of decreasing lengths cavities to improve the rise time and to obtain higher level current modulation. In Fig.3, the beam current waveforms detected by magnetic probes are shown. The cavities with lengths of 300-150-150 mm increased the level of current fluctuation compared with those with the lengths of 150-150 mm. In Fig.3, the level of modulation is still low, so that we develop the devices to obtain higher modulation.

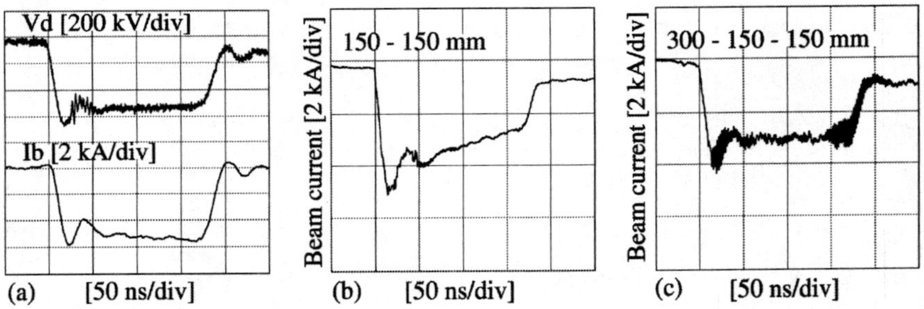

FIGURE 3. (a) Diode voltage and beam current without cavities detected by the magnetic probe. (b) Beam current with 150-150 mm cavities. (c) Beam current with 300-150-150 mm cavities.

ACKNOWLEDGMENTS

A part of this work is supported by a Grant-in-Aid for Scientific Research from Ministry of Education, Science, Sports and Culture, Japan.

REFERENCES

1. N. S. Gintburg, I. V. Zotovan, et.al., Phys. Rev. Lett., vol. 78, pp. 2365--23680, 1997.
2. D. Hasegawa, K. Kamada, et. al., IEEE Trans. Plasma Sci., vol. 28, pp. 1648--1651, 2000.
3. M. Friedmn, V. Serlin, A. Drobot and L. Seftor, J. A ppl. Phys., vol. 56, pp. 2459--2474, 1984.

Suppression of the Transit –Time Instability in Large-Area Electron Beam Diodes*

Matthew C. Myers, [a] Moshe Friedman,[a] Stephen B. Swanekamp,[b] Lop-Yung Chan,[c] Larry Ludeking,[d] and John D. Sethian[a]

[a]*Naval Research Laboratory, Laser Plasma Branch, Washington, D.C. 20375*
[b]*JAYCOR, McLean, VA 22102*
[c]*Science Applications International Corporation, McLean, VA 22102*
[d]*Mission Research Corporation, Newington, VA 22122*

Abstract. Experiment, theory, and simulation have shown that large-area electron-beam diodes are susceptible to the transit-time instability. The instability modulates the electron beam spatially and temporally, producing a wide spread in electron energy and momentum distributions. The result is gross inefficiency in beam generation and propagation. Simulations indicate that a periodic, slotted cathode structure that is loaded with resistive elements may be used to eliminate the instability. Such a cathode has been fielded on one of the two opposing 60 cm x 200 cm diodes on the NIKE KrF laser at the Naval Research Laboratory. These diodes typically deliver 600 kV, 500 kA, 250 ns electron beams to the laser cell in an external magnetic field of 0.2 T. We conclude that the slotted cathode suppressed the transit-time instability such that the RF power was reduced by a factor of 9 and that electron transmission efficiency into the laser gas was improved by more than 50%.

INTRODUCTION

One promising design for a fusion power plant is based on using a modular array of intense krypton fluoride (KrF) lasers to directly compress and heat a small fuel pellet [1]. Such a KrF laser module would be pumped by repetitively pulsed, counter-streaming, large-area electron beams (e-beams). Theory, simulation, [2,3] and experiments [4] performed on the NIKE [5] KrF laser at the Naval Research Laboratory have shown that large-area, e-beam diodes are susceptible to the transit-time instability. The instability modulates the electron beam spatially and temporally, producing a wide spread in electron energy and momentum distributions. This results in inefficient and uneven pumping of the laser gas leading to spatial non-uniformities in the laser pulses. A transmission line analysis and 2-D MAGIC [6] simulations predict that a periodic, slotted cathode structure that is loaded with resistive elements may be used to eliminate the instability and improve efficiency [4]. The following sections investigate the transit time instability and report on the results of experiments using a slotted/loaded cathode on NIKE. We conclude that the slotted cathode suppressed the transit-time instability such that the RF power was reduced by a factor

*Work supported by the U.S. Department of Energy

CP650, *Beams 2002: 14th International Conference on High-Power Particle Beams*,
edited by T. A. Mehlhorn and M. A. Sweeney

of 9 and that electron transmission efficiency into the laser gas was improved by more than 50%. Experimental results and further 3-D Magic simulations indicate that a large-area diode can be designed that will eliminate the instability and meet the efficiency requirements for a power plant module.

TRANSIT-TIME INSTABILITY

A large-area, space charge limited diode can be modeled as a parallel-plate transmission line with distributed conductance and capacitance per unit length, G and C respectively. These parameters are functions of frequency f and the electron transit-time across the anode-cathode (AK) gap τ [2]. Whenever $\tau \approx (n+1/4)/f_n$ where n is an integer, the conductance becomes negative and a TEM wave supported by the gap will grow in amplitude [3]. Electrons emitted from the cathode experience the applied accelerating voltage modulated by the RF voltage from the TEM wave. The electrons alternatively give and receive energy from the field as they are accelerated and decelerated across the AK gap. When the frequency of the wave is such that the electrons give net energy to the field for the duration of their transition across the gap then the wave grows in amplitude and electron flow in the diode is unstable. For a 5 cm AK gap and a 500 kV voltage, the electron transit time is about 0.5 ns, $f_1 \approx 2.5$ GHz, and the predicted growth $\alpha \approx -3.9$ Np/m. Neglecting losses, the gain over a length of diode l=2 m is estimated at $e^{-\alpha l}$=2440. MAGIC simulations of the NIKE diode indicate that the instability has drastic effects on beam generation. A FFT of the simulated diode voltage waveform showed a high amplitude component at 2.5 GHz. The electron momentum distribution across the length of the diode was highly modulated showing the "imprint" of the wave. Moreover, an initially monochromatic 500 keV electron energy distribution had significantly spread from 250 to 750 keV, greatly reducing electron transport efficiency into the laser gas [4].

SUPPRESSION OF THE INSTABILITY

A transmission line analysis was used to design a diode to suppress the transit-time instability [4]. For the KrF laser application, a straightforward solution was to modify the cathode using periodic, resistively loaded, transmission line stubs that would impose a stop-band for the fundamental 2.5 GHz mode of the instability. This was accomplished by constructing a cathode from 3 cm wide vertical strips of emitter separated by 9 mm wide x 2.6 cm deep slots which were periodically loaded with a microwave absorbing material. For this design, G becomes positive at 2.5 GHz, the waves are attenuated, and the instability is suppressed. 2-D MAGIC simulations of this cathode showed 10^3 times less RF amplitude at 2.5 GHz on the diode voltage and no modulation of the electron momentum distribution while the electron energy distribution remained uniform at 500 keV [4]. Note that an additional advantage in electron transport efficiency is gained when openings in the pressure foil support structure (hibachi) are designed to match the "stripped" beam pattern [7].

Figure 1 is a plan view of one side of the NIKE 60 cm aperture laser amplifier. For laser operations, counter-streaming electron beams are emitted from 200 x 60 cm^2 flat velvet cathodes that are immersed in an axial magnetic field of 0.2 T. The beams traverse a 5 cm vacuum AK gap, go through a 33 μm thick Ti anode foil, a 5 cm deep hibachi, and a 43 μm thick Ti pressure foil before entering the laser cell where they excite high pressure KrF gas to produce laser light at 248 nm. For the work discussed here, a single diode is used at 500 kV and a diagnostic plate/beam dump is placed at either the anode or pressure foil position. The diagnostic plate is covered with 1 cm thick carbon tiles and has several, centrally located, 5 cm diameter access ports for diagnostics. These include: i) a calorimeter which measured the average energy loss of electrons in a 50 μm thick Ti foil, ii) magnetic probes which measured dI/dt, and iii) an energy analyzer which consisted of an evacuated tube in which a beamlet propagated through 4 equally spaced, 140 μm thick, carbon foils with 5 Rogowski coils measuring the attenuation of the beam current after each foil. The average electron energy and a rough energy distribution were obtained by comparing the attenuated currents with results from MAGIC simulations. A magnetic probe outside the beam measured the RF noise; the total diode current and generator voltage were also monitored. A standard, flat velvet cathode and the slotted/loaded cathode with velvet strip emitters were compared.

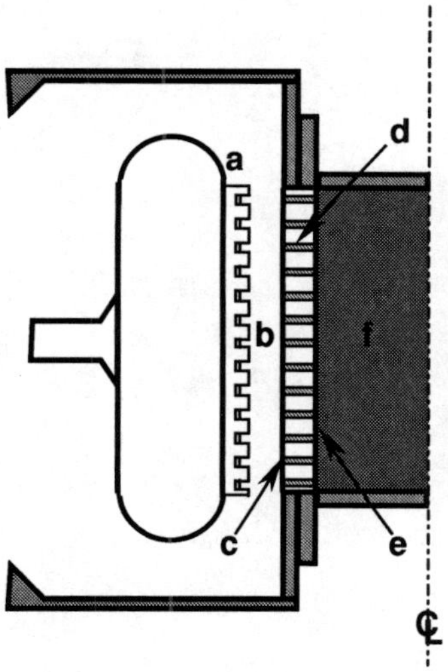

FIGURE 1. Drawing of NIKE diode: a) slotted cathode, b) AK gap, c) anode foil, d) hibachi, e) pressure foil, and f) laser cell.

Figure 2 illustrates the suppression of the instability. The time differentiated RF current in the flat cathode case shows early time, high amplitude oscillations. A FFT of the signal reveals high frequency components spread over 2-3 GHz range. The RF current in the slotted cathode case starts some 150 ns later and is significantly lower in amplitude with a narrow frequency spectrum near 2.2 GHz. When the two signals are corrected for cable attenuation, integrated, and squared, the slotted cathode case shows a factor of 9 reduction in RF power. Figure 3 shows the effect of the instability on the electron energy distribution. Energy analyzer data (•) taken at the pressure foil position with the flat cathode give attenuated currents that are comparable to those seen in simulations using an initial electron distribution that is a Gaussian of 400 keV average energy with a half-width of 150 keV (line A). The instability has caused a spread in electron energy which results in an average loss of ~50 kV/electron. With exacerbated losses in the pressure foil, the total electron transport efficiency was measured at 35%. Using the slotted cathode, the instability growth is reduced and the energy analyzer shows data (■) that closely resembles a 500 keV monochromatic distribution (line B). Suppression of the instability allows beam electrons to maintain

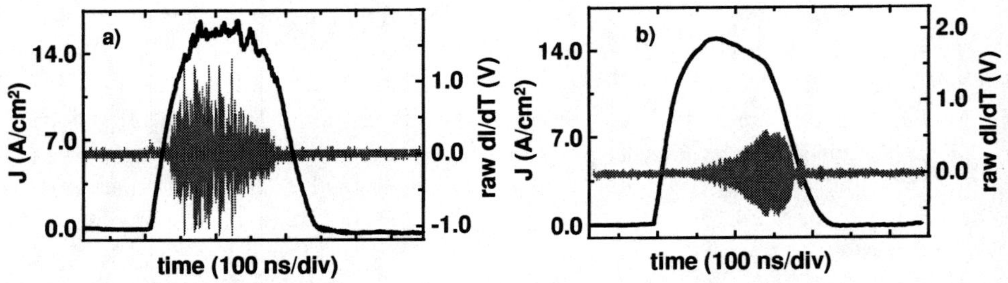

FIGURE 2. Time differentiated RF current superimposed on the beam current density profile measured at the pressure foil position for a a) flat cathode and b) slotted cathode.

their energy in the diode, reducing foil scattering and hibachi losses. The slotted cathode improved the total electron transport efficiency to 55%.

As designed, the slotted cathode suppressed but did not eliminate the transit-time instability. Although the gain is much smaller, it is possible that TEM modes may grow along the other, 60 cm dimension of the cathode. 3-D MAGIC simulations show that with a cathode slotted in both dimensions the instability is completely eliminated. Such a cathode design, used at the size and voltage required for a power plant module, is estimated to achieve an electron transport efficiency of 80%.

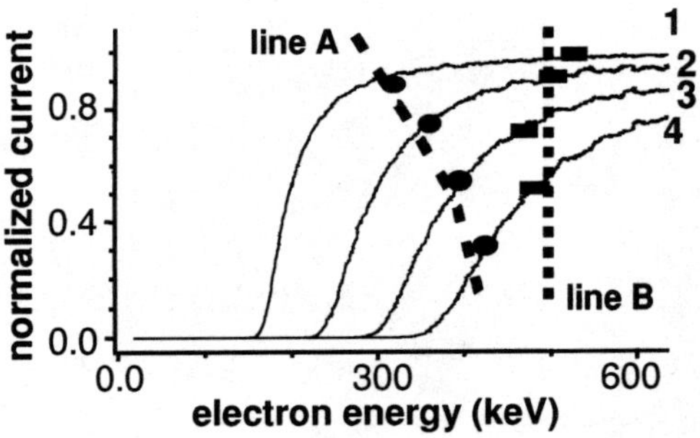

FIGURE 3. Calculated, normalized transmitted current through carbon foils 1-4 as a function of electron energy: line A- Gaussian electron distribution with average energy 400 keV and half-width 150 keV, line B- uniform 500 keV distribution. Energy analyzer data: •- pressure foil position with flat cathode; ■-anode foil position with slotted cathode.

REFERENCES

1. M. McGeoch, et al., Fusion Technology, 32, 610 (1997).
2. C. K. Birdsall and W. B. Bridges, *Electron Dynamics of Diode Regions*, (Academic, New York, 1966).
3. T. M. Antonsen, et al., Phys. Fluids 27, 1257 (1984).
4. M. Friedman, et al., Appl. Phys. Lett. 77, 1053 (2000).
5. J. D. Sethian, et al., Rev. Sci. Instrum. 68, 2357 (1997).
6. B. Goplen, et al., Phys. Comm. 87, 54 (1995).
7. F. Hegeler, et al., "Efficient electron beam deposition for repetitively pulsed krypton fluoride lasers", these proceedings.

On Measurements of the Basic Parameters of High-Current Relativistic Electron Beams

Mikhail I. Fuks,[a] Mikhail B. Goikhman,[b] Nikolay F. Kovalev,[b] Alexey V. Palitsin,[b] and Edl Schamiloglu[a]

[a] *University of New Mexico, Albuquerque, NM 87131 USA*
[b] *Institute of Applied Physics, Nizhny Novgorod, 603600 Russia*

Abstract. The possibilities of performing measurements such as electron energy on the basis of cyclotron absorption of electromagnetic fields by electrons, and current using a Rogowski coil accounting for dispersion, are considered for short pulse relativistic electron beams.

INTRODUCTION

In order to improve the characteristics of high power microwave sources it is necessary to increase the precision of the measurements of electron beam parameters. Difficulties associated with this are connected with the poor shot-to-shot reproducibility of the parameters. This paper describes techniques for measuring electron energy W using the cyclotron absorption of electromagnetic radiation, and measuring electron current using a Rogowski coil accounting for dispersion. These techniques are developed for application to short pulse electron beams.

MEASUREMENT OF ELECTRON ENERGY

In order to measure the energy of electrons moving with velocity v_z in a uniform channel of length L immersed in an homogeneous magnetic field H we use the resonant interaction of fast cyclotron waves of the electron beam with the microwave field when the condition $|\theta| < \pi$ is satisfied. Here

$$\theta = kL_{eff}\left(\beta_z^{-1} - \beta_{ph}^{-1} - \omega_c(\gamma\beta_z\omega)^{-1}\right); \quad (1)$$

$L_{eff} = 1/\nu$, where ν is the decrement of spatial attenuation; $\beta_{ph} = v_{ph}/c = k/h$ is the phase velocity v_{ph} of the external wave related to the speed of light c; h is the longitudinal wavenumber; $k = 2\pi/\lambda = \omega/c$; ω is the angular frequency of radiation; $\omega_c = eH/mc$ is the cyclotron frequency; e and m are the charge and rest mass of an electron, respectively; $\gamma = \left(1-\beta_z^2-\beta_\perp^2\right)$ = 1 + W/mc^2; $\beta_z = v_z/c$; and $\beta_\perp$ is the relative transverse electron velocity. The incident wave with negative group and phase

CP650, *Beams 2002: 14th International Conference on High-Power Particle Beams,* edited by T. A. Mehlhorn and M. A. Sweeney

velocities is reradiated into the cyclotron wave whose energy is scattered in the electron dump in narrow resonant bands. Assuming that $\beta_\perp^2 << \beta_z^2$, we have

$$\Delta H/H \approx \lambda \left|\beta_z^{-1} - \beta_{ph}^{-1}\right|^{-1} / L_{eff} \text{ and } \Delta\gamma/\gamma \approx \lambda\beta_z \left|1-(\beta_z\beta_{ph})^{-1}\right|^{-1} / L_{eff} . \quad (2)$$

To avoid line broadening, ν must satisfy the condition $1/L < \nu < 1/L_k$, where L_k is the spatial extent of the fields. Also, the increment is proportional to bandwidth via Eq. (2). The average electron energy is found from $\gamma_\Sigma \approx 0.5(\omega_c/\omega + \omega/\omega_c)$, $\omega_c/\omega > 1$, which follows from Eq. (1) when $\theta = 0$. Equation (2) is an estimate of the maximum precision for determining the average electron energy. An experimental set-up to perform such a measurement in a waveguide is shown in Fig.1.

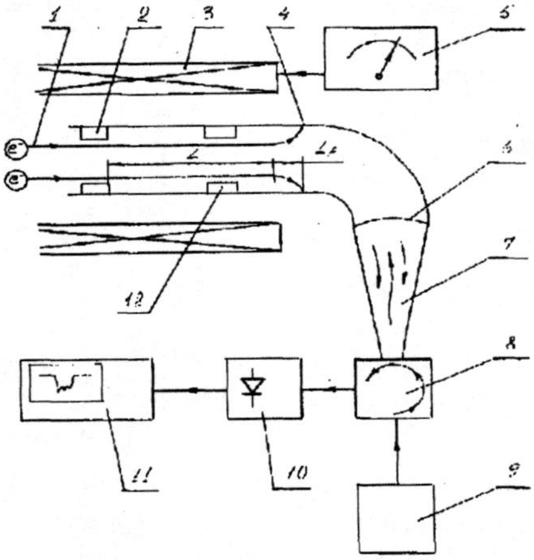

FIGURE 1. Conceptual schematic of an experiment to measure electron energy using cyclotron absortion: 1.- thin-walled electron beam; 2.- reflector-collimator; 3.- solenoid; 4.- electron beam dump; 5.- controlled power supply for solenoid; 6.- curved waveguide with output vacuum window; 7.- mode converter; 8.- circulator; 9.- magnetron; 10.- microwave probe; 11.- oscilloscope; 12.- location of semi-transparent diaphragm

The wave formed by mode converter (7) is reflected by (2) and recorded by detector (10). When the electron beam propagates through the channel, the wave is completely absorbed. To provide the required decrement and, therefore, precision of measurement, it is necessary to select the operating wave and the direction of its rotation relative to the magnetic field. The measurement precision, which depends weakly on the value of the current J_0, is determined by the bandwidth and the form of the reradiated line. Using equations from [1] describing the interaction of a plane linear polarized wave with a thin monoenergetic beam we obtain transmission coefficients (Fig. 2, left), the line form and also expressions (in explicit form) for the bandwidth of the resonant interaction, $\Delta\omega_c/\omega = 4[I(1+\beta_z)/\gamma]^{1/2} / \beta_z$ and for $L_{eff} = 1/\nu_{\max} = \beta_z[\gamma/I(1+\beta_z)]^{1/2}$. For narrower bandwidth, one must decrease the coupling parameter $I = \left(eJ_0/mc^3\right)/4k^2S_\perp$ ($S_\perp$ is cross-section of the channel) and to increase the length L (Fig. 2, left). When it is not possible to achieve greater accuracy the resonant method (placement of diaphragm 12, Fig. 1) is more suitable.

The use of multi-frequency waves and multiple resonances can expand the method, particularly for single pulse measurements. The use of higher order waves leads to multipole interactions, including the magnetic dipole one, facilitating measurement of the initial transverse electron velocities.

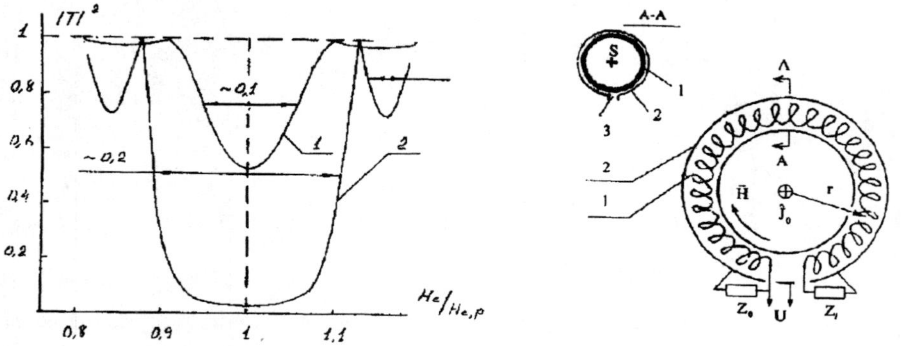

FIGURE 2. **Left:** Lines of cyclotron absorption: $kL = 30$; 1.- I = 10^{-4}; 2.- I = 10^{-3}. **Right:** Rogowski coil with parallel connection to load: 1.- winding; 2.- shielding screen; 3.- narrow slit.

MEASUREMENT OF SHORT CURRENT PULSES

In early publications on Rogowski coils (see, for example, [2]) it was shown that the coils are appropraite for measuring short current pulses down to a few nanoseconds. However, the response of the coils to short and long pulses is marketly different. For short pulses a winding (Fig. 2, right) covered by a shielding screen with a longitudinal narrow slit consitutes a waveguide where slow waves are excited by the self-magnetic field of the current $J_0(t)$. When the pulse duration τ_0 is less than its transit time along the waveguide, we observe a series of pulses on the load with a waveform of the first pulse $U_1(t)$ close to the waveform of the current $J_0(t)$ without additional integration. Usually distortion of the waveform $U_1(t)$ for short τ_0 is explained [2] by the influence of parasitic capacitance and inductance of the load, and losses in the windings. However, for short pulses waveguide dispersion is evident because of the expansions of the spectrum $\left(\omega_{\max} \sim 2\pi/\tau_0\right)$, whereas it is difficult to reject the dispersion of the spiral waveguide at relatively low frequencies.

We solve for the complex amplitudes of the voltage on the load using a rapidly converging series describing the equivalent circuit for the measurement (Fig. 3, right)

$$U = -\frac{\alpha}{2} J_0(\beta_0 + 1)\left[1 - (\beta_l + 1)\sum_{n=0}^{\infty} \beta_0^n \beta_l^n e^{-i(2n+1)hl} + (\beta_0 + 1)\sum_{n=1}^{\infty} \beta_0^{n-1} \beta_l^n e^{-2inhl}\right], \quad (3)$$

where $\beta_0 = (Z_0 - \rho)/(Z_0 + \rho)$ and $\beta_e = (Z_l - \rho)/(Z_l + \rho)$ are the reflection coefficients from the load; $\alpha = \omega\kappa/h$; h and ρ are the wavenumber and impedance of the main wave in the spiral waveguide; $\kappa = N\mu_0 S/2\pi a$ is a coupling parameter; N is density of the windings, and $2a$ is the coil diameter. Taking an inverse Fourier transform of the series in Eq. (3), each term will correspond to separate pulses with different delays according to the exponential factors. This series is convenient for the analysis of the short pulse operation of the Rogowski coil because the first few pulses are separated from each other when the current pulse duration is less than its transit time in the coil and distortions are modest. The characteristic $h = h(\omega)$ of the wave in the screened spiral waveguide can be represented as $h = -i\sigma + (\omega/v) + (q\,\omega'/v)$; $\sigma = \lim_{\omega\to 0} ih$; $v = \lim_{\omega\to 0}(d\omega/dh)$; $q = (v/6)\lim_{\omega\to 0}\left(d^3h/d\omega^3\right)$. The first term in the inverse Fourier series

$$\hat{U}_1(t) + \sigma v\int_0^t \hat{U}_1(t')dt' - q\left(d^2\hat{U}_1/dt^2\right) = -\kappa v(\beta_0 + 1)\hat{J}_0(t)/2 \tag{4}$$

couples the current pulse $\hat{J}_0(t)$ with the first pulse $\hat{U}_1(t)$ on the load. Equation (4) allows one to estimate the difference between these pulses and compare the influence of different factors on forming $\hat{U}_1(t)$. The integral in Eq. (4) provides corrections to losses in the waveguide; the next term is the influence of dispersion. The strongest influence of dispersion is on the pulse edges; the shorter the pulse the larger the distortion. Dispersion limits the ability to measure short pulses. On the other hand, Eq. (4) can be used to reconstruct the current pulse from the observed pulse $\hat{U}_1(t)$ on the load, thereby expanding the applicability of Rogowski coils. Further improvements can be obtained by replacing the screen and narrow slit with other cylindrical conductor configurations.

ACKNOWLEDGEMENTS

This work is supported, in part, by an AFOSR/New World Vistas Grant.

REFERENCES

1. Fuks, M.I., Schamiloglu, E., Kovalev, N.F., and Palitsin, A.V., Proc. SPIE 4371, Intense Microwave Pulses VIII, 146-156 (2001).
2. Vasserman S.B., Devices and Technique of Experiments, 2, 99-103 (1972).

E-Beam Accelerator For Eximer Laser*

Boris M. Kovalchuk, Edward N. Abdullin, Dimitry M. Grishin, Vladimir P. Gubanov, Valery B. Zorin, Alexandre A. Kim, Eugeny V. Kumpjak, Alexandre V. Morozov, Victor S. Skakun, Alexey S. Steptchenko, Viktor F. Tarasenko, Valery S. Tolkachev, Peter M. Schanin, Nikolay V. Tsou

High Current Electronics Institute, Academichesky Ave. 4, 634055 Tomsk, Russia

Abstract. A high current e-beam accelerator for pumping of the 200 liter eximer laser is developed providing electron energy of 550 keV, diode current of 320 kA, e-beam current of 250 kA. The high voltage part of the accelerator consists of two linear transformers with the stored energy of 98 kJ. In order to reduce the influence of the self magnetic field on e-beam formation the vacuum diode is divided into 6 separate magnetically isolated diodes.

INTRODUCTION

The design of the high current accelerators for eximer lasers with large cross section of the e-beam has to solve the problem of the influence of the magnetic fields in the diode on the e-beam formation. In order to reduce the self magnetic field of the beam, it could be produced by several accelerators with reduced current or the diode could be divided into separate magnetically isolated sections. Each section is driven by a separate power supply like pulse forming lines or Marx generators [1, 2].

In the report we present the design and test results of the accelerator consisting of 6 separate magnetically isolated diodes that are powered by 2 parallel linear transformers. The current distribution between the diodes is provided by a cylindrical collector of large diameter and several plates that determine the current flow between the cathodes and the return current.

DESIGN OF THE ACCELERATOR

The voltage pulse on the diode is provided by 2 parallel linear transformers with vacuum insulation of the secondary turn [3]. The voltage to the diodes is delivered by vacuum lines that are secondary turns of the transformers.

The design of the e-beam diode is shown in Fig. 1. The main parts of the diode are: vacuum chamber 1 with side flanges 2, collector 3, cathodes 4 with cathode holders 5, direct current plates 6, magnetically isolated return current plates 7, 8, 9, gas cavity 10, and cavity holders 11. The collector is suspended coaxially with vacuum chamber by two springs 12, that locate at the upper part of the chamber.

* This work is supported by the contract CXSDC – 1A98014b

CP650, *Beams 2002: 14th International Conference on High-Power Particle Beams*, edited by T. A. Mehlhorn and M. A. Sweeney

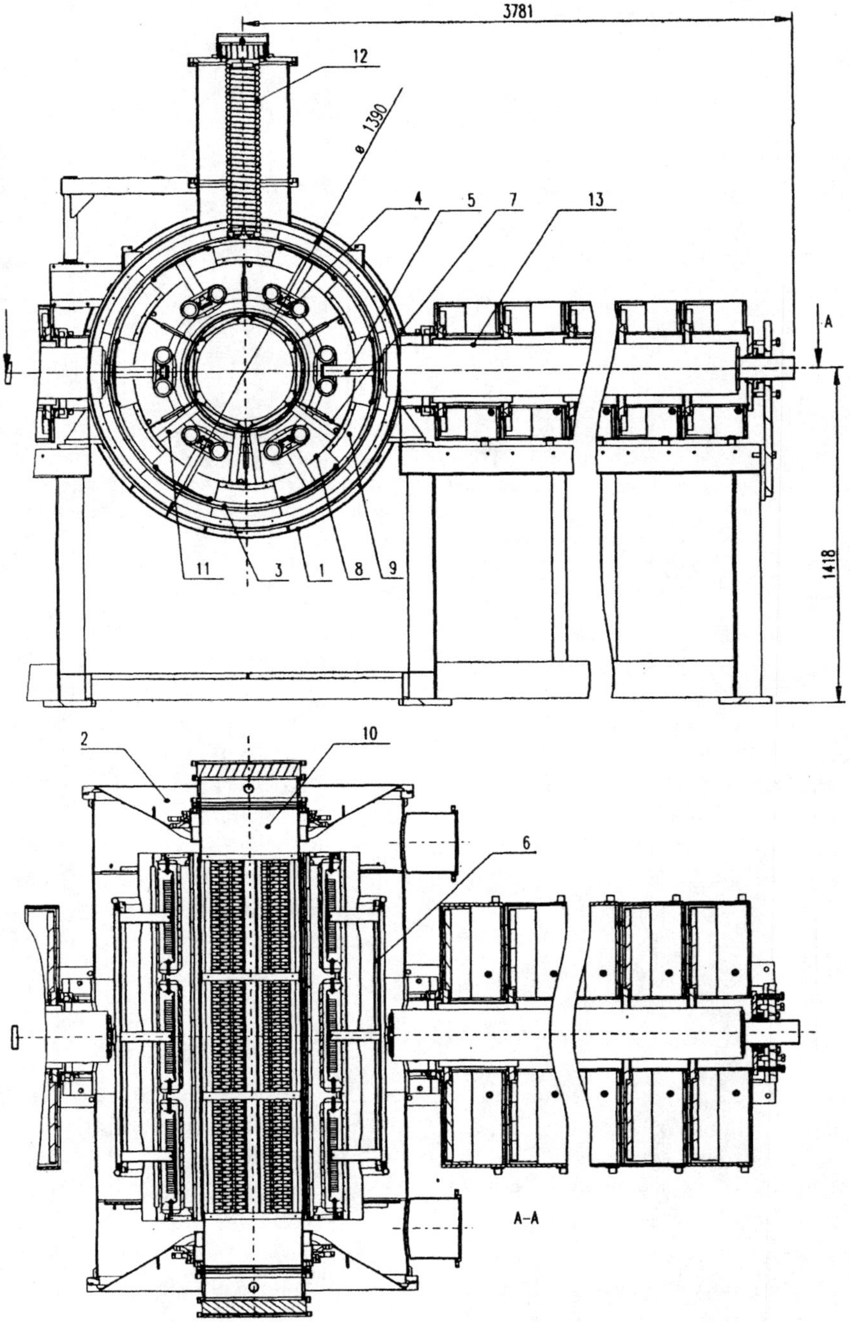

FIGURE 1. Design of the e-beam diode.

There are 18 cathodes in total, they are installed with cathode holders 5 on the plates 6 in such a way that the cathode of each magnetically insulted diode consists of three parts. Such design reduces the current flowing along the outer surface of the cavity in axial direction. The magnetic insulation of the diodes is provided by the collector, cathode plates 6 and return current plates 7-9. The inductance of the e-beam diode is ~40 nH, close to the inductance of the vacuum insulators of the pulse forming lines in usual accelerators.

The shape of the cathodes was defined in simulations performed by using a specially developed 2-D numerical code. The emitting surfaces are covered with velvet. The width of the emitting surface is 120 mm, total emitting area is 0.95 m^2, the AK-gap is 6 cm.

The geometric transparency of the support frame is ~ 75 %. A 40 um Ti foil is used to separate the gas cavity from the vacuum volume. The diameter of the gas cavity is 410 mm, the volume – 200 liters.

The vacuum in the system is provided by two AVDM–250 pumps with the speed of 870 l/s, the residual pressure is (3-4)10^{-5} Torr.

EXPERIMENTAL RESULTS

Fig. 2 shows the curves characterizing the performance of the accelerator. Traces U_1, U_2 represent sum of voltages on primary turns of two transformers, these voltages

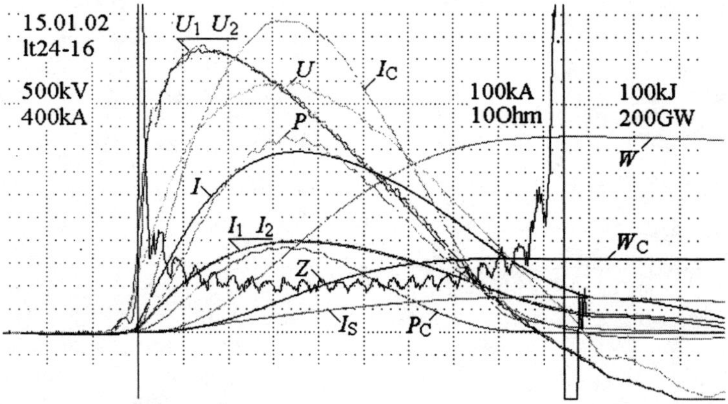

FIGURE 2. Experimental curves at 85 kV charge voltage (time scale is 100 ns/div).

are measured by capacitive dividers installed in each stage. Trace U is the voltage on the collector. It is measured by recording the voltage on the inductance of a part of a spring that holds the collector. The difference in the traces U_1, U_2 and U appears after the explosive emission on the cathodes begins at ~150 kV. The currents I_1, I_2 and I are integrated signals of the B-dot probes in the secondary turns of the transformers, the scale for these currents is 40 kA/div. The probe used to calculate I_1 is installed in one module between the first and the second stages, the probe of I_2 locates in other module at the output of the 10th stage. They coincide indicating no current loss in secondary transformer turns. The trace $I = I_1 + I_2$ is total diode current, W is the energy delivered to the diode, it is calculated by using I and U. Trace I_c (10 kA/div) is the current measured on a 390 mm diameter cylinder that was installed inside the laser cavity. This cavity was pumped to ~ $3 \cdot 10^{-5}$ Torr with Ti foil on the windows. Trace W_c represents estimation of the e-beam energy behind the Ti foil, it is calculated by using U and I_c and does not take into account the electron energy loss in the foil. Trace $Z = U/(I-I_s)$ is the impedance of the diode, I_s is the current in the springs.

Table 1 shows the parameters of the accelerator at two charge voltages. The e-beam energy delivered to the gas, W_g, is measured from the pressure jump in the gas cavity by a probe 6MDX-3B. The laser beam spot on the photopaper indicates that the radiation is uniform within ~10 %.

TABLE 1. Parameters of the accelerator'

Charge voltage, kV	**80**	**85**	**Efficiency**
Diode voltage U, kV	500	550	
Total diode current I, kA	280	320	100%
E-beam current in the diode I_b, kA	214	250	78%
Anode current density, A/cm^2	14	16.5	
E-beam current behind the frame I_f, kA	148	176	
E-beam current behind the foil I_c , kA	115	136	54.4%
Current in the springs I_s (at t = 400 ns), kA	30	32	
Current loss from the collector surface, kA	36	38	
Peak power in the diode P, GW	140	160	
Peak power in the cavity P_c, GW	58	78	
Stored energy W_0, kJ	87	98	100%
Energy delivered to the diode W, kJ	75	86	
E-beam energy W_b, kJ	51	61	62%
E-beam energy behind the frame, kJ	35	45	
E-beam energy behind the foil W_c, kJ	26	32	
Energy delivered to the gas W_g, kJ	16	19	31%
Energy loss in the springs (to t = 1us), kJ	8.6	8.6	
Energy loss from the collector surface, kJ	15.4	16.4	
Total efficiency (W_g / W_0)			**19.4%**

The accelerator operates with the e-beam beginning from July 2001. During this time it demonstrates reliable operation, the voltage and current variation are few %, the shape of the pulses is stable, no any damage of the window foil was observed. Compared with traditional e-beam accelerators on pulse forming lines, the reported accelerator is simple in design, reliable and compact, and provides similar total efficiency of energy transfer from the storage to the gas.

REFERENCES

1. Angood S., Bailly-Salings R., Hirst G. J., at al, *IIIrd Workshop on KrF laser technology*, PP.13, 1992.
2. Abdullin E. N., Bugaev S. P., Efremov A. M., Zorin V. B., Kovalchuk B. M., at al, *IX Intern.Conf. On High-Power Particle Beams*, p. 419, 1992.
3. Kovalchuk B. M., Kim A. A., Kumpjak E. V., Zoi N. V., Zorin V. B., *XXVIII IEEE Intern. Conf. On Plasma Science and XIII IEEE Intern. Pulsed Power Conf.*, p. 1488, 2001.

Characteristics of Ion Induced Secondary Emission Electron Gun

Priya R. Chalise,[a] Masato Watanabe,[a] Akitoshi Okino,[a] Kwang-cheol Ko [b], Eiki Hotta [a]

[a] *Department of Energy Sciences, Tokyo Institute of Technology, Yokohama, 226-8502, Japan.*
[b] *Department of Electrical and Computer Engineering, Hanyang University, Seoul, 133-791, Korea*

Abstract. This paper presents some fundamental characteristics of a secondary emission electron gun using a pulsed glow discharge wire ion plasma source (WIPS). The positive helium ions extracted from WIPS are accelerated in vacuum toward the negatively biased cold cathode surface, which is set oblique to the ion loci in order to inject the emitted secondary electrons perpendicular to the foil window. The physical mechanisms governing the characteristics such as space charge and secondary electron emission have been reviewed. The dependence of such characteristics on the kinetic energy of incident ion and on the ion incidence position have been experimentally investigated. The experimental results are discussed together with the available theoretical models of each characteristic to establish the relative understanding of such phenomena in side-extraction-type secondary emission electron gun. The experimental results are further discussed in the light of a self-developed numerical simulation using the finite element method, which presents a good understanding of particle trajectories as well as potential distribution inside the gun geometry.

INTRODUCTION

Secondary emission electron gun (SEEG) has been extensively used in the area requiring large cross section high-energy electron beams. The device developed in this concept have been reportedly applied in areas such as pre-ionization and excitation of laser [1][2], electron beam processing [3] etc. The SEEG uses a thin wire pulsed glow discharge ion source called wire ion plasma source (WIPS). The energetic singly charged positive helium ions (He^+) extracted from the WIPS are accelerated in vacuum towards a negatively biased stainless steel cathode. When such ions collide on the cathode surface, a numbers of secondary electrons are generated due to a kinetic emission process. The emitted secondary electrons are then accelerated towards an electron window and form a very wide electron beam. This configuration, where the electron window is kept on the side orthogonal to the ion extraction window is referred to as side-extraction-type. Basic characteristics such as space charge limit (SCL), charge exchange and secondary electron emission are important in order to design an efficient SEEG. In this paper, experimental results are compared with relevant theoretical models of each characteristic in order to establish a relative understanding of such phenomena particularly in case of side-extraction type SEEG. In addition, experimental results are explained in the light of a self-developed finite element numerical simulation.

RESULTS AND DISCUSSION

The details of schematic cross-section and discharge characteristics of WIPS have already been reported. [4] A schematic view of side-extraction-type SEEG is shown in Fig. 1. In order to study the basic characteristics, the experimental values of ion beam current and electron beam current were measured by varying the externally controlled acceleration voltage up to 100 kV, a value limited by the power supply. Inter-electrode

CP650, *Beams 2002: 14th International Conference on High-Power Particle Beams*, edited by T. A. Mehlhorn and M. A. Sweeney

distance (d = 50 mm) and the pressure (3 x 10^{-4} Torr) of WIPS were kept constant during the entire experiment. A physical insight of acceleration region shown in Fig. 2 has been presented with the aid of a self-developed numerical simulation using finite element code. The numerical simulation was carried out with the following assumptions: a secondary electron is emitted without initial velocity from the ion collision position on the cathode. The numerical simulations were performed in a Cartesian space with a cathode plate being positioned at an angular position of 45^0, 60^0 and 73^0 with respect to the ion transmission window. (Fig. 1) The electric field vectors are plotted normal to the equipotential contours give fair estimates of actual trajectories. These trajectories show how an average ion and electron motions are directed as a consequence of potential distribution inside the gun geometry. The angular position is referred to as cathode angle. The ion collision position at 73^0 is near top of the cathode plate, where effect of electric field is relatively strong. This is the position, which gives the electron beam trajectory perpendicular towards the center of the electron window.

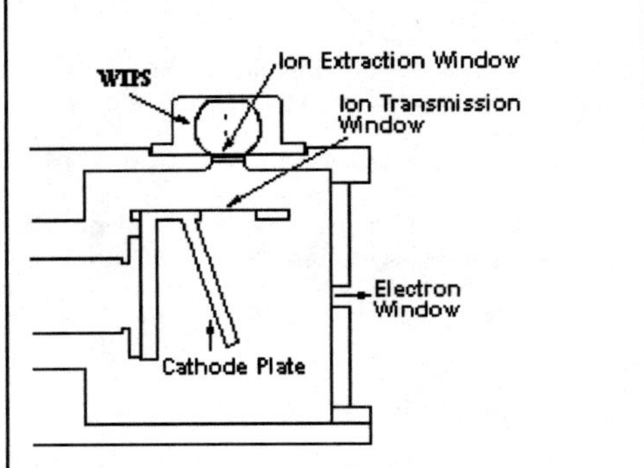

FIGURE 1. Secondary emission electron gun

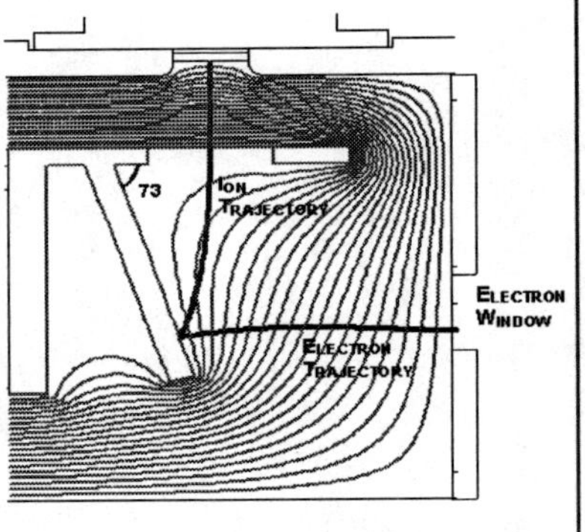

FIGURE 2. Numerical simulation

Space Charge

The space charge limit (SCL) has high practical significance because it determines the maximum electron (or ion) current density that can be extracted from the given electrode geometry. A complete analytical model of space charge accounting a non-parallel electrode and initial ion velocity effect applicable in the similar situation like side-extraction-type SEEG is lacking. The classical Child Langmuir [5] law was derived using parallel electrode of infinite area and the gap filled with charges, where the influence of initial ion velocity was neglected. The influence of the initial particle velocities on SCL has been accounted in the models of Lu Dougal (LD) [6] and Akimov et.al. [7] However, a geometrical end effects are still lacking in their treatments. Since, the extracted ion beam current is independent of the cathode angle in our case, we have considered these models for a comparative study.

The influence of space charge on the ion beam current has been studied on the basis of curves shown in Fig. 3. The measured ion beam current as a function of ion kinetic energy is compared with theoretical values predicted by LD and Akimov model. The curve corresponding to the Child Langmuir model serves as a reference. The experimental values are slightly dependent on the ion kinetic energy at lower energy ranges. A virtual saturation of ion beam current indicates its independence on ion kinetic energy at higher energy ranges. The physical mechanism of this behavior can be explained in terms of ion number densities at WIPS, which contribute to the space charge roughly as much as is necessary to keep the ion beam current constant. Another feature of Fig. 3 is that the experimental values are much higher than the theoretical values. The higher experimental values compared to the Child Langmuir is well expected, but with the other two models, which accounts the initial ion velocity is interesting. This fact is explained in the light of numerical simulation result (Fig. 2), where the equipotential contours at ion extraction region are curved and extend a considerable distance into the extraction grid indicate the extraction of large numbers of initially charged ions. There remain other possibilities that the increased ion beam current is due to a partial neutralization of the ion space charge in the ion extraction region and a charge-exchange process in the acceleration region. A point of intersection between experimental curve and the theoretical curve (LD and Akimov) indicate that the ion beam current has been space charged at the ion kinetic energy between 90 keV to 100 keV.

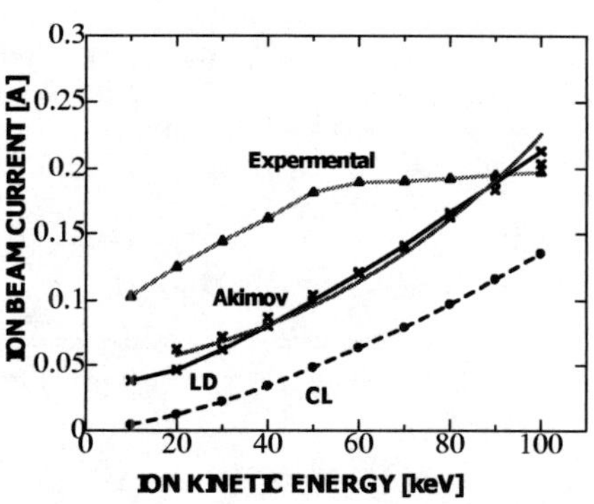

FIGURE 3. Dependence of ion beam current on the kinetic energy of incident ion

The exchange of energetic charges in the acceleration region is an important factor, which influences the secondary electron emission, particularly below the ion energy range of 100 keV. The existence of background helium atom in the space between ion extraction window and the cathode surface causes the charge exchange process to occur. In the charge exchange, the accelerated helium ion collides with natural helium atoms to produce energetic neutral (He^0) and slow speed helium ions (He^+). Since the momentum and energy of such He^0 neutrals are approximately equal to those of the original He^+, the He particle flux incident on the cathode is larger than the ion current. It has been reported that total current of the secondary electrons due to the neutral is 1.1 times the combined current of the positive ions and positive ion secondaries. [8]

Secondary electron emission

The average number of emitted secondary electron per incident ion is referred to as secondary emission coefficient (γ). The γ have been calculated from the ratio of ion beam current and corresponding electron beam current. The calculated γ was corrected to an effective secondary emission coefficient (γ') taking into account the contribution of He^0 particles. There are several different theoretical approaches and numerical models related to the γ have been reported. However, there exists a considerable gap between the theory and experiment in case of low charge state helium ion (below 100 keV) in oblique incidence on the stainless steel target under a moderate vacuum condition. In this context, experimental γ is compared with a more relevant theoretical model of Ghosh Khare (GK). [9]

Fig. 4 shows the dependence of the γ on the kinetic energy of incident He^+ ion. The experimental data has been taken at the cathode angle of 73^0 at the ion kinetic energy between of 40 to 100 keV. The curves corresponds to a least square fitting of experimental γ' and γ caused by incident He^+ ion (γ_{He+}) and He^0 (γ_{He0})

respectively as predicted by GK model. The important significance of the Fig. 4 is the fact that γ is strongly dependent on the ion kinetic energy and experimental yield is much lower than the theoretical one. It is observed that at the ion kinetic energy of 100 keV, the experimental γ' is 6.3, whereas the combined theoretical yield (γ_{He+} + γ_{He0}) is 16.4. The major reason for large discrepancy is the fact that most of the theoretical models, including the one considered here, assume uniform electric field around the acceleration region. In contrast, the numerical simulation (Fig. 2) of the acceleration region shows a non-uniform electric field distribution. The electric filed is shown much weaker around the cathode plate, which is expected to reduce the secondary electron emission substantially.

Fig. 5 shows the dependence of γ' on the cathode angles of 45^0, 60^0 and 73^0 as the ion kinetic energy varies from 40 keV to 100 keV. The profound influence of cathode angle on γ' is observed. The fact can be physically explained in terms of path length of the incident ion within escape zone of the cathode. The prolonged ion path length on increased cathode angle increase the excitation energy and thus the number of emitted electrons. The increased electric field (Fig. 2) at the higher cathode angle is expected to enhance this process. For the ion of same mass and charge states, the dependence of $\gamma(\gamma')$ on the cathode angle has been related by following equation [10][11] $\gamma(\phi)=\gamma(0)\ \cos\phi^{-z}$, where, $\gamma(0)$ is the secondary emission coefficient at perpendicular incidence, ϕ is a cathode angle and z is a fitting parameter. (z=1 for helium ion)[11] A clear deviation from z=1 is noted. The fitting parameter of z=2.77 gives the best fit of the experimental values. Hence, the above equation is modified as $\gamma(\phi)=\gamma(0)\ \cos\phi^{-2.77}$, which is expected to be applicable in similar situation, at least in case of cathode angle below 60^0.

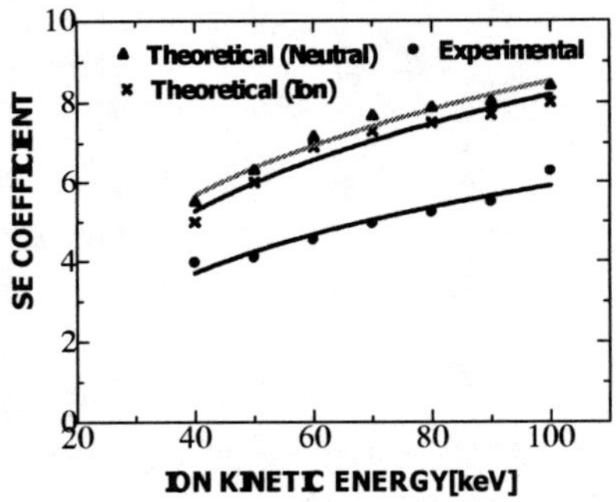

FIGURE 4. Dependence of γ on kinetic energy of incident ion

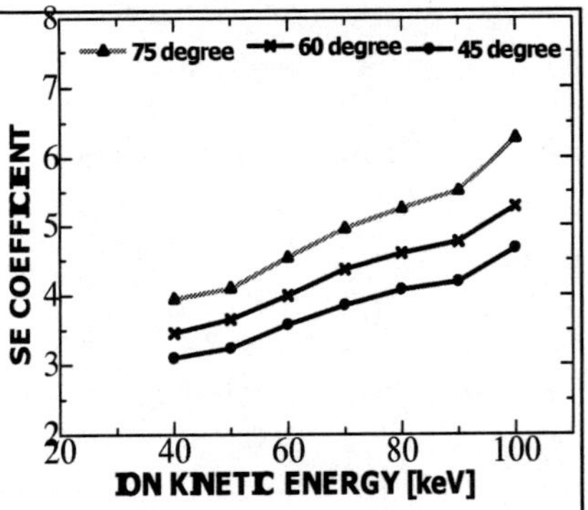

FIGURE 5. Dependence of γ' on cathode angle

SUMMARY

This simple analysis provides an immediate understanding of fundamental characteristics such as space charge limit and secondary electron emission in a side-extraction-type secondary emission electron gun using wire ion plasma source. The agreement between the experimental results and theories is reasonably good in case of space charge analysis, whereas a large discrepancy is seen in case of secondary electron emission coefficient. The excellent correlation between the experimental results and the numerical simulation results presented herein leads authors to believe that the numerical model is able to provide a good physical insight of particle beams inside the electron gun geometry. We hope that presented experimental result and explanations serves as a good reference in the similar situation like side-extraction-type secondary emission gun.

References

1. D. Pigache et. al., J. Vac. Sci. Technol., **12** (6) 1197 (1975)
2. W. M. Clark and G. J. Dunning, IEEE J. of Quantum Electron, **QE-14** (2) 126-129 (1978)
3. H. Suezawa et.al., Vacuum, **47** (6-8), 613-616 (1996)
4. P. R. Chalise et.al., Jpn. J. Appl. Phys., **40**, 1118-1121(2001)
5. C.D. Child, Phys. Rev. **32**, 492, 1911, I. Langmuir, Phys. Rev. **2**, 450, 1913
6. S. Liu and R. Dougal, J. Appl. Phys. **78** (10), 5919-5925 (1995)
7. P. V. Akimov et.al., Phys. Plasmas, **8** (8), 3788-3798 (2001)
8. G. W. Mclure, Phys. Rev., **124** (4) 969-980 (1961)
9. S. N. Ghosh and S.P. Khare, Phys. Rev. **129** (4), 638-1642 (1962)
10. W. O. Hofer in Scanning Microscopy Supplement, **4**, 265-310 (1990)
11. B. Svensson, G. Holmen and A. Buren, Phys. Rev. B, **24** (7), 3749-3755 (1981)

Search for Backstreaming Ion Defocusing During a Single Pulse of a 2 kA Relativistic Electron Beam*

E. J. Lauer, G. J. Caporaso, F. W. Chambers, Y-J Chen, S. Falabella, G. Guethlein, J. McCarrick, R. Richardson, S. Sampayan, and J. Weir

Lawrence Livermore National Laboratory, P.O. Box 808, Livermore, CA 94551

Abstract. Desorption and subsequent ionization of the monolayers from the vacuum wall of an accelerator system can have a detrimental effect on the performance of the beam transport system. Ions extracted from the resultant plasma neutralize the spacecharge and dynamically perturb the net focusing forces within the beam. To study the effect, a transparent first foil, presumably with contaminants on the surface, intercepts the beam. Placing an imaging foil tens of centimeters downstream from the first foil allows observation of minor fluxuations in the envelope. Using conducting foil targets, we see no effect unless the beam radius is small enough to damage the foil. Non-conducting foils produce a strong effect.

INTRODUCTION

The DARHT accelerator will deliver several intense relativistic electron beam pulses to an x-ray conversion target during a few microseconds [1]. Plasma from the target can cause a partial neutralization of the vacuum self-Er field resulting in an unacceptably large beam radius at the target. The Livermore group has been developing barrier foils to block the plasma from moving upstream . Positive ions accelerated upstream from the foil in the self-Ez field during a single pulse could defocus the beam. In May, 2001 LANL used a sensitive "two foil" experiment to search for such effects [2]. They measured significant time dependent effects using conducting foils. In Janurary, 2002, the Livermore group repeated the experiment using the ETA II accelerator. We saw no significant effect from conducting foils unless the beam radius was small enough to damage the foil. The reason for the different results has not been explained and is still being investigated. Possibilities have to do with the longer pulse length at LANL, (60 ns compared to 40 ns) or with the higher energy at LANL, (20 Mev compared to 5.7) We also did some experiments on dielectric targets where there is a strong effect to test our techniques.

APPARATUS

ETA II is an induction linear accelerator that produces 2 kA pulses of 5.7 Mev electrons. The energy is constant within 0.5 % during 40 ns. Fig. 1 shows the apparatus used in these tests. Most of the tests were "two foil" experiments with the

CP650, *Beams 2002: 14th International Conference on High-Power Particle Beams*, edited by T. A. Mehlhorn and M. A. Sweeney

beam passing through a thin upstream foil and the time dependent beam radius measured on the downstream foil.

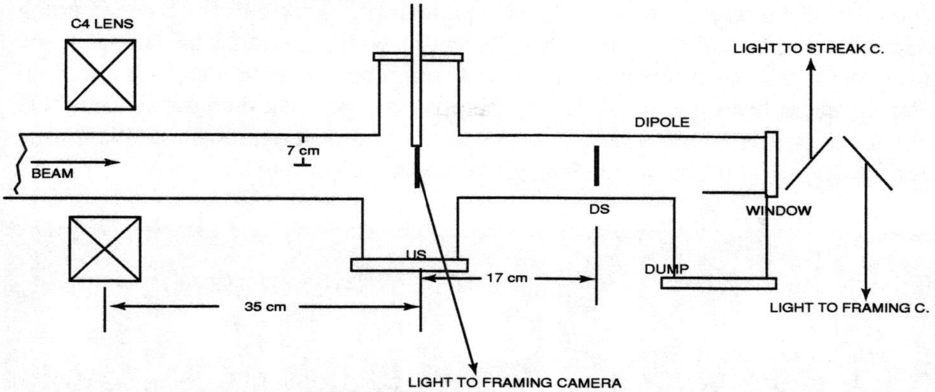

FIGURE 1. Schematic of apparatus

The most important measurements were made by imaging the light produced on the downstream foil onto a streak camera located outside the shield wall. A narrow slit was located at the entrance to the streak camera to form a beam profile and the profile light was streaked with the time axis perpendicular to the profile. The slit width was typically 0.1 cm and the time resolution was estimated to be 1.5 to 4 ns. In one test the slit was removed and the downstream foil was replaced by a thin quartz fiber. There was a graphite foil just upstream of the fiber. Cerenkov light from the fiber produced a beam profile which was imaged onto the streak camera. This was the technique used at LANL. The results were unchanged. Framing camera images of the beam light from the two foils were also recorded. The final magnetic focussing lens called C4 was centered 35 cm upstream of the first foil. In a typical run on a foil, we started with a low current in C4 and increased the current in 10 or 20 A steps until either a hole was produced in the foil or a beam waist of 0.15 cm radius was produced at the foil. The currents in all the other lenses and steering coils were held constant.

The images acquired by the cameras are stored as electronic data files on computers. Our data reduction system starts with the beam profiles from the streak camera or from vertical and horizontal profiles generated from the framing camera images. A Gaussian shape is fitted to the profile data and the half width at half- maximum amplitude is the measured beam radius.

EXPERIMENTAL RESULTS

In the upstream foil location we tested the following conducting foils:1 mil aluminum, 3 mil graphite, 0.5 mil stainless steel, 0.3 mil tantalum, 0.5 mil mylar with aluminum evaporated on one side, 2 mil kapton with evaporated aluminum. We also tested non-conducting kapton. In our experiments, the strength of single-pulse time-dependent focussing effects are strongly correlated with whether the target is a conductor or a non-conductor. Figs. 2 and 3 are for a 1 mil aluminum upstream foil and are typical of all the results on conductors. Fig. 2 presents the streak camera images as the C4 lens current is increased in small steps up to 361 A where we produce a 0.15 cm waist radius and a hole resulted. Fig. 3 presents the corresponding downstream profile widths vs time. On the pulse where the foil was damaged, the beam radius at the downstream foil increased from about 1 cm to about 1.8 cm in 40 ns. For upstream beam radius >0.15 cm there is no significant time dependence. On the figures of profile width vs time, we suggest that the most significant data is for the time range 50 to 100 ns; outside this range the beam current is too low.

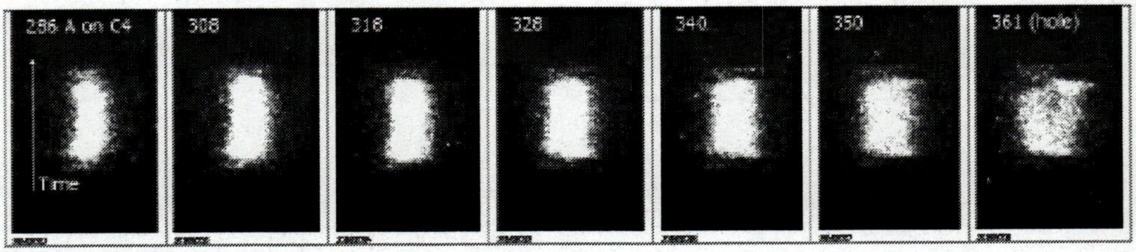

FIGURE 2. Streak images for 1 mil aluminum

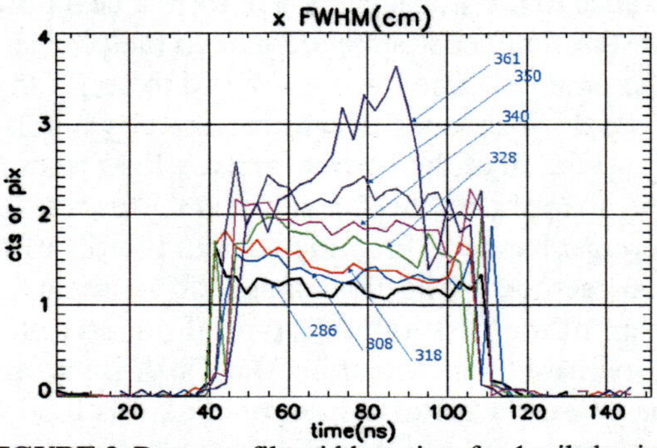

FIGURE 3. Beam profile width vs time for 1 mil aluminum

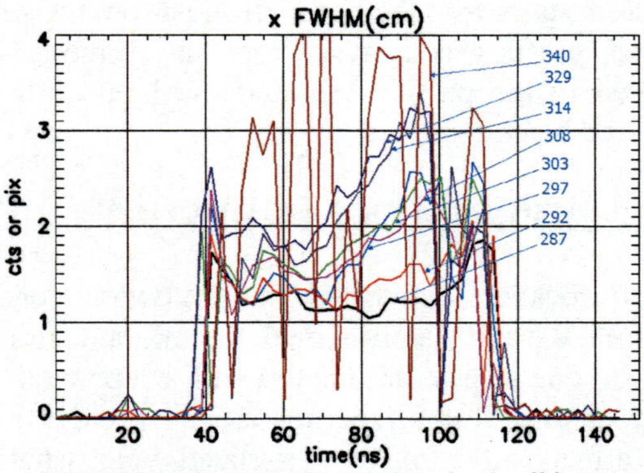

FIGURE 4. Beam profile vs time for 2 mil kapton

Fig. 4 shows the results for a non-conducting upstream foil. As the spot size on the upstream foil is reduced, time dependence starts for about 0.5 cm radius and is very strong for 0.2 cm. Our model for the emission from dielectrics is that at the head of beam the dielectric is a non conductor and there is a strong Er and no Ez. Er is strong enough to cause electric breakdown over the surface of the dielectric with electrons avalanching outward. This creates a plasma which is a conductor and is also a source of ions. Er switches off and Ez turns on accelerating ions.

BEAM ENVELOPE MODEL

Fig. 5 presents calculations using the beam envelope equation and assuming a fractional neutralization f= 0.2 of Er. Protons are predicted to move about 32 cm in 40 ns or 48 cm in 60 ns. The latter case simulates the effect of the longer pulse at LANL.

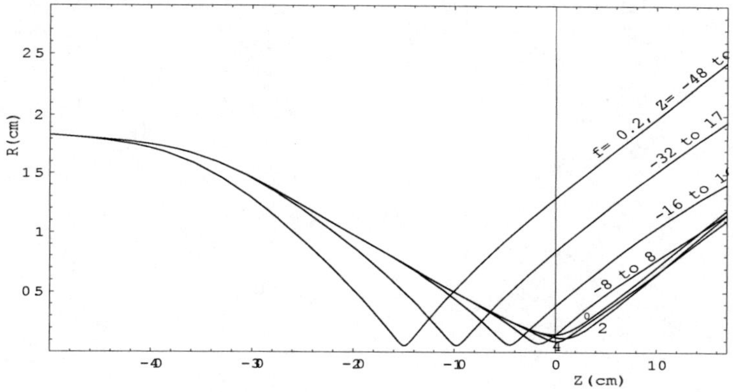

FIGURE 5. Beam envelope model with different lengths of rectangular f-column, 1 mil aluminum

REFERENCES

1. M.J. Burns, "Status of the DARHT Phase 2 Long-Pulse Accelerator," in Proceedngs of the 2001 Particle Accelerator Conference, Chicago, IL; June 18-22, 2001, p. 325.
2. H. A. Davis, "Electron Beam disruption Due to Ion Release Due to Ion Release From Targets," this conference

*This work was performed under the auspices of the US Departmrnt of Energy by University of California, Lawerence Livermore National Laboratory under contract No. W-7405-ENG-48.

HIGH POWER MICROWAVES

S-Band Resonant BWO with 5 GW Pulse Power

Sergei A. Kitsanov, Aleksei I. Klimov, Sergei D. Korovin, Ivan K.Kurkan, Igor V. Pegel, and Sergei D. Polevin

Institute of High Current Electronics SD RAS, 4 Akademichesky Ave., Tomsk 634055, Russia

Abstract. The paper presents numerical and experimental studies of the resonant relativistic BWO. By using reflections of the operating TM_{01} wave from edges of the electrodynamic system, optimum conditions for electron beam interaction with both (-1st) and fundamental harmonics of the wave are achieved. In experiment, 5 GW, 3.6 GHz microwave pulses with 30 % power efficiency were produced.

INTRODUCTION

The basic distinctive feature of the resonant BWO comparing with the conventional relativistic BWO [1] is the increased Q-factor of the electrodynamic system achieved by reflection of the TM_{01} operating wave from the waveguide narrowing at the collector edge of the slow-wave structure (SWS). The resulting improvement of longitudinal distribution of the RF field (both the (-1st) and the fundamental harmonics are important) allowed an increase in the microwave efficiency up to 30%, which is twofold comparing to the conventional BWO. Compare to theory [2] where, for a resonant BWO, accounting the (-1st) harmonic interaction only resulted in 20% efficiency increase. An important merit of the device is its reduced interaction length (~3 λ) allowing smaller energy in the guide magnetic field (which makes 2 – 3 T).

SCHEMATIC AND SIMULATIONS

Configuration of the resonant BWO is represented in Figure 1.The electron beam is produced in a vacuum diode with edge-type cathode in strong magnetic field and is delivered to the tube via beyond cutoff-neck, which serves as a reflector for the backward wave. Slight narrowing of the slow-wave structure at its collector edge provides partial reflection of the wave and increased Q-factor of the tube. A piece of smooth cylindrical waveguide between the beyond cutoff-neck and the beginning of corrugation is used to adjust integer number of oscillations of the fundamental standing wave along the tube. The ratio between amplitudes of fundamental harmonic of the forward wave and (-1st) harmonic of the backward wave in the beam trajectory is determined by the corrugation depth of the SWS. The impedance of the diode was adjusted by varying the anode-cathode gap spacing. The guide magnetic field was produced by pulsed solenoid of rectangular cross section. The generator was optimized numerically using PIC-code KARAT [3].

CP650, *Beams 2002: 14th International Conference on High-Power Particle Beams,*
edited by T. A. Mehlhorn and M. A. Sweeney

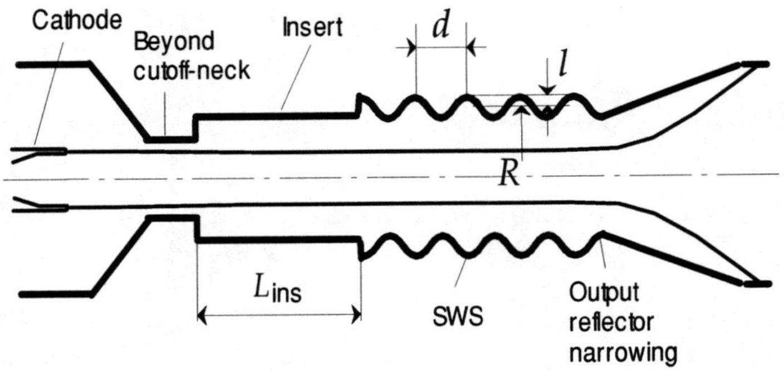

FIGURE 1. Schematic of the resonant BWO

Simulations evidenced that, with optimum geometry of the electrodynamic system ($l \approx \lambda/8$, $R \approx \lambda/2$, $L_{sws} \approx 2.5\lambda$, $L_{ins} \approx \lambda$, $Q \approx 100$ (where L_{sws} and L_{ins} are lengths of the SWS and the insert, respectively and λ is the wavelenght), the microwave power efficiency makes ~30% changing only slightly in a wide range of electron beam powers (from 5 up to 20 GW). The optimum vacuum diode impedance was about 100 Ω. The time for onset of microwave oscillation was ~20 ns (Figure 2).

It was found that changing the length of the cylindrical insert results in periodic jump-wise changes in the longitudinal index of the fundamental standing wave in the tube (from 2 up to 6 variations). At the jump points, multi-frequency operation with substantial drop of power was observed. Within the same standing wave index, the frequency tuning can be 3- 4 %.

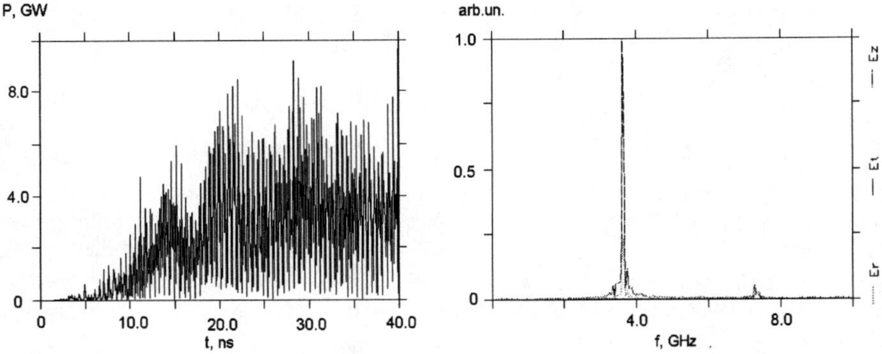

FIGURE 2. Microwave power and spectrum of resonant BWO (PIC code KARAT simulation): electron energy 1.2 MeV, beam current 12 kA, magnetic field strength 25 kOe

EXPERIMENTAL

The experimental study of S-band resonant BWO was made using the SINUS-7 electron accelerator [4]. Optimization of fundamental-harmonic interaction was made by adjusting the length of the cylindrical insert (Figure 3). Variation of the insert length resulted on variation of frequency within 3%. Multifrequency generation with substantial drop in microwave power was observed at insert length of 60 mm. In the optimum regime, microwave generation at 3.6 GHz frequency with ~30% efficiency was obtained. At 1.2 MeV electron energy and 15 kA beam current, the peak microwave power was 5.3 GW with 25 ns pulse width (at half power level). The typical waveforms of diode voltage, beam current, and microwave detector signal are depicted in Figure 4. The microwave pulse energy was ~100 J as measured with a calorimeter. In experiment, like in simulations, the microwave efficiency demonstrated minor changes in the range of electron beam power from 5 up to 19 GW (Figure. 5). The guide magnetic field strength was about 25 kOe. The pattern of radiation from a horn antenna corresponded to the TM_{01} mode.

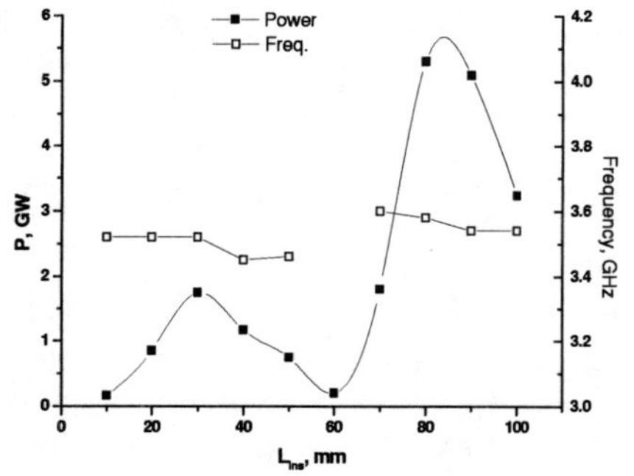

FIGURE 3. Microwave peak power and frequency vs insert length, SWS corrugation period 45mm, diode voltage 1.2MV, beam current 15 kA, magnetic field strength 25kOe.

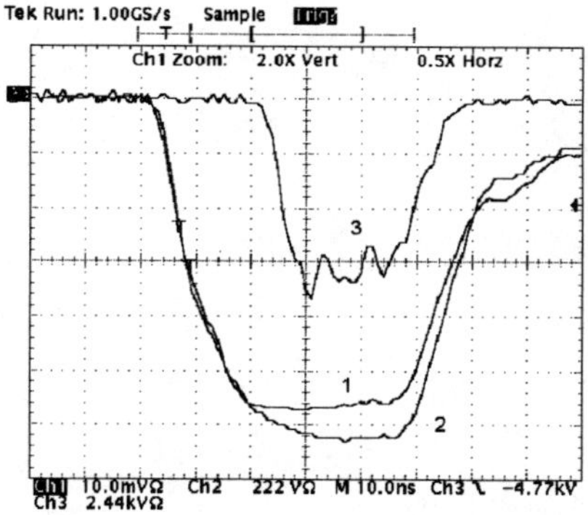

FIGURE 4. Waveforms of (1) diode voltage, (2) electron beam current, and (3) microwave detector signal, diode voltage 1.2MV, beam current 15 kA, magnetic field strength 25kOe.

The magnetic field dependence of the microwave power is depicted in Fig. 6. As found in simulation, the power drop below 15 kOe was caused by partial dumping of beam electrons on SWS surface under the action of the intense RF field. The drop boundary moved to lower magnetic fields as the microwave power decreased.

The lower limit for the guide magnetic field may be also caused by cyclotron absorption of the backward electromagnetic wave by beam electrons:[5]:

$$H > H_r = \frac{mc^2}{e} \beta_e \gamma \bar{h} ,$$

where $\bar{h} = 2\pi / d$; e, m and $c\beta_e$ are electron charge, mass and velocity. For the 3.6-GHz resonant BWO with 1.2 MeV electron energy, $H_r \approx 7$ kOe.

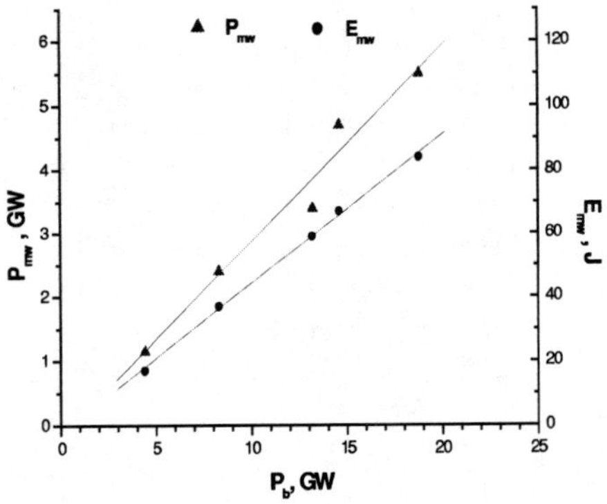

FIGURE 5. Microwave pulse peak power and energy vs electron beam power (diode impedance ~ 80 Ω, magnetic field strength 25 kOe).

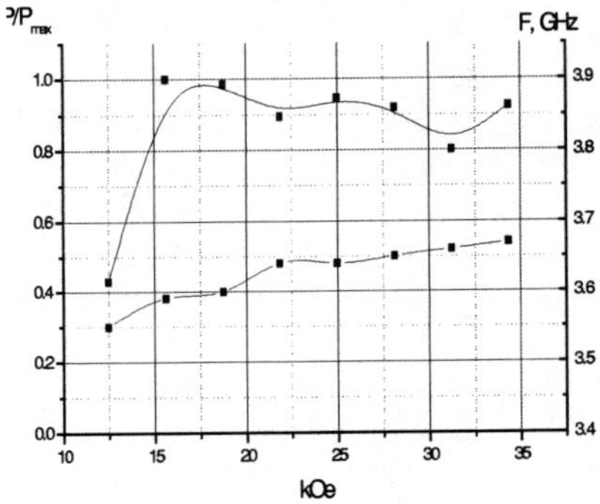

FIGURE 6. Magnetic field dependence of output power and frequency (diode voltage 1.2 MV, beam current 15 kA).

CONCLUSIONS

Thus, the resonant BWO was demonstrated to be an efficient HPM source in S-band. In experiment using SINUS-7 high-current accelerator, 5 GW, 3.6 GHz microwave pulses with 30 % power efficiency were produced. The pulse energy was close to 100 J. An important merit of the device is its reduced interaction length (~3 λ) meaning smaller energy in the guide magnetic field.

REFERENCES

[1] V. L. Bratman, N. S. Ginzburg, N. F. Kovalev, et al., *Relativistic High Frequency Electronics.* Issue 1. Gorky, pp. 249 – 274 (1979) (in Russian).

[2] B. Levush, T. Antonsen, A. Bromborsky, et al., *IEEE Tran. Plasma Sci.*, **20**, no. 3, pp. 263 – 280 (1992).

[3] V. P. Tarakanov, *User's Manual for Code KARAT. Springfield*, VA: BRA (1992).

[4] S. D. Korovin and V. V. Rostov, *Russian Physics J.*, **39**, no. 12, pp. 21 – 30 (1996).

[5] A. S. Elchaninov, F. Ya. Zagulov, S. D. Korovin, et al, *Proc. 3rd Int. Conf. on High Power Electron and Ion Beams*, **1**, pp. 191 – 197 (1979).

Broadband Frequency Tuning in a Gigawatt S-Band Relativistic BWO

Sergei A. Kitsanov, Sergei D. Korovin, Ivan K. Kurkan, Igor V. Pegel, Sergei D. Polevin, Vladislav V. Rostov, and Eugeny M. Totmeninov

Institute of High Current Electronics SD RAS, 4 Akademichesky Ave., Tomsk 634055, Russia

Abstract. Frequency tuning capability was studied for two varieties of pulsed gigawatt relativistic backward-wave oscillator (BWO): the resonant BWO and the BWO with increased cross-section having a resonant reflector. The tuning assumed adjustment of the tube's electrodynamics only, with no variation of electron beam parameters. The studies demonstrated that frequency tuning of as wide as 10–20% is achievable, depending on the tube's design peculiarities.

INTRODUCTION

Recent experiments [1–3] demonstrated efficient single-mode operation of BWO-based HPM sources in X- and S-bands, with peak power of up to 5 GW. The capability for broadband frequency tuning could notably increase the practical value of these devices. In the relativistic BWO, unlike low-voltage BWOs, there is only a very limited possibility to change the frequency by varying the electron beam energy. The goal of this work was to study the frequency variation by adjusting the electrodynamic system parameters only, with constant parameters of the driving electron beam. The two tubes studied were (a) the resonant BWO with single-mode slow-wave structure (SWS) and (b) the BWO with increased SWS cross size having a resonant microwave reflector.

FREQUENCY TUNING IN THE RESONANT BWO

Estimates and Simulation

The basic operation principles of the resonant BWO are described in paper [2] present in these Proceedings and its schematic is shown in Figure 1.

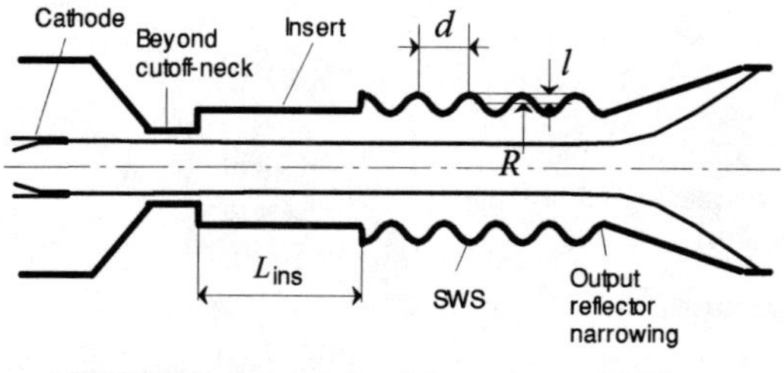

FIGURE 1. Schematic of the resonant BWO.

CP650, *Beams 2002: 14th International Conference on High-Power Particle Beams*, edited by T. A. Mehlhorn and M. A. Sweeney

It is reasonable to suggest that the operation frequency of a BWO (with no exception of the resonant tube) can be changed by shifting the point of Cherenkov synchronism between the (-1st) spatial harmonics of the backward wave and electron beam by varying of the SWS period (keeping the corrugation depth l and the electron beam parameters constant). For a relatively shallow corrugation ($l<<R$, where R is the mean radius of SWS) the condition of Cherenkov synchronism yields

$$\frac{\Delta\lambda}{\lambda} \approx \frac{\lambda}{d} \frac{\beta_e}{\left(1+\beta_e\beta_{ph}\right)} \frac{\Delta d}{d},$$

where $c\beta_e$ is the electron velocity and $c\beta_{ph}$ is the phase velocity of the fundamental harmonic of the forward wave, d is the SWS period, and λ is the open-space wave length. Thus, for a BWO with single-mode SWS and 1 MeV electron energy we have

$$\frac{\Delta\lambda}{\lambda} \approx (0.4 \div 0.5)\frac{\Delta d}{d}.$$

The suggestion was confirmed in simulations using the KARAT code [4]. The simulation evidenced that, using this method, it is possible to vary the frequency from the critical frequency of the SWS up to the critical frequency of the beyond cutoff-neck, which can exceed 25% at half power level. It was found however that the frequency tuning is accompanied by periodic jump-wise changes in the longitudinal index of the fundamental standing wave in the tube (from 2 up to 6–7 variations). At the jump points, multi-frequency operation with substantial drop of power was observed. Within the same standing wave index, the frequency tuning may not be over 5–6 %. For each SWS period, optimization of fundamental harmonic interaction is needed by means of adjusting the length of the cylindrical insert situated between the beyond cutoff-neck and the SWS. Figures 2 and 3 illustrate the simulation results for an S-band BWO tube.

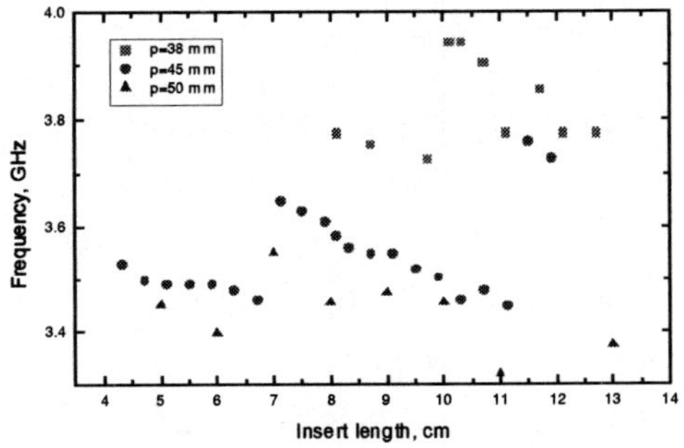

FIGURE 2. Microwave frequency vs length of the insert for a series of SWS corrugation periods. Beam power ~15 GW, magnetic field 20 kOe.

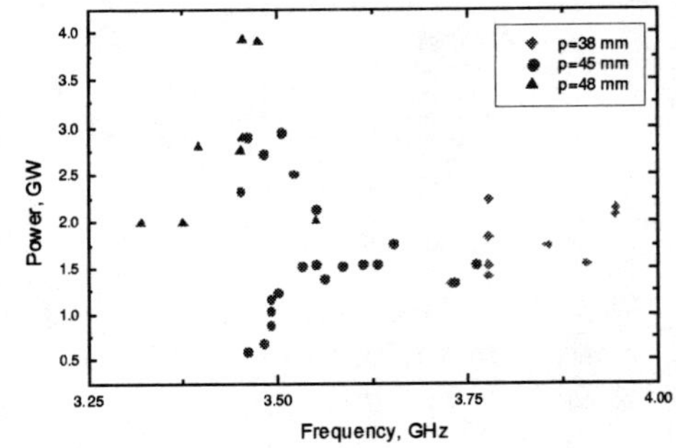

FIGURE 3. Microwave power vs frequency for a series of corrugation periods, each with various insert lengths.

The typical maximum powers simulated were 3 – 5 GW depending on the beam power, with microwave efficiency being about 30%. The optimum vacuum diode impedance was about 90 Ω.

Experimental

The experimental study of the frequency tuning in S-band resonant BWO employed the SINUS-7 electron accelerator [5] (Figure 4). The diode voltage was 1.2 MV, the beam current was 15 kA, and the guide magnetic field strength was 25 kOe.

Since consistent nonlinear variation of two SWS parameters – the period and the insert length – is a sophisticated mechanical problem, we should point out that the frequency can be tuned, within some narrower band, by changing only one of the parameters, namely, the SWS period. In experiment, we fixed the insert length and the corrugation period was discretely changed from 38 up to 48 mm. The electron beam parameters (diode voltage, beam current) and guide magnetic field strength were unchanged. The experimental results are shown in Figure 5. The frequency tuning range obtained was about 10% with 3.9 GW peak power and 22% efficiency.

FIGURE 4. SINUS-7 resonant BWO experiment.

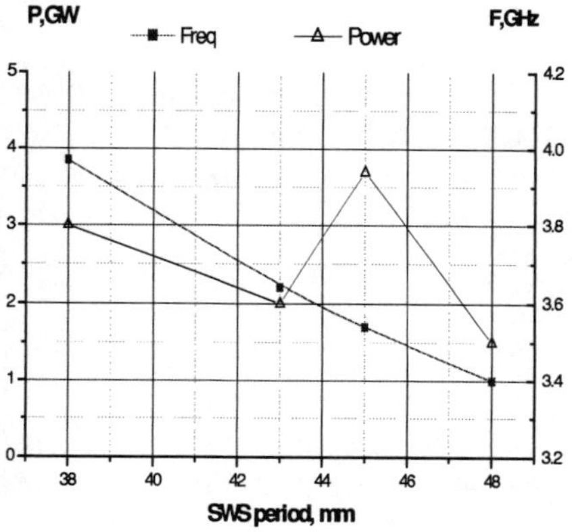

FIGURE 5. Microwave peak power and frequency vs SWS corrugation period at invariant insert length (10 cm).

CAPABILITY FOR FREQUENCY TUNING IN THE LOW-MAGNETIC FIELD BWO WITH RESONANT REFLECTOR

Another kind of the relativistic BWO demonstrating the capability for frequency tuning is the BWO with increased cross size of the slow-wave structure and a resonant reflector [3]. This device is operable in low magnetic field. Reflection of the backward wave occurs due to excitation of critical-frequency locked mode in the resonant reflector, which at the same time serves as the energy modulator for the driving electron beam. The frequency tuning in this device may be done by moving the position of the resonant reflector relative to the slow-wave structure (changing distance L in Figure 6), whose parameters (in particular, the corrugation period) remain unchanged. Moving the reflector also changes to some extent the starting conditions of the generator.

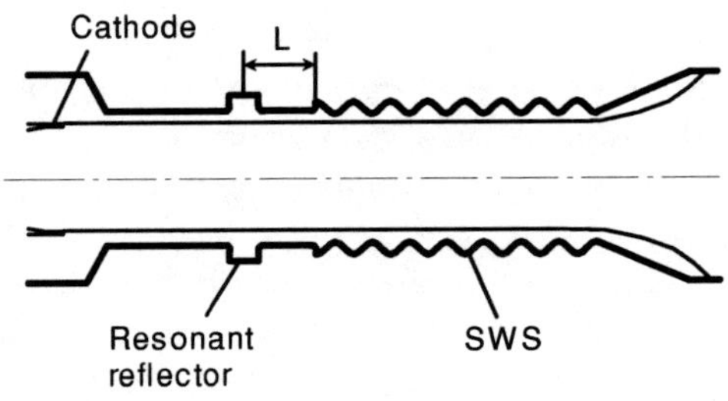

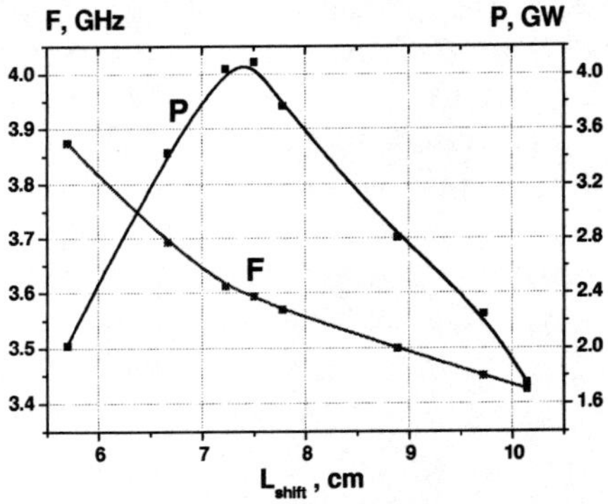

FIGURE 6. Schematic of the relativistic BWO with resonant reflector.

FIGURE 7. Microwave power vs frequency vs position of the resonant reflector (simulation).

Figure 7 illustrates the results of the numerical experiment with the KARAT code. For the guide magnetic field strength of ~ 4 kOe, the frequency tuning was about 12% at the half power level. Clear narrow spectrum with TM_{01} mode was observed throughout this tuning band, with the power efficiency generation between 15% and 30%.

CONCLUSIONS

Thus, the studies demonstrated that, in certain versions of the relativistic BWOs, frequency tuning in excess of 10% is possible at gigawatt power level.

ACKNOWLEDGMENTS

The authors are grateful to Dr. Vladimir Tarakanov, the author of the KARAT code, for collaboration, and to Commissariat à l'Energie Atomique (France).

REFERENCES

1. A. V. Gunin, et al., *IEEE Trans. Plasma Sci.* **26**, no. 3, 326–331 (1998)
2. S. A. Kitsanov, et al., *in these Proceedings*
3. V. V. Rostov, et al. *Proc. 1st Int. Congress on Radiation Physics, High Current Electronics, and Modification of Materials. Tomsk, Russia, Sept. 2000.* **2,** 408-411
4. V. P. Tarakanov, *User's Manual for Code KARAT.* Springfield: BRA, 1992.
5. S. D. Korovin and V. V. Rostov, *Russian Physics J.*, **39**, no. 12, 21–30 (1996).

High Peak Power and High Average Power Subnanosecond Ka-Band Relativistic BWO

Vladimir P. Gubanov,[a] Dmitry M. Grishin,[a] Aleksandr V. Gunin,[a] Sergey D. Korovin,[a] Sergey K. Luybutin,[b] Gennady A. Mesyats,[b] Aleksandr V. Nikiforov,[b] Vladislav V. Rostov,[a] Sergey N. Rukin,[b] Boris G. Slovikovsky,[b] Konstantin A. Sharypov,[b] Valery G. Shpak,[b] Sergey A. Shunailov,[b] Marat R. Ulmaskulov,[b] and Michael I. Yalandin [b]

[a] *Institute of High Current Electronics, Siberian Division, RAS, Tomsk 634055, Russia*
[b] *Institute of Electrophysics, Ural Division, RAS, Ekaterinburg 620016, Russia*

Abstract. We have developed a relativistic Ka-band backward wave oscillator (BWO) capable of generating stable 250-ps pulses of peak power up to 300 MW. The BWO operates in a 1-s burst mode at a PRF of 1-3.5 kHz.

INTRODUCTION

Earlier we have demonstrated [1] that a relativistic BWO operating in a superradiative mode (see, e.g., [2,3]) can produce a narrow (0.25-ns in the Ka-band) spike containing ~10 microwave oscillations with a peak power close to the power of the e-beam (~600 MW). The magnetic field induction $B \leq 5.5$ T was provided by a pulsed solenoid operated with a shot-to-shot interval of 60 s. Experimentally, two important results have been obtained [1]: (i) at a strong B-field (>5 T) the maximum peak microwave power was several times higher ($P_{max} \approx 400$ MW) than that achieved in previous experiments [2,3]; (ii) we have detected, to the left of the wide region of cyclotron absorption, a band of moderate B-fields where the microwave power drops not substantially (280 MW at $B \approx 2$ T) [4]. The above results and the use of a dc solenoid made it possible to realize a highly efficient repetitive regime of microwave generation. This has been done during a specially arranged experiment. The side goal of the experiment described below was to harness the repetitive regime to make microwave measurements more accurate by using a wide-band sampling oscilloscope and a special calorimeter. The slow-wave system (SWS) was identical to that used in [1] and had an increased cross-size ($R \approx 0.7\lambda$) and a wide-band resonant reflector [4] to extract a TM_{01} wave. The coupling impedance was profiled along the SWS.

EXPERIMENTAL SETUP

A hybrid modulator [5] built around an SM-3NS nanosecond driver [6] was used as

CP650, *Beams 2002: 14th International Conference on High-Power Particle Beams*, edited by T. A. Mehlhorn and M. A. Sweeney

a source of stable accelerating voltage pulses. The driver utilized the SOS-switch technology and provided batch operation (~1 s) at PRF's ≤ 3.5 kHz. The SM-3NS driver comprised several energy compression stages equipped with solid-state switches. The output stage was an inductive energy store with a high-current, high-voltage semiconductor opening switch. The modulator output was equipped with a peaking spark gap filled with hydrogen to a pressure of 100 atm and a stepwise coaxial line which increases the amplitude of the nanosecond voltage pulse by a factor of 1.6. Thus, the accelerating voltage at the cathode reached ~270 kV (fig.1,a). It was measured by a capacitive divider with an accuracy of ±4%. In accordance with the data presented in [5], the RMS deviation of the accelerating voltage was not over 1%. The hollow graphite cathode was 8.6 mm in diameter and had an edge of thickness 0.1 mm. The current pulse of the e-beam (fig.1,b) was measured by a Faraday cup with an accuracy of ±5%. The solenoid winding involved 12 serially connected sections which were cooled with transformer oil. During 1-s operation, 50 kJ of energy were dissipated in the solenoid. A magnetic field of induction 2 T was generated with the use of a dc current source which consisted of a molecular capacitor bank (4.6 F) and a powerful transistor switch operated in the mode of stabilized current. The diameter of the vacuum extraction window of the microwave horn was 50 mm. The BWO's vacuum chamber was evacuated down to a pressure of 10^{-4} mbar.

DIAGNOSTICS AND EXPERIMENTAL RESULTS

Numerical simulations (particle-in-cell method, KARAT code [7]) have shown that the microwave pulse may contain an energy of up to 100 mJ (fig. 2,a). A specially designed calorimeter, calibrated in the range 10–100 J, was used to measure the energy of trains of microwave pulses. The estimated accuracy of calorimetric measurements was (−8) ÷ (+3) %. Microwave pulses were detected by a hot-carrier germanium detector and were recorded by a type Tek-TDS820 sampling oscilloscope

FIGURE 1. A voltage pulse at the 45-Ω output of the hybrid modulator (*a*) and e-beam current pulses recorded for a virgin cathode and after 11,000 shots (*b*). Each stroboscopic waveform accumulates 512 sequential samples at PRF of 1 kHz.

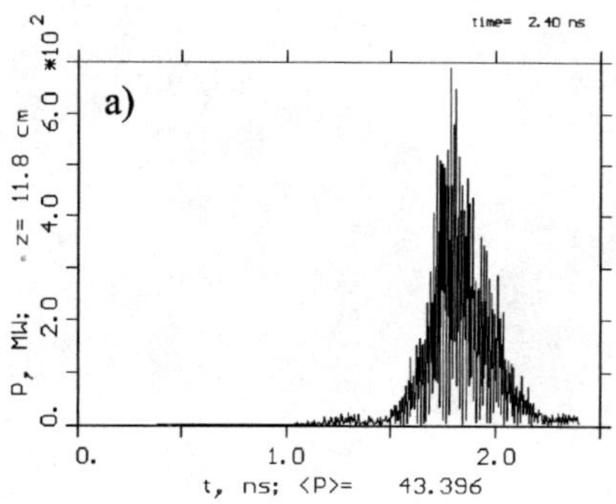

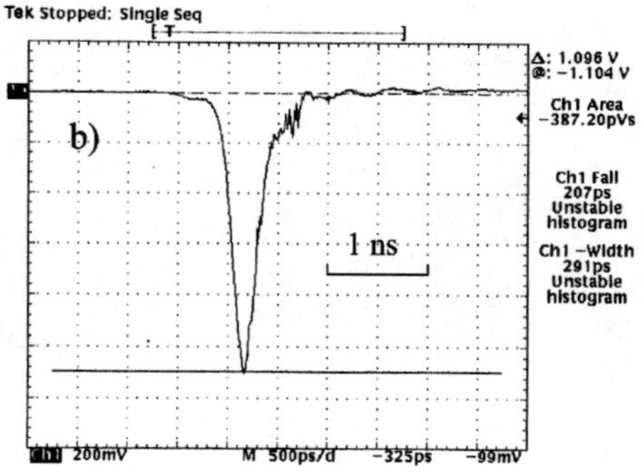

FIGURE 2. The non-averaged microwave power (PIC simulation) (*a*) and the pulse from the microwave detector (*b*).

with a 6-GHz analog band. Thus, the formation of each waveform required accumulation of 512 sequential samples. A magnetron oscillator was employed for the calibration of the microwave detector in the range 0.1–50 kW. Because of the nonlinearity of the detector sensitivity, all waveforms obtained in the experiments were transferred as data files to MathCad code and used to restore the original shape and effective width of microwave pulses. Special measures we taken to minimize probable distortions of the signals. These measures concerned the choice of the detector design and its receiving antenna. The same reason determined the use of time isolation between the detector and the waveguide load and the application of a short (~1.5 m) wide-band cable connected the detector and the oscilloscope (accommodated in a shielded room). The main restrictions in recording subnanosecond signals were associated with the limited transient response of the oscilloscope used (67 ps) and the parasitic capacitance between the miniature Ge sensor and the detector's waveguide case. Time-domain reflectometry has shown an increase in risetime of the test pulse from 100 to 150 ps. The noticeable distortions in the detector – cable – oscilloscope recording duct were typical of pulse widths shorter than 100 ps. The measurement error for the range of pulse widths 200–300 ps can be estimated as $(-5)\div(+10)\,\%$.

A typical signal of the microwave detector is presented in fig. 2,b. Its effective width is about 250 ps. The pulse energy averaged over a great number of shots was 65 mJ. For calorimetric measurements we produced at least 30 bursts, 1000 shots each; the readings fell in the range 55–80 J. Thus, the peak microwave power was in the range 220–300 MW. This is close to results of our previous experiment [1]. The average microwave power was over 200 W at 3.5 kHz PRF.

In the experiments, a decrease in microwave peak power, pulse duration, and stability was observed with increasing number of pulses produced by the same cathode. Measurements of the e-beam current at the inlet of SWS have shown (see fig.1,b) that for ~10^4 shorts a 200-ps-delay of the electron emission with respect to the leading edge of the accelerating voltage pulse and a 10-% decrease in e-beam current took place. Electron microscope photography of a graphite cathode has shown dramatic smoothing of its microrelief. It should be noted that the observed rate of

FIGURE 3. Micrographs of the emitting edge of a graphite cathode, taken before operation (*a*) and after 10,000 and 34,000 shots, respectively (*b*, *c*).

degradation of the emissive property of the graphite cathode does not contradict to the results obtained earlier on the nanosecond scale (10 ns, X-band BWO [8]). From the point of view of the specific material removal from the cathode, it is reasonable to compare the values of the parameter (q/NS), where q is the e-beam charge for a single shot; N is the number of shots, and S is the surface of the cathode emitting edge. To lengthen the lifetime of the developed microwave oscillator probably requires to increase the width of the accelerating voltage pulse.

Thus, the experiments that were started in the mode of single pulses [1] have confirmed our supposition that the peak microwave power of a relativistic BWO operating in the superradiative mode can further be increased by using a nonhomogeneous SWS possessing an extended bandwidth. A high-power subnanosecond microwave pulses have been generated at PRF > 1 kHz for the first time.

ACKNOWLEDGMENTS

The authors are grateful to N. S. Ginzburg for his attention to the work and to I. V. Pegel, A. I. Klimov, V. P. Tarakanov, and A. M. Murzakaev for their helpful assistance.

REFERENCES

1. Korovin S.D., et.al. *Tech. Phys. Letters* **28**, No.1, 76-79, (2002)
2. Ginzburg N.S., et.al. *Izv. Vyssh. Ucheb. Zaved., Radiofizika* **21**, No.1, 1037-1052, (1978).
3. Yalandin M.I., et al. *IEEE Trans. Plasma Sci.* **28**, No.5, 1615-1619, (2000).
4. Kurkan I.K., et al. *Tech. Phys. Letters* **24**, No.5, 388-389 (1998).
5. Lyubutin S.K., et al. *Prib. Tekh. Eksper.* **5**, 80-88, (2001).
6. Lyubutin S.K., et al. *Prib. Tekh. Eksper.* **3**, 52-60, (2000).
7. Tarakanov V.P. User's Manual for Code KARAT. VA, USA: Berkeley Research Associates, Inc. 1992.
8. Gunin A.V., et al. *IEEE Trans. Plasma Sci.* **28**, No.3, 537-541, (2000).

Cyclotron Suppression of Pre- and Post-Pulse Radiation from a Relativistic BWO Pulse

Mikhail I. Fuks[a], Nikolay F. Kovalev[b], Alexey V. Palitsin[b], and Edl Schamiloglu[a]

[a]*University of New Mexico, Albuquerque, NM 87131 USA*
[b]*Institute of Applied Physics, Russian Academy of Sciences, Nizhny Novgorod, 603600 Russia*

Abstract. The possibility of using cyclotron interaction to suppress pre- and post-pulse radiation from a relativistic BWO is considered. It is shown that resonant cyclotron absorption, which depends on electron energy, is as able to suppress unacceptable generation as to eliminate reflected signals.

INTRODUCTION

Successful experiments using a relativistic BWO in a radar [1] provoked interest in the possibility of improving the quality of radiation from this microwave source. In addition to the requirements of high repetition rate, phase and frequency stability, much attention is given to the improvement of the output radiation waveform. In particular, the requirement to decrease the power of the extended tail of the pulse to less than –35 dB is motivated by the desire to increase the contrast of the radar image.

In this paper we present the possibility of using the effect of cyclotron interaction to eliminate the pre- and post-pulse radiation from a relativistic BWO. For illustration the simplest model of the interaction of a magnetized electron beam with a plane electromagnetic wave (EMW) propagating opposite to the electron beam, and with its slow spatial harmonic synchronous with the beam, is described by [1]

$$
\begin{aligned}
C' &= -I \int_0^{2\pi} \left[\kappa e^{-i\theta} - i\varepsilon(\rho/\gamma\beta_z)e^{-i\chi}\right]^* d\omega t_0; \\
\gamma' &= -\operatorname{Re} C\left[k e^{-i\theta} - i\varepsilon(\rho/\gamma\beta_z)e^{-i\chi}\right]; \\
\rho' &= \varepsilon\left(\beta_z^{-1}+1\right)\operatorname{Re}\left(iCe^{-i\chi}\right); \\
\theta' &= \beta_z^{-1} + 1 - \bar{h}/k; \\
\chi' &= \left[(1+\beta_z - \omega_c/\gamma\omega) - (\omega/2\omega_c)(1+\beta_z - \bar{h}/k)\operatorname{Re} Ce^{-i\theta} + (\varepsilon/\rho)(1+\beta_z)\operatorname{Re} Ce^{-i\chi}\right]/\beta_z; \\
\beta_z &= \sqrt{1-\left(1+\rho^2\right)/\gamma^2}; \quad \eta = (2\pi)^{-1}\int_0^{2\pi}\left[(\gamma_0-\gamma)/(\gamma_0-1)\right]d\omega t_0.
\end{aligned}
\tag{1}
$$

Here $k = \omega/c; \omega$ is the angular frequency; (`) is the derivative of the normalized functions: amplitude of the -1st spatial harmonic is $C(\varsigma) = eE_{z,-1}/kmc^2$, $\rho(\varsigma) = \gamma\beta_\perp$, $\gamma(\varsigma) = \left(1-\beta_\perp^2-\beta_z^2\right)^{-1/2}$, which is related to electron energy W, $\gamma = 1 + W/mc^2$ and the phase characteristics of electrons $\theta(\varsigma)$ and $\chi(\varsigma)$ with respect to the normalized

CP650, *Beams 2002: 14th International Conference on High-Power Particle Beams*, edited by T. A. Mehlhorn and M. A. Sweeney

longitudinal coordinate $\varsigma = kz$; $\beta_z = v_z/c$; $\beta_\perp = v_\perp/c$; v_z and $v_\perp$ are the translational and transverse electron velocities, respectively; $\overline{h} = 2\pi/d$, where d is the period of the electrodynamic system; $\varepsilon = -E_{x,0}/2E_{z,-1}$ is relation of the amplitudes of the fundamental and the -1st spatial harmonics; ω_c is the nonrelativistic electron cyclotron frequency; $I = |C|^2 J_0 mc^2/4\pi eP > 0$ is a parameter of the interaction that is proportional to the electron beam current J_0; e and m are the electron charge and rest mass, respectively; and c is the speed of light. The system described in Eq. (1) is applicable for arbitrary initial electron energy γ_0 and allows one to calculate completely until the total deceleration of individual electrons, important for the study of both regimes with strong cyclotron interaction and regimes with maximum efficiency. Equation (1) accounts for two degenerate Langmuir waves of the electron beam (LWB) and a fast cyclotron wave (FCW) with properties analogous to electromagnetic waves on a transmission line. In particular, for synchronous interaction of the FCW with an opposite propagating EMW, we have

$$|1+\beta_z - \omega_c/\gamma_0\omega|/\beta_{z0} < \pi/L \qquad (2)$$

and a stop-band appears (Fig. 1 right) that is responsible for the disappearance of generation (Fig. 1 left) in the region of resonant magnetic fields given by Eq. (2).

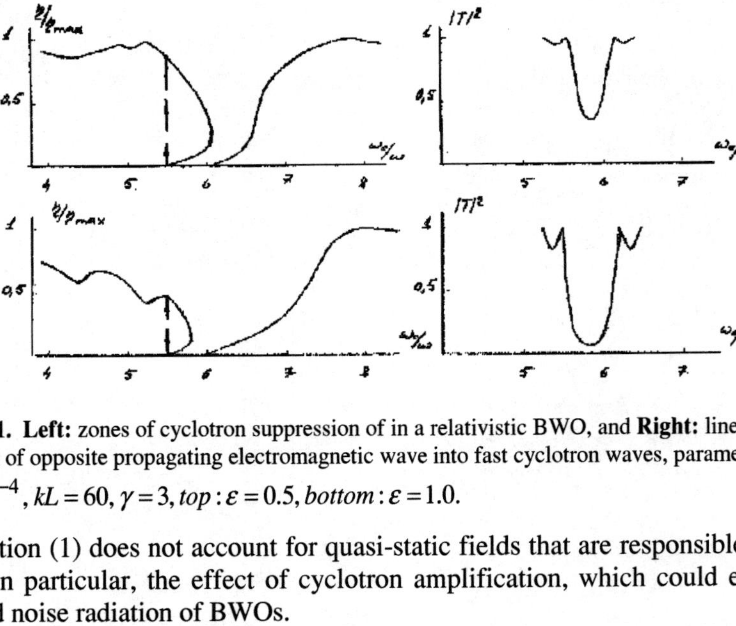

FIGURE 1. Left: zones of cyclotron suppression of in a relativistic BWO, and **Right:** linear bands of conversion of opposite propagating electromagnetic wave into fast cyclotron waves, parameters are $I = 3*10^{-4}, kL = 60, \gamma = 3, top: \varepsilon = 0.5, bottom: \varepsilon = 1.0.$

Equation (1) does not account for quasi-static fields that are responsible for many effects, in particular, the effect of cyclotron amplification, which could explain the increased noise radiation of BWOs.

In order provide regimes of generation that are not sensitive to small variations in the electron beam parameters, the start current must be less than the beam current a few times. In this case the pre- and post-pulses that often accompany the beam current pulses (Figs. 2a,b) can lead to undesirable generation or amplification.

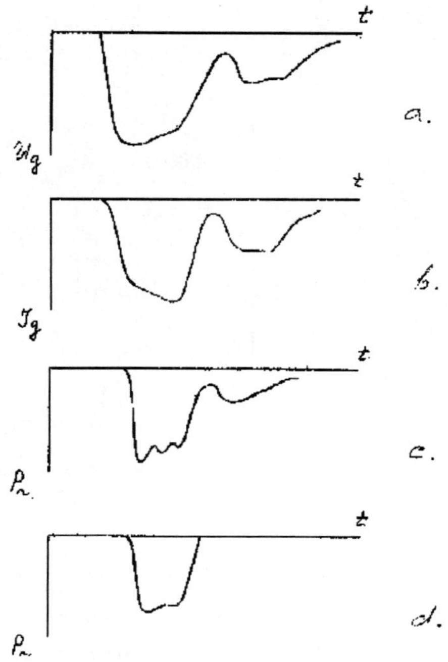

FIGURE 2. Typical pulses from a relativistic BWO operating at high repetition rate: a) pulse voltage; b) electron beam current; c) radiated power; d) radiated power with suppression of trailing edge pulses.

In order to suppress these undesirable pulses, it is possible to use the dependence of the resonant magnetic field in Eq. (2)

$$(\omega_c/\omega)_{res} \approx \gamma + \sqrt{\gamma^2 - 1} \tag{3}$$

and the dependence of the band of suppression of generation on electron energy when the BWO operates with a smaller value of the guide magnetic field H

$$\omega_c < (\omega_c)_{res}; \quad H < H_{res}. \tag{4}$$

When the electron energy W is decreased, for example, by a factor of two (from $\gamma = 3$ to $\gamma = 2$), the value of the resonant magnetic field H_{res} increases by 1.5. It is always possible to find the optimal value H for generation in the region described by Eq. (4) and for the cyclotron absorption of the opposite propagating EMW generated by electron beams with lesser power (Fig. 2c), including the reflected signal in the output channel that is accompanied by the amplification of electron beam noise. A Brillouin diagram depicting both regimes is shown in Fig. 3.

The boundaries of the zone of cyclotron absorption are not symmetric: the right boundary (Fig. 1 left) is gently sloping as the value of the magnetic field decreases towards the resonant value, whereas the left boundary exhibits a hysteresis while transitioning to the regime given by Eq. (4). A fast nonadiabatic transition typical of hysteresis gives the possibility of increasing the steepness of the transition and achieving a more stable leading edge of radiation pulse generation.

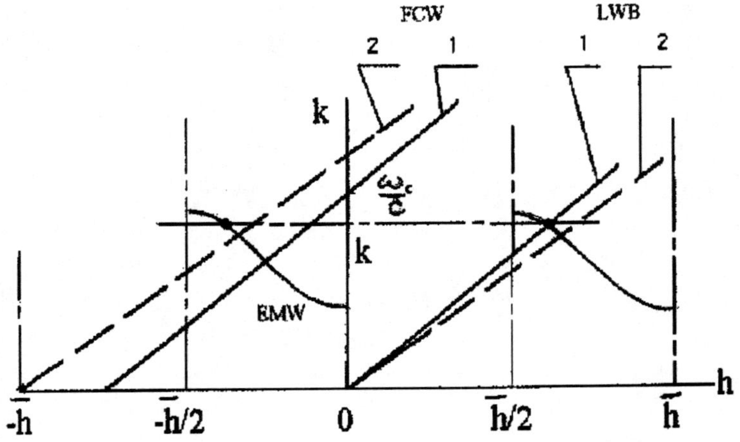

FIGURE 3. Brillouin diagram for a BWO operating using the -1st spatial harmonic of the EMW in the regime given by Eq. (4).

Owing to the hysteresis a disruption of radiation generation for transition in the opposite direction occurs for smaller electron energies that makes it more complicated to suppress the post-pulses. To correct for this we can add small drift sections to both ends of the BWO. We can place a diaphragm with small reflections before the electron dump, and in so doing, we can change the wavenumber of the operating wave in the drift sections (Fig. 1). Elongation of the space of interaction between the electromagnetic and cyclotron waves improves selectivity of transitions.

In order to have experimentally demonstrated (Fig. 2) the calculated solutions (Fig.1) it was convenient to choose the TM_{11} and TM_{02} modes as the operating ones because for these modes there are diameters of the thin-walled electron beam for which cyclotron absorption is absent. The idea of suppressing the pre- and post-pulses is also applicable to a resonant TWT where the FCW of the electron beam interacts with an opposite propagating EMW, which is responsible for the feedback.

ACKNOWLEDGEMENTS

This work is supported, in part, by an AFOSR/New World Vistas Grant.

REFERENCES

1. Clunie, D., Mesyats, G., Osipov, M.I. *et al,. Proc. Int. Workshop on Strong Microwaves in Plasmas*, IAP, Nizhny Novgorod, vol.2, pp. 886-902 (1996).
2. Fuks, M.I., Schamiloglu, E., Kovalev, N.F., and Palitsin, A.V., *Proc. SPIE on Intense Microwave Pulses VIII*, Orlando, vol.4371, pp.146-156 (2001).

Stability of Generation of a Relativistic BWO With Reflections in the Output Waveguide

Nikolay F. Kovalev[a], Mikhail I. Fuks[b], and Edl Schamiloglu[b]

[a]*Institute of Applied Physics, Nizhny Novgorod, Russia*
[b]*University of New Mexico, Albuquerque, NM USA*

Abstract. Effects of the appearance of non-stationary operating regimes of a relativistic BWO with reflections in the output waveguide are considered.

When a relativistic BWO operates far above the threshold of self-excitation, stationary generation is impossible [1]. In this case the BWO operates in a stochastic or self-modulation regime. The threshold of self-modulation depends, in particular, on the value and phase of reflected radiation in the output waveguide. However, reflections not only shift the boundaries of instabilities, but they are also responsible for the appearance of dynamic or non-stationary regimes of generation, even for current levels that barely exceed the start current. This paper describes this regime and elucidates its threshold for onset.

The influence of reflections on BWO operation has been often discussed (see, for example, [2]). We consider here the influence of reflections using a simple graphical method based on the system of stationary ultrarelativistic equations [3] without taking into account the quasistatic fields of the electron beam:

$$\begin{aligned} &C' = I\int_0^{2\pi} e^{-i\theta} d\theta_0, && C(\varsigma_L) = C_L; \\ &\Sigma' = \operatorname{Re} Ce^{i\theta}, && \Sigma(0) = 1; \\ &\theta' = \Sigma^{-2} - B, && \theta(0) = \theta_0 \in [0, 2\pi), \end{aligned} \tag{1}$$

where (') = $d/d\varsigma$; $\Sigma = \gamma/\gamma_0$ is the ratio of electron energy to its initial value at the input (z=0) of the interaction space with length *L*; $\varsigma = kz/2\gamma_0^2$; $\varsigma = kL/2\gamma_0^2$; $\theta = \omega t - hz$ is the phase of the wave acting on electrons; $B = 2\gamma_0^2(h/k - 1)$ is the mismatch that determines the radiation frequency; $C(\varsigma) = (2\gamma_0 e/kmc^2)E_z e^{-i\theta}$ is the normalized complex amplitude of the operating longitudinal electric field E_z; and

$$I = \gamma_0^3 e|E_z|^2 J_0 / \pi k^2 mc^2 P \tag{2}$$

is the parameter characterizing the interaction of the electron beam with a current J_0 and the wave transferring a power *P*. The efficiency of BWO operation is characterized by

$$\eta = \left(\int_0^{2\pi} [1 - \Sigma(\varsigma_L)] d\theta_0\right) / 2\pi = \left(|C_0|^2 - |C_L|^2\right) / 4\pi I . \tag{3}$$

Part of the output radiation returns to the BWO input, given by

CP650, *Beams 2002: 14th International Conference on High-Power Particle Beams*, edited by T. A. Mehlhorn and M. A. Sweeney

$$C_L = C_0 \,\mathrm{Re}^{2ikL_c}, \tag{4}$$

because of reflections in the output waveguide. Here the wavenumber in a homogeneous feedback channel of length $2L_c$ is given by the free space value $k = \omega/c$.

The boundary value problem given by Eqs. (1) and (4) describes regimes of stationary generation with frequency ω and efficiency η depending on a complex reflection coefficient *R*. For a long delay in the feedback (hL_c), that is, for slow processes, the system in Eq. (1) can be used to describe stationary regimes as well as dynamic ones, or at least give the conditions for the appearance of such regimes.

It is convenient to represent solutions of the system in Eq. (1) with the condition given by Eq. (2) in the form of Ricke's diagrams or load diagrams that are isolines for mismatches $B = const$ and amplitudes $C_0 = const$ with $\arg C_0 = 0$ on the plane of complex reflections $R\exp(2ikL_c)$. For $I = 0.15$ and $\varsigma_L = 2.2$ (parameters, which are close to the regime with maximum efficiency) Ricke's diagrams are shown in Fig. 1.

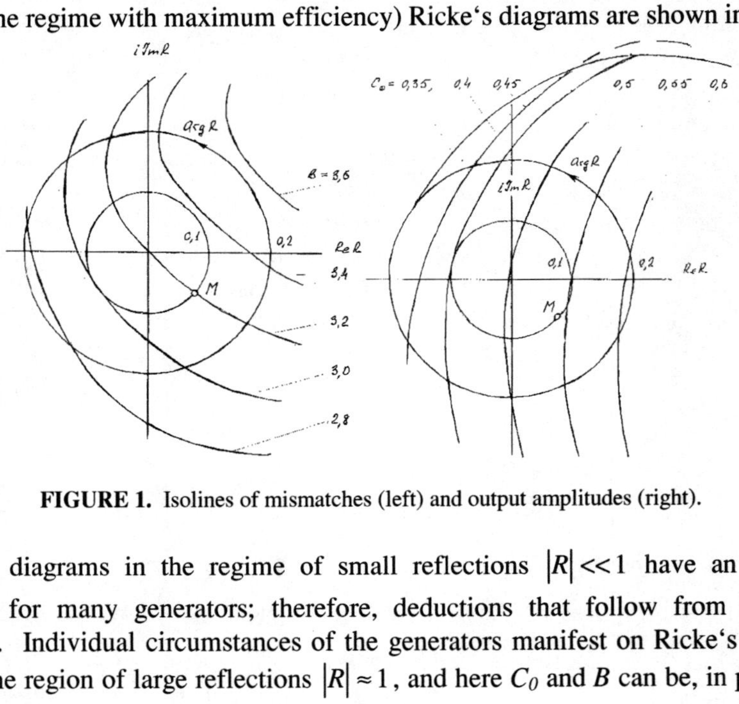

FIGURE 1. Isolines of mismatches (left) and output amplitudes (right).

These diagrams in the regime of small reflections $|R| << 1$ have an identical topology for many generators; therefore, deductions that follow from them are universal. Individual circumstances of the generators manifest on Ricke's diagrams only in the region of large reflections $|R| \approx 1$, and here C_0 and *B* can be, in particular, multi-valued.

For example, when properties of the reflector do not depend on frequency, i.e. $dR_0/dk = 0$, then the geometric operating point for a generator with different phases of reflection from a load ($\varphi_r = 2kL_c + \arg R_0$) is a circle with center coinciding with the center of the diagram (for $|R| = 0.1$ and $|R| = 0.2$ the circles are shown in Fig. 1). It is convenient to show the dependencies of frequencies (*B*) on reflection phases (*arg R*) in a plane (Fig. 2). If one were to also show the dependence of the phase φ_r on frequency, then the point (*M*) of intersection of these lines will determine the operating regime of the BWO, and together with this one can obtain the dependencies of frequency and output power on adiabatically slow variations in *R*.

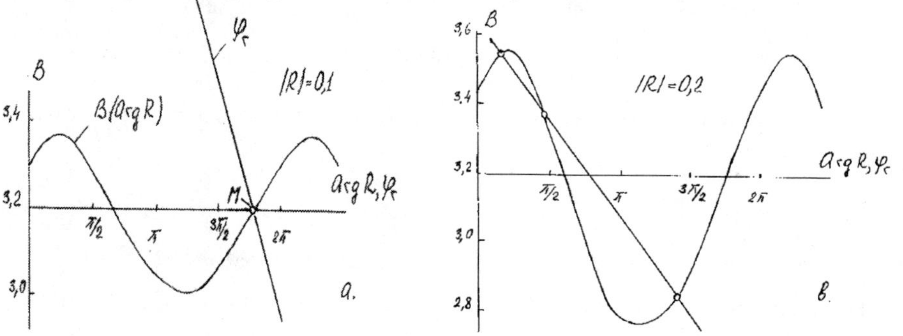

FIGURE 2. Reflection phases along lines (a) $|R| = 0.1$ and (b) $|R| = 0.2$.

Parameters of stationary generation are determined by the position of the point of intersection, whereas the number of intersections is determined by the slope of line φ_r,

$$dB/d\varphi_r = -(\gamma_0^2/kL)(1 - c/v_{ph}). \tag{5}$$

However, the problem of stability of operation of generators is not solved even for the case with one intersection (Fig. 2(a)). To solve this problem, we will use a method similar to the method of Poincare's mapping together with Lameray's diagram [1]. Representative diagrams are shown in Fig. 3. If we take a frequency of generation that differs slightly from the stationary one, then according to curve $\varphi_r(B)$ the phase of the reflected signal also changes. The new setting (with delay) of phase φ_r will correspond to another frequency of BWO generation. If the difference between the new and stationary frequencies is less than the initial deviation, then Lameray's "stairs" leads to stable, stationary generation. When the difference is greater than the initial deviation of frequencies, then the stationary state is unstable. For example, the point M_2 in Fig. 3(b) is unstable, whereas points M_1 and M_3 can be either stable or unstable, depending on the slope of line φ_r. When the slope of line φ_r is greater than the slope of the curve $B(\arg R)$ in the point of intersection,

$$|dB/d\varphi_r| > |dB/d(\arg R)|, \tag{6}$$

that is, when Lameray`s stair converges to the point M_1 (or M_3), it corresponds to a regime of stationary generation. Figure 3(b) shows the hysteresis transitions between difference stationary states that are also possible when the position of the reflector is changed. For inverse inequality (7), when the length L_c is large, the points M_1 or M_3 are unstable. In this case the system evolves to the regime of self-modulation (Fig.3(c)). The self-modulation instability appears when following criteria take place:

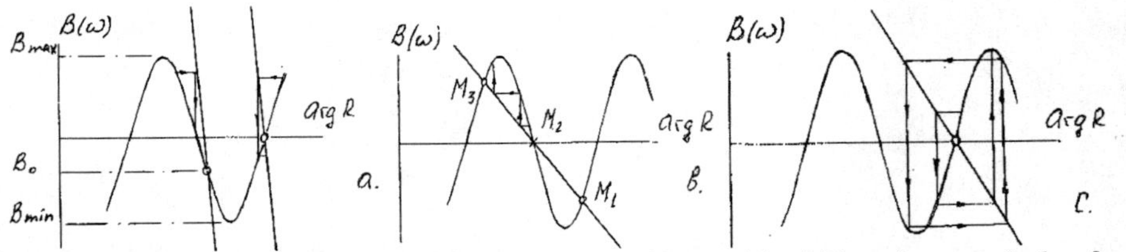

FIGURE 3. Lameray's diagram for (a) stable and (b) unstable (M_2) regimes, and also for (c) appearance of self-modulation instability.

$$|dB/d\varphi_r| < |dB/d(\arg R)|_{\max} \quad \text{or} \quad kL_{c\,\mathrm{lim}} = \gamma_0^2\left(1 - c/v_{gr}\right)/|dB/d(\arg R)|_{\max} \tag{7}$$

For small reflections the derivative $|dB/d(\arg R)|$ is proportional to $|R|$; therefore, in this case the minimal length $L_{c\ lim}$ is large, the transient time to self-modulation increases, and spectrum of output radiation is narrowed.

According to Fig. 3, the self-modulation spectral width cannot exceed a value $\Delta\omega_{\max} = \omega(B_{\min}) - \omega(B_{\max})$ depending on $|R|$. Therefore, when $\Delta\omega_{\max}$ is less than (or comparable to) the natural width of the generation line, the self-modulation does not occur. On other hand, when $\Delta\omega_{\max}$ is greater than the band of self-modulation without reflections, then the results based on stationary Ricke's diagram are invalid.

This method of analysis is also acceptable for reflections, whose properties depend on frequency. For example, for resonant reflections the geometric position corresponding to the operating points is a non-axisymmetric closed curve close to a circle. When the diameter of this curve is small, the dependence $B(\arg R)$ can be close to that in Fig. 2, and the curve $\varphi_r(B)$ gives the set of intersections that corresponds to the range of operating regimes of the BWO.

A shift of the operating regime into a regime of multiple-valued $B(R)$ makes the analysis somewhat more difficult: lines $B(\arg R)$ become multiple-valued, additional stable and unstable states appear, the appearance of hysteresis transitions even for small L_c become possible, and the transitions to dynamic regimes of generation increase.

ACKNOWLEDGEMENTS

This work is supported, in part, by an AFOSR/New World Vistas Grant.

REFERENCES

1. Rabinovich M.I., Trubetskov D.I. "Introduction into theory of oscillations and waves," Moscow, "Nauka," 1984, p. 431.
2. Okress, E., Ed., "Crossed-field microwave devices," Academic Press, New York, 1961.
3. Kovalev N.F., Petelin M.I., Raizer M.D. *et al.*, "O-type devices based on stimulated Cerenkov and transient radiation of relativistic electrons," in "Relativistic microwave electronics," edited by A.V. Gaponov-Grekhov, NIRFI, Gorky, 1979.

Relativistic Backward Wave Oscillators with Distributed Output of Microwave Power

Edward Abubakirov, Andrey Denisenko, Michael Goikhman, Alexey Palitsin, Evgeny Soluyanov, Vladimir Yastrebov

Institute of Applied Physics RAS, 46 Ulyanov street, Nizhny Novgorod, 603950, Russia
e-mail edward@appl.sci-nnov.ru

Abstract. Application of the Bragg reflectors made as periodically corrugated metallic waveguide for microwave output in relativistic BWOs allows to acquire new positive qualities with comparison to traditional system based on a cut-off mirror. In the report various versions of implementation of the Bragg reflector into the microwave system of the relativistic BWO are proposed and results of experimental testing of such BWOs are presented.

INTRODUCTION

Backward wave oscillators (BWOs) with microwave systems made with slightly corrugated periodic metal waveguides are the simplest and the most spread version of relativistic sources of high power microwaves (HPM). In these devices relativistic electron beam interacts synchronically with the (-1)st spatial harmonics of the operating wave

$$\omega \approx (2\pi/d - h_0)v. \quad (1)$$

Here ω is operating frequency, h_0 is longitudinal wave number of the fundamental harmonics, d is a period of corrugation, v is velocity of electrons. The microwave energy in the BWO is transported mainly by the 0-th spatial harmonics, which propagates in the reverse direction towards the electron emitter having high electric potential. To provide output of the microwave energy in this design usually an additional cut-off section situated at the cathode edge of the microwave system is used. This method realize easily the microwave output through a ground potential collector region. However, reflection of the wave adds to the operation of the relativistic BWO some peculiarities:

a) near the cut-off neck a region with a standing wave arises, where high-frequency electric field has double strength with respect to the output wave value. This obviously provokes high-frequency break-down in the region. At the same time double field strength makes essential a role of non-synchronous interactions, which can produce serious changes in the operation of the relativistic BWO;
b) the reverse path of the operating wave through the interaction region gives ground to an appearance of new zones of the cyclotron absorption of the microwave power

CP650, *Beams 2002: 14th International Conference on High-Power Particle Beams,*
edited by T. A. Mehlhorn and M. A. Sweeney

by the magnetized electron beam and by this makes it difficult to work at low guiding magnetic fields;

c) in high power BWOs having oversized microwave systems the cut-off section may disperse the operating wave into other modes of the system. That makes a strong disturbance of spatial structure of the output radiation and decreases efficiency of the energy output.

Obviously, for the relativistic BWO with oversized microwave systems special approaches to energy output have to be used. Essential improvement of the microwave output system in the relativistic BWO can be achieved if the principle of distributed energy extraction based on application of Bragg reflectors (or converters) will be used. Bragg mode converter in the relativistic BWO allows:

- to output energy with a waveguide mode forming simple radiation wave pattern, which is convenient for application;
- to made spectral filtration of the output radiation using selective properties of the converter;
- to get a possibility to choose phase velocity of output wave and, by this, to control zones of cyclotron absorption at low magnetic fields;
- to optimize longitudinal structure of the high-frequency electromagnetic field in the BWO for efficiency enhancement.

VERSIONS OF RELATIVISTIC BWO WITH BRAGG MODE CONVERTER

BWO with Bragg mirror. In the simplest version of the BWO with mode converter a Bragg reflector made as periodically corrugated waveguide can be placed at the cathode edge of the BWO [1]. The corrugation period has to satisfy resonance condition

$$h + h_0 = 2\pi / d , \tag{2}$$

h is longitudinal wave number of the output mode. In this case the Bragg reflector operates similar to the cut-off section and does not change the regime of interactions between electrons and electromagnetic field.

Relativistic BWO with Bragg scattering. In BWO with oversized microwave system an operating wave is usually close to Bragg resonance with some lower mode of the system and corresponding mismatch is small

$$(h + h_0 - 2\pi / d)L \approx (h - \frac{\omega}{v})L \approx \frac{\omega L}{2c}(\frac{k_\perp^2}{k^2} + \frac{1}{\gamma^2}) \sim \pi \tag{3}$$

because of relativistic velocities of electrons $\gamma = (1 - v^2 / c^2)^{-1/2} >> 1$ and proximity of phase velocity of the lower wave $v_{ph} = \omega / h$ to the light velocity. Here, k, h, $k_\perp$ are full, longitudinal and transverse wave numbers of scattered wave accordingly, L is a length of the structure. An essential Bragg scattering of the operating wave to the lower mode can take place in this case just in the microwave system of the BWO itself [2]. Obviously the energy transformation to lower mode is not very high due to mismatch from the exact Bragg resonance. but it can be increased, if an irregular system will be used. For example, it can be done as waveguide with a smooth decrease of period. Its cathode edge corresponds to Bragg reflector, the collector edge provides interaction by BWO mechanism. Due to scattering of the operating wave along the whole microwave system this version of the microwave oscillator realizes distributed output of energy, and by this provides relatively small value of field amplitude at the beginning of the interaction space. That gives ground to efficiency enhancement in this version of the BWO. However, fast wave at the start of interaction groups electrons mostly in its accelerating phase, so the phase velocity profile in this version of microwave oscillator is far from optimal.

Sectioned BWO. To improve the situation a small modulating section should be inserted before the reflector (Fig.2). In the scheme microwave energy generated by the main BWO section is mainly reflected by the Bragg mirror. Its small part penetrates to the first BWO section and produces the initial velocity modulation of the electron beam. So now the reflector simultaneously serves as a drift space and the device in a whole gets a longitudinal distribution of high-frequency field, which allows to increase essentially operation efficiency up to 30-40% The scheme is similar to the BWO with pre-modulation gap [3], but more reliable for powerful microwaves.

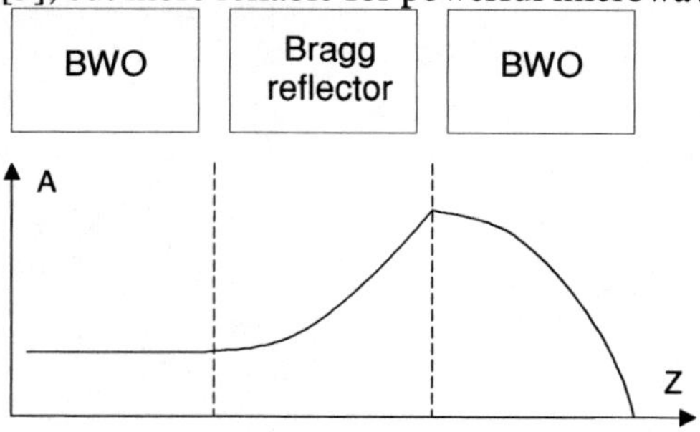

FIGURE 1. Sectioned BWO, scheme (top) and microwave field distribution (bottom).

EXPERIMENTAL RESEARCH OF SECTIONED BWO WITH BRAGG MODE CONVERTER

The main part of experimental researches was carried out with the sectioned BWO, which seems to be the most prospective version of the BWO with Bragg reflector. To make a proof-of-principle experiment the simplest combination of waves was chosen: TM_{11} of a circular waveguide served as operating wave and microwave output was provided with TE_{11} mode. The TE_{11} mode forms easily and effectively gaussian wave beam, so it is very convenient for the energy extraction.. The radial transverse structure of the TM_{11} mode has a region where a cyclotron interaction of the wave and tubular electron beam is minimal. The cyclotron absorption prevents operation of microwave oscillators at relatively low values of the magnetic field, which are used for focusing of the electron beam. In the case of the TM_{11} mode, similar to [4], the absorption can be suppressed essentially.

Three sections of the oscillator were made on a base of circular waveguides with an axis-symmetric periodic corrugation of their walls. Edge sections realized backward wave mechanism of interaction of TM_{11} wave with an electron beam, the middle section was the Bragg converter of TM_{11} to TE_{11} mode with about 90% of the transformation efficiency. TE_{11} mode was radiated to atmosphere by a conical horn through oversized vacuum window.

The oscillator was realized in 3-cm wavelength band and was driven with 0.8 MeV, 6 kA tubular electron beam, formed with explosion emission magnetically insulated gun. Electron beam diameter corresponded to the minimum cyclotron absorption for TM_{11} mode with a rotating transverse structure.

Maximum output peak power of the oscillator (up to 700 MW with 15% efficiency) was observed at magnetic fields below the resonance in the interval of 6-8 kGs (Fig.2). Spectral measurements gave for carrier frequency 10.1 GHz with radiation band of 20-30 MHz. Spatial structure of the radiated wave was close to a gaussian wave beam with circular polarization (Fig.2). As microwave system of oscillator was axis symmetric, a direction of the rotation was strictly determined by orientation of focusing magnetic field.

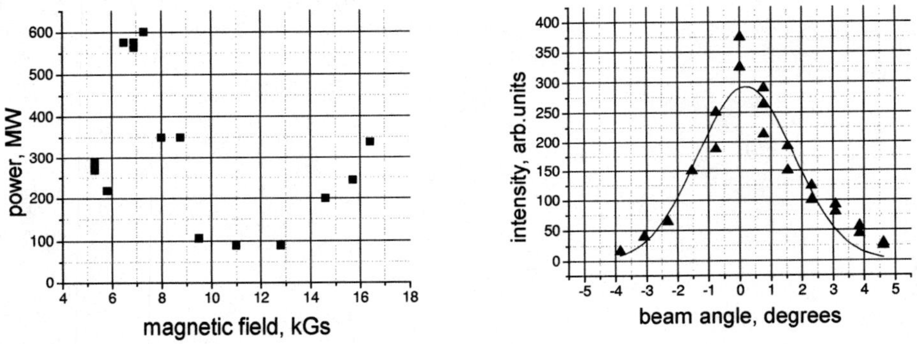

FIGURE 2. Dependence of output power of the sectioned BWO on focusing magnetic field (left) and spatial structure of the output wave beam (right)

ACKNOWLEDGMENTS

The research was supported by the Program of Scientific Research of Russian Academy of Sciences "Pulse High Power Electronics".

REFERENCES

1. Abubakirov E.B., Denisenko A.N., Fuch M.I et al., *Int. Workshop on High Power Microwave Generation and Pulse Shortening. Digest of Technical Papers.* Edinburgh, 1997, p.179-183
2. Abubakirov E.B., Ginzburg N.S. Sergeev A.S., *Radiotekhnika i electronika*, **46**, No.11, 1374-1378 (2001).
3. Kurkan I.K., Rostov V.V., Totmeninov E.M. , *Pisma ZhTF*, **24**, No.10, 43-47 (1998).
4. Abubakirov E.B., Belousov V.I., Zaitsev N.I. et al .*Proc. of 8th International Conf. on High-Power Electron Beams.* World Scientific, 1991, vol.2, 1155-1160.

Highly Efficient Generation of Subnanosecond Microwave Pulses in Ka-Band and Nanosecond Pulses in X-Band

Anton A. Elchaninov,[a] Sergey D. Korovin,[a] Igor V. Pegel,[a]
Vladislav V. Rostov,[a] Valery G. Shpak,[b] and Michael I. Yalandin[b]

[a] *Institute of High Current Electronics, Siberian Division RAS, Tomsk 634055, Russia*
[b] *Institute of Electrophysics, Ural Division RAS, Ekaterinburg 620016, Russia*

Abstract. The paper presents theoretical and experimental studies of the Superradiance (SR) regime of short high-power microwave pulse production in the Cherenkov relativistic BWO. It was found in theory and simulations, that under certain conditions, the peak power of the microwave pulse may be higher than the electron beam power. It is possible for enough lengthy uniform slow wave structure in the case of moderate electron energy and task easy for longitudinal profiling of coupling impedance. In experiment using the RADAN-303SNS mini-accelerator, 420 MW, 200 ps pulses were produced in Ka-band. In X-band experiment using SINUS-150 accelerator, 1.2 GW, 0.5 ns pulses were produced with 140% power conversion efficiency.

INTRODUCTION

As is was found in theory [1] and confirmed in experiment [2, 3], the non-linear stage of transient process in the Backward-Wave Oscillator (BWO) is characterized to short giant pulse with a bandwidth of about 10% if the beam current much exceeds the starting current. This observation promoted origination of idea of short microwave pulse production in superradiance regime using subnanosecond and nanosecond electron beams [4-6].

Below, we demonstrate how superradiance in the relativistic BWO may be realized on the basis of long electron beam in slow wave structure (SWS) with low dispersion of the group velocity.

THEORY AND SIMULATIONS

The questions about saturated power conversion efficiency, pulse width and critical parameters were of interest. Let us find main the regularities by simple models.

Two one-dimensional time-dependent models were developed to study the SR [7,8]. The first one is based on classical theory [1]. It assumes unidirectional motion of electrons and contains a minimum number of independent input parameters (three, for a uniform SWS: C - Pierce parameter, γ_0 – relativistic factor and ξ_k - normalized SWS length). The other PIC-code allows reflection of beam particles and the presence of non-synchronous waves. This results in larger number of input parameters. Both codes assume non-dispersive wave propagation. The third and most complex level of

CP650, *Beams 2002: 14th International Conference on High-Power Particle Beams,*
edited by T. A. Mehlhorn and M. A. Sweeney

simulation employs the 2.5-D code KARAT [9]. Having large number of inputs, it was used for quantitative optimization of concrete tube design.

The slow variable approximation means the requirement $C<<1$ and the following description of synchronous spatial harmonic:

$$E=Re\left(\hat{E}(r_b)E_{z,-1}(z,t)exp\, i\omega_0(t-z/V_0)\right)$$

Here r_b is the beam radius, ω_0 is the frequency of exact synchronism between the wave and particles having velocity V_0. According to [1], the parameters are defined as

$$\xi_k=\frac{\omega_0 CL}{V_0},\; C=\left(\frac{eJ_b\hat{Z}}{2mc^2\gamma_0^3}\right)^{1/3},\left(\hat{Z}\equiv 2\left|\hat{E}(r_b)\right|^2/k^2N\right)$$

Here L is the SWS length, J_b - the beam current, N - the wave norm, $k=\omega_0/c$. To produce intense SR peaks, it is necessary to provide extended length of interaction $\xi_k>>\xi_{k,st}$ (here $\xi_{k,st}\approx 2$ is the BWO starting length in steady-state regime). Further increase in the SWS length ($\xi_k>8$) causes negligible growth of the power conversion, with its saturation. For a uniform SWS of any length the conversion (defining K as ratio microwave peak power to beam pulse power) is small, if parameter $\nu=2\gamma_0^2C\rightarrow 0$. It is so called low signal approximation, where the pulse width may be maximal and close to the specific time: $\Delta t\approx t_S=(1+V_0/V_{gr})/\omega_0 C$, V_{gr} is the group velocity of the wave. Conversion of $K>1$ may be realized in the range of parameter $\nu\approx 0.2\ldots 0.6$ (see Fig. 1) near the limit where reflection of particles is possible. As is seen in Fig.1, the normalized pulse width ($\Delta\tau=\Delta t/t_S$) reduces in the range of high conversion coefficients.

Higher power conversion may be achieved in nonuniform SWS's where the coupling impedance is profiled to grow towards the collector (for example, by gradually increasing the SWS corrugation depth). Such analysis in the case of $\gamma_0>>1$ where reflection of particles doesn't occur was performed (and of any energy by 1-D PIC code, see Fig. 2). So, with no account of group velocity dispersion and non-synchronous interaction, the power conversion coefficient may be as large as 10 ($K\approx\eta_0\xi_k$, η_0 is difference in the beam kinetic energy before and after pulse passage). This phenomenon is explained by spatial accumulation of energy in a short microwave pulse propagating in long SWS. The best type of coupling distribution corresponds to constant amplitude of harmonic and almost linear increase in microwave pulse energy.

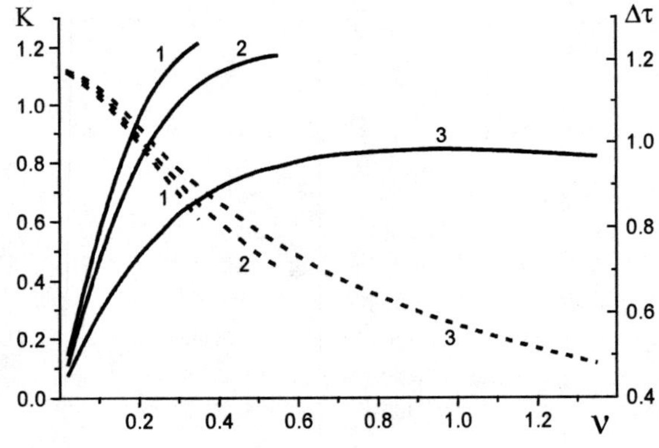

FIGURE 1. Simulated power conversion coefficient for uniform SWS (solid lines) and normalized pulse width (dashed lines) vs parameter $\nu = 2\gamma_0^2 C$; $\xi_k = 6$
1) $\gamma_0 - 1 << 1$; 2) $\gamma_0 = 1.6$; 3) $\gamma_0 >> 1$.

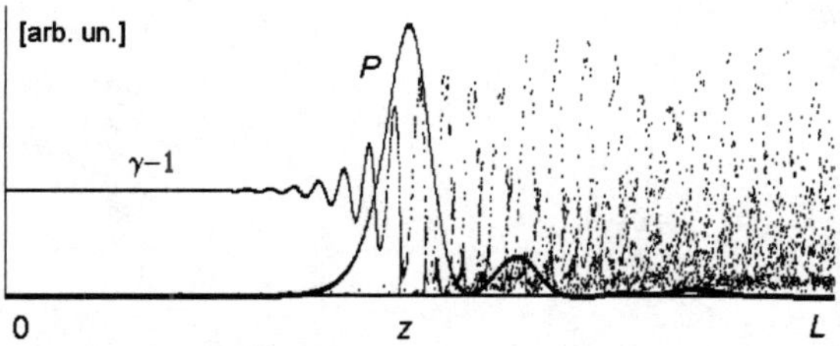

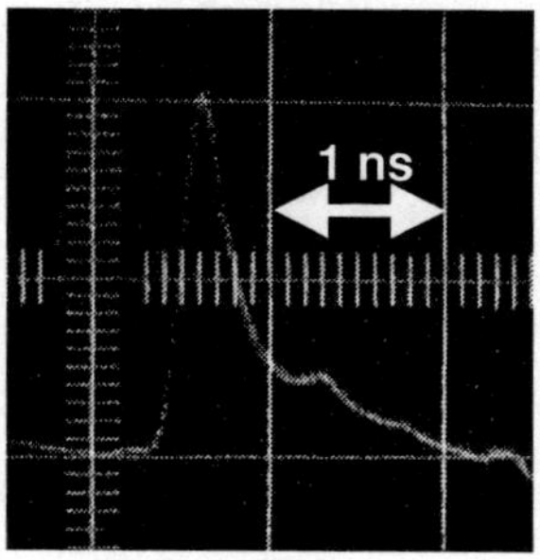

FIGURE 2. Typical shape of superradiance pulse and electron beam phase plot (1-D PIC modeling).
FIGURE 3. Waveform of detected 38-GHz, 400-MW microwave pulse.

The SWS's used earlier with the diameter of corrugation close to the wavelength had a respectively small pass band (30...40%). Strong frequency dependence of the group velocity over the operating band resulted in broadening of the wave envelope, which essentially (in 2...3 times) decreased the peak power. To suppress this effect SWS of increased cross size (in 1.4...1.5 times [10]) with pass band broadened to ≈100% was used. The resonant reflector with low Q-factor [11] provides reflection of pulse instead of beyond cufoff-neck. Using the KARAT code, two superradiance BWO's in Ka-band and X-band were designed.

Ka-BAND EXPERIMENT

The Ka-band experiment [12] was performed with the using a high-current subnanosecond electron accelerator based on RADAN-303SNS pulser with an additional unit of compression and peaking of voltage pulses. The pulse power of the beam with length about 1 ns injected by explosive-emission cathode was 600 MW. The pulse power of the microwave oscillator was determined by special germanium detector and integrating of the radiation pattern. Peak powers ranging from 240 to 280 MW (with $B \approx 2$ T) and 360–420 MW ($B > 5$ T) were obtained. The typical pulse width was 200–250 ps (Fig. 3). Thus, the power conversion efficiency was ≈70% and the energy efficiency was ≈15%.

X-BAND EXPERIMENT

The above result stimulated an attempt to overcome the 100-% threshold in power conversion efficiency. A BWO with longer SWS length (22 corrugation periods: $\xi_k \approx 4.5 \rightarrow \approx 6$) was used. After optimization of SWS nonuniformity and beam parameters (330 kV, 2.6 kA, 4 ns) simulated power conversion of ≈1.4 was achieved. The SINUS-150 compact electron accelerator [4] was used (Fig. 4). The guide magnetic field was 2.4 T. Processing the microwave detector signals (accounting for its nonlinearity) showed that the microwave pulse width was typically of 0.5 ns. The maximum peak power obtained by integration of power density over radiation pattern was 1.2 GW (with beam power less than 0.9 GW). The central frequency was 9.3 – 9.4 GHz, with the spectrum width of ≈10% at (-3...-6 dB) level. The pulse energy measured with a calorimeter reached 0.6 J. The power conversion efficiency was ≈140% and the energy efficiency was ≈17%.

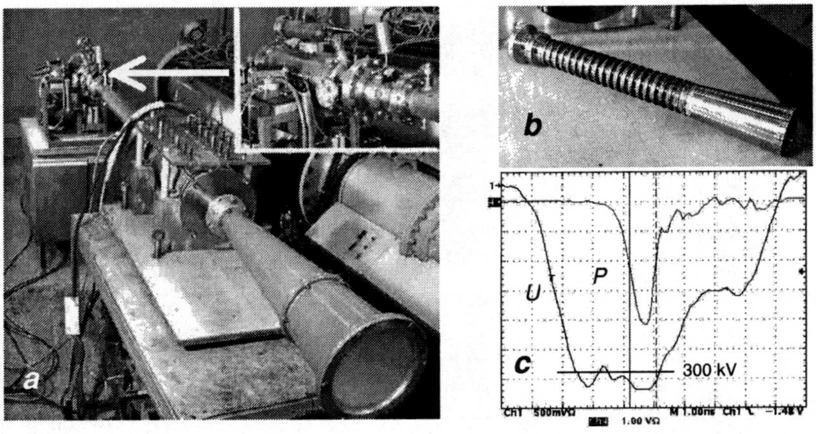

FIGURE 4. (*a*) X-band superradiance BWO experiment (insert shows the SINUS-150 pulser), (*b*) tube with resonant reflector, and (*c*) waveforms of diode voltage and microwave detector signal.

CONCLUSIONS

Thus, for the first time the Superradiance regime of HPM production was demonstrated with microwave power conversion in excess of 100%. In theory and simulations, it was found that the phenomenon is explained by the spatial accumulation of energy in a short microwave pulse propagating in a long slow wave structure with nonuniform coupling impedance. The maximum power conversion efficiency in X-band 140% was obtained.

ACKNOWLEDGMENTS

The authors are grateful to Dr. Naum Ginzburg for his interest to this work and to Dr. Vladimir Tarakanov, the author of the KARAT code, for collaboration.

REFERENCES

[1] N. S. Ginzburg, et al., *Izvestiya vuzov. Radiofizika.* **21**, 7, 1037–1052 (1978).
[2] B. P. Bezruchko, S. P. Kuznetsov, *Izvestiya vuzov. Radiofizika.* **21**, 7, 1053–1059 (1978).
[3] N. S. Ginzburg, et al., *Pis'ma J. Tech. Phys. (Rus.)*, **24**, 20, 66 – 71 (1998).
[4] V. P. Gubanov, et al., *Proc. SPIE «Intense Microwave Pulses 4»*, **2843**, 228 – 237 (1996).
[5] N. S. Ginzburg, et al., *Phys. Rev. E.*, **60**, 3297–3304 (1999).
[6] M. I. Yalandin, et al., *Pis'ma J. Tech. Phys. (Rus.)*, **25**, 10, 19 – 23 (1999).
[7] A. A. Elchaninov, et al., *to appear in Izvestiya vuzov. Radioelektronika (Rus.).*
[8] I. V. Pegel, *Russian Physics J.*, **39**, 12, 1210 – 1228 (1996).
[9] V. P. Tarakanov, *User manual of code KARAT.* Springfield, VA: BRA, 1992.
[10] V. V. Rostov, et al., *Pis'ma J. Tech. Phys. (Rus.)*, **24**, 10, 43 – 47 (1998).
[11] A. V. Gunin, et al., *Proc. 12th Int. Conf. On High Power Part. Beams,* Haifa, **2**, 849-852 (1998).
[12] S. D. Korovin, et al., *Pis'ma J. Tech. Phys. (Rus.)*, **28**, 2, 81 – 89 (2002).

High-current Electron Beam Coupling to Hybrid Waveguide and Plasma Modes in a Dielectric Cherenkov Maser with a Plasma Layer

Anatoli S. Shlapakovski

Nuclear Physics Institute at Tomsk Polytechnic University, 2a Lenina Str., Tomsk 634050, Russia

Abstract. The linear theory of a dielectric Cherenkov maser with a plasma layer has been developed. The dispersion relation has been derived for the model of infinitely thin, fully magnetized, monoenergetic hollow electron beam, in the axisymmetric case. The results of the numerical solution of the dispersion relation and the analysis of the beam coupling to hybrid waves, both hybrid waveguide and hybrid plasma modes, are presented. For the hybrid waveguide mode, spatial growth rate dependences on frequency at different plasma densities demonstrate improvement in gain for moderate densities, but strong shifting the amplification band and narrowing the bandwidth. For the hybrid plasma mode, the case of mildly relativistic, 200-250 keV beams is of interest, so that the wave phase velocity is just slightly greater than the speed of light in a dielectric medium. It has been shown that depending on beam and plasma parameters, the hybrid plasma mode can be separated from the hybrid waveguide mode, or be coupled to it through the beam resulting in strong gain increase, or exhibit a flat gain vs frequency dependence over a very broad band. The parameters, at which the -3 dB bandwidth calculated for 30 dB peak gain exceeds an octave, have been found.

So far, wide-bandwidth microwave amplifiers driven by high-current electron beams remain unexplored. Meanwhile, a development of devices capable of about octave frequency coverage at 1-100 megawatt output power levels would be an important progress in the field of high-power microwaves. At relativistic phase velocities of the rf drive signal, rather weak dispersion, necessary for broadband amplifier operation, takes place in dielectric-lined waveguides; another way to provide weak dispersion is to use plasma as a waveguide structure. A dielectric Cherenkov maser (DCM) operates in a usual electromagnetic waveguide mode. In a plasma maser, the operating mode is the plasma Trivelpiece-Gould mode, which is suitable for extraction as its phase velocity is close to c. At the same time, research performed in the last decade showed that filling a slow-wave structure with plasma can improve amplifier characteristics [1]. In particular, significant power enhancement along with doubling the bandwidth was achieved in the coupled-cavity traveling wave tube [2] due to the formation of hybrid waves, which combine the properties of plasma and waveguide modes. Hence, for keeping broad bandwidths typical for low-voltage helix tubes at much increased voltages, where a helix is inapplicable, it would be promising to apply plasma filling to another smooth slow wave structure such as dielectric-lined waveguide.

Recent experiments on the wide-bandwidth DCM amplifier [3] showed that the plasma layer is formed in its interaction space at the dielectric surface. The analysis of the hybrid waves in the dielectric-lined waveguide with the near-surface plasma layer

---

The work was supported in part by the Russian Foundation for Basic Research (grant #00-02-17720).

CP650, *Beams 2002: 14th International Conference on High-Power Particle Beams,*
edited by T. A. Mehlhorn and M. A. Sweeney

given in [3] revealed that at the certain parameters, the hybrid plasma mode dispersion can be very weak with phase velocities just slightly exceeding the speed of light in the liner material and, at the same time, its coupling impedance is strong enough and the portion of rf power contained in the dielectric region approaches 80% – hence, the possibility of the interaction between a mildly relativistic (200-250 keV) electron beam and hybrid plasma mode of such a structure seems very interesting. In addition, the configuration with the plasma layer may allow a DCM operation at longer beam pulse durations, since plasma presence prevents from charge build-up at the dielectric surface.

In this work, the linear theory dispersion relation has been derived for the DCM with the plasma layer, and the results of its numerical solution have been analyzed depending on the plasma parameters. We consider the cylindrical geometry that can be depicted as follows. The conducting wall is at radius b, a dielectric fills the region between a and b, an annular plasma layer fills the region between r_p and a, and an infinitely thin electron beam is at radius r_b. Plasma is assumed uniform, collisionless, fully magnetized, and cold, ion motion is neglected. The beam is also assumed fully magnetized and monoenergetic. Using the one-dimensional fluid equations for the beam and plasma and matching the solutions of the wave equations for E_z in various regions, one can finally obtain the following dispersion relation for axisymmetric TM-modes of the system:

$$D(\omega,k)\left[(\omega-ku)^2-\frac{2q^2c^2I_b}{\beta\gamma^3I_A}\Phi(qr_p,qr_b)\frac{I_0(qr_b)}{I_0(qr_p)}\right]=$$

$$=\frac{2q^2c^2I_b}{\beta\gamma^3I_A}\frac{I_0^2(qr_b)}{I_0^2(qr_p)}\frac{\Phi'(\kappa a,\kappa r_p)-\dfrac{\varepsilon q^2}{\kappa p}\dfrac{F'(pa,pb)}{F(pa,pb)}\Phi(\kappa r_p,\kappa a)}{\kappa r_p\Phi'(\kappa r_p,\kappa a)-qr_p\dfrac{I_0'(qr_p)}{I_0(qr_p)}\Phi(\kappa r_p,\kappa a)}. \tag{1}$$

Here, $D(\omega,k)=\dfrac{\varepsilon q^2}{\kappa p}\dfrac{F'(pa,pb)}{F(pa,pb)}+\dfrac{\Phi''(\kappa r_p,\kappa a)+\dfrac{q}{\kappa}\dfrac{I_0'(qr_p)}{I_0(qr_p)}\Phi'(\kappa a,\kappa r_p)}{\Phi'(\kappa r_p,\kappa a)-\dfrac{q}{\kappa}\dfrac{I_0'(qr_p)}{I_0(qr_p)}\Phi(\kappa r_p,\kappa a)}$,

$D(\omega, k) = 0$ is the dispersion relation for the no-beam system, ω is the frequency, k is the axial wavenumber, q, p, and κ stand for the transverse wavenumbers in vacuum ($-q^2=\omega^2/c^2-k^2$), dielectric ($p^2=\varepsilon\omega^2/c^2-k^2$), and plasma ($\kappa^2=q^2(1-\omega_p^2/\omega^2)$) regions, ε is the dielectric constant, ω_p is the plasma frequency, $u=\beta c$ is the velocity of electrons, γ is Lorentz factor, I_b is the beam current, $I_A = mc^3/e$. Helper functions are defined as: $F(x,y)=J_0(x)Y_0(y)-J_0(y)Y_0(x)$, $F'(x,y)=J_0'(x)Y_0(y)-J_0(y)Y_0'(x)$, $\Phi(x,y)=I_0(x)K_0(y)-I_0(y)K_0(x)$, $\Phi'(x,y)=I_0'(x)K_0(y)-I_0(y)K_0'(x)$; J_0, Y_0, I_0, and K_0 are Bessel functions; $\Phi''(x,y)=I_0'(x)K_0'(y)-I_0'(y)K_0'(x)$.

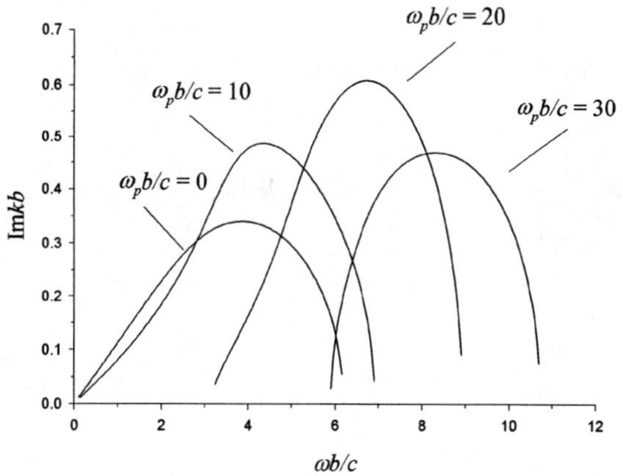

FIGURE 1. Spatial growth rate vs frequency for different plasma frequencies. $\varepsilon = 2.25$, $a/b = 0.5$, $r_p/b = 0.42$, $r_b/b = 0.4$, $\gamma = 1.8$, $I_b = 3.4$ kA.

Some results of the numerical solution of Eq. (1) are presented below. We analyze the beam coupling to the hybrid waveguide and hybrid plasma modes as defined in [3]. The case of interaction with the hybrid waveguide mode is illustrated in Fig. 1 for beam and geometrical parameters close to those of the experiment [3]. The improvement in gain is seen for moderate plasma densities compared to the no-plasma case, as well as strong shifting the amplification band and narrowing the bandwidth. At the frequency given, the dependence on the plasma frequency is shown in Fig. 2. Increasing the plasma density results, first, in the gain increase; then, it abruptly decreases and amplification fails, this is well known [1], but further, it appears again due to the beam coupling to the hybrid plasma mode. This could take place in the experiments [3]. However, the gain in this case is much less; in addition, for these parameters, the rf power is concentrated mostly in the plasma region, that results in big extraction losses. Interaction with the hybrid plasma is more interesting for less beam energy (see Figs. 3, 4).

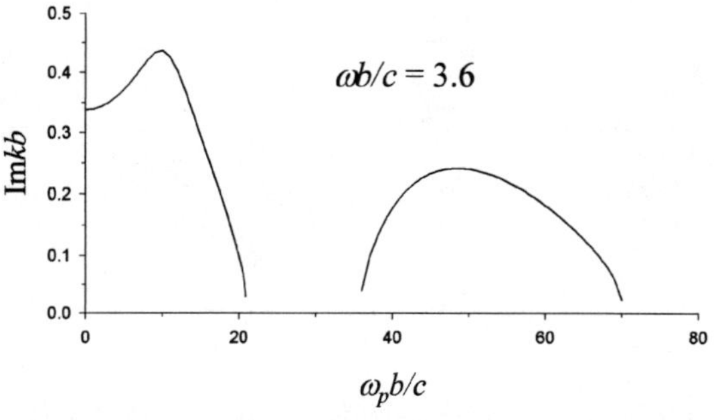

FIGURE 2. Growth rate vs plasma frequency at the frequency given. Parameters are the same as in Fig. 1.

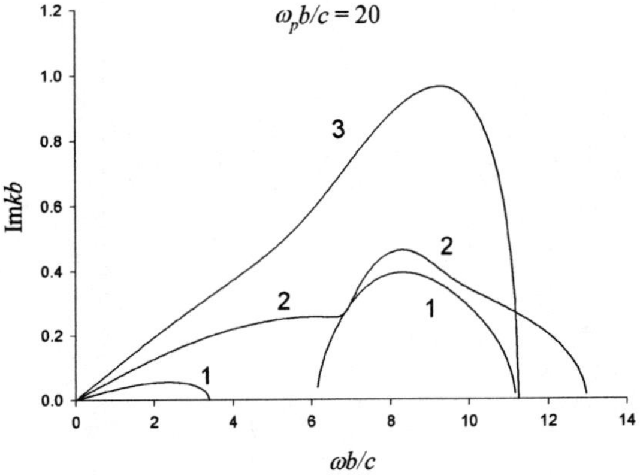

FIGURE 3. The same as Fig. 1 (except for $\gamma = 1.5$) for different beam currents: I_b =: 1 – 500 A; 2 – 1 kA; 3 – 2 kA.

For the parameters of Fig. 3, maximum phase velocity of the hybrid plasma mode is less than the beam velocity, so that the interaction occurs in Raman regime, at large enough currents, due to synchronism with the beam slow space charge mode. As the current increases, its phase velocity decreases, and interaction occurs over a wider frequency range. The hybrid plasma and waveguide modes are coupled through the beam resulting in strong gain increase. At higher plasma density (Fig. 4), the hybrid waveguide mode is separated, and a superwide bandwidth becomes possible in the hybrid plasma mode. The -3 dB bandwidth value calculated for the parameters of the curve 3 in Fig. 4 at 30 dB peak gain is of about 74%, i.e., exceeds an octave. The output power of such amplifier (the beam of 250 keV and 3 kA) could be of tens megawatts.

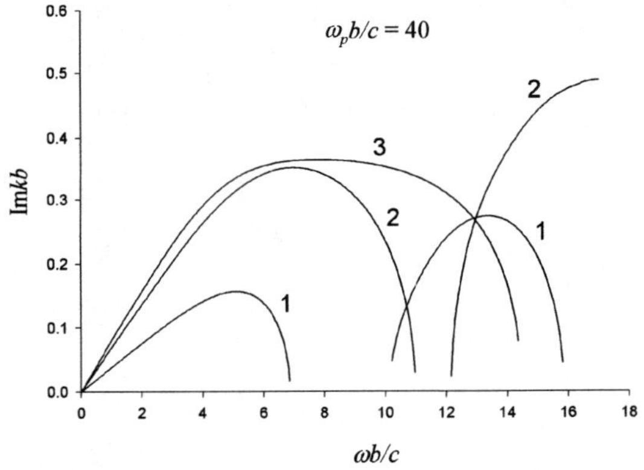

FIGURE 4. The same as Fig. 3 for higher plasma frequency. I_b =: 1 – 170 A; 2 – 1.5 kA; 3 – 3 kA.

REFERENCES

1. Nusinovich, G.S., Carmel, Y., Antonsen Jr., T.M., et al., *IEEE Trans. Plasma Sci.* **26**, 628-645 (1998).
2. Zavialov, M.A., Mitin, L.A., Perevodchikov, V.I., et al., *IEEE Trans. Plasma Sci.* **22**, 600-607 (1994).
3. Mel'nikov, G.V., Petrov, A.V., and Shlapakovski, A.S., *Plasma Physics Reports,* **26**, 1015-1026 (2000).

Multiwave Cerenkov Generator with Nonuniform Slow-Wave Structures

Michail P. Deichuly, Vladimir I. Koshelev, Viktor M. Pikunov, Vladimir A. Popov, Andrey A. Petkun

Institute of High Current Electronics, RAS, 4 Akademichesky Ave., 634055 Tomsk, Russia

Abstract. Influence of a distributed reflector on parameters of microwave radiation of Cerenkov generators with overmoded slow-wave structures (SWS) was investigated both experimentally and theoretically. It was shown that reflector using results in the change of frequency dependence of the SWS reflection coefficient, stability and generation efficiency increase. Radiation pulses with the power of ~1.4 GW at the efficiency of ~30% were obtained at the electron energy up to 500 keV.

INTRODUCTION

A significant direction in microwave electronics is radiation power increase due to the beam current rise (~10^4A) at comparatively low electron energies (~10^5 eV). Development of this direction of investigations will allow creating compact sources of gigawatt power level radiation that is important for applications. A promising device for this purpose is a multiwave Cerenkov generator (MWCG) on the basis of the overmoded multisectional SWS [1, 2]. Preliminary investigations [2] at the electron energy of ~1 MeV have shown that use of nonuniform SWS and a distributed reflector at their input allows increasing the efficiency and stability of generation. Results of experimental investigations at the electron energy up to 500 keV are presented below. To analyze a linear stage of the processes in generators, a MULTIWAVES-5.2 code is used [3]. Special consideration was given to the investigation of the reflector influence on radiation parameters.

INFLUENCE OF REFLECTOR ON GENERATION IN A ONE-SECTIONAL SLOW-WAVE STRUCTURE

The experiments were carried out at the Sinus-7M accelerator. A hollow beam with the average diameter $D_b = 2r_b = 100\div116$ mm was formed at the diode voltage $U_d = 460\div500$ kV. Current pulse of the value $I_b = 9\div10$ kA has duration $\tau_p \approx 40$ ns (fig. 1). Magnetic field in the SWS was changed in the limits of B = 3÷15 kG. The SWS diameter in all experiments was D = 130 mm ($D/\lambda \approx 4$). Radiation power was determined by the measured patterns for two polarizations. Radiation spectrum was measured by means of a bandpass filter with the bandwidth of 0.5%.

Initially, influence of a distributed reflector on operation of the generator with a one-sectional SWS was investigated both theoretically and experimentally. The

CP650, *Beams 2002: 14th International Conference on High-Power Particle Beams,*
edited by T. A. Mehlhorn and M. A. Sweeney

reflector (a corrugated waveguide) consisted of four diaphragms presenting a half-tore with the radius ρ_{ref} = 4.8 mm at a pedestal with the height h_{ref} = 0.9 mm having a period d_{ref} = 18 mm (fig. 2). The distance between the reflector and SWS was L_t = 11 mm.

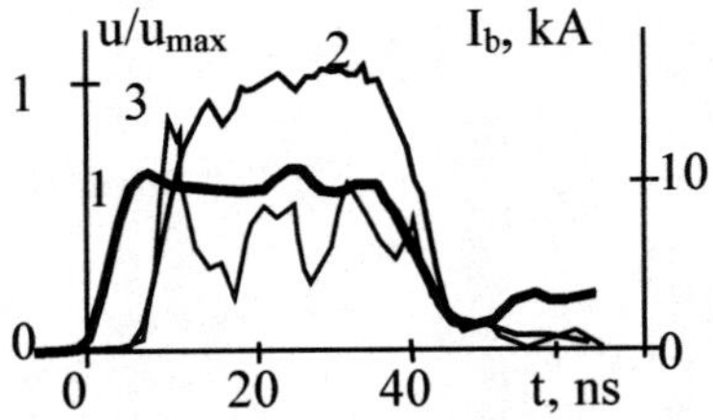

Figure 1. Waveforms of the beam current (1) and the signals from a microwave sensor for the SWS with a reflector (2) and without the one (3).

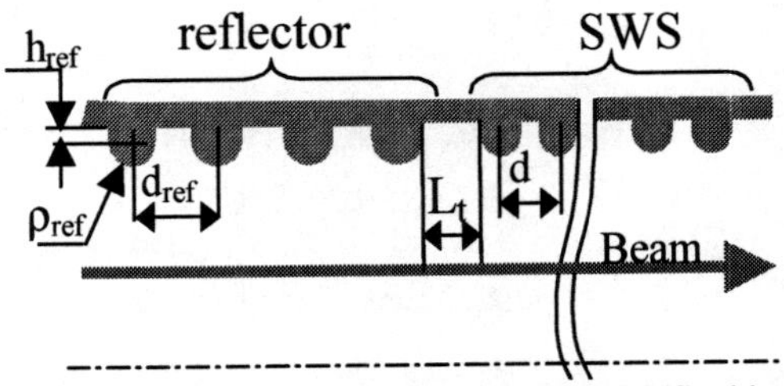

Figure 2. A schematic of a one-sectional SWS with a reflector.

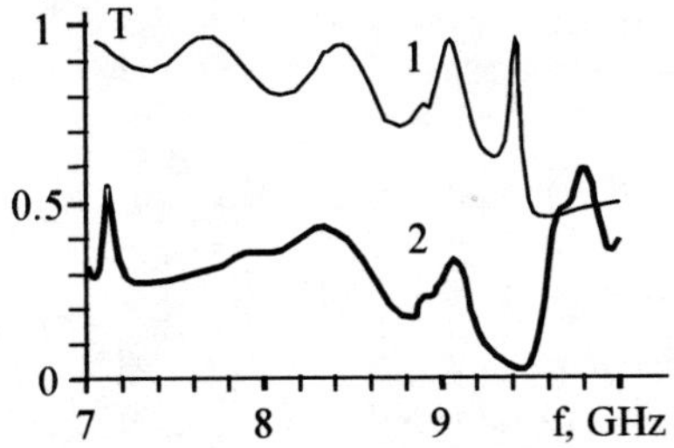

Figure 3. Transmission coefficient of the operating mode E_{01} for the SWS without a reflector (1) and with the one (2).

The reflector geometry was chosen from the condition that the frequency of the π-type oscillations of the reflector mode E_{01} would be lower than generation frequency. Calculations have shown (fig. 3) that use of the reflector results in increase of the frequency interval between longitudinal resonances and decrease of their Q-factor. Starting currents and frequencies of the E_{01}-mode increase and parasitic mode HE_{11} is not excited.

Experimental investigations have shown that use of the reflector results in radiation pulse waveform smoothing (fig.1). The power of radially polarized radiation P_θ increased from 120÷180 MW to 280 MW. The power of cross-polarized radiation $P_\perp$ didn't exceed 60 MW in both cases. Results of experimental investigations of a one-sectional generator with a reflector are presented in figs.4-7.

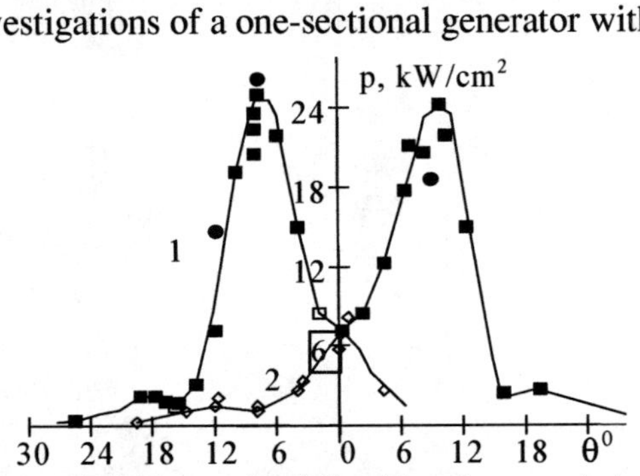

Figure 4. A pattern of radially- (1) and cross-polarized (2) radiation at r_b=55.4 mm and B = 5.8 kG.

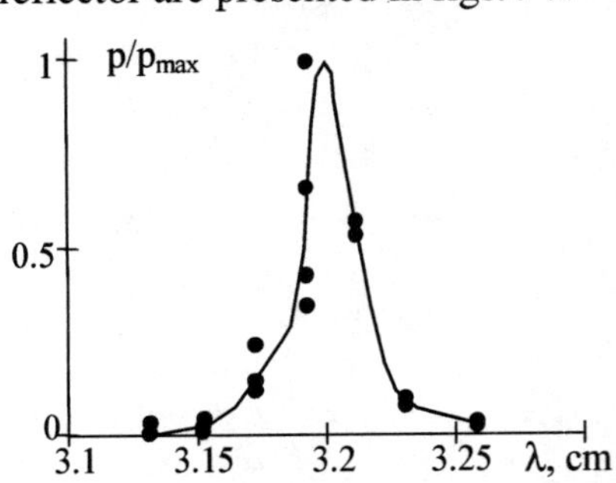

Figure 5. Radiation spectrum at r_b = 55.4 mm and B = 5.8 kG.

Reflector using results in weak increase of the starting radius of the beam (~1MM) as well as in the wavelength decrease (~0.2 mm) in comparison with calculations. Note, that in the calculations reflections from the cathode and the output were not taken into account.

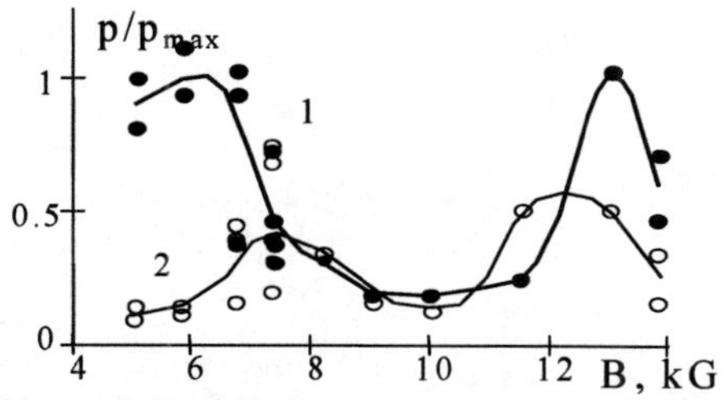

Figure. 6. Radiation power versus magnetic field for radially- (1) and cross-polarized (2) radiation.

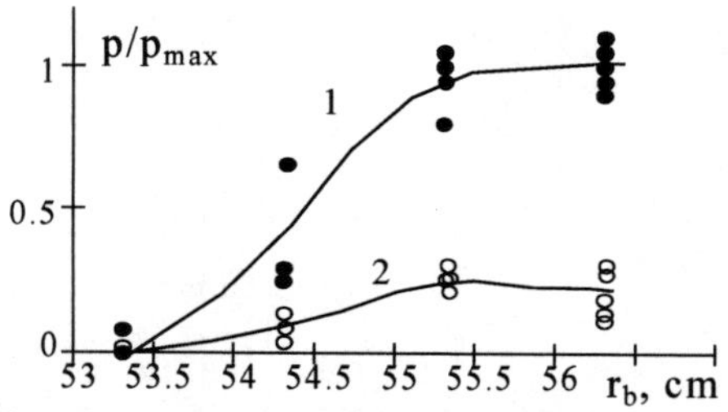

Figure. 7. Radiation power versus beam radius for radially- (1) and cross-polarized (2) radiation.

GENERATION IN MULTISECTIONAL STRUCTURES

Two multisectional SWS were investigated experimentally. The first structure contained three sections: $L_1 = 5d_1$, $L_2 = 6d_2$, $L_3 = 5d_3$, $d_1 = d_2 = d_3 = 13$ mm, $\rho = 4.8$ mm, $h = 0.9$ mm. The distance between the sections was of 8.4 mm. The second structure was two-sectional. The first section had the length $L_1 = 6d_1$, $\rho = 3.2$ mm, $h = 1.6$ mm, $d_1 = 12.4$ mm. The second section was biperiodic and contained 6 diaphragms with different geometry and the period $d_2' = 12.4$ mm ($\rho = 3.2$ mm, $h = 1.6$ mm) and $d_2'' = 13$ mm ($\rho = 3.2$ mm, $h = 1.2$ mm). The distance between the sections was of 12.6 mm.

Comparative investigations have shown that use of a reflector for both SWS results in increase of stability and generation efficiency. The pattern and radiation spectrum for a three-sectional SWS with a reflector ($L_t = 8.4$ mm) are presented in figs. 8, 9, respectively.

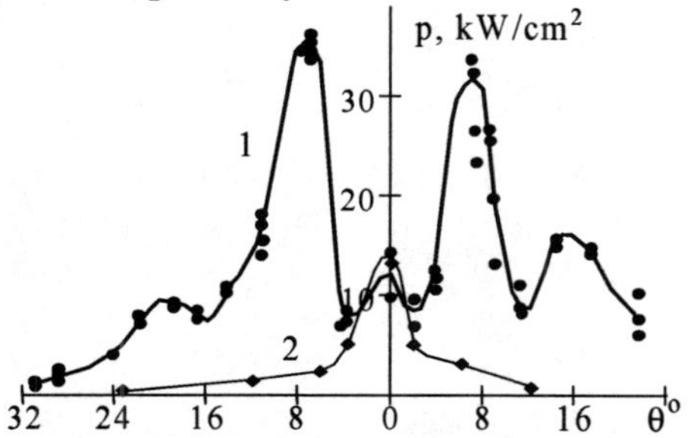

Figure. 8. Pattern of radially- (1) and cross-polarized (2) radiation at $r_b = 53.8$ mm and B = 13.8 kG.

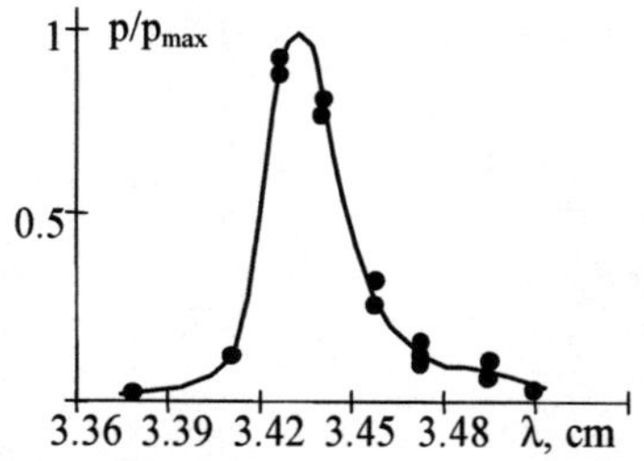

Figure. 9. Radiation spectrum at $r_b = 53.8$ mm and B = 13.8 kG.

The powers in both polarizations were $P_\theta = 670$ MW and $P_\perp = 70$ MW at $U_d = 460$ kV and I_b=9 kA. Generation efficiency at the wavelength $\lambda = 3.43$ cm made up ~16%. Numerical calculations in this case also have shown that reflector using results in the increase of the starting current generation at the E_{01}-mode at the same beam radii and generation at the HE_{11}-mode isn't excited.

Results of investigations of the MWCG with the output biperiodic section are presented in figs.10-13. Use of the reflector ($L_t = 11.6$ mm) resulted in the stability increase of radiation parameters. The power increased by ~2 times at B >12 kG. At $U_d = 470$ kV and $I_b = 9.8$ kA radiation pulses with the power $P_\theta = 1.1$ GW and $P_\perp = 0.3$ GW were obtained. Radiation spectra for two polarizations practically coincide.

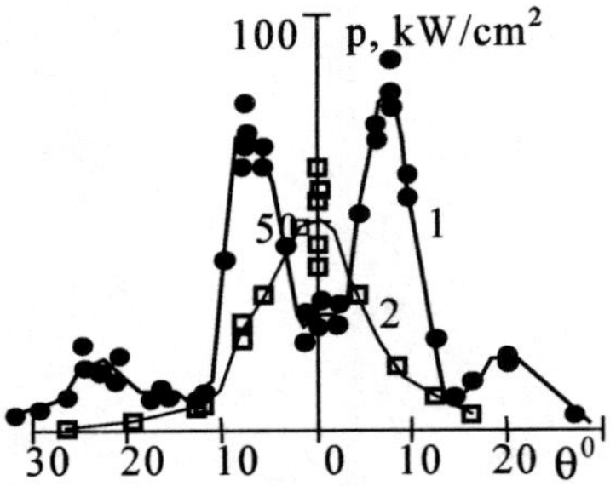

Figure 10. Pattern of radially- (1) and cross-polarized (2) radiation for the SWS with a reflector at $r_b = 56.8$ mm and B = 13.7 kG.

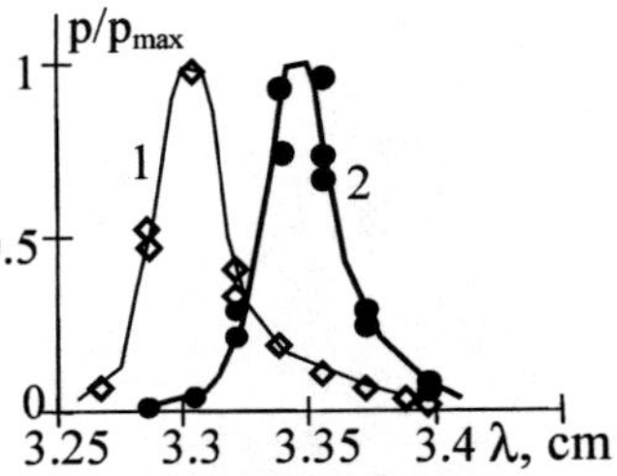

Figure 11. Radiation spectrum for the SWS with a reflector at $r_b = 56.8$ mm and B = 5.8 kG (1), B = 13.7 kG (2).

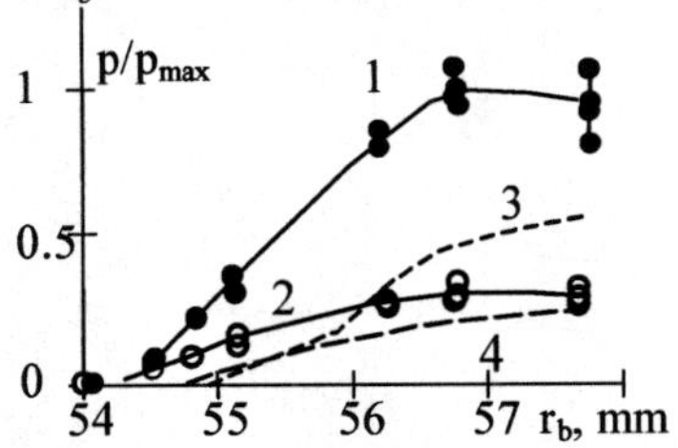

Figure 12. Radiation power versus beam radius for the SWS with a reflector at B = 13.7 kG (1) and at B= 5.8 kG (2) and without a reflector at B = 13.7 kG (3) and at B = 5.8 kG (4).

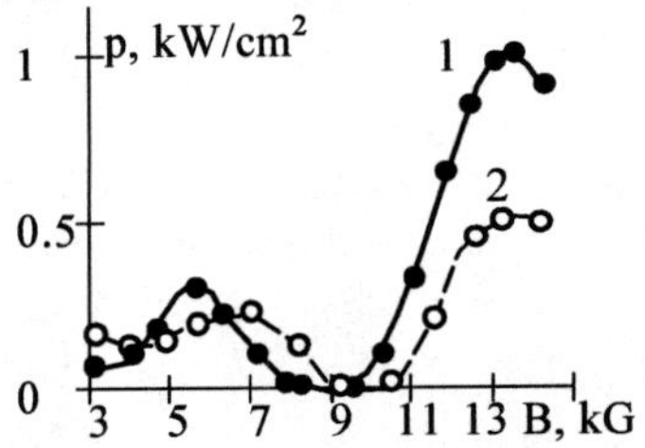

Figure 13. Radiation power versus magnetic field for the SWS with a reflector (1) and without a reflector (2) at $r_b = 56.8$ mm.

Generation efficiency of radially polarized radiation made up 24%, and generation efficiency of total power (~1.4 GW) was of ~30%.

CONCLUSION

The investigations that were carried out have demonstrated possibility to increase stability and efficiency of radiation generation at using a distributed reflector. The highest power and efficiency of generation was obtained for the MWCG with a biperiodic output section.

ACKNOWLEDGMENTS

This work was supported through the ISTC No.2005p / EOARD # 00-7017 Grant.

REFERENCES

1. Bugaev, S.P., Cherepenin, V.A., Kanavets, V.I., Klimov, A.I., Kopenkin, A.D., Koshelev, V.I., Popov, V.A., and Slepkov, A.I., *IEEE Trans. Plasma Sci.* **18**, 525-536 (1990).
2. Koshelev, V.I., Deichuly, M.P., "Optimization of electron beam – electromagnetic field interaction in multiwave Cerenkov generators", in *High Energy Density Microwaves*, edited by Robert M. Phillips, AIP, Woodbury, New York, 1999, pp. 347-359.
3. Koshelev, V.I., Pikunov, V.M., Deichuly, M.P., Popov, V.A., Petkun, A.A., "Cerenkov generators with overmoded one-sectional slow-wave structures", in *Proc. IEEE Pulsed Power Plasma Science Conf.,* edited by R. Reinovsky and M. Newton, 2001, pp. 1634-1637.

Increasing of Peak Power of Superradiation Pulses by Variation of Accelerating Voltage

N. S. Ginzburg, I. V. Zotova, R. M. Rozental, A. S. Sergeev, M. Kamada[a], K.Sugawara [a], K.Kurihara [a], H. Shirasaka [a], R. Ando [a], K. Kamada [a]

Institute of Applied Physics, Russian Academy of Science, Nizhny Novgorod, Russia
[a] Department of Physics, Faculty of Science, Kanazawa University, Kanazawa, Japan

Abstract. Modeling carried out demonstrates possibility of generation of gigawatt power level S band microwave pulse with duration of several nanoseconds using superradiation of short electron beam moving along slow-wave periodical structure. A 10 ns / 500 keV / 5 kA accelerator of Kanazawa University can be used in such experiments. It is shown that significant increasing peak power can be obtained by optimization of voltage and current pulses waveforms.

INTRODUCTION

Stimulated emission from electron bunches – superradiance (SR) can be considered as effective method of generation ultrashort electromagnetic pulses. Different types of SR emission (bremsstrahlung, Cherenkov, cyclotron) were studied theoretically [1-5] and were recently observed experimentally [6-9] at millimeter wavelength bands. RADAN accelerator producing intense subnanosecond electron bunches (250-300 keV, 1-2 kA, 0.3-1 ns) has been used in these experiments [10].

The maximal peak power of SR pulses was associated with Cherenkov emission from an electron bunches moving through a periodical slow-wave structure and interacting with backward wave. Based on these mechanisms, generations of SR pulses have been observed experimentally at all millimeter waveband through the frequencies 35-150 GHz. Theoretical and experimental studies of superradiance should be continued at centimeter (X and S) wave bands. An electron pulse duration should be increased in correspondence with increasing wavelength and nanosecond accelerators can be used for production of driving electron pulse. Experiments on observation of SR at the S band based on 10 ns /500 keV /5 kA high-current accelerator are planning at Kanazawa University. This paper devoted to increasing of peak power of SR pulses by optimization of an accelerating voltage and current pulse profiles.

In the previous theoretical works [1-5] it was assumed that the electron pulses with a top plate waveform are the most suitable for SR (similar to the case of steady state generation). However when a short SR pulse occurs inside interaction space and propagates inside interaction space, it is appropriate to feed this pulse by electrons with longitudinal velocities more and more exceeding phase velocity of the wave. Due to increasing in time SR pulse amplitude, the SR pulse can effectively extract energy from electrons with velocities strongly different from a synchronous value. Based on

CP650, *Beams 2002: 14th International Conference on High-Power Particle Beams*,
edited by T. A. Mehlhorn and M. A. Sweeney

average time-domain model and PIC code simulations it is shown that optimization of profile of voltage and current pulses provides possibility for increasing peak power of SR pulses in several times.

MODEL BASED ON AVERAGED EQUATION

For a model based on averaged equations the longitudinal electric field of the synchronous wave can be presented in the form $E_z = \mathrm{Re}\left[E_z^s(\mathbf{r}_\perp)A(z,t)\exp(i\omega(t - z/V_0))\right]$ where $E_z^s(\mathbf{r}_\perp)$ describes the transverse field distribution and $A(z,t)$ describes a temporal evolution of the longitudinal distribution. The interaction of the electrons with radiation can be described by the following equations

$$\frac{\partial a}{\partial \tau} - \frac{\partial a}{\partial \zeta} = -f(\tau)\frac{GI}{\pi}\int_0^{2\pi} e^{-i\theta} d\theta_0$$

$$\frac{\partial \theta}{\partial \zeta} = \frac{1}{\sqrt{1-\gamma^{-2}}} - \frac{1}{\sqrt{1-\gamma_0^{-2}}} \qquad \frac{\partial \gamma}{\partial \zeta} = \mathrm{Re}\left(a(\zeta,\tau)e^{i\theta}\right)$$

Here we are using dimensionless variables: $a = eE_z^s(r_b)A/mc\omega$, $\tau = \omega(t - z/V_0)\left(1/\beta_0 + 1/\beta_{gr}\right)^{-1}$, $\zeta = \omega z/c$, $\theta = \omega t - h_s z$ is the electron phase, $h_s = hc - h$, $G = Zc/2\beta_0^2$, Z is the coupling impedance of the TM$_{01}$ mode, $I = eJ_0/mc^3$, J_0 is the electron current, $\gamma = \left(1-\beta_0^2\right)^{-1/2}$ is relativistic mass factor, $V_{gr} = \beta_{gr} c$ is the electromagnetic wave group velocity. The function $f(\tau)$ describes the unperturbed electron density $f(\tau) = 1$, $\tau \in [0,T]$, where $T_e = \omega t_{e.pulse}\left(1/\beta_0 + 1/\beta_{gr}\right)^{-1}$ is the dimensionless duration of the electron bunch. The boundary conditions when emission start up related with electron density fluctuations q<<1 can be presented in the form:

$$\theta\big|_{\zeta=0} = \theta_0 + q\cos\theta_0, \theta_0 \in [0,2\pi], \gamma\big|_{\zeta=0} = \gamma_0$$

$$a\big|_{\zeta=L} = 0, a\big|_{\tau=0} = 0$$

where $L = \omega l/c$ is the dimensionless length of the interaction region.

In FIGURE.1, temporal dependence of radiation power in the case of voltage and current pulses of constant amplitude (top plate pulses) is shown by curve 1. Curve 2 corresponds to the situation when accelerating pulse amplitude grows from 300 keV to 900 keV. Curve 3 correponds to the case when electron current also grows in time from 4 kA to 12 kA. We see that for the last case a peak power of SR pulse increases in 2 times in comparison with SR pulse generating from top plate electron pulse with current 8 kA and electron energy 600 keV. Note that total energy in the both electron pulses is approximately the same. It is also seen that optimization of driving electron pulse results in shortening of microwave pulse.

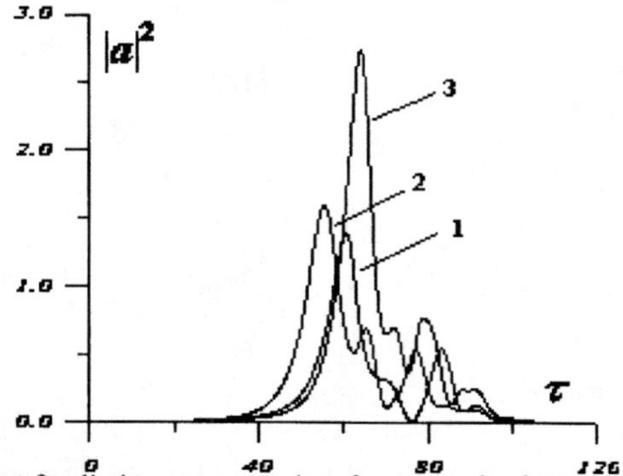

FIGURE 1. Dependence of radiation power on time for rectangle electron pulse curve 1, for electron pulse with only voltage variation curve 2, for pulse with variable both voltage and current curve3.

PIC CODE SIMULATION

The results on enhancement of amplitude of SR pulse obtained in the frame of the model based on averaged equations agree with result of PIC code simulation. In FIGURE 2a SR pulse presented in the case of voltage and current pulse variation shown in FIGURE 3. Maximal power amounts 4 GW for average power of electron beam 4.8 GW that corresponds to instant efficiency > 80%. For comparison, FIGURE 2b corresponds to SR pulse emitted by top plate electron pulse with mean voltage 600 kV and current 8 kA (the kinetic energy in such a beam equal to optimal one). The gain in the SR pulse peak power exceeds factor of 2. The pulse duration 1.5 ns is much shorter than in the case of top plate electron beam. In FIGURE 4 shown dependence on time energy of electrons injected in consecutive moments. The most of electrons are decelerated in the case of the pulse of optimal waveform.

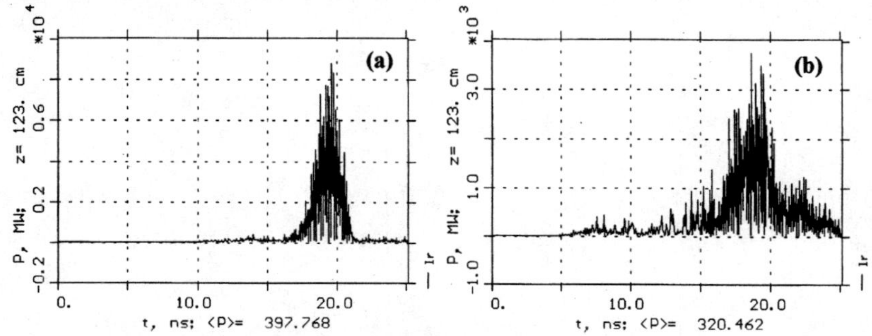

FIGURE 2. (a) SR pulse for the case of optimal electron pulse, (b) SR pulse for the case of rectangle electron pulse of the same energy

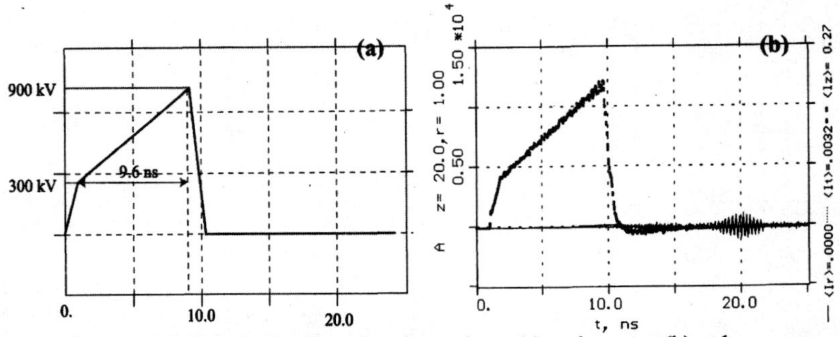

FIGURE 3. Optimal accelerating voltage (a) and current (b) pulses

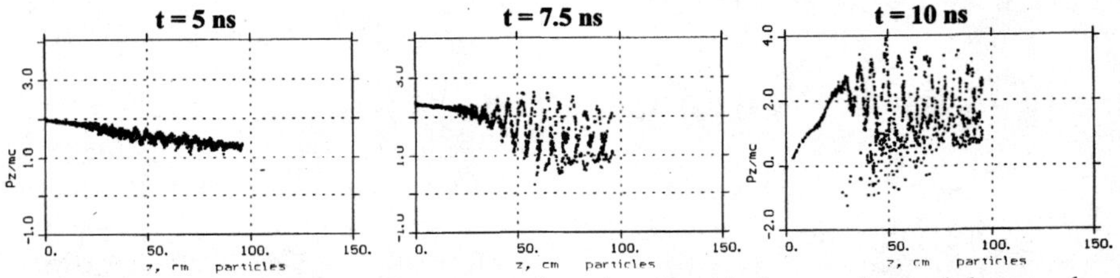

FIGURE 4. Evolution of positions of electrons at phase plane (p_z, z) in the case of optimal electron pulse

CONCLUSION

Modeling carried out demonstrates possibility of generation of gigawatt power level S band microwave pulse with duration of several nanoseconds using superradiation of short electron beam moving along slow-wave periodical structure. It is shown that significant increasing peak power can be obtained by optimization of voltage and current waveforms. Note that required increasing of electron energy and current by the end of electron pulse can be achieved by using self-acceleration of a short beam passing through a system of passive cavities [11].

REFERENCES

[1] R. H. Bonifacio, C. Maroli and N. Piovella, *Opt .Comm.* **68**, 369 (1988).

[2] N. S. Ginzburg, *Sov. Tech. Phys. Lett.* **14**, 197 (1988).

[3] N. S. Ginzburg, I. V. Zotova and A. S. Sergeev, Pis'ma ZhTF **15(14)**, **83** (1989).

[4] N. S. Ginzburg, Yu. V. Novozhilova, and A. S. Sergeev, *Sov.Tech.Phys.Lett.* **22**, 359 (1996).

[5] N. S. Ginzburg, I. V. Zotova, A. S. Sergeev, et al., *Phys. Rev. Lett.* **78**, 2365 (1997).

[6] N. S. Ginzburg, I. V. Zotova, A.S.Sergeev, et al., *IEEE Trans.Plasma Sci.* **27**, 462 (1999).

[7] N. S. Ginzburg, I. V. Zotova, Yu. V. Novozhilova, et al., *Nucl.Instr.and Meth.in Phys.Res.A* **393**, 352 (1997).

[8] N. S. Ginzburg, N. Yu. Novozhilova, I. V. Zotova, et al., *Phys.Rev. E* **60**, 3297 (1999).

[9] M. I. Yalandin, V. G. Shpak, S. A. Shunailov, et al., *IEEE* **28**,1615 (2000).

[10] M. I. Yalandin, V. G. Shpak, S. A. Shunailov, and M. R. Ulmaskulov, *Proc.SPIE* **2557**, 289 (1995).

[11] D. Hasegawa, K. Kamada, K. Shimizu, R. Ando and M. Masuzaki, *IEEE Trans. Plasma Sci.* **28**, 2275 (2000).

Vircator Operation at Plasma Assistance

V. A. Balakirev, I. I. Magda, I. N. Onishchenko, S.S. Pushkarev

National Science Center "Kharkov Institute of Physics and Technology"
Academic St. 1, Kharkov 61108, Ukraine
E-mail: onish@kipt.kharkov.ua

Abstract. The influence of plasma on the processes of virtual cathode (VC) formation, high frequency (HF) generation, and plasma ions acceleration were studied. The theoretical consideration of the problem was analytically undertaken in one-dimensional approximation, that has shown the principal ability of low frequency (LF) oscillations of VC potential well and ion flow due to the periodic space charge neutralization by plasma ions. The experiments were carried out at the pulsed electron accelerator "AGAT", which is a microsecond diode with a magnetic insulation. The accelerator produced an annular relativistic electron beam (REB). Plasma was produced by the beam electrons. It was observed that relaxing VC with low frequency modulation of 20 MHz was formed. The current of REB was temporary modulated at this frequency. The temporal modulation of intense REB is now proposed for excitation of slow space charge wave when further REB is being spatially modulated by periodic magnetic field. It constructs the scheme of the collective ion accelerator.

THEORY

At first we consider the mechanism of the LF oscillations excitation by an ion flux, that bases on the dependence of ion emission current from the plasma emitter (layer of plasma) upon the electric potential of a space charge in the VC area.

The physical picture of LF oscillations excitation is the following. Under action of the electrical field created by a space charge REB an emission from the plasma emitter appears. In process of arrival of ions to VC area (minimum of negative electric potential) the electric potential will be increased, i.e. the depth of the potential well will become less. Thus the emission current of ions will be decreased selfconsistently. The departure of ions from the area of electric potential minimum leads to the renewal of this phenomena and to its periodical repeating.

The consideration is made in the approximation of one-dimensional model. REB drift area is confined between two grounded conductive planes. One-dimensional REB is injected into the drift area. On an internal surface of the first plane ($z=0$) a thin layer of plasma is located, that serves as ion emitter.

The ions emission is limited by a space charge. The electric potential in the drift area ($L>z>0$) is described by Poisson equation with boundary conditions

$$\Phi(z=0)=\Phi(z=L)=0\,.$$

CP650, *Beams 2002: 14th International Conference on High-Power Particle Beams,*
edited by T. A. Mehlhorn and M. A. Sweeney

i.e. the electric potential becomes zero on the grounded planes. Considering, that the time of electron transition through the drift area is much less than the period of LF oscillations, we describe REB in a static approximation.

In this approaching the equation for electric potential becomes

$$\frac{d^2\Phi}{dz^2}=\frac{4\pi j_e}{c}\frac{\gamma_0+e\Phi/mc^2}{\sqrt{\left(\gamma_0+e\Phi/mc^2\right)^2-1}}-4\pi e\, n_i(z,t), \tag{1}$$

where j_e is REB current density, γ_0 is initial value of a relativistic factor.

Below we consider a case $e\Phi/mc^2\gamma_0 \ll 1$. Then, taking into account the main term of expansion, we have the following expression, instead of (1)

$$\frac{d^2\Phi}{dz^2}=\frac{4\pi j_e}{V_0}-4\pi e\, n_i(z,t),$$

V_0 is initial velocity of REB electrons.

Firstly let's define the potential of an infinitely thin sheet of ions (macro particle) with density of particles (Green function)

$$dn_i=\frac{j_i(t_0)dt_0}{e}\delta(z-z_L(t,t_0)),$$

where $j_i(t_0)$ is the ion density to be found on the entrance of the system, $j_i(t_0)dt_0$ is charge of the sheet which has flown into drift area at the time point t_0, $z_L(t,t_0)$ is Lagrangian coordinate of a sheet. To determine the potential Φ_G of macro particle we use the equation

$$\frac{d^2\Phi_G}{dz^2}=\frac{4\pi j_e}{V_0}-4\pi\, j_i(t_0)dt_0\,\delta(z-z_L(t,t_0))$$

The solution of this equation looks like

$$\Phi_G=\frac{2\pi j_e}{V_0}(z-L)z+4\pi\, j_i(t_0)dt_0\, G(z,z_L(t,t_0)), \tag{2}$$

where

$$G(z,z_L)=\begin{cases} z(1-z_L/L), & z\le z_L, \\ z_L(1-z/L), & z\ge z_L. \end{cases}$$

The potential produced by the whole ion flux can be found by integrating of an second term in (2) over entry times

$$\Phi=\frac{2\pi j_e}{V_0}(z-L)z+4\pi\int_0^t j_i(t_0)dt_0\, G(z,z_L(t,t_0)).$$

The electrical field on the emitter $z=0$ is determined by the expression

$$E(z=0)=\frac{2\pi j_e L}{V_0}-4\pi\int_0^t j_i(t_0)\left[1-\frac{z_L(t,t_0)}{L}\right]dt_0\,.$$

As the ions emission is limited by a space charge, the condition $E(z=0)=0$ is fulfilled on the emitter. From this condition we receive the integral equation for determination of the ion current density

$$\frac{j_e L}{2V_0}=\int_0^t j_i(t_0)\left[1-\frac{z_L(t,t_0)}{L}\right]dt_0\,.$$

For the determination of Lagrangian coordinate value it is necessary to use the equation of motion for ions, from which we find

$$z_L(t,t_0)=\frac{L}{2}(1-\cos\,\omega_i\,(t-t_0)),\qquad (3)$$

where

$$\omega_i=\left(4\pi n_0 e^2/M\right)^{1/2},$$

M is ion mass. With taking into account (3) the integral equation takes the form

$$\frac{j_e(t)L}{V_0}=\int_0^t j_i(t_0)\,[1-\cos\,\omega_i\,(t-t_0)]dt_0\qquad (4)$$

Let's suppose, that the REB current density changes in time by the law

$$j_e(t)=j_{e0}\left(1-e^{-t/t_b}\right),$$

where t_b is REB front time duration.

The solution of the Volterrs equation of II kind (4) can be found by Laplace transformation. In result for emission current density we receive the following expression

$$j_i(t)=j_{e0}\frac{\omega_i L}{2\sqrt{2}}\sin\frac{\omega_i t}{\sqrt{2}},\quad \frac{\pi}{\omega_i}\geq t.$$

From this expression it follows, that in the system, considered the nonstationary mode of ion emission takes place. The frequency is proportional to the square root of REB density and is inversely proportional to the square root of ions mass.

EXPERIMENT

The experiments were carried out at AGAT accelerator, which is a microsecond diode with a magnetic insulation [1]. The accelerator produces an annular electron beam with the following parameters: energy 280 keV, current 3.5 kA, pulse duration 1 μ s, and diameter 32 mm. The beam was injected into a drift chamber 4.9 cm in diameter which was placed in an external magnetic field of 4-4.5 T. It was found that, when the magnetized IREB with a current higher than the limiting current was injected into the drift chamber filled by a gas at a pressure of 2×10^{-5} - 4×10^{-4} Torr, the relaxing VC was formed. Simultaneous measurements were made by a set of the capacitive probes located along the drift chamber. The dimensions of the region occupied by the VC were determined by comparing the signals from the probes located at distances of 3, 5,7, and 11 cm from the cathode. It was shown that, at a distance of 3 cm from the cathode, the potential is equal or higher than the voltage applied to the diode. As the distance increases, the potential drops rapidly, and then, at a distance of 11 cm, it becomes constant. The magnitude of magnetic field in the region of the IREB injection was chosen such that a portion of the beam escapes to the walls of the chamber. Thereby, a plasma was produced near the anode providing the ion flow.

The structure of the ion pulses for different geometry of the diode and also for different values of the electron beam diameter, magnetic field magnitude and gradient, and gas pressure were studied. To measure the ion-beam parameters, we used collimated and wide-aperture collectors, negatively biased Faraday cylinders, Rogowski loops, magnetic analyzer, and track detectors made from the cellulose nitrate. It was found that the accelerated ions appeared behind the VC when the injection current J was either comparable or higher than the limiting current. The ion current reaches its maximum of 200 A when the radius of the beam propagating through the drift chamber is 14 mm. When the injection current is smaller than the limiting current, the ion current was noticeably smaller. Most of the nitrogen ions of the wall plasma, which were accelerated by the potential well of the VC, had energies of the order of 300 keV. From the energy spectrum and potential distribution, we found that ion energy was governed by the depth of the potential well of the VC.

In the VC regime, we observed LF modulation of the electron beam; the modulation characteristics were determined from the measurements of the bremsstrahlung X radiation and beam-collector current and from the signals of the magnetic loop probes installed along the drift chamber. From the correlation analysis of the measured signals, we could determine the fluctuation parameters. It was found that the frequency spectra of the signals, which were measured by three independent diagnostics, coincide quite well over the entire transportation channel. The frequency spectrum has a maximum at a frequency of 20 MHz; the modulation amplitude reaches 40%.

Therefore, virtual cathode at plasma assistance executes three functions: produces of the LF temporal modulation of REB, serves as a ion source, and provides the preliminary acceleration of ion flow. Vircator at plasma assistance can be considered as the first section of a two-staged collective ion accelerator based on simultaneous temporal and spatial modulation of REB [2].

The results obtained inspire to elaborate the collective ion accelerator based on slow space charge wave of REB, subjected consequently to the temporal and spatial modulation. Besides plasma assistance can govern the generation spectrum of the vircator.

ACKNOWLEDGEMENTS

This work was partly supported by STCU grant #1569

REFERENCES

1. Balakirev V.A., Gorban A.M., Magda I.I., Novikov V.E., Onishchenko I.N., and Pushkarev S.S. Plasma Phys. Reports Vol.23, No.4, 1997, pp323-327
2. Lymar A.G., Khizhnyak N.A., Belikov V.V., VANT series High Energy Physics, 1973, No. 3(5) p.78.

The Direct Generation of High Power Microwaves with Compact Marx Generators

Jon R. Mayes and William J. Carey

Applied Physical Electronics, L.C., P.O. Box 341149 Austin, Texas 78734 USA

Abstract. High Power Microwave energy may be directly generated with ultra-fast voltage pulses driving an antenna. Recent efforts with the wave erection Marx generator have seen the production of voltage pulses in excess of several hundred kV, with rise times as fast as 200 ps. This generator has been used to source Ultra Wideband antennas as well as Narrow Band antennas, each resulting in high electric field strengths. This paper describes the Marx generator and explores its use for generating UWB and NB energy. Experimental results are presented.

INTRODUCTION

Compact Marx generators have been employed for many years as trigger generators for larger systems. These generators are typically designed with pulse characteristics of more than 200 kV peak voltage, 3 – 4 ns rise times, 1 – 3 ns rms jitter and pulse widths of 10's of ns. Unfortunately, these characteristics make the compact Marx generator an impractical source for directly generating RF energy by driving an antenna due to the slow rise time.

However, recent efforts [1] with compact Marx generators have brought faster rise times, higher peak voltages and extremely low jitter values, making these sources viable candidates for compact Ultra Wideband (UWB) sources for both single and multiple generator systems.

While efforts continue the successful development of these generators, deployability will be limited by compact, high voltage, impulse radiating antennas. Typical impulse antennas, with any respectable amount of gain and directivity are large and cumbersome. Furthermore, high voltage antennas capable of handling voltages in excess of 100 kV are simply not available.

BACKGROUND

The most efficient, compact and economical method of generating a repetitive, large magnitude, electromagnetic impulse is the wave erection of a spark gap-switched Marx circuit. Wave erection is necessary to obtain the fast voltage rise times from the Marx circuit that generates the ultra-wideband of frequencies necessary for high resolution radar or the interdiction of flight controls and computer memories for electronic warfare. Wave erection is made possible through the proper design of the stray capacitance and the inter-stage capacitance, in concert with coupling the spark

CP650, *Beams 2002: 14th International Conference on High-Power Particle Beams*, edited by T. A. Mehlhorn and M. A. Sweeney

gaps via ultra-violet energy, resulting in a sub-ns rise time for output voltages of several hundred kV at moderate per pulse energies.

A 17-stage Marx generator is fabricated for the antenna testing. This generator provides a peak voltage of 170 kV, and a rise time of approximately 250 ps, as shown in Figure 1.

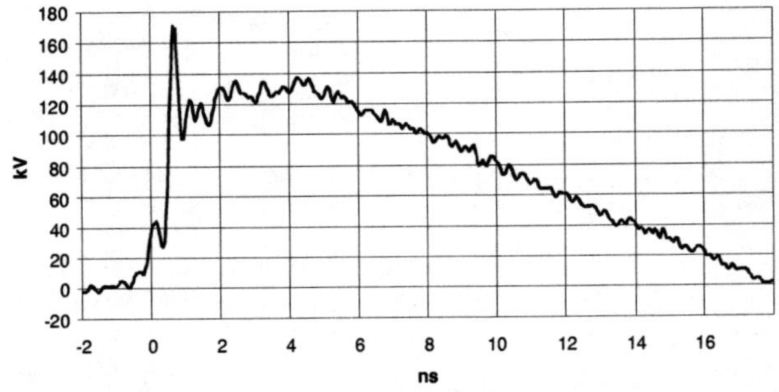

FIGURE 1. Output waveform from the Marx generator.

Several antenna designs are pursued including a linear half TEM horn and a spiral antenna. The initial focus is to gain an understanding of pulsed high voltage antenna characteristics, and to establish antenna baseline parameters for gain, radiation pattern, optimal impedance transformation requirements, and field polarization over the frequencies of interest. Future designs will build on these results.

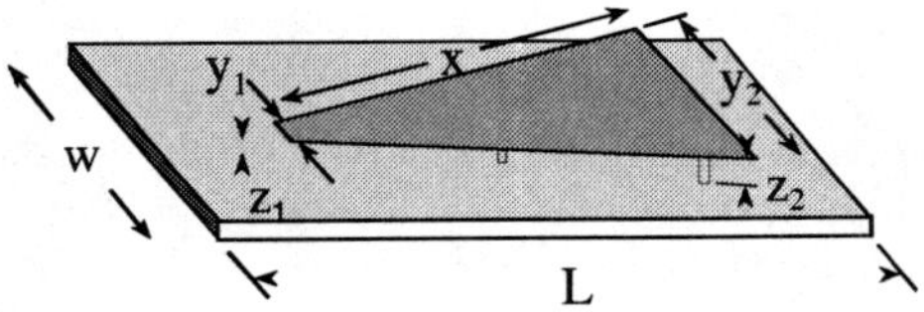

FIGURE 2. Geometry of the linear half TEM horn antenna.

The linear half TEM-styled horn antenna is illustrated in Figure 2. This antenna is characterized by a single TEM plane extending over a large ground plane. The TEM plane linearly expands in both the E and H directions as the wave propagates toward the output. The basic transmission line equation for a parallel plate line is used for the design; however, since the antenna is to be operated at very high voltages, the input portion of the antenna has an insulator other than air. As the wave propagates toward the output, the insulator tapers to a zero thickness. Thus, at the input, the width of the TEM plate is defined as $w = 377\ h/(z_o\, \varepsilon_r^{1/2} - 2)$. Likewise, the width of the TEM plate at the output is defined as $w = 377\ h/(z_o\, \varepsilon_r^{1/2})$.

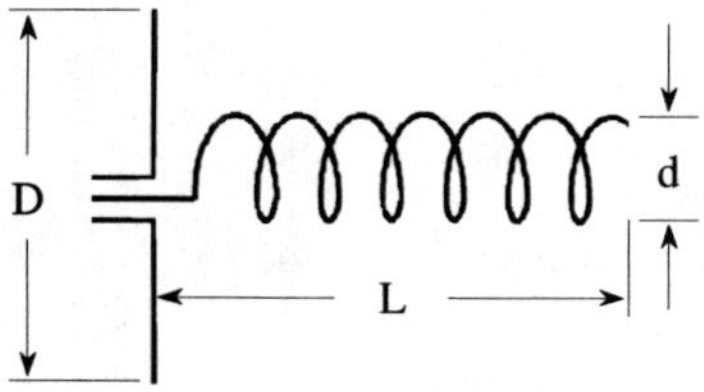

FIGURE 3. The spiral antenna.

Spiral antennas offer more control over the radiated frequency. These antennas offer bandwidths of 2:1, and up to 5:1 is possible. They are unidirectional and are circularly polarized. As illustrated in Figure 3, once the center conductor parts from the ground plane, it spirals around an imaginary cylinder whose circumference is approximately one wavelength in diameter. Choosing the number of coils, which defines the gain, the length, L is derived from a coil spacing definition of $\lambda \sin \theta$, where θ is typically 12 – 14 degrees. The minimum ground plane diameter is ¾ λ.

EXPERIMENTAL ARRANGEMENT

Antennas

The linear half-TEM horn antenna is designed for easy connection to RG-220 cable and is insulated at the feed point for high voltage hold-off. The insulating material continues in the direction of the wave propagation and gradually tapers into the ground plane. Matching the 50 Ω cable, the input side of the antenna has dimensions of y_1 = 40 mm and z_1 = 13 mm. The output side of the TEM has dimensions of y_2 = 290 mm and z_2 = 290 mm for a 377 Ω impedance match to free space. The length, x, of the TEM is chosen to be 910 mm for a gradual transition. Finally, the ground plane has dimensions of w = 600 mm and L = 1220 mm.

The spiral antenna, like the half-TEM ground plane antenna, is fabricated for easy connection with the RG-220 cable. The antenna is designed for an operating frequency of 1 GHz, and 10 turns. Therefore, the coil diameter, d, is set at 95 mm, with a length, L, of 770 mm. The ground plane diameter, D, is chosen at 230 mm.

EXPERIMENTAL RESULTS

The radiated waveform from the half-TEM ground plane antenna is shown in Figure 4(a). The resulting waveform is very nearly a monopulse, with an electric field of 1100 V/m at 100 m. The FFT response, shown in Figure 4(b), shows that the signal is very broadband, with most of the energy in the hundreds of MHz range. The 7 ns delayed spike coincides with the test range's calculated first ground bounce signal.

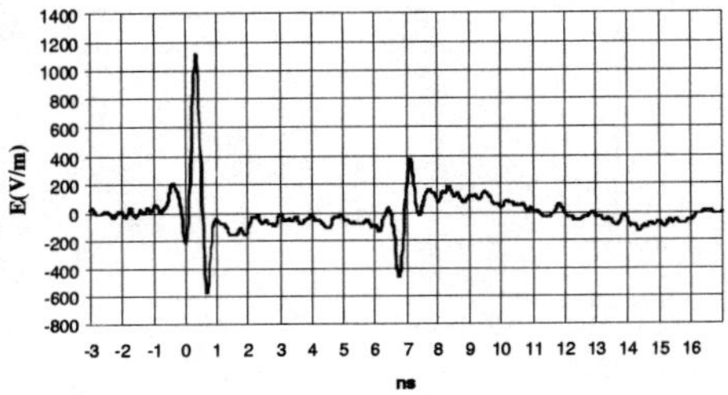

(a) Radiated impulse

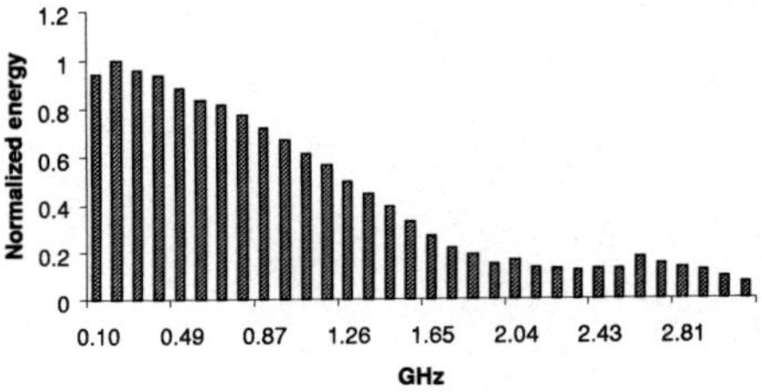

(b) FFT response

FIGURE 4. TEM antenna results.

The range results of the coil antenna are shown in Figure 5. As predicted, this antenna produced a narrowband signal centered at 1 GHz, as shown in Figure 5(b). The electric field strength is approximately 300 V/m for both the horizontal and vertical polarizations. The energy is spread over several cycles in a narrow bandwidth.

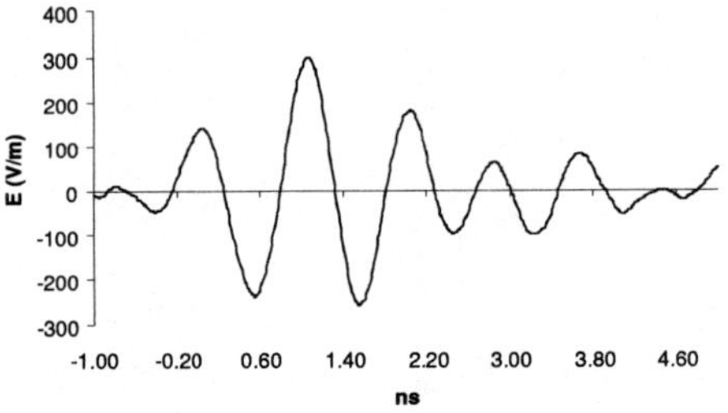

(a) Radiated impulse

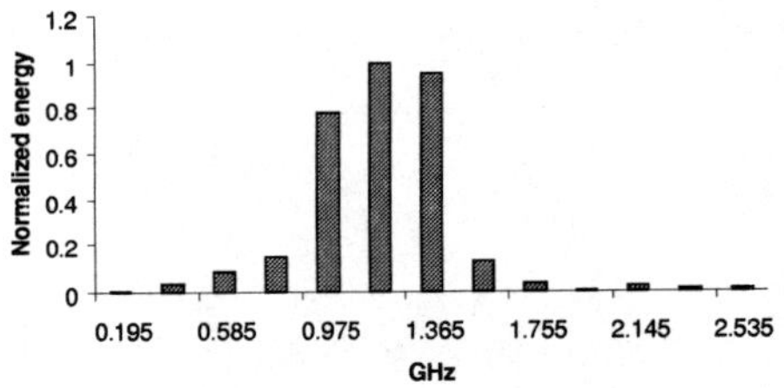

(b) FFT response

FIGURE 5. Coil antenna results.

CONCLUSION

This paper has presented preliminary results of a wave-erection Marx generator driving several impulse-type antennas. The generator's power was reduced so as to minimize breakdown problems with the antennas.

The radiation results do provide a basis for future work. The TEM horn antennas demonstrated UWB behavior. The coil antenna demonstrated narrowband operation and the ability to focus energy into microwave frequencies. Each of the antennas demonstrated the ability to successfully hold off the high voltage, with no apparent breakdown problems. However, impedance matching issues were observed and the gain and directivity measurements indicate large margins for improvement.

REFERENCES

1. Mayes, Jon R., and Carey, William J., *The Marx Generator as an Ultra Wideband Source*, 13th IEEE International Pulsed Power Conference, Las Vegas, NV, July 2001.

Project of Powerful 4-beam Planar FEM with Distributed Feedback for 75 GHz Band

Andrei V. Arzhannikov, Vladimir B. Bobylev, [a]Naum S. Ginzburg, Petr V. Kalinin, Sergei A. Kuznetsov, [a]Nikolai Yu. Peskov, [b]Petr V. Petrov, [a]Aleksandr S. Sergeev, Stanislav L. Sinitsky, Vasilii D. Stepanov

Budker Institute of Nuclear Physics, Novosibirsk, 630090, Russia
[a]*Institute of Applied Physics, N-Novgorod, 603600, Russia*
[b]*RFNC-VNIITF, Snezhinsk, 456770, Russia*

I. INTRODUCTION

In [1,2] we proposed a novel concept to achieve a few GW power level in mm-wave band that is based on creation of a multichannel planar FEM. Such generator consists of several planar masers with a moderate (~20cm) width, which operate as oscillators connected together as modules in a single device. The device has a multilayer electrodynamic structure using 2-D Bragg gratings to produce 2-D distributed feedback in each layer and special wave-guides to connect feedback between all layers. In this case the space of the beam-wave interaction can be developed in two transverse directions.

As to operation of a single channel of planar FEM, it is successfully investigated in experiments at the "ELMI"- device [3]. But there are a few problems for realisation of the multichannel concept. The key problem is to provide co-phase operation of module-oscillators in the multichannel generator. The other problems are to generate several sheet beams with appropriate parameters, to design and to construct multilayer magnetic and electrodynamic systems. Some results on solutions of the pointed problems for 4-channel generator are presented in this paper. These problems are solved by an analytical description, computer simulations, engineer and technology studies, and experimental investigations.

II. GENERAL ARRANGEMENT OF FOUR-CHANNEL FEM

The general arrangement of four-channel FEM is shown in a Fig.1. The geometry of a single module was chosen close to the conditions of the experiments at the ELMI-device. Each of modules operates as an oscillator with 2-D distributed feedback. This feedback is realised by a planar resonator composed two reflectors consisting of Bragg gratings pairs.

Each FEM-module is driven by a sheet beam generated by a magnetically insulated diode. The transversal velocity of electrons ($\beta_\perp \approx 0.2$) is pumped by the magnetic field of a planar undulator. The undulator field is created by five parallel planes of coils. Each plane consists of coils connected consequently with alteration of polarity. The number of turns in the coils at the beginning and at the end of each plane is smoothly varied from 6 turns up to 20 ones on the length 12 cm. At the spatial period of the

CP650, *Beams 2002: 14th International Conference on High-Power Particle Beams,*
edited by T. A. Mehlhorn and M. A. Sweeney

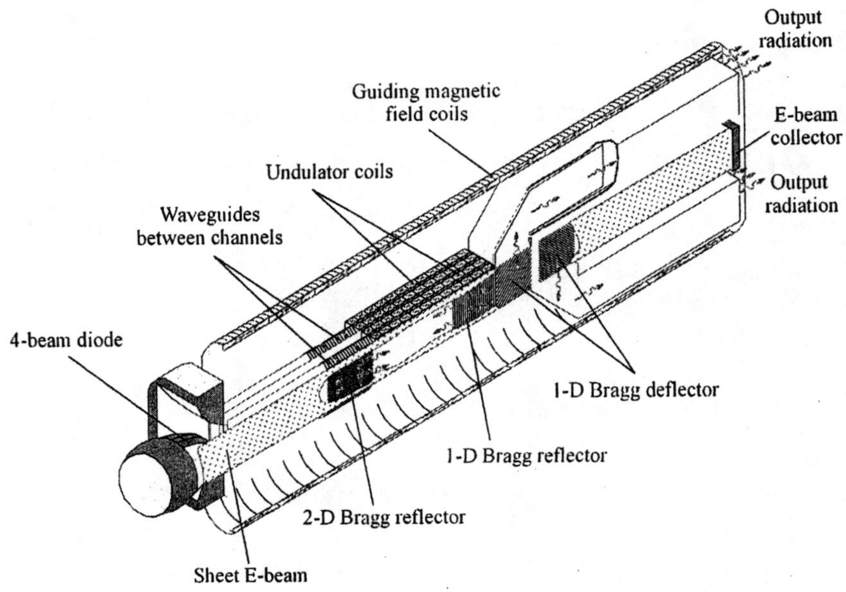

Fig.1. Schematic of 4-channel FEM-generator.

undulator 4 cm the maximum value of the transverse magnetic field is 0.2T and the longitudinal one is up to 1,5T. According to computer calculations, the difference in the magnetic field strength between various slit channels is less than 3% in the regions with passing sheet E-beams. This value is appropriate enough to provide synchronous pumping of the waves. Width of the sheet beams is 15 cm at the current per unit width ~ 200 A/cm.

The geometry of the resonators permits to use different combinations of 1-D and 2-D structures for upstream and downstream reflectors. Synchronisation of modules is provided by connection of resonators with bent waveguides located in the places of the 2-D resonator reflectors through microwave energy cross flows produced by scattering in 2-D Bragg gratings. To form a circle of a feedback the first module should be connected to the fourth one by the special waveguide.

After each resonator exit there is a Bragg deflector that serves for inclining the electromagnetic flux from the beam transport channel into output waveguide. In order to prevent the influence on the wave generation process by plasma from the beam collector, the collector is shifted from the exit of the resonator on a distance about 0.5 meter.

III. COMPUTER SIMULATIONS AND EXPERIMENTAL TESTING OF VARIOUS PROCESSES

Synchronisation of the channels.

In the first series of computer simulations, the radiation synchronisation process was considered on the base of solution of averaged equations of the beam-wave interaction for 6 modules of the multichannel FEM-generator (see [3]). These simulations have shown that at appropriate parameters the processes of the wave excitation in the modules are realised under a strong synchronising, and a good coherence of the radiation at the maser exit is provided.

Then we carried out simulations for 4-module FEM generator with geometry presented in Fig.1 and parameters close to the ones required for the experiments, which are planed at the ELMI-device (see above). The beam parameters were the following: the electron energy - 1MeV, the current per unit of width – 200A/cm, the electron transverse velocity – (0.2-0.25)c. At a generated wave frequency 75 GHz the Peirce parameter was $C \approx 4\times10^{-3}$. Width of an interaction space was $l_x = 20$ cm and length of regular part of the resonators was $l_z = 35$ cm. The entrance 2-D Bragg reflector had the length 16cm and the exit 1-D Bragg reflector 10cm. The coefficient of Ohm losses in the wave guides connected channels was varied in interval 5-50%. The computer simulations show that the frequency of oscillations in a steady-stated mode is close to the frequency of a precise Bragg resonance of the grating reflectors and a stationary regime of generation is established in all modules simultaneously in time 300ns. Moreover, the amplitudes and the phases of the wave in the modules are practically the same. It means that the radiation generated in all modules is strongly synchronized. The dependence of electron efficiency averaged over 4 modules as a function of RF-field losses in the waveguides between the FEM-modules is shown in Fig.2.

It should be also noted that synchronization takes place even at a spread of initial mismatches from the resonance in different modules (for example due to difference in the electron energies) exceeding halfwidth of self-excitation band. Further auto-oscillations were broken. Situations, in which the synchronization state was disturbed and the separate modules generated on miscellaneous frequencies, was not carried out in simulations.

Generation of E-beams.

Four sheet electron beams are generated by magnetically insulated diode with four parallel strongly elongated cathodes (there is a picture for a half of the diode in Fig.3). Geometry and location of the cathodes, as soon as those of the anode slits are chosen from self-consistent simulations of the beam generation and propagation in such diode,

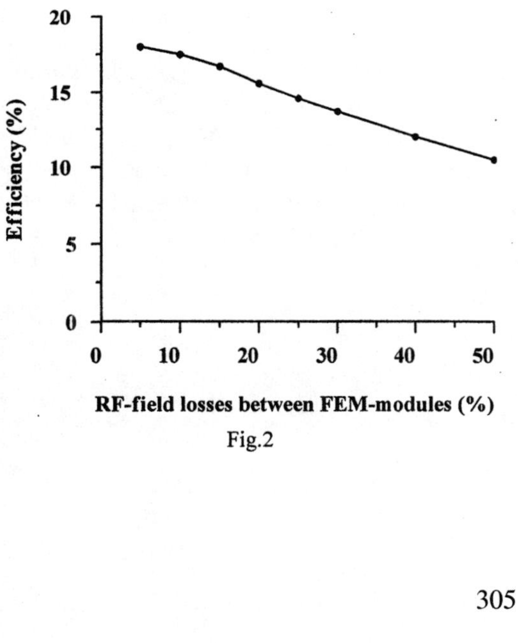

Fig.2

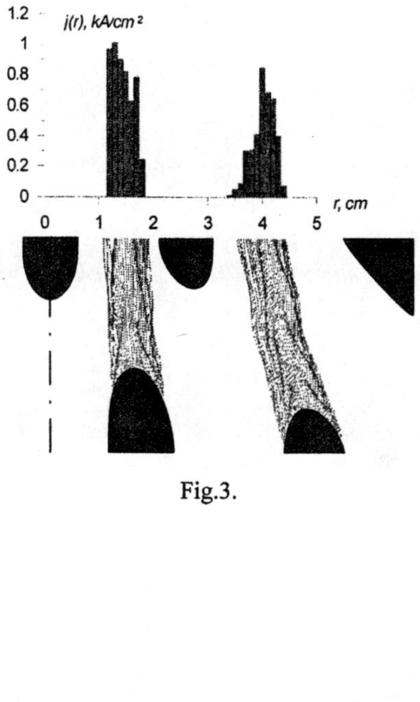

Fig.3.

to provide close values of the current density of the beam electrons in all channels. Example of results obtained in one of the simulation runs is presented in Fig.3. Current density distribution on the beam thickness for two channels is shown for the case after passing the beams through the anode slits. It should be pointed that graphite formers additionally cut out the thickness of these beams in the channels down to 4mm.

Output radiation.

In order to test units of the multichannel FEM we made resonator for present experiments at the ELMI-device with geometry like the one of a single channel of the multichannel maser. In particularly, the Bragg RF-deflector inclining the electromagnetic flow from the beam pass region was made tested in "cold" and "hot"

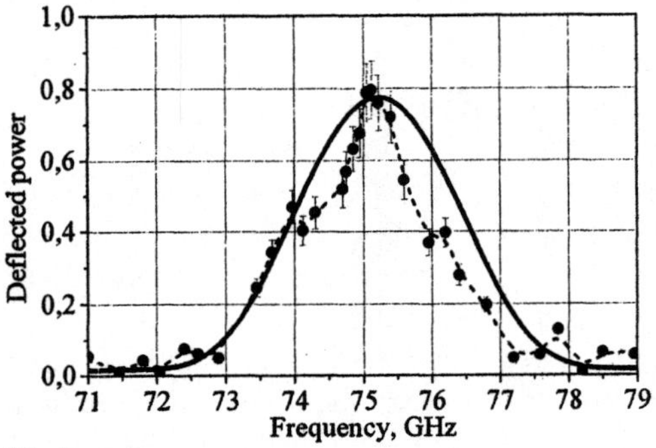

Fig.4a.

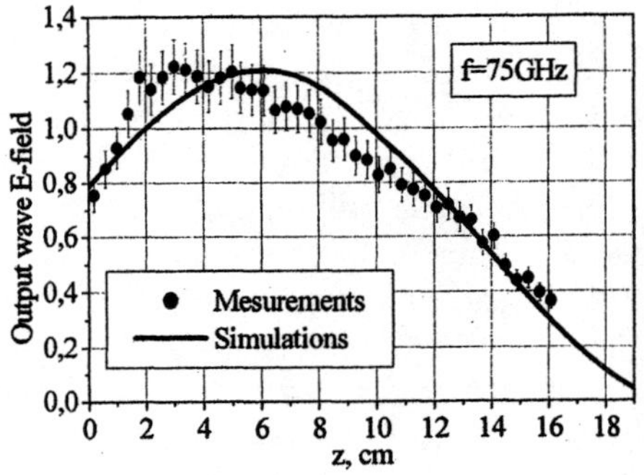

Fig.4b.

experiments. Results of computer simulations and "cold" measurements at a special test bench are presented in Fig.4 for the RF-deflector with following geometry: the angle between wave-vector of the input wave and the ripples of the thread – 45°, the depth of the thread - 0.2mm, the spatial period – 2,82mm. The thread was made as a ribbon with a width 15cm from a corner to corner in a square region with the sizes 19x19cm. Fig. 4a shows spectral properties of this RF-deflector and Fig. 4b demonstrates the electric field distribution of the output wave over the width of electromagnetic flow at the exit of the deflector.

IV. CONCLUSION

Theoretical and numerical considerations have shown that a planar FEM with 2-D distributed feedback is suitable to be used as a module for constructing multichannel devices to produce microwave radiation of GW level. Computer simulations for operation of 4-modules FEM–generator have demonstrated a perfect synchronization of all modules even at a little bit different energies of the electron beams. The project of 4-channel FEM–generator is designed on the base of the "ELMI"-device.

1. Arzhannikov A.V. et. al. Digest of Technical Papers of PPPS-2001, Las Vegas, USA, June 2001, p. 561-564.
2. Ginzburg N.S. et. al. Nucl. Instr. and Meth. in Phys. Res. A 475 (2001) 173-177.
3. Agarin N.V. et. al. Nucl. Instr. and Meth. in Phys. Res. A 445 (2000) 222-229.

Recent Results in Generation of Powerful Pulses of 75GHz Radiation at ELMI-Device

Andrei V. Arzhannikov, [a]Naum S. Ginzburg, Viktor G. Ivanenko, Peter V. Kalinin, Alexander S. Kuznetsov, Sergei A. Kuznetsov, [a]Nikolai Yu. Peskov, Stanislav L. Sinitsky, Vasilii D. Stepanov

Budker Institute of Nuclear Physics, 630090, Novosibirsk, Russia
[a]Institute of Applied Physics, 603600, Nizhnii Novgorod, Russia

INTRODUCTION

The main aim of research on generation of microwave radiation carried out at BINP (Novosibirsk) and IAP (Nizhnii Novgorod), is to create a powerful planar free electron maser (FEM) with 2-dimensional distributed feedback driven by sheet electron beam. Attractiveness of this device is concerned with a possibility to increase one of transverse dimensions of electrodynamic system and the beam and hence to get more powerful and energetic microwave pulses. The coherence of radiation in such planar system is provided by 2-D distributed feedback. It is realized in Bragg resonator due to scattering of the wave pumped by the beam, in transverse direction and then backward on 2-D corrugation of lateral surface of Bragg reflectors. Due to this feedback one-mode one-frequency regime can be established in such device [1]. One of the main problems of this FEM common for all microwave generators, is pulse shortening, which is caused by HF discharge in vacuum channel at a certain level of the wave energy density and by coming plasma from the beam collector to the resonator. To solve this problem in our case we have done two things. First, we increased the length of downstream reflector at keeping its reflection coefficient on the same level. By this way we decreased the density of the transverse flow of radiation. For the second we have suggested and realized a novel scheme to separate the radiation flow from the beam electrons and to dump electrons to a remote collector. In this paper we present the results of recent experiments carried out at the ELMI-device.

EXPERIMENTAL DEVICE AND MM-WAVE DIAGNOSTICS

The ELMI-device [2] consists of following basic units: a magnetically-insulated ribbon diode, a former of the sheet beam, a slit vacuum channel with an undulator and a resonator, a deflector of mm-wave from E-beam and a diagnostic system to measure the parameters of the generated radiation.

The magnetically insulated diode generates the sheet beam with E_e~1MeV, I~10kA, τ ~5μs and a cross section 1x20cm. Then a central part of the beam cross section containing the electrons with minimal angular spread is cut out by the special former. After passing through the former the sheet beam with a cross-section

CP650, *Beams 2002: 14th International Conference on High-Power Particle Beams,*
edited by T. A. Mehlhorn and M. A. Sweeney

0.4x8cm comes into a planar resonator. The former and the undulator are located inside a slit vacuum channel, which coils create a uniform longitudinal magnetic field up to 14 kG.

The undulator magnetic field used for pumping the transverse electron velocity in the maser has the spatial period 4cm, the length of the field increase is 12cm. The amplitude of the perpendicular component of the field on the axis is up to 2kG.

After passing the resonator the beam is transported in a slit channel at a distance 0.5m to a graphite collector, where it is dumped on a large area to eliminate a creation of a dense plasma here. The remoteness of the collector from the resonator serves for delaying this plasma.

Schematic of the resonator and the deflector used in the experiments, is shown in the Fig.1. The resonator consists of two Bragg reflectors (upstream and downstream), connected together by a piece of a regular waveguide with cross-section 0.93x8.8cm and the length 32cm. The first reflector is formed by two parallel Bragg gratings with chessboard thread. The length of corrugated area is 19cm, the period is 0.4cm and the depth is 0.025cm. The downstream reflector also consists of two plates, but the only one of them has the chessboard thread. The length, period and depth of this Bragg grating coincide with the ones for the gratings of the upstream reflector. Spectral properties of these reflectors measured in "cold" experiments, are shown in the Fig.2. It is seen that both reflectors have a narrow band of reflection 74-76GHz with maximal values 0.9 and 0.7.

Just after the downstream reflector we put a wave deflector. It is a wave transformer separating the wave from the beam by means of scattering it in perpendicular direction. This wave transformer is formed by two 1-D Bragg gratings with parallel rectangular grooves mechanised at the angle 45° to the direction of incident wave. The length of the corrugation is 10cm, its period is 0.282cm and the depth is 0.03cm. Dependence of the transformation efficiency on the frequency of the incident wave H_{10} and the scattered wave amplitude profile are shown in the Fig.3. After scattering in the wave transformer the radiation is reflected by a special mirror at the angle 90° into a rectangular waveguide, then pass through the output window into atmosphere.

To measure the absolute value of microwave radiation power in the experiments we used calibrated semiconductor detectors on hot electrons. For spectral analysis of

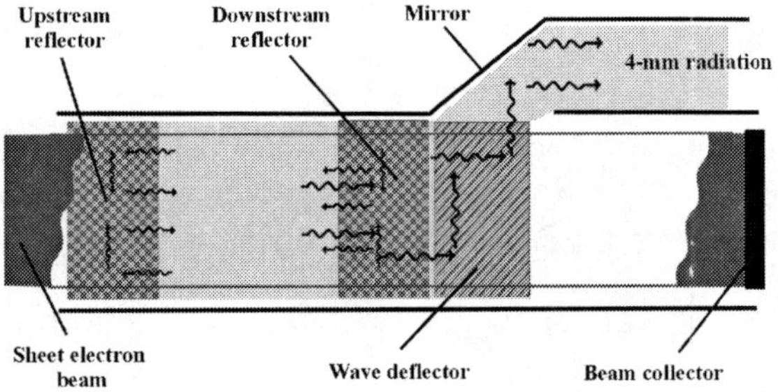

FIGURE 1. Shematic of the Bragg resonator and deflector

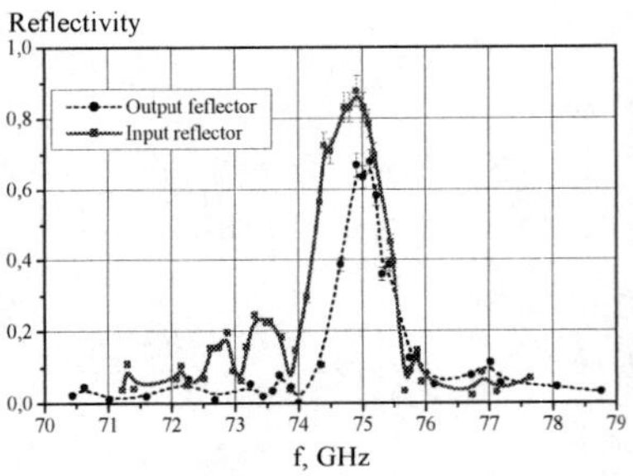

FIGURE 2. Selective properties of 2-D Bragg reflectors.

the radiation a set of detectors consisting of receivers, attenuators, narrow-band resonance filters and semiconductor diodes of 4mm band, was used. In experiments it was located in a distance 1m from the maser exit. For visual observation of the radiation power distribution a panel of neon lamps was used.

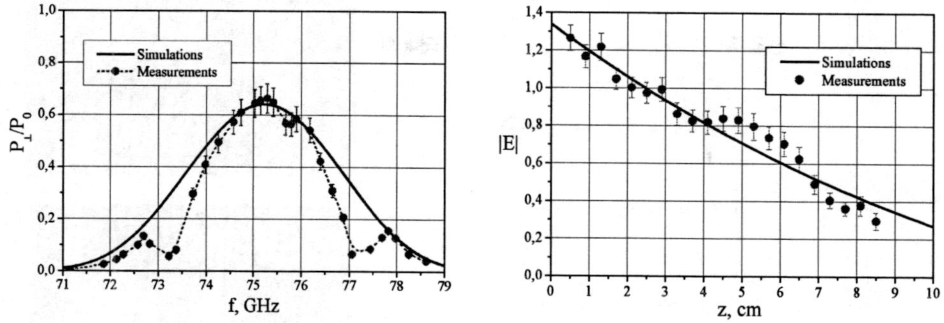

FIGURE 3. The efficiency of transformation (left graph) and the scattered wave amplitude profile (right one)

RESULTS OF EXPERIMENTS

In recent series of the experiments on generation of 4-mm radiation in the FEM with the novel scheme of separation the strength of transverse component of the undulator magnetic field was varied at a fixed longitudinal one 10kG. The obtained dependence of the radiation power on the amplitude of the undulator field is close to the one obtained in previous series of experiments [3], in which the dump of the beam electrons to collector was made near the region of the radiation pass. The results of one typical shot for the amplitude of the undulator field 0.8kG at which the maximum of the radiation power is obtained, are shown in the Fig.4. The diode voltage and the beam current in the collector are presented in its upper graph. The lower one illustrates by thin blue line the radiation power signal measured by semiconductor detector on hot electrons and by thick red line the signal of the semiconductor diode supplied by attenuator and the resonant filter with transmission band 74.7-75.8 GHz. As it is seen these two signals are in good correlation. If to compare with the results of the previous experiments, the duration of the obtained signals of the radiation power is increased and reached the value 300ns.

The visual picture of the radiation angular distribution after the window is registered by the panel of 100 neon lamps, standing at a distance 2cm one from each other. This panel located in the experiments at a distance 0.5m from the maser window, was irradiated by the flow of mm-wave radiation going out from the maser through a sheet of the radiation absorber. The upper photo in the Fig.5 was taken in the case when the amplitude of the undulator field was equal to 0.8kG. The lower one shows the geometry of the panel. If the undulator field was equal zero, no luminescence was registered. It is evident that in the upper photo the intensity of luminescence is increasing to the bottom of the panel and is left - right symmetric. This behaviour of intensity is qualitatively close to the distribution of the radiation power obtained in the "cold" measurements and in numerical simulations for deflection of H_{10} wave.

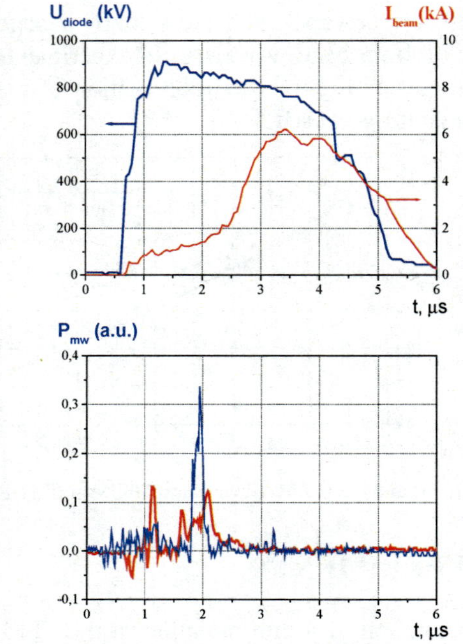

FIGURE 4. Signals of the diode voltage and the beam current (upper graph). Oscilloscope traces of the radiation power (lower one).

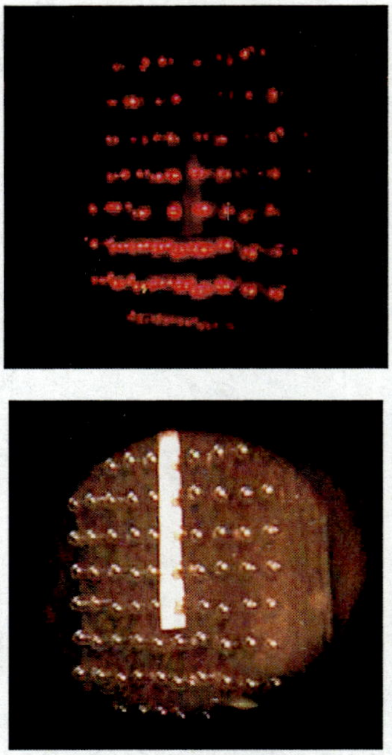

FIGURE 5. Visual picture of the radiation angular distribution

CONCLUSION

The novel scheme of separation of the mm-wave radiation from the intense electron beam using planar Bragg wave transformer, is suggested and realized. In the experiments at the planar FEM with the new scheme of separation, the generation of powerful 4-mm radiation pulses was observed. The duration of these pulses is increased in two times in comparison with the old scheme of the beam dump and has reached 300ns. By means of the panel with neon lamps the angular distribution of the generated radiation at the exit of the maser is registered. The distribution of the luminescence intensity qualitatively agrees with the results of "cold" measurements and computer simulations.

Work was supported by grant 01-02-16749 of RFBR and grant INTAS 2192.

ACKNOWLEDGEMENTS

We are grateful to Prof. V.S.Koidan for his support of the work and Dr. V.S.Nikolaev, A.F. Rovenskikh and S.V.Polosatkin for help in preparation of experiments.

REFERENCES

1. N.S.Ginsburg et al. Pis'ma v ZhTF, vol.18, no.9, 23-28 (1992) (in Russian).
2. N.V.Agarin et al. Proc. of 12-th International Conf. On High-Power Particle Beams (Haifa, Israel, 1998), v. II, p. 687.
3. A.V.Arzhannikov et al. Proc. of 13th Int. Conf. on High-Power Particle Beams, Nagaoka, Japan, 2000, v.1, pp.244-247.

Full Power Operation of an Ultra-Compact, Low Voltage Reltron Microwave Source

Song Hun Choi and Edl Schamiloglu

University of New Mexico, Albuquerque, NM 87131 USA

Abstract. This paper presents the results of a characterization study performed on the pulsed power system of the UNM Reltron, an ultra-compact, low voltage (~100 kV) high power microwave source. The results presented include vacuum system operation, full power operation of the pulsed power system, and initial operation of the tube. Since the electron beam modulator is essentially the heart of the Reltron, responsible for the efficient generation of the electron bunches in a short distance, we also present cold test measurements on the electron beam modulator and microwave extraction system. The main objective of the research was to characterize the electron beam modulator and identify a regime of stable operation.

INTRODUCTION

The phenomenon of transit time oscillations being generated in the cavities located along the path of an electron beam was discovered early in the development of electron beam microwave devices [1]. The early discoveries did not lead to a widespread development of a device due to the fact that the amplitude of the oscillations generated were at levels too low to be of practical use with the technology available at the time. The practical aspect of this phenomenon was rediscovered with the advent of intense relativistic electron beams ("IREBs"). The amplitude of oscillations generated using IREBs proved that the phenomenon could be utilized to generate high power microwaves [2,3]. The Reltron is an electron beam-driven source that uses the phenomenon of transit time oscillations to generate a stable narrowband pulse of high power microwave radiation in the L to C frequency bands. The novel feature that distinguishes the Reltron from other high power microwave sources is the simple design of the device and its compact size (relative to other HPM sources).

The Reltron produces microwave radiation through the formation of electron bunches from the electrons passing through a compact modulating cavity, similar to a Klystron. However, unlike the Klystron the Reltron requires no input feed signal and no guide magnetic field for the bunching of the electron beam (see Fig. 1). The bunching of the electron beam is accomplished by coupling to the mode generated by electron transit through a multi-cavity oscillator. The modulated beam exiting the modulating cavity can be post-accelerated to preserve the electron bunches and to increase the output power. The bunched electron beam is then converted to microwave radiation through an extraction cavity consisting of a simple mode converter connected to a rectangular TE_{10} waveguide to direct the microwave output.

CP650, *Beams 2002: 14th International Conference on High-Power Particle Beams,*
edited by T. A. Mehlhorn and M. A. Sweeney

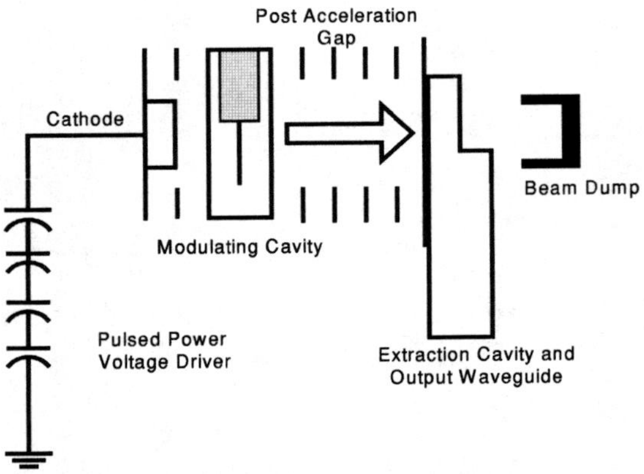

FIGURE 1. Schematic drawing of the University of New Mexico's ultra-compact Reltron.

The unique geometry of the multi-cavity oscillator provides for efficient beam modulation in very short distances, thus eliminating the need for a guide magnetic field in most cases. (For higher voltage levels, permanent magnets have been installed in the cavity to provide beam compacting [4].) Commercial Reltron sources with peak output power levels of several hundred megawatts have been reported. The commercial applications of the Reltron source is mainly as an HPM source for survivability and vulnerability testing of sensitive electronic equipment.

The Reltron source at the University of New Mexico ("UNM Reltron") is a scaled down version of the commercial high power microwave source sold by Titan Inc. The Reltron source at UNM preserves the essential operating principles of the commercial source: modulation by transit time effect, post acceleration, and no guide magnetic field, with the reduction in voltage driver levels to about 120 kV. The UNM Reltron is designed to generate microwaves in S band at 10's MW power levels. The extraction of beam bunches to microwaves is accomplished through a single extraction waveguide connected to a TM-to-TE mode converter. The UNM Reltron was designed to be an educational device for conducting pulsed power and high power microwave experiments at the graduate level. The device can also serve as a platform for testing cathode and grid materials, and tube designs. The device can be taken apart and reassembled in short time with minimal difficulty.

SYSTEM DESCRIPTION

The UNM Reltron device comprises three main sections: the voltage driver section, the tube section, and the output section. The voltage driver section consists of a Marx bank and pulse forming line, gas flow system, power supplies, and control electronics. The tube section consists of the cathode, modulating cavity, post acceleration gap, and the tuners. The output section consists of the mode converter, output waveguide, and the waveguide components. The parameters that can be directly monitored from the Reltron high voltage pulser through the monitoring points on the system are summarized in Table 1. (Considerable details on the entire system can be found in [5].)

TABLE 1. Summary of directly monitored parameters.

	High Voltage Pulser Parameter	Voltage Division Ratio (into 50 Ω)
1	Marx Output Voltage	2550 to 1
2	Marx Trigger Signal	100 to 1
3	Marx Trigger Generator Output	1000 to 1
4	Crowbar Trigger Signal	100 to 1
5	Crowbar Trigger Output	1000 to 1
6	HVPS Voltage Monitor	3000 to 1

PULSER PERFORMANCE

The pulser was tested using a 770 Ω load resistor. The ideal voltage driver output would deliver a pulse with sharp voltage rise and fall characteristics and with consistency. The main parameters that determine these characteristics are the "tube voltage" (charge) and the gas pressure of the spark gap switches. The tube voltage is an important parameter since it determines the performance of the switch in terms of closing time and consistency. One major problem of concern with the switches is not with the initiation of the pulse, but rather with the termination of the pulse. If the crowbar switch fails to close, the voltage pulse takes the form of a decaying pulse that can last for 10's μs. This increases the beam current in the tube and cause damage to the modulating cavity by melting the grids. The gas flow in the sparkgap switches is fixed to the rate recommended by the switch manufacturers. Typical output pulses are shown in Fig. 2.

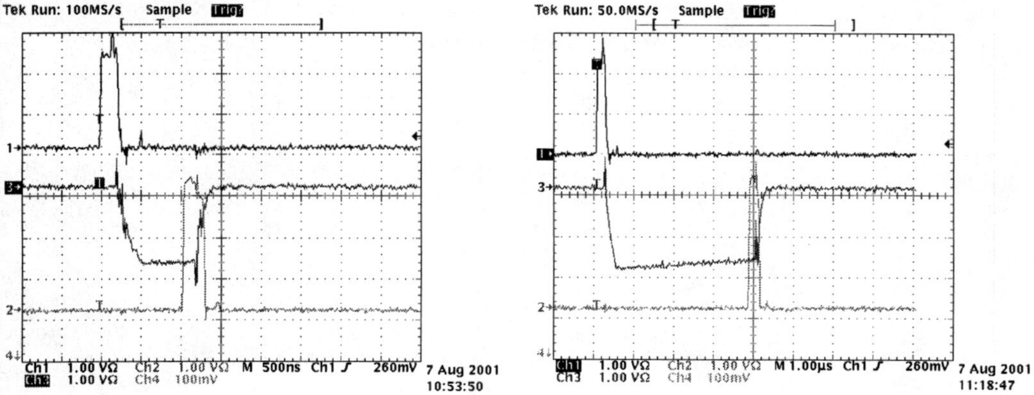

FIGURE 2. Typical output pulse from the pulser for a 0% pulse width setting (left) and 100% pulse width setting (right). The voltage traces are in the middle. The top and bottom are trigger pulses.

MODULATOR CHARACTERIZATION

In order to characterize the electron beam modulating cavity, both B-dot and E-dot loops were used to excite the TM_{01} mode. S_{12} measurements were performed using an HP 8720 network analyzer, as indicated in Fig. 3.

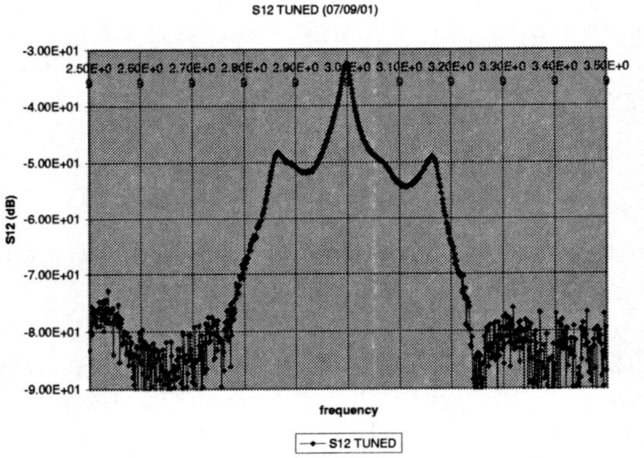

FIGURE 3. Cold test results of the modulating cavity (f_0 = 3.0 GHz, f_1 = 2.85 GHz, f_2 = 3.15 GHz).

Initial hot tests indicated a beam current between 200 and 300 A, with a bulk impedance of about 550 Ω. The FFT of the B-dot probe signal was inconclusive. It appears that instrumentation modification is needed for improved measurements.

ACKNOWLEDGEMENTS

This work was supported by an AFOSR and AFOSR/DOD MURI grant. Useful discussions and assistance from R. Bruce Miller are greatly appreciated.

REFERENCES

1. Marcum, J, "Interchange of energy between an electron beam and an oscillating electric field," J. Appl. Phys., vol. 17, p. 4, 1946.
2. Friedman, M., Serlin, V., Drobot, H. and Seftor, L., "Self modulation of an intense relativistic electron beam," J. Appl. Phys, vol. 56, p. 2459, 1984.
3. Friedman, M. and Serlin, V., "Parametric interaction of modulated intense relativistic electron beams with high-voltage gaps," Phys. Rev. Lett., vol. 54, p. 2018, 1985.
4. Miller, R.B. McCullough, W.F., Lancaster, K.T., and Muehlenweg, C.A., "Super-Reltron theory and experiments," IEEE Trans. Plasma Sci., vol. 20, p. 332, 1992.
5. Choi, S.H., "Characterization of electron beam modulation in a compact reltron high power microwave source," M.S. Thesis, University of New Mexico, Albuquerque, NM, 2002.

Vircator Experiments on Repetitive Pulsed Power Generator "ETIGO-IV"

Weihua Jiang, Masaki Sato, and Kiyoshi Yatsui

Extreme Energy-Density Research Institute, Nagaoka University of Technology
Nagaoka, Niigata 940-2188, Japan

Abstract. A new configuration of virtual cathode oscillator is proposed and tested experimentally. A resonant cavity is used to enhance the field feedback to the virtual cathode oscillation. The experiments were carried out on repetitive pulsed power generator "ETIGO-IV" (400 kV, 13 kA, 120 ns, 1 Hz). The diagnostic results have shown the peak microwave output power of ~ 450 MW, giving peak-to-peak microwave power efficiency of ~ 9 %.

INTRODUCTION

High power microwaves have potential applications in many fields such as controlled nuclear fusion, high-energy particle acceleration, material processing, and space development. Among various high-power microwave sources, virtual cathode oscillator (VCO) is of special interest because of its simplicity and high-power capability. In addition, most VCO configurations do not require the external magnetic field, which greatly saves volume, weight, and cost.

However, the VCO has a drawback which is its low conversion efficiency from the electron beam to microwave. So far, the most typical energy efficiency obtained in the experiments is on the order of several percents. It is clear that, without significant improvement in its conversion efficiency, the VCO can hardly be considered for practical applications.

It has been realized that the microwave efficiency of VCO depends very much on the intensity of the electromagnetic field around the virtual cathode [1,2]. The low efficiencies obtained in most of the experiments were the results of weak interaction between the microwave field and the electron beam, as the interactions usually happened in part of a waveguide where the microwave field is week. Since each electron interacts with the field only during a few cycles of movement, to obtain high energy-extraction efficiency, one has to increase the field intensity to the same order as that of the space-charge field of the virtual cathode. Therefore, theoretically, enhancing the oscillating field by using a resonant cavity would significantly increase the microwave efficiency of VCO.

By injecting high-current electron beam into a resonant cavity, a virtual cathode can be formed inside the cavity. The cavity field modulates the electron beam and extracts energy from it. However, in order to build the virtual cathode in the cavity with a given beam current, the cavity must be large enough so that the space charge of the

CP650, *Beams 2002: 14th International Conference on High-Power Particle Beams*, edited by T. A. Mehlhorn and M. A. Sweeney

electron beam takes effect. For this reason, it is usually difficult to avoid mode competition and to develop strong field of cavity oscillation.

In this paper, we propose a new configuration of VCO where a small cavity is used. Two electron beams are injected into the cavity from opposite directions, which allows both cavity walls to be located close enough to the virtual cathode. In addition, the transmitted electrons of one of the beams join the reflected electrons of the other beam resulting in efficiency enhancement of microwave extraction.

We have tested this new VCO configuration on the repetitive pulsed power generator "ETIGO-IV" which generates high-voltage (400 kV) and high-current (13 kA) pulses of 120 ns, at the repetition rate of 1 Hz. The initial experimental results on output microwave power are reported.

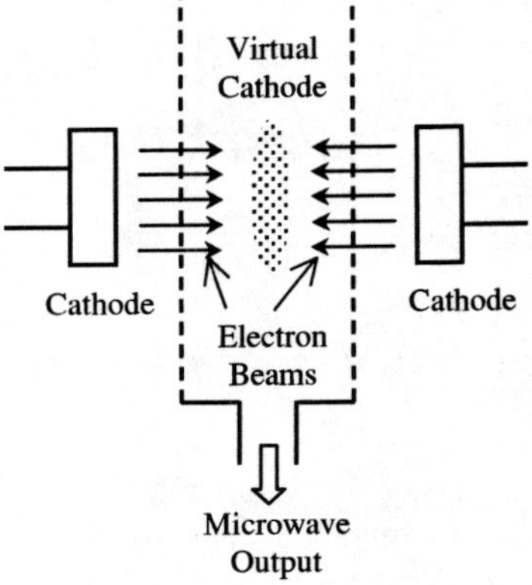

FIGURE 1. New configuration of virtual cathode oscillator.

EXPERIMENTAL SETUP

Figure 1 shows the VCO configuration used in this experiment. Two electron beams are injected into a pillbox cavity from both sides. In case of independent oscillation, each beam produces an individual VCO. However, if the TM_{010} cavity mode dominates, the oscillations on both sides are separated by a constant phase of π. As a result, the transmitted electron beam from one side can join the reflected electron beam of the other side, given the right timing. Therefore, during operation, the virtual cathode oscillates back and forth rotationally reflecting one of the electron beams and transmit the other. The resulting current oscillation feeds energy to the cavity mode and drives the microwave output.

The experimental setup is shown in Fig. 2. The positive high-voltage pulse is applied to the cavity by the repetitive pulsed power generator "ETIGO-IV". Electron beams are extracted from the cathodes on both sides of the cavity. There is an opening on the cavity wall that allows microwave extraction through the cylindrical waveguide.

Figure 3 shows the typical waveforms of the diode voltage and the diode current. The peak voltage of ~ 400 kV and peak current of ~ 11.5 kA have been obtained. The pulse width is ~ 150 ns.

The microwave power is measured by using both microwave antennas and E-field probes. They were calibrated by using network analyzer HP8753C. Figure 4 shows the relative microwave output power for different A-K gap lengths. The results shown in Fig. 4 can be explained by the following. When the A-K gap is large, the space charge

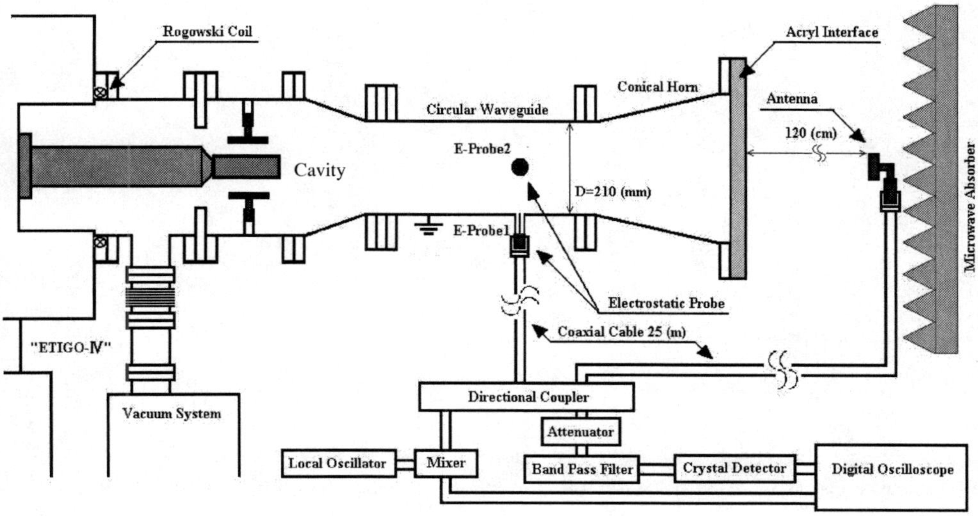

FIGURE 2. Experimental setup.

is not strong enough to form a virtual cathode in the cavity. When the A-K gap is small, on the other hand, the two interaction areas are separated so that the transmitted electrons from one beam can not join the reflected electrons of the other beam at the right time.

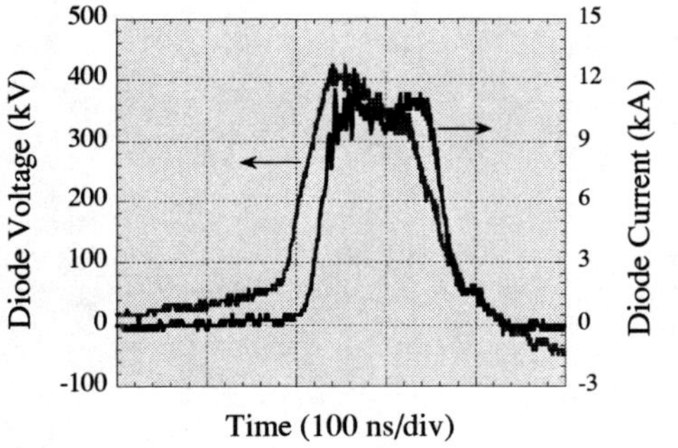

FIGURE 3. Typical waveforms of diode voltage and current.

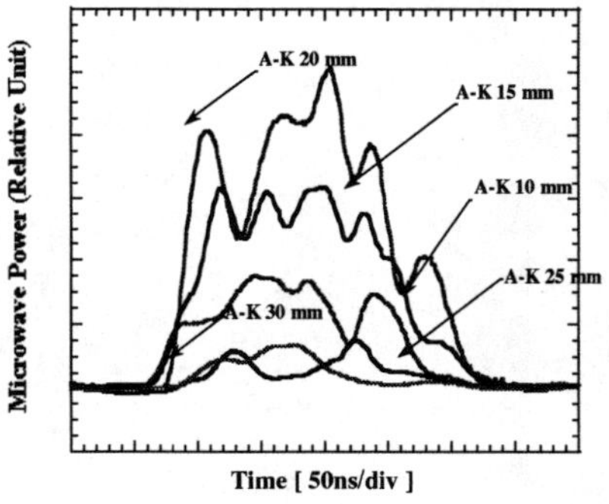

FIGURE 4. Relative microwave output power for different A-K gaps.

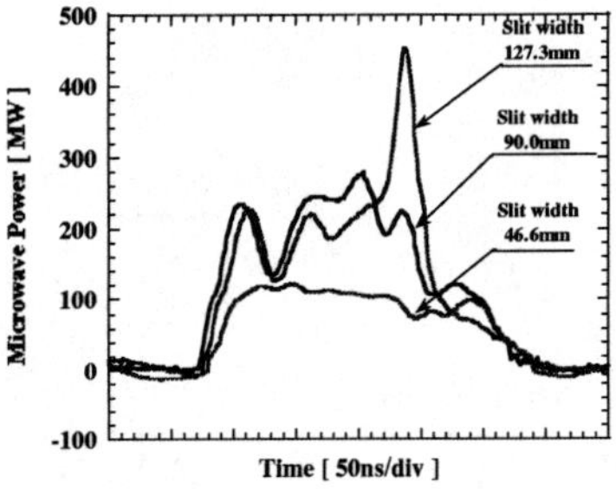

FIGURE 5. Microwave output power obtained with different sizes of cavity opening.

Figure 5 shows the waveforms of microwave output power obtained with different sizes of cavity opening. It is seen that a very sharp peak is obtained with the slit width of 127.3 mm. The maximum power reaches ~ 450 MW, giving peak-to-peak microwave power efficiency of ~ 9 %. However, as the pulse width of the maximum output power is relatively narrow, the energy efficiency is considered to be much lower.

ACKNOWLEDGMENTS

This work is supported by Asian Office of Aerospace Research and Development (AOARD).

REFERENCES

1. W. Jiang, K. Masugata, and K. Yatsui, *Physics of Plasmas* **2**, 982-986 (1995).
2. W. Jiang, J. Dickens, and M. Kristiansen, *IEEE Transactions on Plasma Science* **27**, 1543-1544 (1999).

ELECTRON BEAM APPLICATIONS

Progress on Channel Spark Development and Application of Pulsed Electron Beam Deposition (PED) in the Field of Medical Coating Work

Christoph Schultheiss, Lothar-H.-O. Buth, Wolfgang Frey, Hanns-G. Mayer*, Hansjoachim Bluhm

*Forschungszentrum Karlsruhe GmbH, Institute for Pulsed Power and Microwave Technology, *Technology Transfer and Marketing, Postfach 3640, D-76021 Karlsruhe, Germany*

Abstract. A promising source for Pulsed Electron Beam Deposition (PED) is the channel spark. Recent improvements helped to reduce beam instabilities which up to now have limited the life time of the system. The beam power could be increased and because of better beam quality the transport length of the beam is increased from 1 to several centimeters (up to 10 cm). Together with other improvements on the triggering system and beam transport in dielectric tubes, the channel spark approaches industrial standards. An overview of actual applications in research and industry will be presented. An attractive feature of the pulsed electron beam thin film deposition is the conservation of stoichiometry even during deposition of multi-component earth-alkali and alkali glasses. Specially developed glasses like BIOGLAS® [i] have the ability to anchor soft living tissue at the surface. In form of a bulk material bio active glasses are brittle limiting its applications. Contrary to brittle bulk material a thin layers on medical implants exhibits reliable bio-functionality. Coating of implants with this category of materials is subject of the European INCOMED project (Innovative Coating of Medical Implants with Soft Tissue Anchoring Ability) which just has started.

INTRODUCTION

This paper deals with the channel spark (CS)[1] which is a further development of the well known pseudospark described in 1978 by Christiansen and Schultheiss[2] initially. Both systems generate pinched electron beams which can be used to ablate target material for coating (PED) purposes. The application spectrum of these sources is similar to that of pulsed laser deposition (PLD), however investment and operation efforts in comparison with laser systems like Excimer Laser are magnitudes lower. Therefore PED and especially channel spark meets the interest of the industry because many coating processes which up to now fail at the cost barrier of PLD seem now to be possible[3]. However the quality and performance of pulsed electron systems has to be improved. The main goals for the channel spark development were i) to decrease the complexity of the pseudospark source by reduction to essential building groups, ii) the increase of power of the pulsed electron beam by coupling to an external capacitance. Actual goals are i) the operation of the source at repetition rates in the order of 20 to 50 Hz and ii) to realize technical standards i.e. reliability, pulse to pulse stability etc. and fulfil industrial needs like sufficient power per source.

---

[i] Bioglas® is a registered product, developed at the University of Florida

CP650, *Beams 2002: 14th International Conference on High-Power Particle Beams*, edited by T. A. Mehlhorn and M. A. Sweeney

IMPROVED CHANNEL SPARK SYSTEM

The CS-system consists of three functional parts: the gas discharge trigger, the hollow cathode and the dielectric channel which range into a vessel on anode potential at a gas pressure of 1 Pa. As can be seen in fig. 1 a trigger plasma is generated by a glow discharge which takes place in a glass tube after charging the capacitors and after the ignition of a gas switch to ground. The plasma density is enhanced by the axially field of a permanent magnet. The middle section of the CS is a conically shaped hollow cathode which supports the glow discharge to develop more current and allows the current amplification of the channel spark discharge to some 100 Amperes.

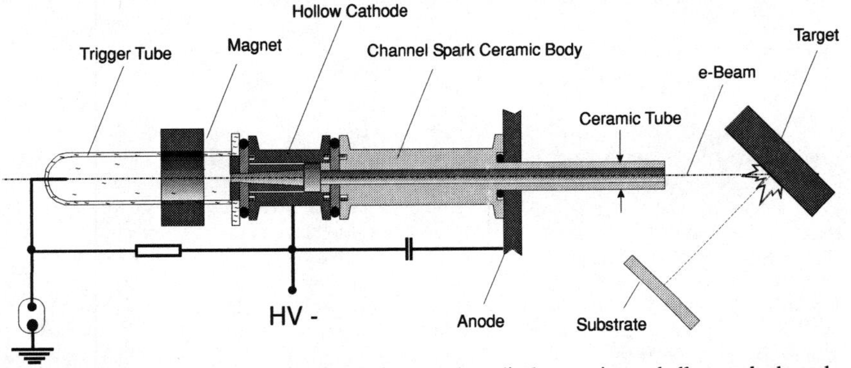

FIGURE 1. Channel spark device consisting on glow discharge trigger, hollow cathode and ceramic tube arrangement as used for coating substrates

The glow discharge trigger allows repetition rates up to 30 Hz. More promising is a trigger method basing on the Penning discharge effect where an electronically controlled current generator establishes magnetic fields via field coils at a trigger discharge tube and induce powerful trigger plasmas[4]. The trigger reliability as well as the performance of the CS beam in terms of ablated material per shot is improved decisively.

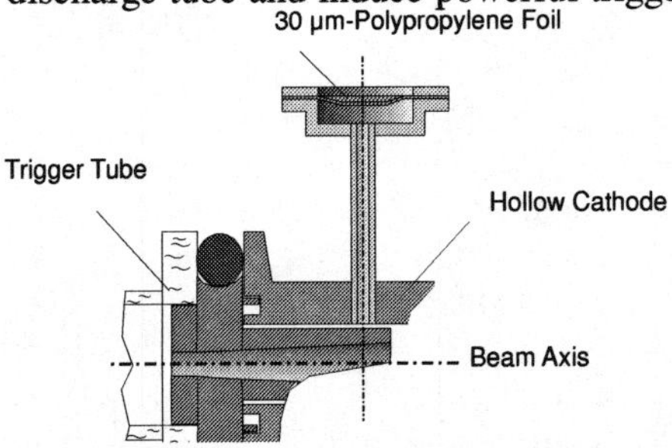

FIGURE 2. Hollow cathode provided with a “gas noble leak” to improve beam stability

The trigger method also influences a kind of hose instability in the discharge which is responsible that the beam touches the inner wall of the ceramic tube in a chaotic manner. Sooner or later this instability leads to a drilling of the beam though the channel wall. It can be demonstrated that a lets say “gas noble leak” in the vicinity of the hollow cathode suppresses the beam instability. A tiny pressure difference between the hollow cathode and the tube end of a factor of only 10^{-4} is necessary. This is magnitudes lower then usual micro valves can meter out. By means of a “gas noble leak” realized by air permeation though a 1 cm^2 area polypropylene foil at a pressure difference of 1 hPa a proper gas flow could be established (see fig.2). The elimination of the beam instability increases the range of the beam from 1 to several centimeters. It also improves is the beam to beam stabitlity measured by target bremsstahlung. The x-ray signal deviates not more then 10% from shot to shot. This may be the consequence of a quiet plasma development in the conically shaped hollow cathode without sharp edges etc. which suppresses undesired explosive electron emission (EEE) events.

MEDICAL COATING WORK

The energy transported in the electron beam is 1-2 J and each single ablation event heats up the substrate surface to some centigrade corresponding to the heat absorption of 0.1 J/cm^2. By controlling the repetition rate even temperature sensitive hydrocarbons can be coated. Targets used in PED can be from dielectric. This opens the way to coat arbitrary medical implants with bio active glasses[5]. Normally very brittle, bio active glasses in form of thin coatings i.e. 10 µm thick are flexible and gives implants the desired bio active surface. The perspective is not only hard tissue bonding (bone implants) but also soft tissue anchoring since the phase diagram of bio active glass shows a small area with this behavior (see fig.3). To meet the desired

FIGURE 3. Test substrates from TiAlV used in in-vitro and in-vivo experiments. Substrates (right) are coated with 14 µm thick bio active glass; the left one is uncoated.

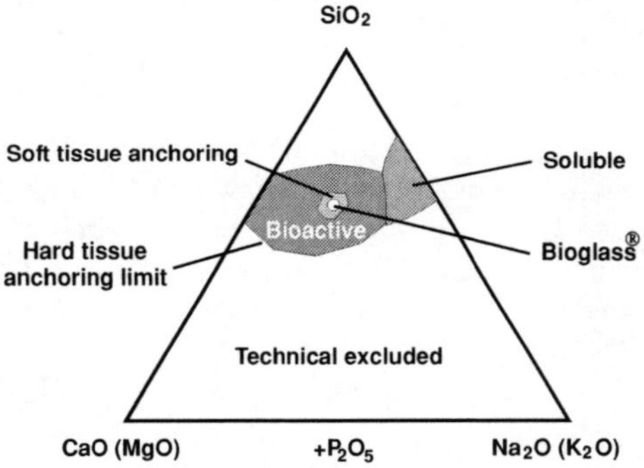

FIGURE 4. Phase diagram of bio active glass

stoichiometry deviations of the coating procedure must be very limited. Fortunately ablation in PED takes place in form of small molten droplets with an characteristic diameter of 40 nm. In the bulk of the droplets the stoichiometry does not change. In table 1 stoichiometry of target and coating is compared. The error of SEM is estimated to 1% and in the frame of errors no deviation between target and film composition can be stated. The conservation of stoichiometry gives hope to reproduce at film side the same tissue anchoring potential as for the bulk material shown in the phase diagram of bio active glasses (see fig. 4).

TABLE 1. Comparison between target and film composition

%	Bio Active Glass Target	Deposited Film
Na_2O	27,62±1	27,1±1
Al_2O_3	2,7±1	3,7±1
SiO_2	43,4±1	41,2±1
P_2O_5	7,0±1	5,9±1
CaO	19,3±1	22,1±1

The thickness of the coating in fig.3 is 14 µm. The coating has to be sufficient thick because in body liquid a solution attack occurs and a accompanying hydroxyapatite (HA) crystal growth take place. Phosphorous groups at the surface of HA are responsible for the tissue anchoring. The adhesion of the coating is measured to 200 N/cm^2. This result is a consequence of a pretreatment of the substrate by means of Argon sputtering.

OUTLOOK

PED open the possibility to coat nearly any of the implant materials with bio active or bio neutral material opens the desired to wide field of medical application. The whole spectrum of reliable tissue anchoring or transcutaneous and subcutaneous bio compatible or bio inert implants can be covered.

ACKNOWLEDGMENTS

The authors would like to thank the European Union for the support of this project in the frame of the 6th framework of Competitive and Sustainable Growth. The authors also like to thank the partners of the INCOMED project, especially Sigfried Jaecque who delivered first bio active targets. We also like to express our gratitude to REHAU AG+Co, Germany which supply us with medical plastic material.

REFERENCES

[1] G. Müller, M. Konijnenberg, G. Krafft, C. Schultheiss, "Deposition by Means of Pulsed Electron Beam Ablation" in "Science and technology of thin film" edited by World Scientific Publishing Co. PTE.LTD 1995

[2] J. Christiansen and C. Schultheiss, Z. Phys. **A 290**, 35 (1979)

[3] See: http://www.codeangel.com/neocera/html/display.asp?recno=70

[4] W. Frey, C. Schultheiss, H.-J. Bluhm, published in this conference

[5] June Wilson and D. Nolletti, "Bonding of Soft Tissue to Bioglass", in Handbook of Bioactive Ceramics, Edited by T. Yamamuro, L. L. Hench, and J. Wilson. CRE Press. Boca Raton, FL (1990) and http://www.bg.ic.ac.uk/Lctures/Hensch/BioGlass/call3.htm

Application of Pulsed Electron Beams for Improvement of Material Surface Properties

Georg Mueller[a], Vladimir Engelko[b], Hansjoachim Bluhm[a], Annette Heinzel[a], Gustav Schumacher[a], Dirk Strauss[a], Alfons Weisenburger[a], Frank Zimmermann[a], Vyacheslav Shulov[c], Nadegda Notchovnaia[c]

a- Forschungszentrum Karlsruhe, Institut für Hochleistungsimpuls- und Mikrowellentechnik, Postfach 3640, D-76021 Karlsruhe, Germany
b- Efremov Institute of Electrophysical Apparatus, 189631, St. Petersburg, Russia
c- Moscow Aviation Institute, 125 993 Moscow, Russia

Abstract. The pulsed electron beam facilities GESA I and GESA II were developed in cooperation between Efremov Institute St. Petersburg, Russia and the Research Center Karlsruhe (FZK), Germany for large area surface treatment with beam diameter of 4 – 10 cm. It melts the material surface down to a depth of 10-100 µm. Technological applications are improvement of high temperature oxidation resistance of MCrAlY and of its bonding properties to TBC's, prevention of liquid metal attack on steels, surface hardening and increase of wear resistance and lowering of sea water corrosion and high cycle fatigue of aircraft engine blades. An overview is given on the status of the current applications.

INTRODUCTION

It is important for an effective surface treatment by a pulsed electron beam to heat and melt the whole area during one single pulse to achieve the structural conditions that cause the property changes of the affected surface layer. The structures produced in this process are very fine grained down to nanometer size or amorphous ones and sometimes exist in metastable phases. They are of high homogeneity and with a smooth surface. The facilities GESA I (1996) and GESA II (2000) are developed for this purpose in cooperation between the Efremov Institute St. Petersburg, Russia and the Research Center Karlsruhe (FZK), Germany who runs both facilities in its laboratories. Table 1 gives an overview on the most important data of the 2 GESA facilities at FZK which are described elsewhere [1].

TABLE 1. Main parameters of GESA-I and GESA-II.

Parameters	GESA II	GESA I
accelerating voltage	200 - 400 kV	50- 150 kV
power density	≤ 6 MW/cm^2	≤ 2 MW/cm^2
beam diameter	4 – 6 cm	5 - 10 cm
pulse duration	5 - 250 µs	4 - 40 µs

Since starting work with the GESA facilities several applications have been conducted and started which mainly concern wear and corrosion problems that can be solved just by changing the surface properties.

CP650, *Beams 2002: 14th International Conference on High-Power Particle Beams*,
edited by T. A. Mehlhorn and M. A. Sweeney

One main application is the improvement of steel behaviour in liquid lead and lead-bismuth. This problem has to be solved for the realization of liquid metal cooled accelerator driven subcritical nuclear reactor systems (ADS) [2]. It is known that oxide layers at the steel surface prevent or postpone the dissolution attack if decomposition of the oxide is prevented by an appropriate concentration of dissolved oxygen in the liquid lead alloys [3, 4, 5]. The protection behaviour can be improved if a stable oxide former like Al is alloyed into the surface of the steel [6, 7].

Another extended application is the improvement of the high temperature oxidation resistance of gas turbine blades which are protected against oxidation at high temperatures (around 900 °C) by coating them with a Ni or Co basis Cr-Al alloy (MCrAlY). It was shown recently that treatment of the coating surface favors the formation of α-alumina layers with optimal protection properties [8]. GESA treatment of the MCrAlY surface before deposition of a TBC also improves the life time of a thermal barrier coating (TBC) which depends on the density and bond strength of the alumina layer that develops at the MCrAlY surface below the TBC [9].

Compressor and turbine blades are the most critical parts of turbine engines in air planes. They are exposed to cyclic loads, corrosive environment at high temperatures and to dust and fluid erosion. High cycle fatigue and sea water corrosion are two processes that influence life time of compressor blades strongly. The effect of both processes can be diminished by changing the structure at the surface using the pulsed electron beam.

The present paper gives a status of the recent and current applications and their results.

STATUS OF APPLICATIONS

Prevention of liquid metal corrosion. The design study of an accelerator driven subcritical reactor system (ADS) considers lead-bismuth as a target and cooling material. One of the important problems to be solved is the compatibility between the structural steel and the liquid metal. Alloying Al into the steel surface by melting a 18 µm Al foil placed on top of it improved the compatibility with liquid Pb/Bi decisively as it was shown in experiments up to 550 °C [6]. The result of corrosion testing in Pb/Bi containing 10^{-6} wt% of oxygen at 650 °C for 2000 hours is presented in Fig. 1.

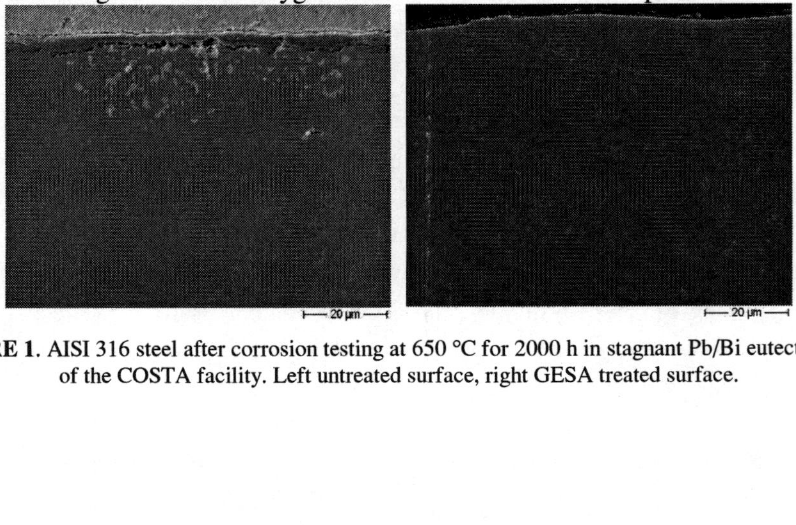

FIGURE 1. AISI 316 steel after corrosion testing at 650 °C for 2000 h in stagnant Pb/Bi eutectic alloy of the COSTA facility. Left untreated surface, right GESA treated surface.

The SEM cross section to the left shows the interface between the molten Pb/Bi and the untreated AISI 316 steel with 20 µm deep dissolution attack. To the right a cross section of an Al-alloyed surface is depicted without any attack although at 650 °C.

The tests have shown that there is a minimum Al concentration that must be exceeded to obtain selective Al_2O_3 formation at the surface. Below 5 – 8 wt% of Al there will be no protective Al_2O_3 scale formation. However the Al concentration in the surface layer should also not exceed 18 – 30 wt%. At higher concentrations there is the danger of crack formation because of embrittlement that depends on steel composition.

Gas turbine blades. It was shown recently [9] that electron beam treatment of the MCrAlY coating on turbine blades significantly reduces the growth rate of the protective Al_2O_3 scale developed on the surface exposed to the hot oxidizing atmosphere of gas turbines. This leads to a higher life time of the Al_2O_3 scale and to a better bonding property if the MCrAlY coating is employed as a bond coat.

Experiments to confirm these results are still running. There is however an effect of the electron beam treatment which must be prevented. If there is more Y in the alloy than can be dissolved melt pearls of an Al-Y alloy appear at the surface which can disturb the growth process of columnar TBC´s when they reach a size above 5 – 10 µm. This can be avoided by alloying less than 0.2 wt% Y into the MCrAlY or by applying multiple electron pulses.

High cycle fatigue strength. Compressor blade samples of an Ni-base superalloy (GS6U) and of a titanium alloy (VT8M) were irradiated by electron pulses to improve the high cycle fatigue strength. The intention was to remelt the surface material layer to remove impurities and defects that can act as spots that inintiate cracking during vibration of the blade. Different energy densities were required to achieve this goal in the two materials. For the Ni-base alloy 40 J/cm^2 were necessary at the surface to obtain an optimal results, while on the titanium alloy the optimal energy density amounted to 26 J/cm^2. Both materials were exposed to a subsequent annealing, the Ni-alloy for 2 h at 950 °C and the Ti alloy for 2 h at 530 °C. For both materials there was a massive increase in the number of cycles up to fatigue cracking of the material. The number of cycles increased for the Ni-alloy by a factor of 15 and that for the Ti-alloy even by a factor of 1000.

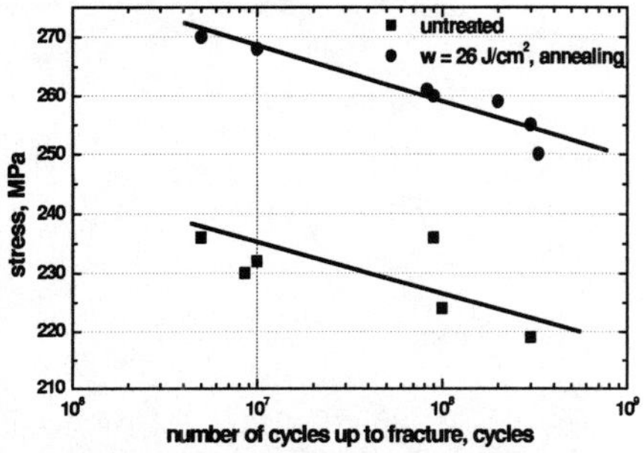

FIGURE 2. Fatigue curves of VT8M titanium alloy samples before and after irradiation and annealing

Fig. 2 shows the fatigue curves for the Ti-alloy with and without electron pulse irradiation. It should be noted, that annealing of the specimen after treatment is necessary because of the stresses which remain on the surface due to the rapid cooling after the irradiation.

Salt corrosion resistance. Compressor blades of an NiCrAlY and a FeCrAlY superalloy were irradiated by 5 electron beam pulses of 16 – 30 J/cm^2 energy density. They were exposed together with an untreated blade to 20 cycles of heating to 600 °C and quenching in sea water at 20 °C. The appearance of the surface after cycling shown in Fig. 3 indicates a much better performance of the irradiated blade surface. The blade in the middle of Fig. 3 exhibits severe corrosion while the surface of the two irradiated blades are not much affected.

FIGURE 3. EP866sh steel blades after corrosion tests under thermocycling conditions (T= 600 °C, sea water at 20 °C, 20 cycles): 1-irradiated blade (w=18 J/cm^2, n= 5 pulses), 2- initial blade, 3-irradiated blade (w=23 J/cm^2, n= 5 pulses).

REFERENCES

1. V. Engelko, G. Mueller, H. Bluhm, Vacuum, Vol 62/2-3, June 2001, pp. 97-103.
2. C. Rubbia, J.A.Rubio, S. Buono, F. Carminati, CERN/AT/95-44 (ET), September 29, 1995
3. V. Markov, Seminar on the Concept of Lead-Cooled Fast Reactors, Cadarache, September 22-23, 1997.
4. B. F. Gromov, Yu. I. Orlov, P. N. Martynov, K. D. Ivanov and V. A. Gulevski, in Liquid Metal Systems, Edited by H. U. Borgstedt and G. Frees, Plenum Press (1995) 339
5. H. U. Borgstedt and H. Glasbrenner, Fusion Eng. Des., 27 (1995) 659
6. G. Müller, G. Schumacher, F. Zimmermann, J. Nucl. Mater., 278 (2000) 85-95.
7. Müller, G., Heinzel, A., Konys, J., Schumacher, G., Weisenburger, A., Zimmermann, F., Engelko, V., Rusanov, A., Markov, V.; J. Nucl. Mater., Vol. 301 (1) (2002) 40-46
8. Mueller G., Schumacher G., Strauss D., Surf. Coat. Technol., 108&109, (1998) 43.
9. D. Strauss, G. Müller, G. Schumacher, V. Engelko, W. Stamm, D. Clemens, W. J. Quadakkers, Surface and Coatings Technology 135 (2001) 196-201.

Measurement of Angular Distribution of Beam Electrons at GESA Facilities

Vladimir Engelko[a], Georg Mueller[b], Vladimir Kavaljov[a], Oleg Komarov[a]

[a] *Efremov Institute of Electrophysical Apparatus, 189631, St. Petersburg, Russia*
[b] *Forschungszentrum Karlsruhe, Institut für Hochleistungsimpuls- und Mikrowellentechnik, Postfach 3640, D-76021 Karlsruhe, Germany*

Abstract. In the paper results of measurements of an angular distribution of intense beam electrons are presented. The measurements were performed at the GESA-1 and GESA-2 facilities. At both facilities multipoint explosive emission cathodes are used as an electron source. The beam is formed and transported with the help of external magnetic field. Measured angular distribution is in a good agreement with calculated one. The influence of the magnitude and distribution of magnetic field, cathode points density, accelerating voltage on the angular distribution was investigated.

INTRODUCTION

When an intense electron beam is used for treatment of materials the knowledge of an angular distribution of beams electrons is necessary for providing effective beam transportation and focusing and for better understanding of details of the treatment.

Here we present results of the angular distribution measurements, performed at the GESA-1 and GESA-2 facilities [1]. At both facilities multipoint explosive emission cathodes are used as an electron source. The beam is formed and transported with the help of an external magnetic field. The method of the measurements is based on the study of electron beam passing through a collimator having a small hole with a radius of about the Larmour radius of the electrons. The current of electrons passed through the collimator decreases when the length of the collimator is increased. The extent of the process is the greater, the greater is the angular spread of electrons [2].

METHOD OF MEASUREMENT

The interrelation between the currents of electrons passed through the collimator I_{pass} and deposited on its side wall I_{side} (that is the passage function: $P=I_{pass}/(I_{pass}+I_{side})$) depends on a number of factors: magnetic field inductance B, electron kinetic energy E, root-mean scattering angle of the electron beam Ω, collimator length L and diameter of its hole ∅. For experimental determination of the angular distribution for given magnetic field B and electron energy E the collimator with a varied length L can be used. By means of variation of the collimator length the passage function P(L) is

CP650, *Beams 2002: 14th International Conference on High-Power Particle Beams,*
edited by T. A. Mehlhorn and M. A. Sweeney

obtained. From comparison the measured passage function with the set of calculated ones the angular distribution is determined.

The function P(L) is calculated, using the following expression for the passed part of the electron current [2]:

$$J = e\pi a^2 \int_0^\infty V_{par} \int_0^\infty T(V_{par}, V_{per}) f_0(V_{par}, V_{per}) \cdot 2\pi V_{per} dV_{per} dV_{par}, \tag{1}$$

where V_{par}, V_{per} are direct and perpendicular components of the electron velocity, $T(V_{par}, V_{per})$ is the analytical function depended from magnetic field inductance B, electron kinetic energy E, collimator length L, and diameter of the collimator hole ∅, f_0 is the angle distribution function of beam electrons determined by its value of the root-mean scattering angle Ω, a=∅/2 is the collimator radius. The analytical function $T(V_{par}, V_{per})$ is expressed by the following equations [1]

$$T(V_{par}, V_{per}) = \begin{cases} 0; \delta \geq 1, \\ (1-\delta)^2; \delta < 1, \end{cases} \quad \text{if} \quad \beta \geq \pi, \tag{2a}$$

and

$$T(V_{par}, V_{per}) = \begin{cases} 0; \delta \geq 1, \beta > \alpha_{max} \\ \frac{2}{\pi}\left[\arccos(\delta \sin\beta) - \delta \sin\beta \sqrt{1-\delta^2 \sin^2\beta}; \delta > 1, \beta < \alpha_{max}\right] \\ \frac{1}{\pi}\left[\pi + \delta(\delta-2)\beta - \arcsin(\delta \sin\beta) - \delta \sin\beta\left(\delta\cos\beta + \sqrt{1-\delta^2 \sin\beta}\right)\right]; \delta \leq 1 \end{cases}$$

$$\text{if} \quad \beta < \pi, \tag{2b}$$

where $\delta = \frac{\rho}{a} = \frac{V_{per}}{aw}$ is the relation between the Larmour radius and the radius of the collimator hole, $\beta = \frac{wd}{2V_{par}}$ is the angle at which electron turns after passing a distance d, $\alpha_{max} = \arcsin\left(\frac{a}{\rho}\right) = \arcsin\frac{1}{\delta}$, w is the cyclotron frequency.

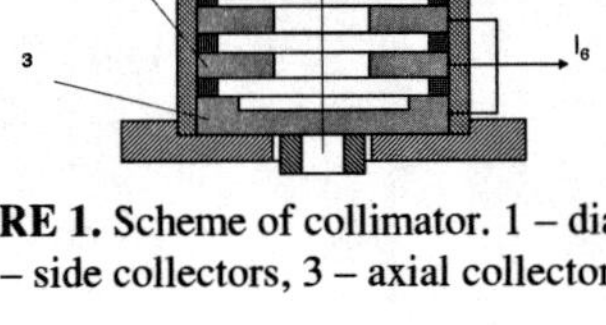

FIGURE 1. Scheme of collimator. 1 – diafrag, 2 – side collectors, 3 – axial collector.

In Fig. 1 the scheme of the collimator is shown. Its design is similar to that described in [3]. The detector consists of the diaphragm made of Ta foil of 1 mm thickness and of 1 cm diameter; twelve graphite side collectors (width 0.6 cm, diameter 1 cm), isolated one from another; central collector.

The appropriate radius of the collimator hole is determined by Larmour radius of electrons. Its length corresponds to saturation of the passage function.

RESULTS

In Fig. 2, 3 measured and numerically calculated passage functions for GESA-1 and GESA – 3 facilities are shown. Here α is the angle, corresponding to maximum of the passage function. Calculations were done under assumption that the majority of electrons are emitted from the surface of spheres of 2 mm diameter which centres are situated on the tops of the cathode points.

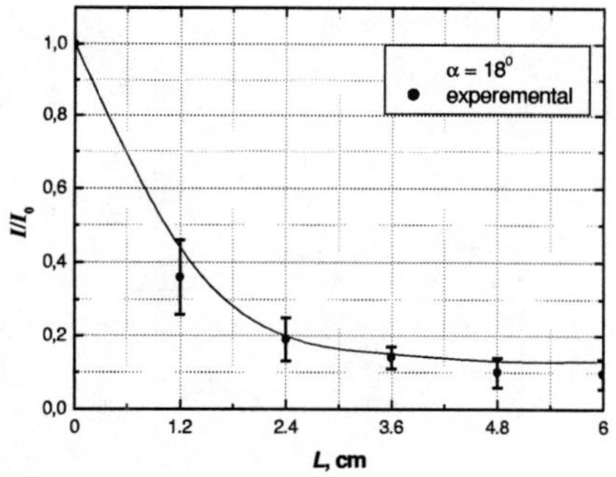

FIGURE 2. Calculated and measured passage functions for GESA-1 facility (U = 120 kV).

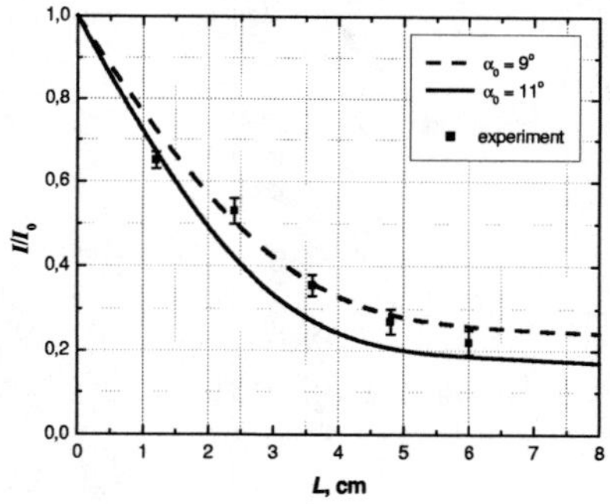

FIGURE 3. Calculated and measured passage functions for GESA-2 facility (U = 160 kV).

In Fig.4 examples of the calculated angular distributions for GESA –2 facility are presented. They are of Gaussian type.

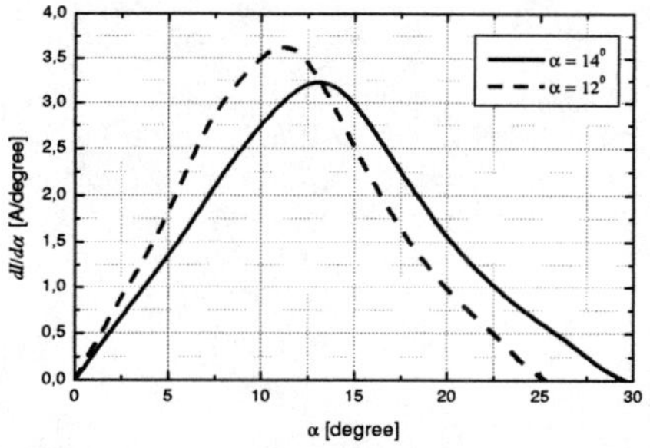

FIGURE 4. Angle distribution functions for GESA 2 facility for U = 160 and 180 kV.

It is seen in Fig.2,3 that calculated and measured passage functions are in a good agreement. Therefore, one can conclude, that the method of the angular distribution calculation is correct. Fig. 5 illustrates the dependence of α on k, where k is the relation of the magnetic field at the cathode to one in the treatment chamber. An increase k from 18 to 30 leads to increasing α from 10 to 18 degree.

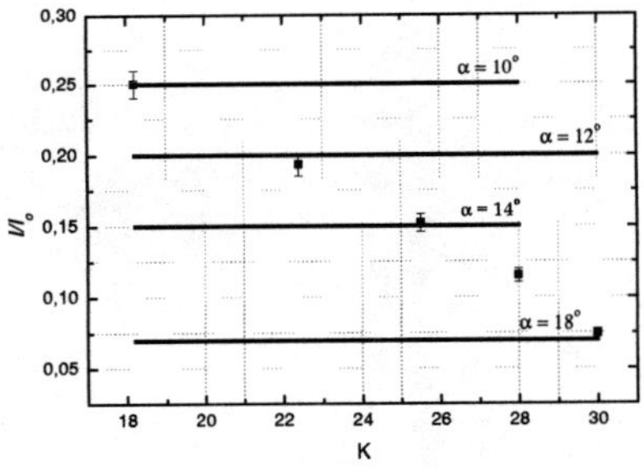

FIGURE 5. Dependence αon k. GESA 2. U=160 kV.

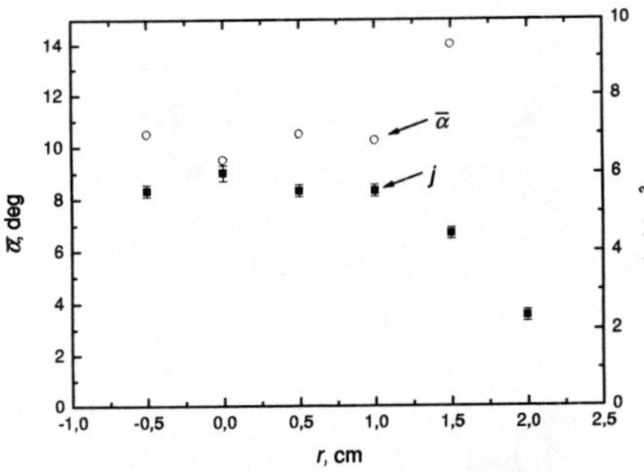

FIGURE 6. Dependence α on the distance from the beam axis

Fig.6 shows the dependence α on the distance from the beam axis. One can see that at the beam periphery electrons have higher angles then at the beam axis.

Measurements performed with the cathodes containing deferent numbers of pins showed that an increase of the cathode pins density from 1/cm2 to 4/cm2 results to decrease of the average angle of electrons from 24 to 15 degree.

CONCLUSION

The angular distribution of electrons was measured at the GESA facilities. It agrees with that, obtained from numerical simulations. Dependencies of the average angle of the distribution on accelerating voltage, magnitude and distribution of the applied magnetic field, distance from the beam axis, density of the cathode pins were obtained.

REFERENCES

1. V. Engelko, B. Yatsenko, G. Mueller, H. Bluhm, Pulsed electron beam facility (GESA) for surface treatment of materials, Vacuum, Vol. 62/2-3, June 2001, pp. 211-216.
2. V.D. Aleksin, B.G. Bocharov. Diagnostics of Plasma. Atomisdat, 1973, p. 345 (in Russian).
3. A.V. Arzhannikov, V.S. Koidan, S.V. Loginov. Preprint INP 81-10, Novosibirsk, 1981 (in Russian), Pribory i Technika Experimenta, v.4, 1983, p. 36-39 (in Russian).

Plasma Dynamic on a Target Irradiated by Intense Electron Beams

Vladimir Engelko[a], Georg Mueller[b]

[a] *Efremov Institute of Electrophysical Apparatus, 189631, St. Petersburg, Russia*
[b] *Forschungszentrum Karlsruhe, Institut für Hochleistungsimpuls- und Mikrowellentechnik, Postfach 3640, D-76021 Karlsruhe, Germany*

Abstract. Theoretical consideration shows, that under the influence of an intense electron beam first an ion flux and then a plasma is formed on the surface of a target as a result of gas desorption and evaporation of target material. The target plasma is created when the density of the ion flux achieves a certain upperlimit. The time necessary for plasma formation depends on the beam current density and the efficiency of gas desorption and ionization. This time becomes a few microseconds under typical vacuum conditions of about $5.0 \cdot 10^{-5}$ mbar, electron beam current density in the range of 10 A/cm^2 and kinetic energy of electrons in the range of 100 keV. When the density of the ion flux reaches a limiting value the beam potential decreases to a level, which is half of the initial one. A transient layer is formed between the plasma boundary and the electron beam. For conditions mentioned above its length is a few centimeters. The target plasma expands into the beam drift region, with a velocity that increases in time. The expansion of the target plasma is the main reason for neutralization of the electron beam space charge.

RESULTS

Under the influence of an intense electron beam at the surface of a target first an ion flux and than a plasma are formed. The investigation of these processes is necessary because both ion flux and plasma influence the electron beam parameters. The ion flux changes the beam current density distribution and beam dynamics in a drift space. The target plasma penetrating into the beam formation region influences the electron source operation. Plasma expansion also changes the potential distribution in the drift space. The last circumstance can be used for obtaining the information about the target plasma and ion flux dynamics.

It is known that the potential on the axis of a long tube, along which a cylindrical electron beam is transported, is determined by the following expression

$$\Phi = \frac{\mathrm{I}}{4\pi\varepsilon_0 v_z} + \frac{\mathrm{I}}{2\pi\varepsilon_0 v_z}\ln\frac{R}{r} \qquad (1)$$

here v_z is the longitudinal velocity of electrons, R is the tube radius, r is the beam radius. The first term in this equation describes the potential drop inside the beam. The second one – the potential drop between the tube and the beam.

Since the target is grounded but the beam has a certain potential a transient region between the beam and target should exist. This is illustrated in Fig.1.

CP650, *Beams 2002: 14th International Conference on High-Power Particle Beams,* edited by T. A. Mehlhorn and M. A. Sweeney

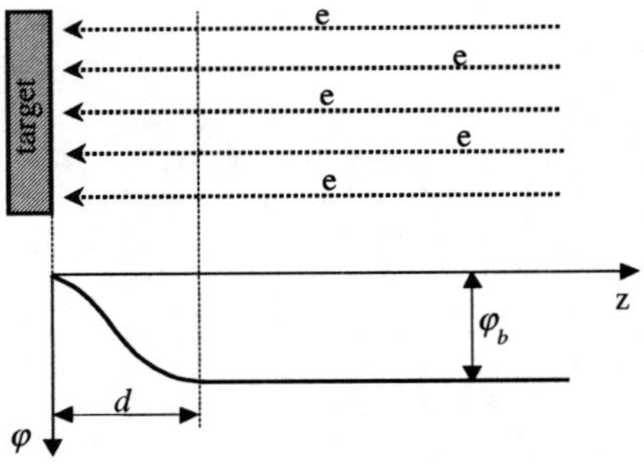

FIGURE 1. Potential distribution along beam axis

The length of this region is d. When the target plasma is formed the transient region exists between the plasma boundary and the beam.

The formation of the target ion flux and plasma and the plasma dynamics comprises the following steps: target irradiation by an electron beam with the current density j_e and kinetic energy of electrons eU; the desorption of a gas from the target surface with the yield a takes place under the influence of the beam electrons; desorbed atoms are ionized with the cross-section σ corresponding to the whole energy of electrons; ionized atoms create the ion flow with the current density j_i moving towards the electron beam; the positive space charge of the ions decreases the electric field at the target surface; when the ion current density achieves a certain limiting value $j_i^{\lim}$ the electric field on the target drops to zero; with zero electric field on the target the ion current density does not increase anymore, further production of ions leads to the plasma formation and its expansion into the transport channel.

In the presence of the ion flow from the target the potential distribution in the transient layer is described by Poisson's equation in the following form

$$\frac{d^2\Phi}{dz^2} = -\frac{1}{\varepsilon_0}\left(\frac{j_e}{\sqrt{\frac{2eU}{m}}} - \frac{j_i}{\sqrt{\frac{2e\Phi}{M}}}\right) = -\frac{j_e}{\varepsilon_0}\sqrt{\frac{m}{2eU}}\left(1 - \frac{j_i}{j_e}\sqrt{\frac{MU}{m\Phi}}\right). \tag{2}$$

Having solved this equation one obtains the following results.
When the electric field on the target surface drops to zero, the beam axis potential is equal to

$$\Phi_b = 4\frac{j_i^{\lim 2} M}{j_e^2 m} U , \tag{3}$$

the length of the transient layer is equal to

$$d = 4.4\frac{j_i^{\lim}}{j_e^{3/2}} U^{3/4}\sqrt{\frac{M}{m}\varepsilon_0\sqrt{\frac{2e}{m}}} = 2.2\sqrt{\Phi_b \varepsilon_0 \frac{v_e}{j_e}} , \tag{4}$$

and the limiting ion current density is equal to

$$j_i^{\lim} = \frac{j_e}{2}\sqrt{\frac{\Phi_b m}{UM}} . \tag{5}$$

Φ_b can be determined from the radial potential well in the drift space. Assuming that Φ_b is the potential at the beam axis, the ions have a velocity corresponding to $e\Psi_b$ and the electrons – to eU, one can write for Φ_b:

$$\Phi_b = \frac{1}{2\pi\varepsilon_0}\left(\frac{I_e}{v_e} - \frac{I_i}{v_{i\,i}}\right)\left(\frac{1}{2} + \ln\frac{R}{r}\right) = \frac{I_e}{4\pi\varepsilon_0 v_e}\left(\frac{1}{2} + \ln\frac{R}{r}\right). \tag{6}$$

Here I_e and I_i are the total electron and ion current correspondingly.

From comparison (1) and (6) one can see that, when the electric field on the target drops to zero, the beam potential decreases by a factor of 2.
For typical parameters of the GESA-1 facility (U=120 keV, j_e=10 A/cm2, R=7.5 cm, r=4cm) one can obtain from (3) – (5):

$$\Psi_b = 25kV,\ j_i^{\lim} = 57.8\, {}^{mA}\!/_{cm^2},\ d = 4.3cm.$$

Let us evaluate now what time is necessary for j_i to reach the limiting value $j_i^{\lim}$. The number of atoms, desorbed from the target surface during the time period t, is equal to

$$N = \frac{j_e at}{e}. \tag{7}$$

The rate of ion production is

$$\frac{dN_i}{dt} = \frac{j_e N\sigma}{e} = \frac{j_e^2 \sigma at}{e^2}. \tag{8}$$

We assume that until the electric field strength on the target drops to zero all ions produced are removed by the electric field to form the ion flow with the density, equal to

$$j_i = e\frac{dN_i}{dt} = \frac{j_e^2 \sigma at}{e}. \tag{9}$$

From (9) one can simply obtain the following expression for the time t_p, which is necessary for the target plasma formation

$$t_p = \frac{j_i^{\lim} e}{j_e^2 \sigma a}. \tag{10}$$

The dependence of the number of target plasma ions on time (when $t > t_p$) is determined by the expression

$$\frac{dN_i}{dt} = \frac{j_e^2 \sigma at}{e^2} - \frac{j_i^{\lim}}{e}. \tag{11}$$

Here the first term describes the income of the ions due to the desorption and ionization of absorbed and evaporated atoms and the second one describes the ions outcome, which is necessary to maintain the target ion flow with the density of $j_i^{\lim}$.
From (11) it follows

$$N_i = \int_{t_p}^{t} \left(\frac{j_e^2 \sigma at}{e^2} - \frac{j_i^{\lim}}{e} \right) dt. \tag{12}$$

Taking into account that $j_i^{\lim} = 0.5 e n_i \sqrt{\frac{kT_e}{M}}$, where n_i and T_e are the target plasma density and electron temperature, and using (10), (12) one can obtain for the distance s(t) on which the target plasma expands during the time period $(t - t_p)$:

$$s(t) = 0.5\sqrt{\frac{kTe}{M}} \left(0.5\, {}^{t^2}\!/_{t_p} - t + 0.5 t_p \right). \tag{13}$$

The velocity of the target plasma expansion is equal to

$$v(t) = 0.5\sqrt{\frac{kTe}{M}}\left({}^{t}\!/_{t_p} - 1 \right). \qquad (14)$$

The time dependencies s(t) and v(t) for H are shown in Fig. 2. (T_e = 1 eV). One can see, if hydrocarbons are formed on the target surface, the plasma expands at a distance of ~1m already within ~ 30 microseconds.

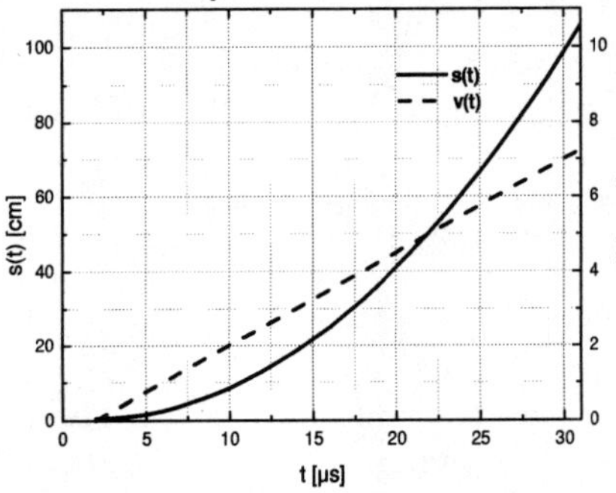

FIGURE 2. Length of expanded plasma s(t) and plasma expansion velocity v(t)

The experiments are performed at the GESA-1 facility [1] with the accelerating voltage U decreased to 60 keV since under higher voltage electrical breakdowns along the glass tube of the potential probe take place. Measured values of the beam potential and length of the transient layer are in good agreement with calculated ones. In Fig.3 one can see the examples of the oscillograms of the accelerating voltage, controlling grid voltage, beam current and probe signal.

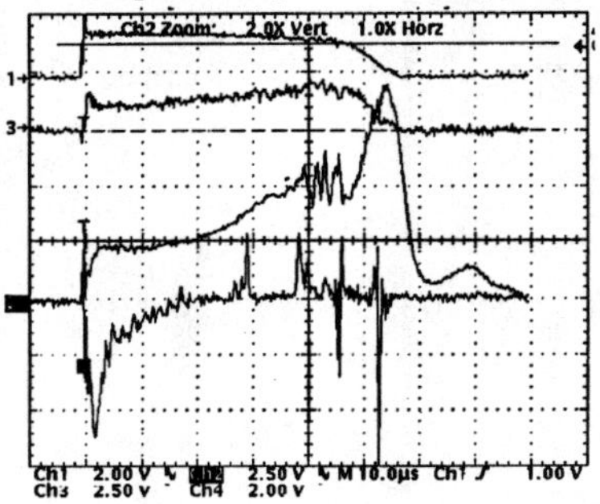

FIGURE 3. Accelerating voltage, grid voltage, beam current and probe signal

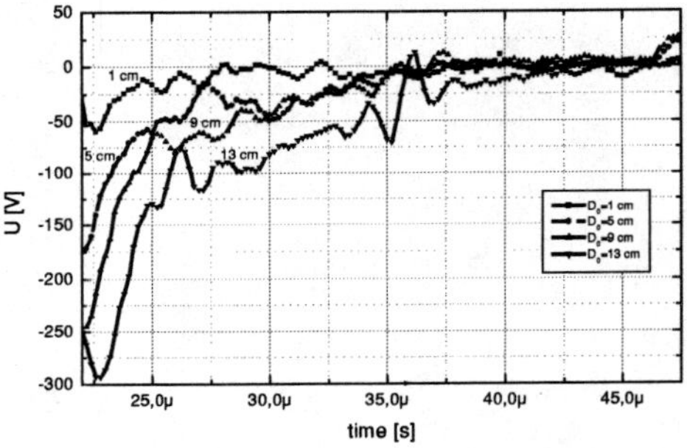

FIGURE 4. Measured probe signals

Calculated Probe signals calculated using (13) and probe calibration curves are given in Fig. 3. They look very similar to those obtained in the measurements. Therefore, one can conclude that the model described above can be used for target plasma formation and expansion analysis.

REFERENCES

1. V. Engelko, B. Yatsenko, G. Mueller, H. Bluhm, Pulsed electron beam facility (GESA) for surface treatment of materials, Vacuum, Vol. 62/2-3, June 2001, pp. 211-216.

Precision of Intense Electron Beams Transported in Vacuum by Applied Magnetic Field

Georg Mueller[a], Vladimir Engelko[b], Hansjoachim. Bluhm[a], Vladimir. An[a], Vladimir Kuznetsov[b], Galya Viazmenova[b]

a Forschungszentrum Karlsruhe, Institut für Hochleistungsimpuls- und Mikrowellentechnik, Postfach 3640, D-76021 Karlsruhe, Germany
b Efremov Institute of Electrophysical Apparatus, 189631, St. Petersburg, Russia

Abstract. Measurements of the spatial distribution of the beam current density in the GESA-facilities [1] shows that the current density oscillates in time although the total beam current is stable. To investigate the nature of these oscillations results from different diagnostics for current and power density measurements have been analyzed (Faraday cups, small metal plates, x-ray detectors, etc.). It was found that oscillations of the beam current density are due to spatial movement (rotation) of the whole beam. The amplitude and appearance of this beam precision strongly depends on magnitude and distribution of the applied magnetic field and also on the target material. The onset of oscillations occurred several μs after the beginning of the pulse. The time of oscillations onset depends on the target material. The period of the oscillation is practically independent of the magnetic field magnitude and distribution and target material, and is always equal to about 1.5 μs. In this paper we discuss the results of current density and x-ray measurements and the possible reasons for the appearance of beam precision during the pulse.

DIAGNOSTICS AND RESULTS

The nature of current density oscillations is studied simply by current measurements on metal plates moved under the beam above the collector. The position of the metal plates to each other can be chosen ad lib. A sketch of the measurement process is given in Fig. 1.

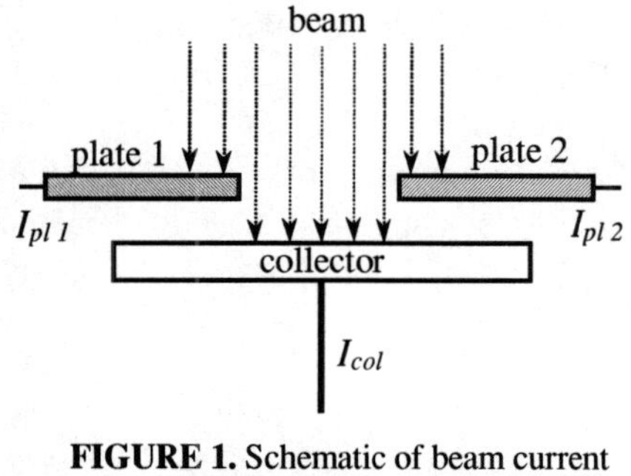

FIGURE 1. Schematic of beam current measurement with two moveable metallic plates.

The second diagnostics used here is a x-ray diagnostics imaging the x-ray emission from a thin metallic target disk with the help of a scintillator and a streak camera, Fig 2. It is possible to detect simultaneously two radial slit areas perpendicular to each other.

CP650, *Beams 2002: 14th International Conference on High-Power Particle Beams,* edited by T. A. Mehlhorn and M. A. Sweeney

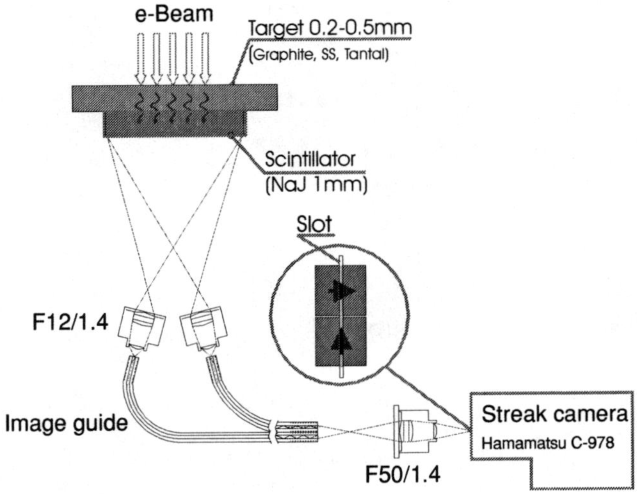

FIGURE 2. x-ray diagnostic to measure the beam current oscillations.

Typical oscillograms of the plate currents are presented in Fig. 3. Oscillations of the current on a stainless steel collectors appear about 5-7 μs after the beginning of the pulse. No oscillations are observed if a large plate is used, capable to cover the whole beam. The results of multi fold measurements demonstrates that plate current oscillations are in anti-phase at any rotation angle and inter-plate distance. Frequency of current oscillations of the both plates at any rotation angle and inter-plate distance is practically the same and corresponds to T ~ 1.5 μs period of oscillations. The period of oscillations do not depend on magnetic field magnitude and distribution. Close to the end of the pulse the oscillations are random.

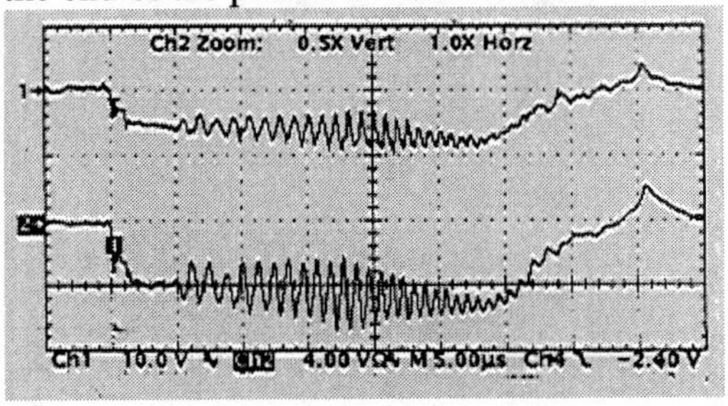

FIGURE 3. Current oscillograms. 1 - the 1-st plate, 2 - 2-nd plate. B_c = 47 Gs, ν= 12.5.

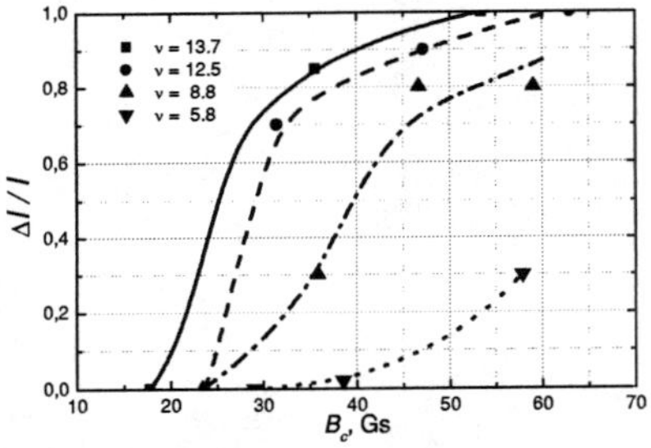

FIGURE 4. Beam current oscillations versus B_c for different $\nu = B_t/B_c$.

Fig. 4 presents relative current oscillations versus magnetic field magnitude at the cathode B_c as a function of the mirror ratio $\nu = B_t/B_c$. The dependence was obtained by treatment of the current oscillograms. Here B_t is the magnetic field magnitude at the target. Amplitude of the oscillations on the metal plate is increased with B_c. At mirror ratios of 8.8, 12.5, 13.7, onset of oscillations occurs only at $B_c > 20$ Gs; and at a mirror ratio of 5.8, at $B_c > 40$ Gs. At large mirror ratios (12.5 and 13.7) the amplitude of oscillations is practically the same. Maximum beam current oscillations on the plate are observed at $B_c \approx 30$ Gs, $B_t = 0$ (at $B_c = 0$, $B_t = 0$ no oscillations are observed.

Oscillations of the beam current density can be connected with spatial movement of the beam. Fig. 5 shows streak images from simultaneously detection of two radial slit areas perpendicular to each other. One can see the time shift between the oscillations which corresponds to ¼ of the time period T, achieved from current measurement (T= 1.5 μs).

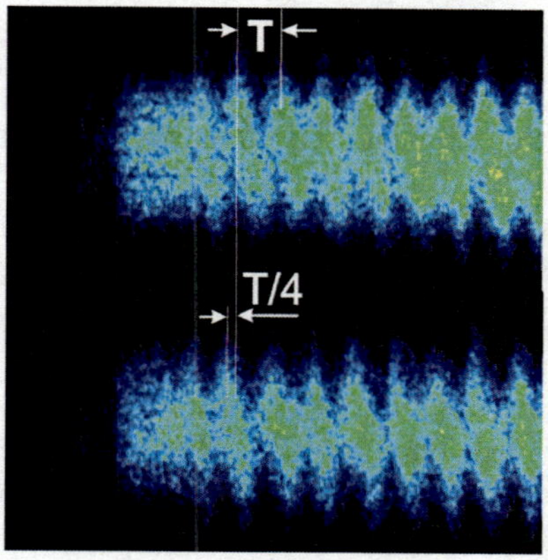

FIGURE 5. Streak images from simultaneously detection of two radial slit areas perpendicular to each other

Further investigations show that the appearance and amplitude of current density oscillations depend not only on the configuration and magnitude of magnetic field but is also strongly influenced by the target material itself. In Fig. 6 x-ray images for graphite, stainless steel and tantalum are shown.

It can clearly be seen, that the onset of oscillations is decreased with the charge number of target material. It starts after about 2-3 µs for Ta, 5-7 µs for SS and about 20 µs for C. Furthermore the amplitude of oscillation increases significantly from C to Ta.

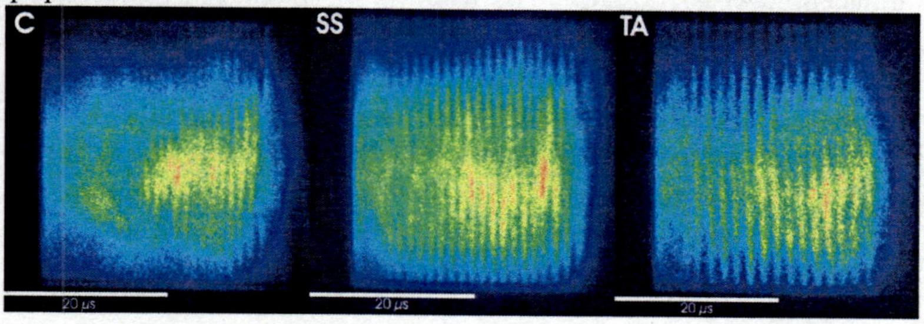

FIGURE 6. x-ray streak images for different target materials (C, SS, Ta), B_c=28 Gs, $\nu = 16.4$.

DISCUSSION

From the results of the experiment, the conclusion can be drawn that the beam oscillates relative to some center. This is illustrated in Fig. 7.

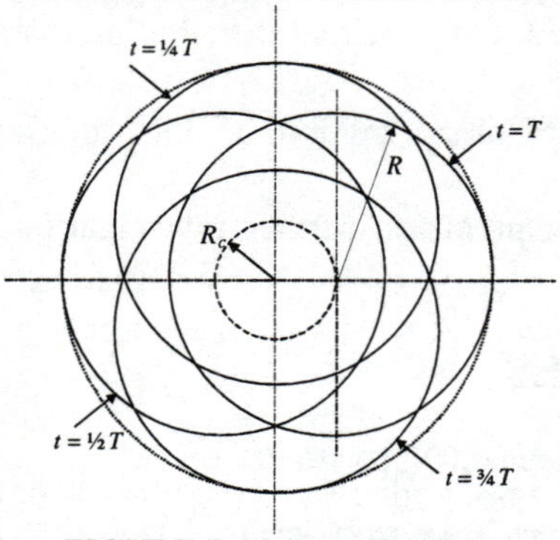

FIGURE 7. Illustration of beam precision

From the oscillograms and x-ray images it is possible to determine the radius of the rotation of beam center axis R_c . For a stainless steel target it is equal to ~ 1/3 of the beam radius , i.e. at the beam radius $R \approx 3.6$ cm, $R_c \approx 1.2$ cm. R_c is somewhat smaller for C and somewhat larger for Ta. The most probable cause of the beam rotation is non-coincidence of the beam center axis with the magnetic field axis and presence of radial electric field that must result in an azimuth beam drift. The drift velocity $V_d=2\pi R_c/T \cong 5{\cdot}10^6$ cm/s corresponds to rotation radius R_c = 1.2 cm and rotation period T = 1.5 µs .With longitudinal magnetic field and radial electric field, the azimuth drift velocity is found from the ratio $V_d=E/B$.

At magnetic strength of 600 Gs and $V_d = 4 \cdot 10^6$ cm/s we have $E \cong 25$ V/cm. Under the assumption of uniform distribution of space charge over the beam cross section, this corresponds to the beam potential well of ~ 40V. Such a low value, sufficient to explain the observed velocity of beam rotation about the magnetic field axis, makes rather plausible the above explanation of the cause of the beam center axis rotation. However, to make finally sure, more accurate simultaneous measurements of current oscillations and beam potential should be carried out at different magnetic field values.

Oscillations of the current density of an electron beam extracted from a plasma source, qualitatively similar to ours, were observed in [2]. Current oscillation frequency was several MHz, which is close to that observed in our experiments. In this paper, the oscillations are attributed to the existence of a potential well in the transient layer of penetrating plasma and oscillations of the cathode plasma ions in this well. By solving relevant dispersion equations, one obtains the following discrete set of possible frequencies $f= V_{0i}/\Delta L \cdot m \cdot \sqrt{5}/2$; m=1,2,3,... where V_{0i} is the velocity of ions defined by the potential well depth, ΔL is the width of the potential well.

From data of the work [3], the depth of the potential well of the penetrating plasma transient layer is 100-200 V, its width is ~ 1-2 cm. For this case the formula gives maximum value of $f \approx 10$ MHz that corresponds to the period of oscillations of 0.1 μs. This value exceeds that observed in the experiment. However, it takes into account that this formula serves only for estimation. Depth and width of the potential well were defined rather approximately and for conditions different from ours. In [2] the oscillation frequency was shown to be inversely proportional to the extracting gap. This fact can be used to clarify the question whether this mechanism can be responsible for the current density oscillation in our case.

CONCLUSION

The analysis of performed experiments made it possible to identify two probable factors responsible for the observed oscillations.

- Rotation of the beam center axis about the magnetic field axis. In this case no oscillations of the beam total current are observed. Only current distribution over the collector is changed in time.
- Double stream instability appearing under oscillatory motion of ions in the potential well of the penetrating plasma of emitter.

To clarify the reason for the appearance of beam precision further investigations are necessary.

REFERENCES

1. V. Engelko, G. Mueller, H. Bluhm, Vacuum, Vol 62/2-3, June 2001, pp. 97-103.
2. M.V. Nezlin. Beam dynamics in plasma. M. Energoizdat, 1982, P.20.
3. P.E. Belensov, A.T.Kapin, A.A.Pljutto, V.N. Ryzhov. ZHTF, 1964, №12, p.2120-2128.

Transportation of Intense Microsecond Electron Beams in Vacuum and their Interaction with Matter

Vladimir Engelko[a], Georg Mueller[b], Hansjoachim Bluhm[b]

[a] *Efremov Institute of Electrophysical Apparatus, 196641, St. Petersburg, Russia*
[b] *Forschungszentrum Karlsruhe, Institut für Hochleistungsimpuls- und Mikrowellentechnik, Postfach 3640, D-76021 Karlsruhe, Germany*

Abstract. Application of pulsed intense electron beams for material surface modification to improve wear, corrosion and oxidation resistance, fatigue strength etc. is an attractive technology with promising industrial perspectives. To get stable, reproducible results of the treatment it is necessary to investigate the formation and transportation of pulsed intense electron beams and their interaction with matter. A short review is presented of such investigations performed at the GESA-1 and GESA-2 facilities, producing the beams with electron kinetic energy 50-400 keV, current 200-500 A, pulse duration 5-250 µs, maximum power and energy density at the target of 6 MW/cm2 and 500 J/cm2, respectively.

INTRODUCTION

When an intense pulsed electron beam interacts with matter the following processes take place.

- A part of electrons are reflected back to the source.
- Heating the target results in melting and evaporation of its material and to gas desorption from its surface.
- Hydrodynamic processes in the melted layer take place which are responsible for material mixing and surface relief formation after irradiation.
- Ionisation of desorbed gases and vapours of the target material leads to formation of the target plasma and ion fluxes.

The application of pulsed intense electron beams for material surface modification to improve wear, corrosion and oxidation resistance, fatigue strength etc. is an attractive technology with promising industrial perspectives. For promotion of this technology it is necessary to investigate the above mentioned processes in order to clarify their influence on the beam characteristics the results and of material treatment.

In the report, we present a short review of such investigations performed at the GESA-1 and GESA-2 facilities [1], which produce beams with a electron kinetic energy 50-400 keV, current 200-500 A, pulse duration 5-250 µs and maximum power and energy density at the target of 6 MW/cm2 and 500 J/cm2 respectively.

CP650, *Beams 2002: 14th International Conference on High-Power Particle Beams*, edited by T. A. Mehlhorn and M. A. Sweeney

EXPERIMENTAL

The scheme of the facility is shown in Fig.1.

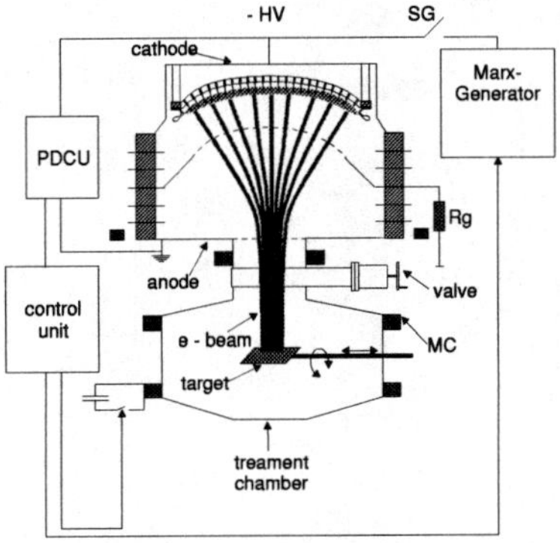

FIGURE 1. Scheme of the GESA facility

The beam is formed in an electron source of triode type with multipoint explosive emission cathode. Then it is transported through the transport channel to the treatment chamber. An external magnetic field is used for beam formation and transportation. In the course of experiments, the following values were measured: accelerating voltage U, beam current I, control grid current I_g and potential U_g, beam current density distribution j(r), intensity of x-ray radiation from the target and its temporal and spatial distribution, dynamic of the beam potential in the drift channel, and so on. Theoretical and numerical investigations related to the mentioned topics were also conducted.

RESULTS

Effects related to electrons back-reflected from a target for the conditions of GESA facilities were considered in [2,3]. Main results are the following. Reflected electrons appreciably decrease the limiting emission current density of the cathode. This decrease is higher in the triode scheme of the electron source. The calculated and measured dependence of the beam current density on the charge number of the target material is presented as an example in Fig. 2. Reflected electrons also change the potential distribution in the grid-anode gap. The consequence of this is a reduction of the beam current value for which a potential well and a virtual cathode are formed in this gap.

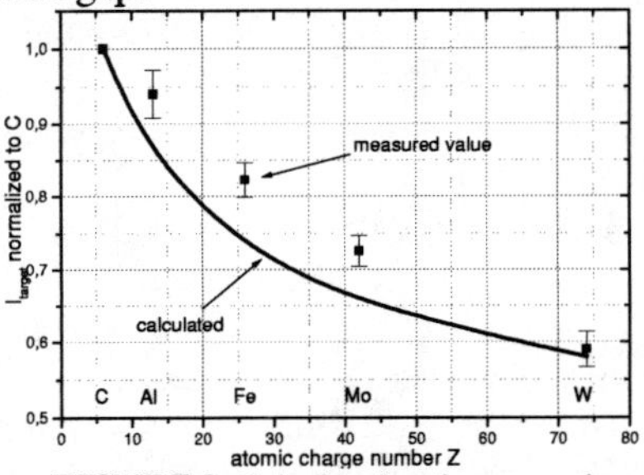

FIGURE 2. Calculated and measured dependence of the beam current density on the charge number of the target material

Reflected electrons do not only change the energy delivered by the electron beam to the target, as described above, but also the distribution of this energy inside the target [4]. The penetration depth of the electrons in matter is energy dependent. It decreases with the energy of the electrons. Therefore, the low energy electrons deposit their energy in the near surface region of the target. This effect increases with the atomic number of the target material. In Fig. 3,4 examples for the distribution of the beam energy inside a matter with and without reflected electrons are given.

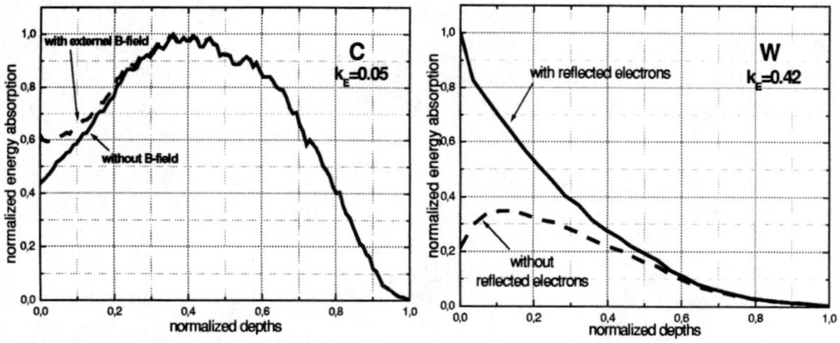

FIGURE 3. Energy absorption profile for carbon and tungsten, normalized to maximum values

Melting and evaporation of the target material was investigated numerically and experimentally. For typical GESA facility conditions the rate of metal evaporation is in the order of 0.1 m/s or 10mg/kJ, the depth of the melted layer – 10-100μm.

As a result of hydrodynamic processes in the melted layer the mass transfer from the surface into the layer takes place. This effect can be used for surface alloy formation. In this case a thin film of any material is deposited on the surface of a base material and irradiated by an electron beam with an energy density sufficient to melt film and base material together within some tens of microseconds. Under such conditions penetration of the deposited material into the whole depth of the melted layer is observed. Concentration of implanted material achieves tens of percents [5]. In Fig. 4 ,as an example, the calculated and measured depths of the layer alloyed with Al are shown. The mechanism of the mass transfer inside the melted layer is not clear up to now. It is possible, that a gradient of a surface tension plays an important role here.

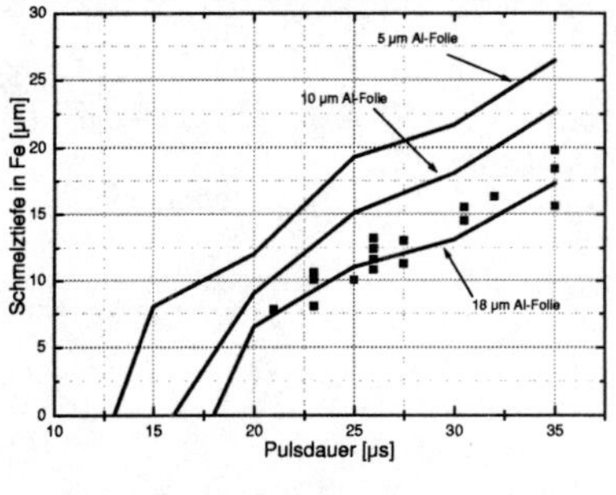

FIGURE 4. Dependence of the depth of Al alloyed layer on beam pulse duration. ■ measured, - calculated

Under the influence of an intense electron beam at the surface of a target first an ion flux and than a plasma is formed. The formation of the ion flux and plasma comprises the following steps: electron beam stimulated desorbtion of a gas from the target surface; ionization of the desorbed atoms; formation of the ion flow; formation of the target plasma. The ion flux from the target with saturated current density is formed within few microseconds. It decreases the beam potential by a factor of 2. The expanding target plasma neutralizes the beam potential and influences the electron source operation [5].

Multifold measurements of the radial distribution of the intense electron beam current density demonstrate that it commonly has a maximum near the beam axis. The reason for this can be related to the influence of target ions on the distribution of the emission current density of the cathode operating in space charge limited mode [2]. The target ions are accelerated towards the cathode by the electric field resulting from the electron beam space charge. At the region of electron beam formation these ions disturb the electric field distribution. Since ions are not magnetized their trajectories do not coincide with the electron trajectories. That is why the distortion of the electric field takes place mainly near the central part of the source. This leads to a change of the electron beam current density distribution. Fig.5 demonstrates results of calculations and measurements certifying this conclusion. To achieve more stable and homogeneous beams, one has to prevent the ion flux from reaching the electron source region.

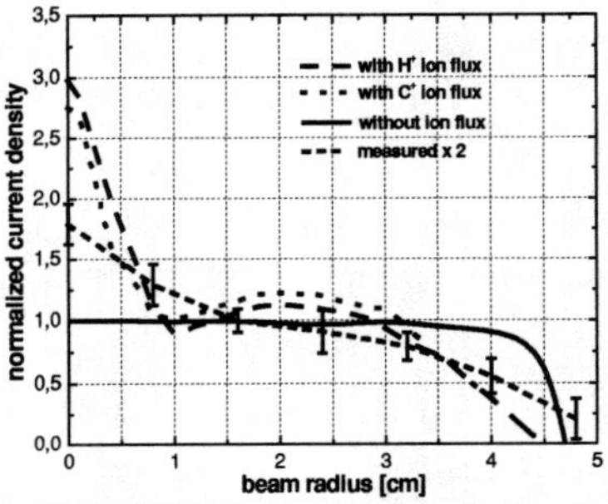

FIGURE 5. Measured and calculated electron beam current density normalized to the calculated maximum value without ion flux from the target.

Measurement of the spatial distribution of the beam current density with the help of different current probes and investigation of the temporal and spatial resolved x-ray image of the beam at the target showed that under certain conditions the precision of the beam takes place (rotation of the beam axis around the axis of transport channel) [6]. The amplitude of the precision increases with increasing magnetic field at the cathode B_c. Maximum deviation of the beam axis is equal to 1 cm. The time of the precision onset reduces with higher B_c. The onset of oscillations mainly occurs 5-7 μs after pulse beginning. When the target is made of graphite the precision is absent. Last circumstance shows that the reason of the beam precision can be in the beam interaction with target plasma or ion flux.

REFERENCES

1. V. Engelko, B. Yatsenko, G. Mueller, H. Bluhm, Pulsed electron beam facility (GESA) for surface treatment of materials, Vacuum, Vol. 62/2-3, June 2001, pp. 211-216.
2. V. Engelko, G. Mueller, H. Bluhm, Influence of particle fluxes from target on characteristics of intense electron beams, Vacuum, Vol 62/2-3, June 2001, pp. 97-103.
3. V. Engelko, V. Kusnetsov, G. Viazmenova, G. Mueller and H. Bluhm, Influence of electrons reflected from a target on the operation of triode-type electron sources, J. Appl. Phys., Vol. 88 No.7 1 October 2000, pp. 3879 – 3888.
4. V. Engelko, G. Mueller, G. Schumacher, D. Strauss, F. Zimmermann, H. Bluhm, Improvement of the corrosion resistance of steels and MCrAlY coatings by application of a pulsed electron beam, 13th Int. Conf. on High-Power Particle Beams, June 25-30, 2000, Nagaoka, Japan, Vol.1 pp. 216-221
5. V. Engelko, G. Mueller, Plasma dynamic on a target irradiated by intense electron beams, this conference
6. V. Engelko, G. Mueller, Precision of intense electron beams transported in vacuum by applied magnetic field, this conference

Generation of Ozone by Ns-width Pulsed Power

Naoyuki Shimomura[a], Masaya Wakimoto[a], Yosuke Shinke[a], Masayoshi Nagata[a], Takao Namihira[b], Hidenori Akiyama[b]

[a]*The University of Tokushima, 2-1 Minimi-Josanjima, Tokushima, 770-8506 Japan*
[b]*Kumamoto University, 2-39-1 Kurokami, Kumamoto, 860-8555 Japan*

Abstract. The demand of ozone will be increasing for wholesome and environment-conscious sterilizations. The generation of ozone using the pulsed power discharge will apply electron accelerations around the head of streamer discharge principally. The breakdown in reactor often limits the efficient generation. Therefore, the pulse shape should be controlled for dimension of the reactor. It is clear that a pulse shortening is one of effective approaches. Pulsed power voltage with ns-width applies for ozone generation. The effects, on concentration and efficiency of generation, of pulse shape, repetition rate of pulse, flow rate of oxygen gas, and dimension and configuration of reactor, are discussed. The dimension and configuration of the reactor are optimized for the pulse width.

INTRODUCTION

Generation of ozone is one of the most typical industrial and commercial applications of electrical discharge. The demand of ozone will be increasing for wholesome and environment-conscious sterilizations. With an expectation of higher efficiency of generation and higher concentration, the generation of ozone by pulsed power discharges has been studied[1]-[8].

The generation of ozone using the pulsed power discharge will apply electron accelerations around the head of streamer discharge principally. The streamer by the pulsed power voltage can sweep large volume and the large and efficient reactor is expected. However, the breakdown in reactor often limits the efficient generation because the streamer would traverse across the electrodes and breakdown occurs. Therefore, the pulse shape should be controlled for dimension of the reactor. It is clear that a pulse shortening is one of effective approaches and the higher field for over-voltage can be obtained.

Here, pulsed power voltage with ns-width applies for ozone generation. The short pulse is generated from a combination of a short pulse forming line and a short gas gap with high pressure, and supplies to the reactor with a wire-to-cylinder configuration. The effects, on concentration and efficiency of generation, of pulse shape, repetition rate of pulse, flow rate of oxygen gas, and dimension and configuration of reactor, are

CP650, *Beams 2002: 14th International Conference on High-Power Particle Beams*,
edited by T. A. Mehlhorn and M. A. Sweeney

discussed. The dimension and configuration of the reactor are optimized for the pulse width.

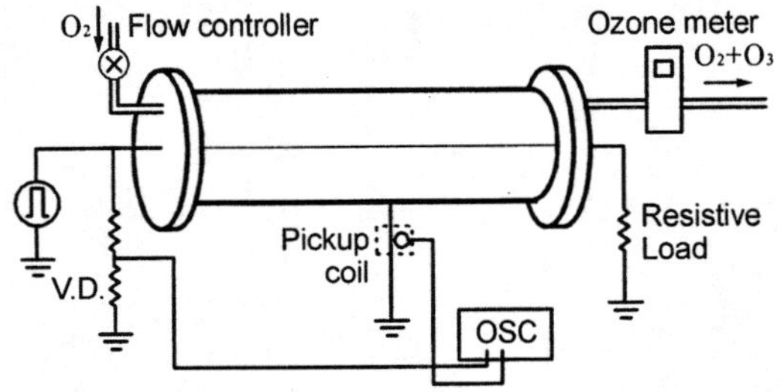

FIGURE 1. The experimental configuration.

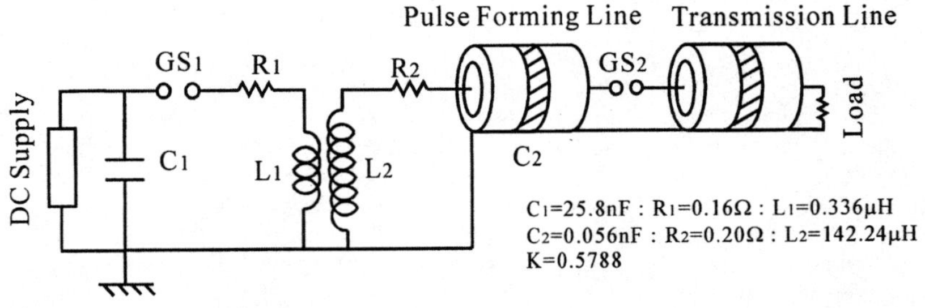

FIGURE 2. The equivalent circuit of the pulsed power generator.

EXPERIMENTAL CONFIGURATION

The experimental configuration is shown in Figure 1. The pulsed power is applied on the inner electrode of a coaxial reactor. The reactor consists of the copper cylindrical outer electrode of 58 mm or 108 mm in diameter and of 500 mm in length and the inner electrode that is selected from several figurations. The shape of the inner diameter is fundamentally a simple copper wire (wire reactor) or a copper spiral winding on polyvinyl chloride (PVC) cylinder (spiral reactor). Oxygen gas regulated by a flow controller axially flows in the coaxial reactor and the density of ozone in oxygen is measured with an ozone meter at an end of the reactor. The applied voltage is observed by a resistive voltage divider and a pickup coil observes the reactor current. In some condition, a resistance is connected to the electrodes at the end of reactor for the prevention of pulse reflection.

Figure 2 represents the equivalent circuit of pulsed power generator. The generator consists of a capacitor bank, a pulse transformer, a considerable short pulse forming line and a gas gap switch with a short separation. The capacitor bank (25.8 nF) is charged by DC power supply and a gap switch self-breakdowns. The pulse transformer (1:24) and the capacitors comport with a tesla transformer and the pulse forming line, whose capacitance is evaluated to be 56 pF, is multiplicatively charged. The gas gap switch with a short separation applies to fast rise of voltage so that the output voltage pulse has sub nanosecond rise and pulse width of ns. The typical waveform of output voltage pulse for 81 Ω load is shown in Figure 3. In this condition, the peak voltage is 130 kV and the pulse width (FWHM) is approximately 1 ns. The DC power supply is controlled by a personal computer and 100 pps of the repetitive ratio of the generator is possible at present.

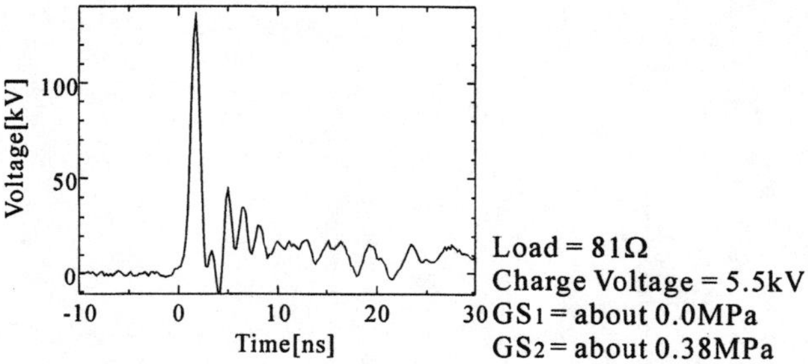

FIGURE 3. The typical waveform of output voltage pulse for 81 Ω load.

EXPERIMENTAL RESULTS AND DISSCUSSION

In the case of the spiral reactor, the dependences of the ozone density and the efficiency of ozone production on the flow rate of oxygen and the repetition rate of the pulse are shown in Figure 4. The efficiency is based on the initial energy of the capacitor bank of the generator. The density increases as the flow decreases and as the repetition rate increases. On the other hand, the efficiency decreases in the same case. The repetitive application of streamer on the same volume increases the concentration of ozone, while the application of streamer for mixture imply depression of production of ozone. Figure 5 is obtained by normalization on deposited energy of the data in Figures 4. The trade off between density and efficiency is confirmed in this figure. In this experiment, the reactor is terminated by a resistive load of 500 Ω owing to the prevention of breakdown. The matching between the generator and the reactor is not good, and the reduction of impedance in reactor is desirable.

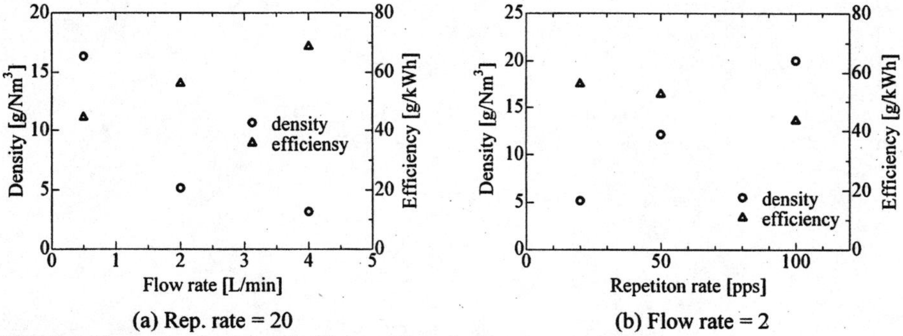

FIGURE 4 The ozone density and efficiency for spiral reactor.

Figure 6 shows the effect of the winding rate of the inner spiral electrode. The density decreases as the winding rate increases. It seems that the increase of impedance of reactor by increase of the winding rate brings the reflection of pulse at connection of the reactor. In this figure, the condition of winding-rate = 0 represents the wire coaxial reactor. While the impedance is lower, the effective reactive volume should decrease owing to the shorter wire. The maximum density of 20.0 g/Nm^3 and the system efficiency of 73.84 g/kWh is obtained respectively at present.

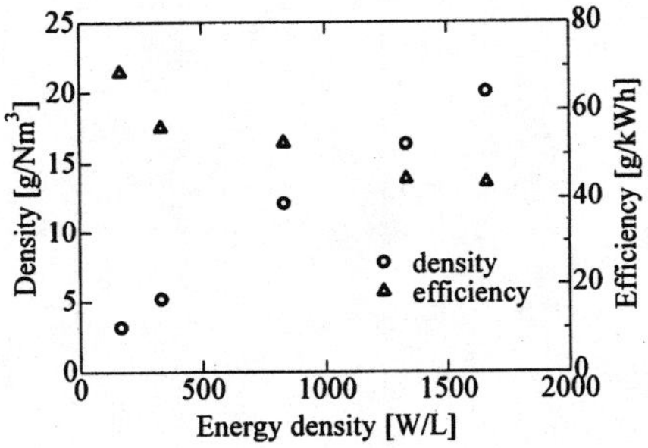

FIGURE 5 Normalization of Figures 4.

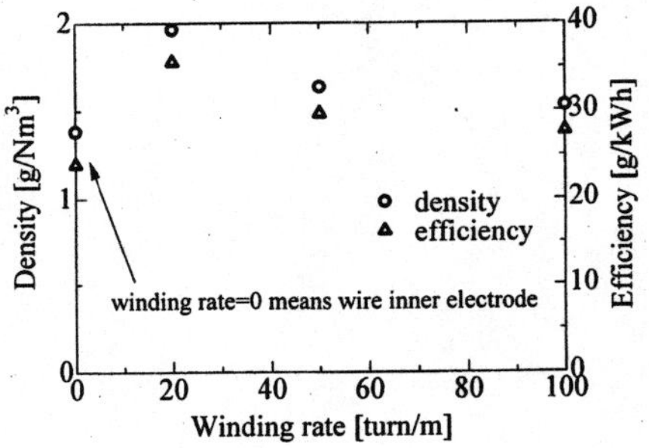

FIGURE 6 Effect of winding rate.

SUMMARY

Pulsed power voltage with ns-width applied to ozone generation. The effects, on concentration and efficiency of generation, of pulse shape, repetition rate of pulse, flow rate of oxygen gas, and dimension and configuration of reactor, were discussed. While the pulse shortening and sharpening require the modification of reactor, the improvement of efficiency is expected for applying overvoltage and high energy density.

REFERENCES

1. Hegeler F., and Akiyama H., *Japanese Journal of Applied Physics* **36**, 8, 5335-5339 (1997)
2. Hegeler F., and Akiyama H., *IEEE Transactions on Plasma Science* **25**, 5, 1158-1165 (1997)
3. Akiyama H., Miyahara Y., Samaranayake J., and Katsuki S., *Journal of Plasma and Fusion Research* **74**, 10, 1139-1143 (1998)
4. Samaranayake W.J.M., Miyahara Y., Namihira T., Katsuki S., Sakugawa T., Hackam R., and Akiyama H., *IEEE Transactions on Dielectrics and Electrical Insulation* **7**, 2, 254-260 (2000)
5. Samaranayake W.J.M., Miyahara Y., Namihira T., Katsuki S., Hackam R., and Akiyama H., *IEEE Transactions on Dielectrics and Electrical Insulation* **7**, 6, 849-854 (2000)
6. Samaranayake W.J.M., Miyahara Y., Namihira T., Katsuki S., Sakugawa T., Hackam R., and Akiyama H., *IEEE Electrical Insulation Magazine* **17**, 4, 17-25 (2001)
7. Samaranayake W.J.M., Miyahara Y., Namihira T., Katsuki S., Hackam R., and Akiyama H., *IEEE Transactions on Dielectrics and Electrical Insulation* **8**, 4, 687-697 (2001)
8. Samaranayake W.J.M., Miyahara Y., Namihira T., Katsuki S., Hackam R., and Akiyama H., *IEEE Transactions on Dielectrics and Electrical Insulation* **8**, 5, 826-831 (2001)

APPLICATION OF CONCENTRATED ELECTRON BEAMS IN EXTRA VACUUM TECHNOLOGIES

GORSHKOV O.A., ILYIN A.A., LOVTSOV A.S., RIZAKHANOV R.N.

Keldysh Research Center.
125438 Russia, Moscow, Onezhskaya street 8, Keldysh Research Center.
Phone: 4566465 Fax: 4568228

Abstract

Results of first experimental investigation in the area of nitric oxides elimination from gas flow by application of concentrated e-beam are represented in this paper. The application of concentrated e-beams instead of distributed ones (injected through foil windows) for gas flows purification is caused by the possibility of development of high unit power accelerator (up to 1 MW) with relatively low accelerating voltage (200-350 kV). The M-2 facility parameters: electron energy – 100 keV, beam power – up to 2 kW – allows us to treat up to 0,05 m^3/s of gas with radiation doses up to 100 kGy.

INTRODUCTION

Recently a raise in the area of concentrated electron beams technology, particularly in the area of the beam extraction into gas of the atmosphere pressure is observed. In addition to the traditional vacuum technologies linked with the surface modification by concentrated energy fluxes (welding, cutting, surface metal hardening), interest toward the plasma chemistry problems is also increased. These processes apply the electron beams for initiation of plasma chemistry reaction.

One of such technologies is gas flow purification from toxic impurities (nitric oxide, sulfur oxide). The electron beam purification of impurities (so-called Ebara-process) is based upon the bounding of the toxic substances stimulated by radiation. The disadvantage of such a method is high level of consumed power during the process and difficulties at the development of the foil ejection facility with the power of 1MW.

It is possible to solve these problems by application of devices using the ejection of the concentrated electron beams into the atmosphere. In this case the electron beam is transported through the sequence of independent lock chambers pumped separately. The disadvantage of the concentrated electron beam is temperature, plasma concentration and dimensional non-uniformity of radiation fields. In this paper the facility for investigation of e-beam gas purification is described. The results of their first investigation are also presented.

A number of e-beam facilities with concentrated beam extraction into the dense gaseous medium (in particular, in the air atmosphere) were developed in Keldysh Research Center. The comparative technical characteristics of these facilities are represented in table 1. All these facilities have the practically similar principal scheme.

CP650, *Beams 2002: 14th International Conference on High-Power Particle Beams,*
edited by T. A. Mehlhorn and M. A. Sweeney

TABLE 1.

Facility	Date of development	Beam power, kW	Accelerating voltage, kV	Possible areas of technological application	Reference
«Onega»	1980	1000	280	gas purification, hardening, investigation, industrial application	[1,2]
M-1	1988	35	100	hardening, welding, industrial application	[3,4]
TUR-3	1993	80	200	gas purification, pilot facility	[5]
M-2	2001	2	100	gas purification, investigation	

1. THE FACILITY FOR E-BEAM PURIFICATION OF GAS FLOWS

In order to investigate the process of electron beam purification (EBP) of exhaust gases of the gaseous toxic impurities, the facility on the basis of the M-2 equipment was developed. The main parameters of the facility: voltage - 80...100 keV, current- 0...20 mA, gas productivity - 0...0,05 m^3/s, initial gas temperature - 20…150 ^oC, radiation dose – up to 100 kGy. The scheme of the facility is presented in fig.1.

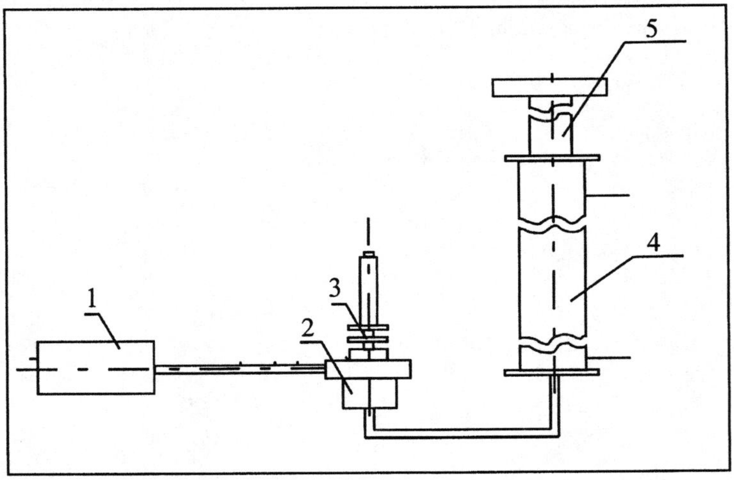

FIG.1. The scheme of EBP facility.

Let us consider main elements of the facility. Gas-heating and preparation unit (1) is designed to simulate gas fed from chimney. Here gas is heated up to the temperature required, and also the studied toxic impurities and necessary reagents are put in it. Radiochemical reactor (2) is a stainless steel reservoir, in which the gas treatment with e-beam plasma occurs. Beam plasma generator (3) is a device, which is designed to form and accelerate the concentrated e-beam with subsequent beam extraction into dense gas through the differential pumping system. It was developed on the basis of updated facility for extraction of continuous concentrated e-beam into gas with the atmospheric pressure [4]. Gas withdrawal system (4) is a heat-insulated pipe with the diameter of 400 mm and length of 4 m. The major task of this unit – gas confinement in the system till the completion of the main chemical reactions. The proper simulation of the processes carried out in the real chimney, requires the constant temperature within the whole system. For that all gas withdrawal system is equipped with isolation. Ventilating system (5) is designed for gas pumping through the gas path. The main measured parameters in this system are: gas flow rate, gas temperature, e-beam parameters, reagents' consumption, as well as gas composition before and after the treatment. Gas flow rate is measured by means of Venturi tube. Measurements of the static pressure are realised in the atmosphere and in the cylindrical part of Venturi tube. Conclusion concerning gas flow rate is drawn proceeding from the pressure difference between them. Flow rate measurement error does not exceed 5%. Gas temperature is measured in several points: ambient temperature, in the radiochemical reactor inlet and at its outlet. The measurements are conducted be means of semiconductor temperature sensors. The e-beam parameters (accelerating voltage and current) are measured by means of the facility measuring instruments; gas composition and humidity are determined via TESTO-350 gas-analyzer. This gas analyzer allows to measure the concentration of the following substances: NO – 0...3000 ppm; NO_2 – 0…500 ppm; SO_2 – 0...5000 ppm, as well as gas temperature and humidity. Reagents mass flow was obtained by the measurement of gas pressure and temperature before the washer of mass flow controller. The radiation doze was calculated proceeding from the difference of gas flow temperature before and after the reactor.

2. Results of the Investigations

The facility has passed the preliminary tests on gas flow purification of nitric oxides. The preparation of toxic impurities was realized through the thermal decomposition of nitric tetraxide into NO and NO_2. The mixture obtained consisted of NO and NO_2 in 1:1 proportion. Water vapor was also introduced into the heater unit. Water vapors presence is very important factor, which impacts on efficiency on nitric oxide removal because water molecules stroke by fast electron produce most chemically active radicals OH and H. Besides, water is necessary for resolution of produced higher nitric oxides NO_2. Acid can be effectively neutralized by different reagents (like NH_3, NH_4OH, or KOH) with generation of easily removing solid salts, which can be used as mineral fertilizer.

Initial NO and NO_2 concentrations were $4{,}3 \cdot 10^{14}$ cm^{-3}. The sense of the electron radiation is in NO conversion (which extremely hard dissolves in water) into NO_2 (which dissolves well). The direct fast electrons influence on NO_2 is weak. The results of conducted studies are presented in table 2.

TABLE 2.

# of experiment	Radiation dose, kGy	Concentration of additional water vapor, cm^{-3}	NO concentration, cm^{-3}	NO_2, concentration, cm^{-3}	Degree of purification %
Initial gas	0	0	$4{,}3 \cdot 10^{14}$	$4{,}3 \cdot 10^{14}$	–
1	10	0	Δ	$3{,}8 \cdot 10^{14}$	56
2	3,0	1,4	Δ	$2{,}9 \cdot 10^{14}$	66
3	10	1,4	Δ	$2 \cdot 10^{14}$	77

The Δ symbol here denotes that the concentration value is lower the threshold of sensor sensitivity, which is equal 10^{13} cm^{-3}.

From the data obtained follows that at radiation dose of 10 kGy the transformation of NO into NO_2 occurs, at that, the former partially dissolved in residual water vapors (experiment #1). At purposeful addition of water into the gas flow the NO_2 concentration considerably decreases (experiment #3). It is possible to obtain good purification effectiveness (66% - experiment #2) at relatively low power contribution (3,0 kGy). From the economical point of view, the mode #2 is more effective than mode #3, although the purification effectiveness is higher in the mode #3 (77%).

On the M-1 facility the investigations of surface hardening of different grades of steel were carried out. As the results show the hardening of a layer 70…320μm thick for grades containing more then 0,4% of carbon occurs. At that, the layer hardness increases in 3...3.5 times. The layer structure is martensite. The results of e-beam surface hardening in detail are stated in [6].

CONCLUSION

A number of facilities with concentrated e-beam ejection in power range from 2 to 1000kW are presented in the paper.

The results of e-beam gas flow purification of the toxic impurities (nitric oxides NO_x) are shown. It is shown that with initial NO_x concentrations at the level of 10^{15} cm^{-3} and radiation doses of 3…10 kGy the purification efficiency runs up to 60...80% depending upon gas humidity.

This work is carried out at the support of the International Scientific-Technical Center (Grant #932).

REFERENCE:

1. V.M. Ievlev, A.S. Koroteev. "Ejection of high-power e-beams into the atmosphere and their investigations."// *Izv. AN SSSR, ser. Energetika i Transport.* 1981. #63. p.3 (In Russian).
2. V.M. Belogrivtsev, A.S. Koroteev, R.N. Rizakhanov, I.I. Shishkanov . A.M. Yartsev. "The application of e-beam technology in systems of power plant smoke gases purification." *Izv. AN SSSR; ser. Energetika i Transport.* 1991. #3. p.26 (In Russian).
3. A.S. Koroteev, V.M. Belogrivtsev, R.N. Rizakhanov, A.M. Yartsev. "Ejection of a concentrated e-beam into dense gaseous media" // *Thesisy dokl. VII Vsesuyzn. Symposium po Silnotochnoy Electronike*, 1988 Tomsk, P.2. pp. 162-163 (In Russian).
4. O.A. Gorshkov, A.A. Ilyin, A.S. Lovtsov, R.N. Rizakhanov. "The Facility for ejection of a concentrated e-beam into gaseous media of atmospheric pressure." // *Pribory i Tehnika Experimenta*, 2002, #2, pp.162 – 163 (In Russian).
5. "The application of e-beams and impulse discharges for purification of smoke gases". *//Materialy seminara. Moskva, IVTAN, oktyabr' 1991.* M. IVTAN, 1991 (In Russian).
6. V.V. Abashkin, O.A. Gorshkov, A.A. Ilyin, A.S. Lovtsov, R.N. Rizakhanov "Extra-vacuum Electron Beam Technologies and Methods of Their Realization." to be published in the Proceedings of EMRS-2002 Spring Meeting *"Progress in plasma Processing of Materials"* Strasbourg, France, June 2002.

Removal of NO_X by Pulsed, Intense Relativistic Electron Beam in Distant Gas Chamber

Go Imada, Tomoki Sakurai, Weihua Jiang, and Kiyoshi Yatsui

Extreme Energy-Density Research Institute, Nagaoka University of Technology, 1603-1 Kamitomioka, Nagaoka, Niigata 940-2188 JAPAN

Abstract. Removal of NO_X has been studied using a pulsed, intense relativistic electron beam (IREB). The dependence of NO_X concentration and the removal efficiency of NO_X on the number of IREB shot have been investigated within a distant gas chamber spatially isolated from the electron beam source. The distant gas chamber is filled up with a dry-air-balanced NO gas mixture with the pressure of 270 kPa, and is irradiated by the IREB (2 MeV, 30 A, 35 ns) passing through a 1.6-m-long atmosphere. With the initial NO concentration of 88 ppm, ~ 70 % of NO_X is successfully removed by firing 10 shots of IREB. The NO_X removal efficiency has been found to be 50-155 g/kWh.

INTRODUCTION

The flue gases produced by the electric power plants and diesel engines have caused serious environmental problems in the earth. The injection of a pulsed, intense relativistic electron beam (IREB) into a flue gas is expected to be effective for pollutant removal. In the removal of NO_X by IREB, NO_X reacts with the radicals. The radicals are formed by the collision of the electrons with the atoms or molecules in the flue gas. The IREB is able to produce a lot of radicals simultaneously due to the high current density. It is also reported that the removal efficiency of NO_X can be improved by utilizing a pulsed electron beam [1]-[3]. In addition, the electrons of IREB have large kinetic energy, resulting in long *range* in the gas. For example, the *range* of IREB at 2 MeV is calculated to be ~ 9 m in an atmosphere [4]. Therefore, the IREB irradiation is believed to be a promising candidate for NO_X removal in the elongate chimney at electric power plants and in the tunnels for automobiles.

We have studied the NO_X removal by IREB of the beam energy of the order of 1 MeV using a pulsed power generator, "ETIGO-III" (8 MV, 5 kA, 50 ns). In this study, the dependence of NO_X concentration and the removal efficiency of NO_X on the number of IREB shot have been investigated within a distant gas chamber spatially isolated from the electron beam source. The gas chamber is filled up with a dry-air-balanced NO gas mixture with the pressure of 270 kPa, and is irradiated by the IREB passing through a 1.6-m-long atmosphere.

CP650, *Beams 2002: 14th International Conference on High-Power Particle Beams*, edited by T. A. Mehlhorn and M. A. Sweeney

EXPERIMENTAL SETUP

Figure 1 shows the schematic diagram of the IREB generator, "ETIGO-III" [5]. It consists of a Marx generator, a pulse forming line, and induction acceleration cells of four stages. The capacitors in Marx generator (450 nF × 34) are charged up to 20 kV with the total stored energy of ~ 3 kJ. The output voltage of the pulse forming line is ~ 670 kV, which is fed to the four acceleration cells through the transmission lines in parallel. Three amorphous-metallic cores are installed in each acceleration cell, producing approximately 2 MV per cell (670 kV × 3). The electron beam with the energy up to 2 MeV is generated by an electron-beam diode placed at the first acceleration cell. The field-emission foilless diode with a hollow cathode is used to generate the electron beam. This electron beam is accelerated by each acceleration cell, and the maximum energy of the IREB increases up to ~ 8 MeV [6].

Figure 2 shows the schematic of the experimental apparatus for NO_x removal by IREB. It consists of an electron-beam diode, drift tube, distant IREB-gas-interaction chamber, external magnetic field, gas mixer, and gas system. The IREB generated by "ETIGO-III" is injected into the distant interaction chamber through not only the two bulkheads but also the 1.6-m-long drift tube filled up with room air. The bulkheads consist of a titanium foil with the thickness of 40 μm and a stainless-steel honeycomb flange with the optical transparency of ~ 50 %. At the IREB energy of 2 MeV, the energy loss of the electron within the titanium foil and the 1.6-m-long atmosphere are calculated to be ~ 26 keV and ~ 330 keV, respectively [4]. The IREB is guided by an external magnetic field (~ 0.5 T) through the drift tube and the distant interaction chamber. The IREB-gas-interaction chamber is made of a stainless-steel pipe with the length of 150 mm. After the interaction chamber is evacuated to ~ 20 Pa by the rotary vacuum pump, it is filled up with a dry-N_2/O_2-balanced NO gas mixture with the total pressure of 270 kPa to simulate the flue gas. The concentrations of NO and NO_2 are measured by a controlled potential-electrolysis gas analyzer after mixing up the gas mixture for 3 minutes.

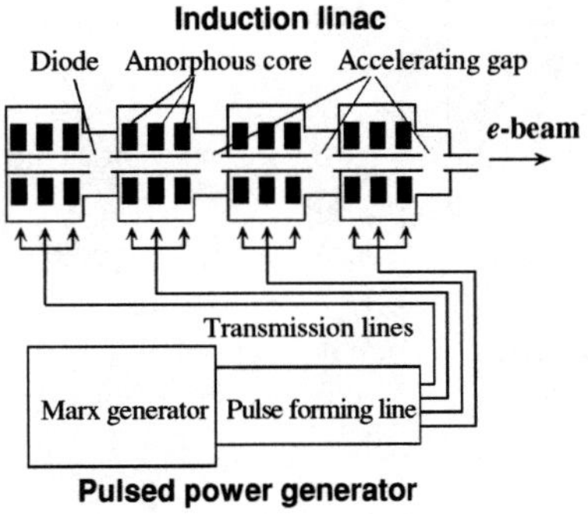

FIGURE 1. Schematic of "ETIGO-III".

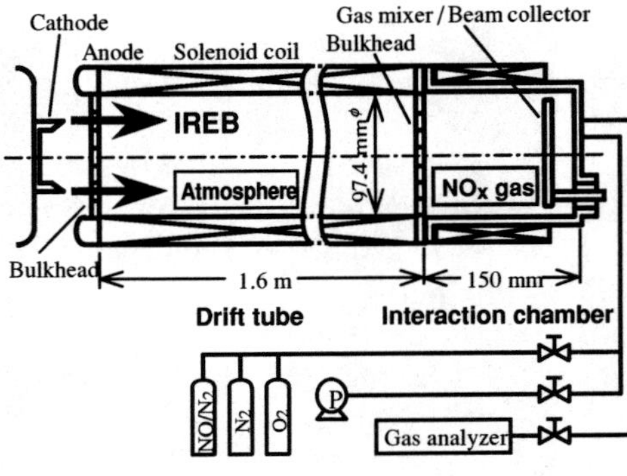

FIGURE 2. Schematic of experimental apparatus.

RESULTS AND DISCUSSION

Figure 3 typically shows the time evolution of the electron beam diode voltage (V_d) and IREB current (I_c), where I_c is measured by a Faraday cup with a pick-up coil placed at the end of the interaction chamber. The peak currents of I_c is observed to be ~ 30 A at V_d ~ 2 MV. The pulse width of I_c is approximately 35 ns (full width at half maximum). The injected IREB energy into the interaction chamber (ε) is estimated to be ~ 4.2×10^{-7} kWh from the time integral of the product of V_d^* and I_c, where V_d^* is the reduced IREB voltage ($V_d^* = V_d - \{26 \times 2 + 330\}$ kV).

Figure 4 shows the NO, NO_2, and NO_x concentration as a function of the number of IREB shot for the initial NO concentration of 88 ppm in the dry-N_2/O_2-balanced NO gas mixture with the composition of $N_2:O_2 = 8:2$, where the NO_x concentration is defined as the sum of NO and NO_2 concentrations. A certain amount of NO is oxidized

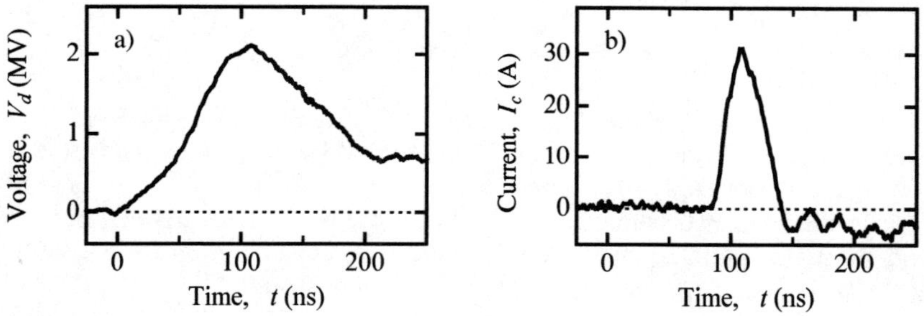

FIGURE 3. Time evolution of a) *e*-beam diode voltage and b) IREB current at end of interaction chamber.

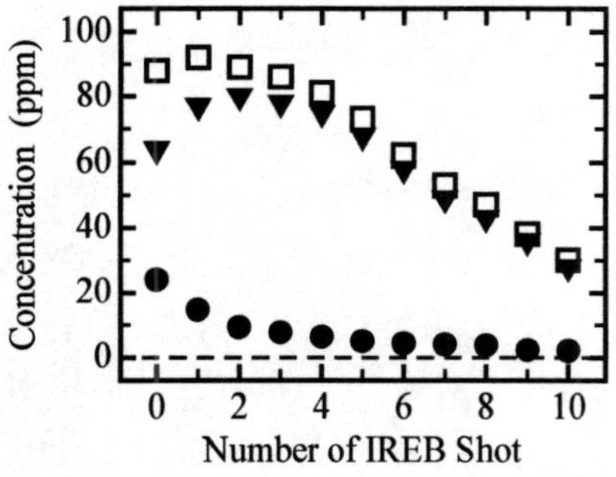

FIGURE 4. NO, NO_2, and NO_x concentrations as a function of number of IREB shot: ● NO, ▼ NO_2, □ NO_x.

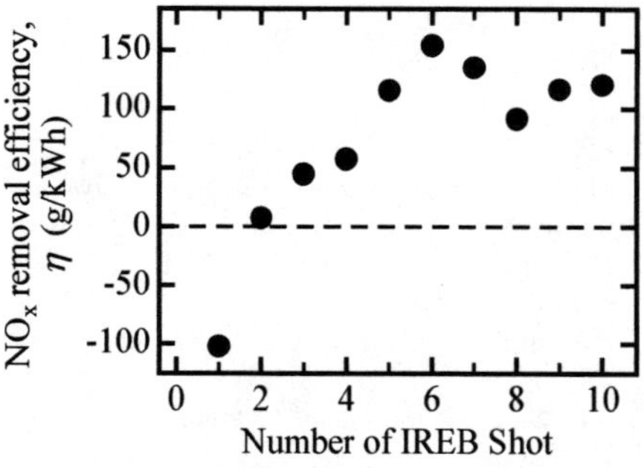

FIGURE 5. NO_x removal efficiency as a function of number of IREB shot.

by ambient O_2 without firing the IREB, and therefore, the NO concentration equilibrates to ~ 24 ppm, and then ~ 64 ppm of NO_2 is produced. It is found from Fig. 4 that the NO is removed by the irradiation of IREB. Until the second shot of IREB, on the other hand, NO_2 is generated by the irradiation of IREB because of high NO concentration. The NO_2 concentration decreases in proportion to the number of IREB shot after the third shot of IREB because the reduction of NO_2 may exceed the oxidation of NO. We found that ~ 70 % of NO_x is removed by firing 10 shots of IREB. Thus, NO has been almost treated by the irradiation of IREB in the distant IREB-gas-interaction chamber.

Figure 5 shows the NO_x removal efficiency (η) as a function of the number of IREB shot, where the experimental conditions are the same as those in Fig. 4. Here, η is defined as $\eta = q/\varepsilon$, where q is the amount of NO_x removal in each shot. Until the second shot of IREB, η is negative because of the generation of NO_2. It is seen from Fig. 5 that η is estimated to be 50-155 g/kWh which is of the same order of magnitude as or somewhat higher than that obtained by discharge treatments [7][8]. It is also found that η tends to increase with decreasing the NO_x concentration (cf. Fig. 4)

CONCLUSIONS

Nitrogen oxide within the distant interaction chamber 1.6-m isolated from the electron beam source has been successfully removed by the irradiation of IREB. The following results can be drawn from the present investigation.

1) In the initial NO concentration of 88 ppm in dry-N_2/O_2-balanced NO gas mixture with the pressure of 270 kPa, ~ 70 % of NO_x is removed by firing 10 shots of IREB.
2) The NO_x removal efficiency of 50-155 g/kWh is obtained.

ACKNOWLEDGMENTS

This work was partly supported by Grant-in-Aid for Encouragement of Young Scientists from the Ministry of Education, Culture, Sports, Science, and Technology in Japan.

REFERENCES

1. B. Penetrante, *Digest of Technical Papers on 11th IEEE Int'l Pulsed Power Conf.*, **I**, 91-96 (1997).
2. Y. Nakagawa and H. Kawauchi, *Jpn. J. Appl. Phys.*, **37**, 5082-5087 (1998).
3. T. Ikegaki, S. Seino, Y. Oda, T. Matsuda, G. Imada, W. Jiang, and K. Yatsui, *Jpn. J. Appl. Phys.*, **40**, 1104-1107 (2001).
4. M. Berger, S. Coursey, and A. Zucker: *ESTAR*: *Computer Programs for Calculating Stopping-Power and Range Tables for Electrons* (ver. 1.21), Online available: http://physics.nist.gov/Star [2001, July 19], National Institute of Standards and Technology, USA.
5. A. Tokuchi, N. Ninomiya, K. Yatsui, G. Imada, Q. Zhu, W. Jiang, and K. Masugata, *Proc. 12th Int'l Conf. on High-Power Particle Beams*, **1**, 175-178 (1998).
6. Y. Oda, T. Matsuda, S. Seino, T. Ikegaki, G. Imada, W. Jiang, and K. Yatsui, *Proc. 13th Int'l Conf. on High-Power Particle Beams*, **2**, 544-547 (2001).
7. Y. S. Mok and I. Nam, *IEEE Trans. Plasma Sci.*, **27**, 1188-1196 (1999).
8. K. Takaki and T. Fujiwara, *IEEE Trans. Plasma Sci.*, **29**, 518-523 (2001).

Efficient Electron Beam Deposition for Repetitively Pulsed Krypton Fluoride Lasers

F. Hegeler,[b] M. C. Myers,[a] M. Friedman,[a] J. D. Sethian,[a]
S. B. Swanekamp[c], D. V. Rose[d], and D. R. Welch[d]

[a] *Naval Research Laboratory, Plasma Physics Division, Washington, DC 20375 USA*
[b] *Commonwealth Technology, Inc., Alexandria, VA 22315 USA*
[c] TITAN/JAYCOR, *McLean, VA 22102 USA*
[d] *Mission Research Corporation, Albuquerque, NM 87110 USA*

Abstract. We have demonstrated that we can significantly increase the electron beam transmission efficiency through a pressure foil structure (hibachi) by segmenting the beam into strips to miss the hibachi support ribs. In order to increase the electron beam transmission, the cathode strips are adjusted to compensate for beam rotation and pinching. The beam propagation through the hibachi has been both measured and simulated with 1-D and 3-D codes.

INTRODUCTION

Electra is a repetitively pulsed, electron beam pumped krypton fluoride (KrF) laser that will develop the technologies that can meet the Inertial Fusion Energy (IFE) requirements for durability, efficiency, and cost. The Electra laser is pumped with two opposing electron beams each with parameters of 500 kV, 100 kA, and with a 100 ns flat-top pulse duration. They are emitted from 27 x 97 cm^2 cathodes in a vacuum diode that is immersed in an external magnetic field of 0.14 T (about 2.5 times the beam self-field). The repetition rate is 5 Hz. Fig. 1 shows the key components of a KrF laser. They include the pulsed power driver, cathode, hibachi, laser cell, gas recirculator, and optical components. As part of the program we must develop a long life, efficient, foil support structure.[1,2] The e-beam propagates through a hibachi structure, which supports a thin foil that isolates the vacuum diode from the high-pressure (>1 atm) laser gas. The ultimate goal for a full-scale 750 kV system is an energy transmission efficiency of 80% into the laser gas during the flat-top portion of the diode pulse. This paper describes hibachi developments on Electra at 500 kV that are used to benchmark simulation codes. These codes are then used to understand the complex beam physics in the diode and to predict the efficiency for a full-scale system.

CP650, *Beams 2002: 14th International Conference on High-Power Particle Beams,*
edited by T. A. Mehlhorn and M. A. Sweeney

ELECTRON BEAM TRANSMISSION INTO LASER CELL

One of the key challenges for a long-lived KrF laser system is the development of a hibachi that allows reliable and efficient injection of the electron beam into the laser gas. It is obvious that to maximize transmission efficiency the electron beam needs to be segmented so that it will miss the hibachi ribs. Fig. 2 shows the basic configuration of the diode. The hibachi ribs are 5 mm in width and 28 mm deep, and they are spaced 4 cm apart. For gas deposition measurements, the laser cell, 30 cm in depth, is filled with Krypton at 1.2 atm, which stops about 99% of 500keV electrons. The cathode consists of 24 strips, each 25 mm in width, and they are placed at an A-K gap of 37.5 mm. Electron beam halos of each strip cathode have been eliminated with "floating" electric field shapers.[3] Instead of preventing electron emission from a field shaper as is standard practice [4], the field shaper is electrically isolated from the cathode, and its capacitance and charge are minimized. Therefore, a floating field shaper should not emit significant currents unless the cathode plasma expands and electrically connects the field shaper to the cathode. The "floating" field shaper concept has been tested under repetitive operation at 1 pulse per second for 3000 shots at 500 kV, with an A-K gap of 42 mm, and 160 ns FWHM pulse duration. The field shaper did not show any visible surface damage.

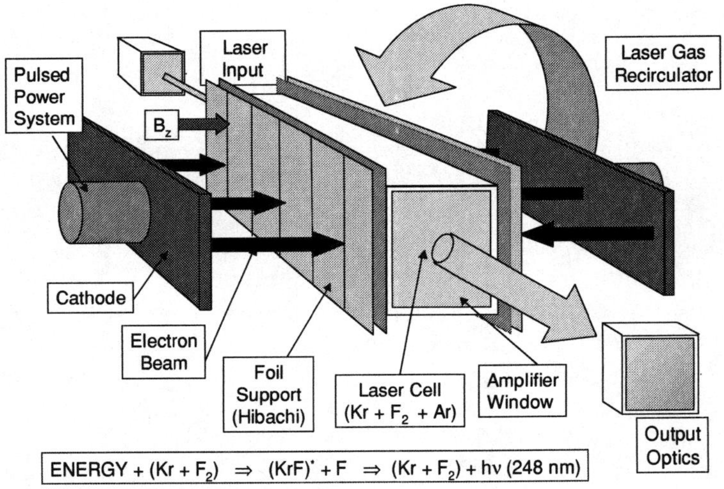

FIGURE 1. Key components of a krypton fluoride (KrF) laser.

The high energy transmission efficiency was achieved with two innovations: 1) eliminating the anode foil on the diode side of the hibachi structure, and 2) patterning the electron emitter into strips so the beam "misses" the hibachi ribs. While these are conceptually simple, they are difficult in practice: Eliminating the anode results in highly non-uniform electric fields, and the beam rotates and shears due to its interaction with the applied magnetic field. Thus the cathode strips need to be "counter-rotated" so the beam will be propagating parallel to the ribs when it gets to the hibachi. The cathode strips were rotated by 6° and the strip-to-strip spacing increased by 0.5 mm compared to the hibachi rib-to-rib spacing to compensate for beam pinching. Typical diode voltage and current waveforms are shown in Fig. 3. During the rise, 100 ns flat-top, and fall time of the pulse, 0.7 kJ, 4.2 kJ, and 1.2 kJ are measured in the diode, respectively. The energy deposited in the gas was determined with a Baratron that detects the pressure rise in the laser cell. 1-D Tiger simulations showed that the rise and fall of the beam contribute 25% and the flat-top 75% of the energy in the laser cell.

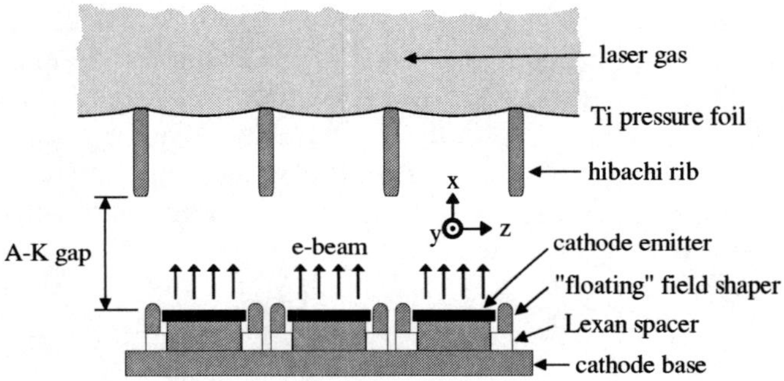

FIGURE 2. Diode configuration with 28 mm deep hibachi and 25 mm wide strip cathode with "floating" field shapers. The A-K gap is measured from the cathode emitter surface to the hibachi front surface.

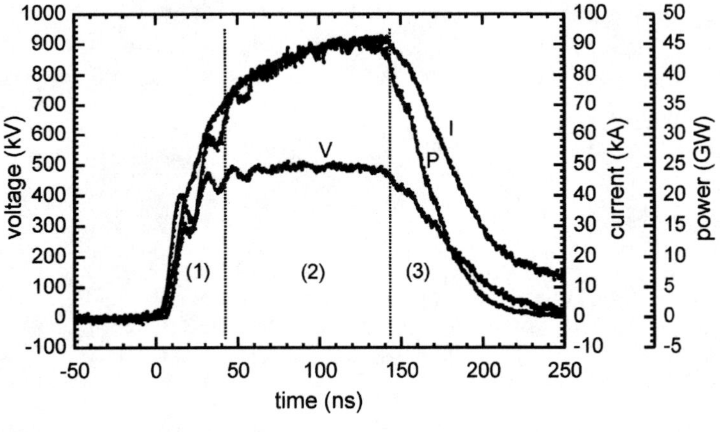

FIGURE 3. Diode waveforms: voltage, current and power are labeled V, I, and P, respectively. The diode energy is 0.7 kJ during the rise (1), 4.2 kJ during the 100 ns flat-top (2), and 1.2 kJ during the fall (3) of the pulse.

Measurements showed that the energy transmission efficiency, defined as the ratio of the deposited energy divided by the beam energy, is 67% during the flat-top portion of the pulse, at a diode voltage of 500 kV and with a 2 mil (50 μm) Ti pressure foil in a configuration as shown in Fig. 2. 1-D simulations without hibachi ribs (e.g., simulation considers e-beam, 2 mil Ti foil, and laser gas only) predict a transmission efficiency of 72%. These results show that the segmented e-beam propagates efficiently between the hibachi ribs with only minor losses, thus, patterning the beam imitates operation of a nearly " rib-less hibachi".

3-D LSP [5] simulations, which include the actual diode geometry, external magnetic field, hibachi ribs, and backscattering, showed that the transmission efficiency can be increased to 74% for a 500 keV beam by decreasing the thickness of the Ti pressure foil from 50 μm to 25 μm. Accounting for the thicker 50 μm pressure foil used in the experiments, the predicted efficiency would be 68%, or quite close to the experimentally observed 67%. A shallower hibachi with a rib depth of 13 mm instead of 28 mm will further increase the transmission efficiency. This new hibachi, which is currently under construction, with create a more uniform electric field at the anode, and thus, minimize e-beam spreading losses. More importantly, preliminary simulations showed that the ultimate goal of >80% hibachi efficiency is achievable at in a full-scale (750 keV) system.

SUMMARY

Measurements and simulations show that the electron beam can propagate efficiently through the hibachi. This is accomplished by patterning the electron beam into strips and by operating the diode without an anode foil, i.e., the hibachi ribs serve as an anode. 3-D simulations agree well with the experiments performed at 500 kV and with a 50 μm pressure foil. The energy transmission efficiency can be further increased by changing the thickness of the pressure foil to 25 μm and by using shallower hibachi ribs.

REFERENCES

1. Sethian, J.D., Myers, M., Smith, I.D., Carboni, V., Kishi, J., Morton, D., Pearce, J., Bowen, B., Schlitt, L., Barr, O., and Webster, W., IEEE Trans. Plasma Sci. **28**, 1333 (2000).
2. Sethian, J.D., Hegeler, F., Myers, M., Friedman, M., Obenschain, S., Lehmberg, R., Giuliani, J., Kepple, P., Swanekamp, S.B., Smith, I., Weidenheimer, D., Morton, D., Schlitt, L., Smilgys, R., and Searles, S., Proc. IEEE Pulsed Power Plasma Sci. Conf., Las Vegas, NV, June 17-22, 2001, 232 (2001).
3. Hegeler, F., Friedman, M., Myers, M.C., Sethian, J.D., and Swanekamp, S.B., submitted to *Phys. Plasmas* (2002).
4. Pierce, J.R., *Theory and design of electron beams*, 2nd ed., D. Van Nostrand Company, New York, 1954.
5. Welch, D. R., Rose, D. V., Oliver, B.V., and Clark, R. E., *Nucl. Instum. Meth. Phys. Res. A* **464**, 134 (2001).

Fragments and Debris Generation Using a High Power Pulsed Electron Beam

Bruno Cassany, Roger Courchinoux, Isabelle Bertron, Frederic Malaise, David Hebert

CEA/CESTA, BP 2, 33114 Le BARP, FRANCE

Abstract. The high power Laser Megajoule (LMJ) will be constructed at CEA/DAM/CESTA near Bordeaux, in the south west part of France. Among the problems encountered in the LMJ experimental chamber, there is the impact of the debris produced after a laser shot on the silica optical windows. The production of debris as well as the behavior of optical materials under their influence can be simulated and studied with a pulsed electron beam. We present in this paper the first experimental results obtained by this original technique.

INTRODUCTION

The utilization of a high current pulsed electron beam is a very practical way to deposit a high energy density in materials. If the energy of the electrons is not too large (< 1MeV) and if the current density is sufficiently large (>10 kA/cm²), most of the materials can be vaporized in times less than 100 ns and on a depth of a few tens of micrometers. Two kinds of experiments are performed. In the first one, a metallic foil is fragmented by a fast energy deposition. The emitted fragments are collected for analysis. Their distribution in size are compared with code predictions. In the second one, a plastic foil which contains a calibrated powder is vaporized. The particles of powder are then launched at high velocity on silica samples. The evolution of the resulting damages in the silica samples are then ready to be studied with a high energy laser beam.

PULSED POWER GENERATOR

The CESAR generator uses a Marx generator associated with a water-filled low impedance electrical line. A stainless-steel cold cathode produces a 300 kA, 60 ns electron beam, which is accelerated to a thin transparent anode up to 600 keV. Behind the anode the beam is neutralized in a low-pressure chamber and focalized by longitudinal magnetic field Bz.

The diode voltage and current are measured around the cathode by using Ddot and Bdot respectively. An inductive correction is applied in order to obtain the anode-cathode voltage. The typical voltage and current signals are displayed in figure 1.

CP650, *Beams 2002: 14th International Conference on High-Power Particle Beams*, edited by T. A. Mehlhorn and M. A. Sweeney

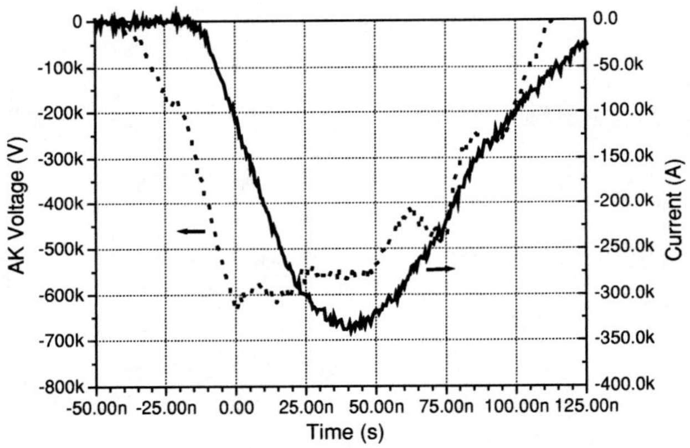

FIGURE 1. Typical voltage and current waveforms.

From these measurements, we are able to compute the power, energy, and impedance of the electron beam as shown in figure 2a. The energy spectrum of the electrons, deduced from the forementioned data, is given in figure 2b. It shows a small contribution of low energies in such a way that we can neglect absorption or scattering in the anode foil. This spectrum and the power are used to compute the energy deposition in the target.

The 1D-CEPXS code [1] is used to predict the ratio of internal energy/electron fluency as a function of the depth inside the foil. The fluency can be adjusted in the range 50–200 cal/cm² by adjusting the magnetic coils in order to modify the focalization of the e-beam.

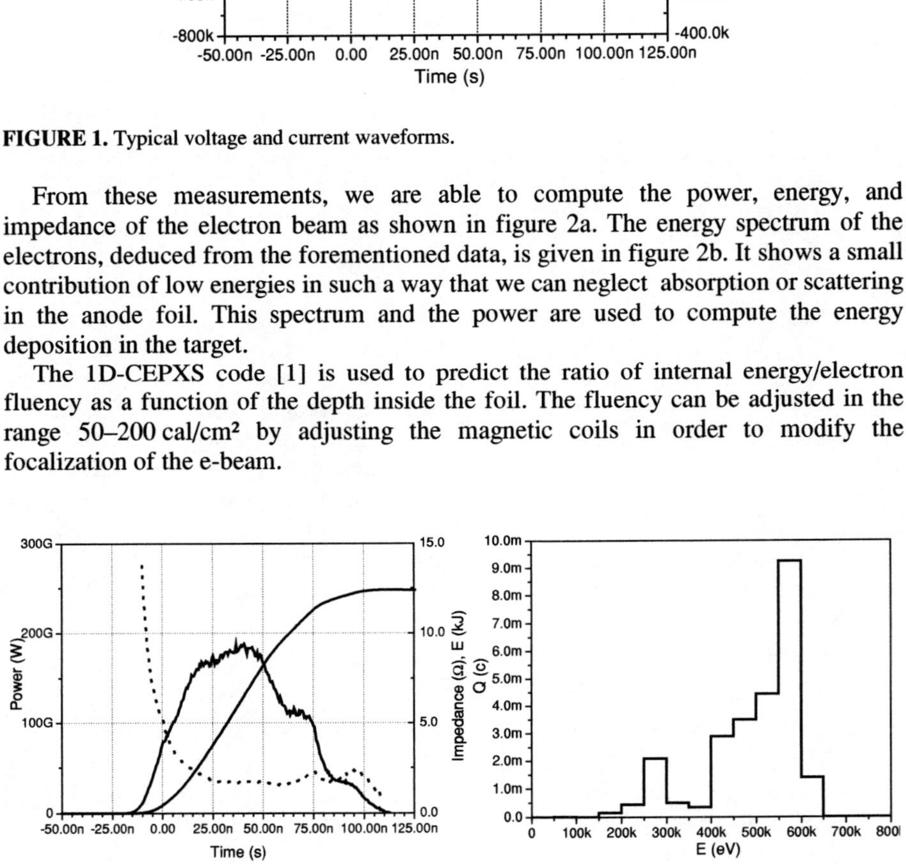

FIGURE 2. (a)-Calculated power, energy and impedance.
(b)- Calculated electron spectrum from -25 ns to 100 ns.

EXPERIMENTAL CONFIGURATION

The experimental set up is illustrated in figure 3 for the particle launching experiment. The electron beam is shown entering on the right. A 50mm-diameter graphite diaphragm is used to protect the silica samples against direct exposure to the e-beam. For the fragment generation experiment, the experimental configuration layout is identical: the silica sample is replaced by the polyethylene collector and the mylar foil is replaced by a metallic foil.

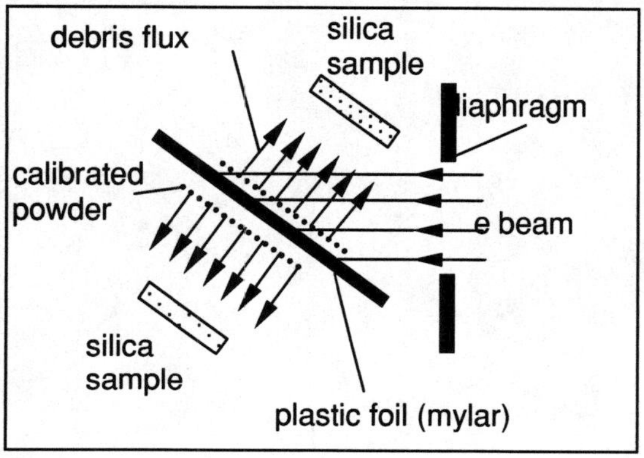

FIGURE 3. Schematic of the experiment.

Fragment Generation

We have performed experiment with a 80 µm thick copper foil submitted to 100 cal/cm2. The fragments are collected by a polyethylene plate located on the back side of the target holder. The shape of the copper fragments are visible in the photograph shown in figure 4. Their spherical shape is due to the liquefaction of copper which appears during the fragmentation process. The distribution of the diameter of the fragments corresponds to a Poisson law with an average value of 9 µm. These results give us data on fragmentation process [2] which can be used to benchmark our simulation codes.

Particle Launching

The thickness of the Mylar foil is in the range 300-1000 µm. Two silica samples are placed in parallel to the foil and the angle between the foil and the beam is 30°.

The foil is coated with calibrated powders (20-75 µm) made of synthetic diamond, tantalum or tungsten.

The silica samples are impacted by particles of powder which have been accelerated by the expansion of the plasma produced by the electron beam energy deposition. An estimation of the plasma velocity expansion of 2 km/s is obtained from hydrodynamic calculations.

When the fragments impinge the silica plates, craters are produced as it can be shown in figure 5 for the case of diamond.

After this kind of experiment, the damaged silica samples are suitable for further analysis, in order to progress on laser damage knowledge. They are submitted to a high fluency laser beam in order to observe the evolution of their optical properties.

FIGURE 4. Photograph of the fragments collected by the polyethylene plate (100 cal/cm^2).

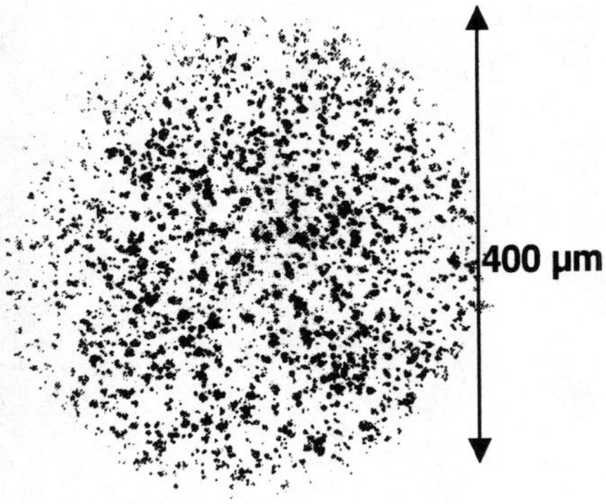

FIGURE 5. Damages induced by 20 µm diamond powder on a silica sample.

CONCLUSION

To study hardening of the optical materials which are used in the LMJ experimental chamber, the production of debris and the damages they can produced on an optical material can be studied by using an electron beam. In the next series of experiment, we will test different materials powder Al, silica.

ACKNOWLEDGEMENTS

The authors would like to thank T. Desanlis, A. Galtié and P. Modin for their experimental support during the experiments. They are pleased to acknowledge stimulating discussions with J. Labrouche.

REFERENCES

1. CEPXS/ONELD One-Dimensional Coupled Electron-Photons Multigourp Discrete Ordinate Code System. RISC COMPUTER CODE COLLECTION CCC – 544

2. High-Pressure Shock Compression of Solids II. Dynamic Fracture and Fragmentation. L. Davison, D.E. Grady, M. Shahinpoor. Publisher Springer.

Features of High-Power E-Beam Application for Plasma Heating in Long Open Trap GOL-3

R.Yu. Akentjev, A.V. Arzhannikov, V.T. Astrelin, A.V. Burdakov, I.A. Ivanov, V.S. Koidan, S.A. Kuznetsov, K.I. Mekler, S.V. Polosatkin, V.V. Postupaev, A.F. Rovenskikh, A.A. Shoshin, S.L. Sinitsky, V.D. Stepanov, Yu.S. Sulyaev

Budker Institute of Nuclear Physics, 630090 Novosibirsk, Russia

Abstract. Recent results of experiments on high-energy content E-beam application for plasma heating in a multimirror trap GOL-3 are presented. Conditions for optimal neutralization of injected beam and its stable transport through long plasma column are defined. Effective electron and ion heating of plasma is observed.

INTRODUCTION

High-energy-content microsecond electron beam is used for plasma heating in a long solenoidal facility GOL-3. This facility is an open trap intended for studies of heating and confinement of a relatively dense ($10^{15}\div10^{17}$ cm^{-3}) plasma in axially-symmetrical magnetic system [1]. Production of ~ 1 keV plasma in the facility gives the possibility to study the multimirror and «wall» confinement of such a plasma according concept of a pulsed multi-mirror fusion reactor [2].

Some important problems of E-beam physics are connected with our activity. The first one is neutralization of the electron beam in the region of its compression before injection into the plasma. The second problem is macroscopically stable transportation of the E-beam with supercritical current through the long (12 m) plasma column. Third one is effective collective relaxation of the beam in the plasma with corrugated magnetic field and fast plasma heating.

GOL-3 FACILITY

Layout of the GOL-3 facility is shown in Fig.1. The magnetic system consists of coils for transport and shaping of the electron beam, a 12-meter-long main solenoid and an exit unit (which includes plasma creation system, expander and exit receiver of the beam and plasma). Main solenoid operates usually with 4.7 T field and 8-9 T in end mirrors. Recently the GOL-3 facility was upgraded to configuration with a long multimirror parts of the magnetic system. Current design of the magnetic system enables operation in the multimirror mode with B_{max}/B_{min}~1.5, 22 cm cell length and

CP650, *Beams 2002: 14th International Conference on High-Power Particle Beams*, edited by T. A. Mehlhorn and M. A. Sweeney

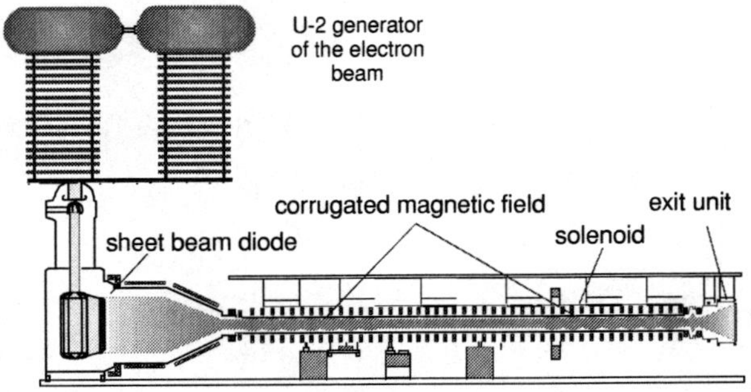

FIGURE 1. Layout of the GOL-3 facility.

with ~4 m sections at the plasma ends (20 cells each). The plasma has an average diameter 6 cm and 12 m length, its density can be varied in $10^{14}\div10^{17}$ cm^{-3} range and can be as of fixed density distribution along the device length and varied one. The plasma heating is provided by a high-power electron beam generated by U-2 device [3].

OPTIMIZATION OF E-BEAM NEUTRALIZATION

Initial electron beam is generated by a ribbon magnetically insulated diode at the U-2 device (Fig.2). Main parameters of the generated beam are: E_e=1 MeV, I_b~50 kA, τ_b~10 μs, θ_s~2°, beam cross section 3.5x130 cm^2. Then the ribbon beam is transformed to circular one and compressed 20-fold before its injection into the plasma chamber of the GOL-3. In order to keep angular spread of magnetized beam electrons at a low level these two processes should be done in adiabatically converging guiding magnetic field and under electron space charge neutralization. Puffing a gas cloud with high density to provide dense plasma in a few hundreds of nanoseconds can solve the problem of space charge neutralization of the beam.

Aim of experiments was to find an appropriate sort of gas, its initial pressure and the beam injection delay from the gas puffing. In the experiments we measured net currents at the end of transformer I_t and in the center of compressor I_c by Rogowsky coils. Four gases were tested: H_2, N_2, Kr and Xe at the puff pressures up to 12 atm and the delay between the beam injection and gas puffing in range 0-12ms. Ratio I_t/I_0 as a

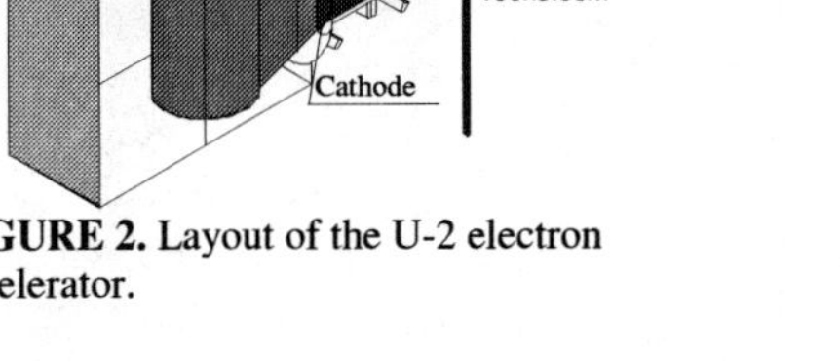

FIGURE 2. Layout of the U-2 electron accelerator.

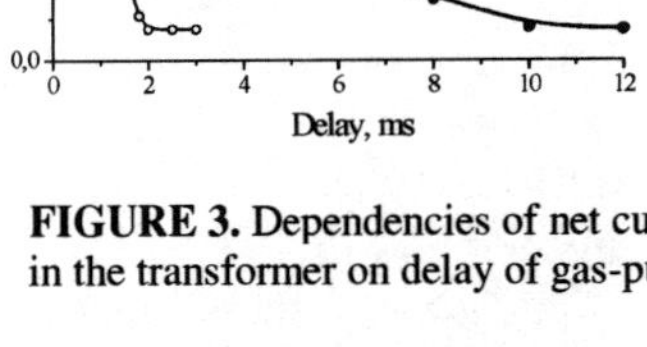

FIGURE 3. Dependencies of net current in the transformer on delay of gas-puff.

function of delay for various gas pressures are shown in Fig.3 (I_0=40kA is the current at the beam dump). In case of hydrogen, the minimal delay for the charge neutralization is about 1 ms and for xenon this minimum is about 4 ms (3 ms for krypton). So, gas puffing provides conditions for high efficiency of the beam compression.

MACROSCOPICAL STABILITY OF BEAM-PLASMA SYSTEM

Macroscopically stable transportation of a beam requires usually that value $q=(H_z/H_\phi)\cdot(2\pi r/L)$ should exceed unity (where H_z and H_φ - longitudinal and azimuthal magnetic field, r and L- plasma radius and column length). In a new regime of GOL-3 operation with a corrugated magnetic field the plasma can be unstable even when Kruskal-Shafranov criterion is formally satisfied.

For prevention of disruption events and achieving of stable plasma a series of experiments on search of an optimum radial structure of net current density in a plasma was carried out. Macroscopically stable beam transport was achieved with specially designed source of the preliminary plasma [4]. The exit receiver enables operation under near-to-floating potential during the short period of the beam injection. The discharge current flows through the entire chamber along the magnetic force lines. Dense gas puffed in the beam compression region provides some limited transverse conductivity to the wall near first (input) magnetic mirror, thus closing the circuit.

It is performed the following scenario of the experiment. At 30 μs before the beam injection the system of preliminary plasma creation is switches on and the current up to 5 kA, opposite to a beam current, is created. Then the electron beam is injected into the prepared plasma. Though the injected beam current achieves ~25 kA, net plasma current in the system practically does not changes. The value and the waveform of the plasma current depends on a delay of discharge start prior to gas puff operation moment. Examples of currents in the plasma, received for various gas conditions in the compression chamber are given in Fig.4. The waveforms corresponds to puffing of Xe and H_2 with two different delays. In case of small delay of gas puffing (small length of a cloud in the compression chamber) the plasma current not exceeds 1 kA. Disruption (bottom curve in the figure) is observed in this case. In a stable operation regimes (top curve on Fig.4) changes of a net plasma current are minimal during the beam injection with current exceeding 25 kA.

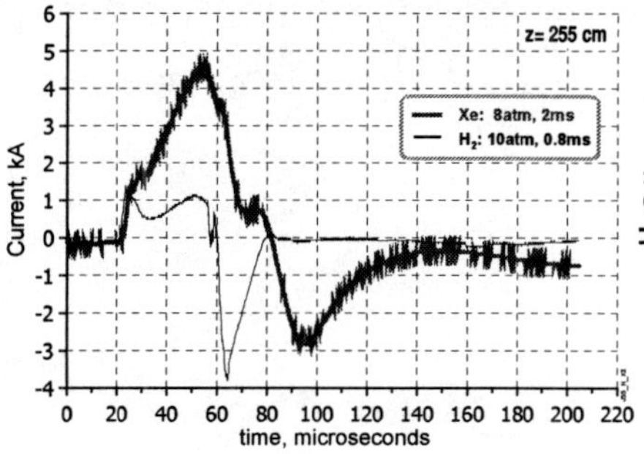

FIGURE 4. Waveforms of net current in plasma.

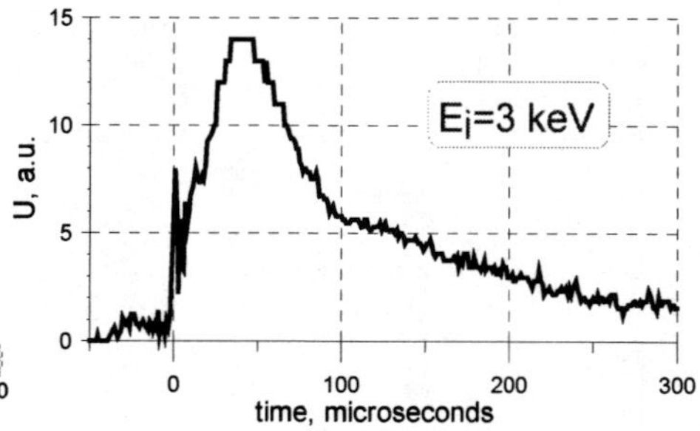

FIGURE 5. Waveforms of CX neutrals.

PLASMA HEATING DUE TO E-BEAM RELAXATION

After beginning of the beam injection a pressure rises and reaches (1-2)x10^{15} keV/cm^3, then quickly falls down by half and then slowly decreases with characteristic time of 100-200 μs. Dynamics of plasma pressure along the column length depends on distribution of initial gas pressure. As measurements by Thomson scattering have shown, peak electron temperature reaches 1-2 keV. In ~10 μs after the ending of a beam heating the electron temperature becomes of 200-300 eV at density of (0.8-1)x10^{15} cm^{-3}. Thus, in contrast with a conditions without a multimirror field, the plasma pressure is determined not only by heated of Maxwellian electrons. Ions have also to be heated.

First experiments with new ion diagnostics are carried out. The scintillation neutron detectors records neutrons and gamma radiation. Emission of particles begins at the time of beam passage. Duration of a radiation pulse is 100-300 μs. Total number of neutrons during a shot determined by activation of the silver is (0.5-1)x10^6 m^{-1} at Z~2 m. Neutron flux obtained from measurements with the scintillation detectors is (1-5)x10^9 s^{-1}m^{-1} and time integration of this data is in agreement with activation detector data. Doppler broadening of D_α line was measured in a place with lowered density (Z=2.15 m). Time evolution of ion temperature was obtained.

Emission of last charge exchange (CX) neutrals starts during the beam injection and lasts some hundreds of microseconds. The signal from channel corresponding to ion energy of 3 keV is given in Fig.5. The ion temperature to be estimated is in the range 0.5-1.2 keV.

CONCLUSION

Conditions for optimal neutralization of compressed E-beam are defined. In the 12-m multimirror trap GOL-3 macroscopical stability of the system electron beam – plasma of 10^{15} cm^{-3} density was achieved. First measurements of ion temperature have shown that ion temperature reaches of 0.5-1.2 keV and exists in the trap for more than 100 μs. Emission of fusion neutrons is observed.

Work is supported by RFBR within the framework of project 00-02-17649.

REFERENCES

1. *A. V. Arzhannikov, et al.*, Transactions of Fusion Technology, **35**, No.1T, 112 (1999).
2. *G. I. Budker, V. V. Mirnov, D. D. Ryutov*, JETP Letters, 14 (1971) 320 - in Russian.
3. *A. V. Arzhannikov, V. B. Bobylev, V. S. Nikolaev, et al.*, Proc. 10th Int. Conf. On High-Power Particle Beams (San Diego), 1, 136 (1994).
4. *V. T. Astrelin, A. V. Burdakov, V. S. Koidan, et al.*, Proc. 25th ICPIG (Nagoya), 2, 183 (2001).

Preliminary Modeling Results of a Thermal Neutron Source Driven with an Electron Linac

Gregory E. Dale and John M. Gahl

Department of Electrical Engineering, University of Missouri-Columbia, Columbia, MO 65211

Abstract. An electron linac, fitted with a standard x-ray converter, can produce high neutron yields in materials with low photonuclear threshold energies, such as D and ^{9}Be. This paper discusses the preliminary MCNP modeling of a thermal neutron source driven with a 10 MeV electron linac operated in the CW mode. Results indicate that a 10 MeV, 10 kW electron linac can produce on the order of 10^{12} n/sec in a heavy water photoneutron target. The thermal neutron flux in an unreflected heavy water target is nearly 10^{10} n/cm^2/sec. Current modeling results are discussed.

INTRODUCTION

There are several potential applications requiring a high flux of thermal neutrons. Potential industrial applications include neutron radiography, Prompt Gamma Neutron Activation Analysis (PGNAA), as well as polymer hardening. Examples of medical applications include neutron radiotherapy, various forms of Neutron Capture Therapy (NCT), and the production of short lived medical radionuclides for nuclear medicine. A further discussion of these applications can be found in [1].

An electron linac driven thermal neutron source is being designed at the University of Missouri-Columbia to meet the need for a low-cost, compact, high-flux thermal neutron source suitable for installation in industrial and clinical environments. Electron linacs with energies in the MeV range and kW power levels are commercially available and are less expensive than comparable ion accelerator systems. The use of electron linacs has a very long history in hospitals. Electron linacs are currently favored over radioisotope systems (^{60}Co and ^{137}Cs) for photon radiotherapy and are found in virtually every major oncology department.

This paper describes preliminary modeling results of the electron accelerator-driven photoneutron source. The neutron source is modeled with the Monte Carlo for N-Particle (MCNP) transport code [2].

ELECTRON LINAC DRIVEN PHOTONEUTRON SOURCE

The electron accelerator driven thermal neutron source is based on the photonuclear dissociation of deuterium. Deuterium has a very low photonuclear threshold energy (2.23 MeV) and is available in a very convenient form, that of heavy water. Heavy

CP650, *Beams 2002: 14th International Conference on High-Power Particle Beams*,
edited by T. A. Mehlhorn and M. A. Sweeney

water is also an excellent neutron moderator. A schematic diagram of the system is shown in Figure 1.

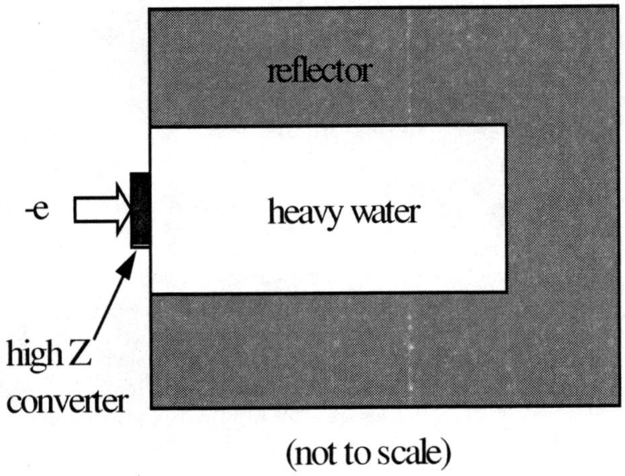

Figure 1. Schematic Diagram of the electron accelerator driven photoneutron source.

Electrons of moderate energy (5 to 10 MeV) from the electron linac are directed onto a target of high atomic number (Z), such as tantalum or tungsten, in a standard x-ray converter configuration. Very large high-energy photon fluxes can be generated in this manner. Photons from the x-ray converter impinge on the photoneutron target where the neutron generation takes place. A further discussion of the modeling of the x-ray converter can be found in [3]

The neutrons are emitted with an energy spectrum similar to that of fission neutrons. The heavy water photoneutron target also serves as a neutron moderator for thermalization of the neutrons. The heavy water target/moderator can be surrounded by a neutron reflector to improve neutron economy.

PRELIMINARY MODELING RESULTS

The production of photoneutrons, or photoneutron yield, in a heavy water target was modeled with MCNP using the photodissociation cross section of deuterium [4]. The photoneutron production in a cylindrical heavy water photoneutron target of various sizes using a 10 MeV electron beam is shown in Figure 3. In this figure the photoneutron production rate is plotted as a function of target radius, *R*, and target axial length, *Z*. It can be seen in the figure that the photoneutron production rate is fairly insensitive to target size. For a 6.28 L (1.66 gal) cylindrical heavy water target of radius 10 cm and a axial length of 20 cm the neutron production rate is 2.64×10^{12} n/s/mA. For a 6283 L (1660 gal) cylindrical heavy water target of radius 100 cm and a axial length of 200 cm the neutron production rate is 7.16×10^{12} n/s/mA. Most of the neutron production occurs near the x-ray converter. Additionally, the heavy water target does not have to be very large to have an appreciable neutron production rate.

The thermal neutron flux in an unreflected heavy water target of radius 100 cm and an axial length of 200 cm using a 10 MeV electron beam is shown in Figure 3. The thermal neutron flux is plotted as a function of radial position, r, and axial position, z. The thermal neutron flux is taken as all neutrons with energy less than 1 eV.

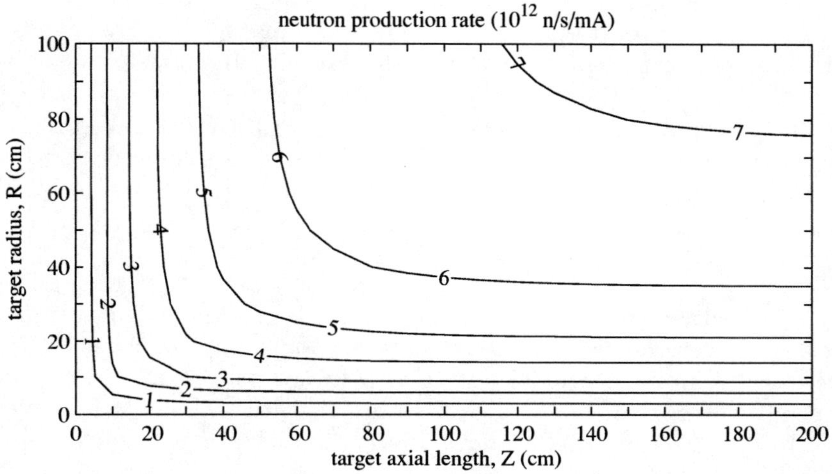

Figure 2. Neutron production rate as a function of target radius, R, and axial length, Z.

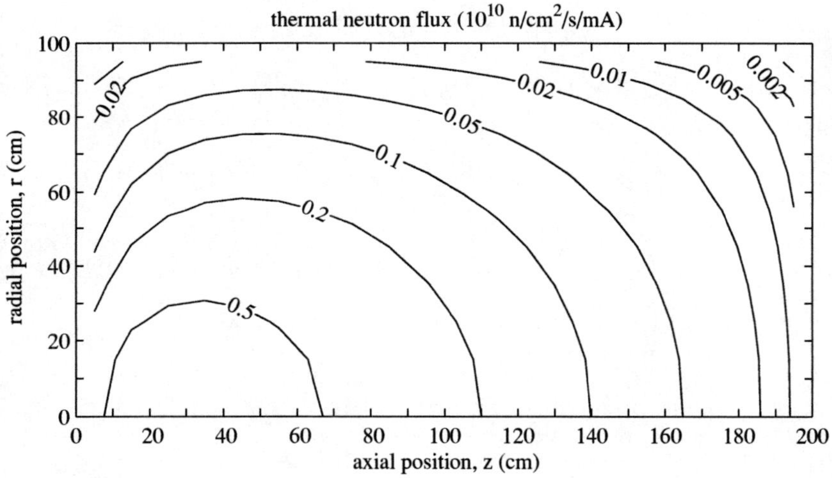

Figure 3. Thermal neutron flux in an unreflected heavy water target of radius 100 cm and axial length 200 cm using a 10 MeV electron beam.

The maximum thermal neutron flux in an unreflected heavy water target with radius 100 cm and axial length 200 cm is 9.01×10^9 n/cm^2/s/mA. It is apparent from the figure that the thermal neutron flux is fairly constant over a large volume. Over half the volume of the heavy water target has a thermal neutron flux greater than 10^9 n/cm^2/sec/mA. This is due to the low neutron absorption in heavy water. The high neutron flux over such a large volume would be very beneficial for applications such as materials modifications.

Modeling results indicate that the peak thermal neutron flux is fairly insensitive to tank radius over a certain range. The maximum thermal neutron flux in an unreflected heavy water target of radius 60 cm and axial length 200 cm was calculated to be 8.29×10^9 n/cm^2/s/mA. This shows that the peak thermal neutron flux decreases by only 8% with a 64% reduction in tank volume. This is because heavy water is an excellent moderator.

CONCLUSIONS

An electron linac driven thermal neutron source is being designed to meet the need for a low-cost, compact, high-flux thermal neutron source suitable for installation in industrial and clinical environments. The photoneutron target of an electron accelerator driven photoneutron source has been modeled with the MCNP transport code. A 10 kW, 10 MeV electron linac driven system has a photoneutron yield in excess of 10^{12} n/s. In an unreflected heavy water target the peak thermal neutron flux approaches 10^{10} n/cm^2/s. The unreflected target/moderator is the base case for the next phase of modeling, the neutron reflector design. A great enhancement in neutron economy should be realized with the use of a neutron reflector.

ACKNOWLEDGMENTS

The authors would like to thank Morgan C. White of Los Alamos National Laboratory for providing the photonuclear cross sections for use with MCNP.

REFERENCES

1. Dale, G.E. and Gahl, J.M, "Applications of Thermal Neutron Generators Driven with a Repetitively Pulsed Electron Linac," *Proceedings of the IEEE Pulse Power Plasma Science (PPPS) Conference*, Las Vegas, NV, June 17-21, 2001.
2. J.F. BRIESMEISTER, Ed., *MCNP-A General Monte Carlo N-Particle Transport Code: Version 4C*, LA-13709-M, Los Alamos National Laboratory (2000).
3. Dale, G.E. and Gahl, J.M., "Modeling the Neutron Yield of a Therapeutic Thermal Neutron Source Driven with a Repetitively Pulsed Electron Linac," *Proceedings of the IEEE Pulse Power Plasma Science (PPPS) Conference*, Las Vegas, NV, June 17-21, 2001.
4. T. MURATA, "Evaluation of the D(γ,n) Reaction Cross Section," *Proceedings of the 1993 Symposium on Nuclear Data*, Tokai, Japan (1993). Data available at the IAEA Photonuclear Data Library (www-nds.iaea.org/photonuclear/).

The Propagation of High Power CW Scanning Electron Beam in Air

Sergey Korenev[a], Ivan Korenev[b]

[a]*STERIS CORPORATION, STERIS Isomedix Services, 2500 Commerce Drive, Libertyville, IL 60048, USA*
[b]*Northern Illinois University, DeKalb, IL 60115, USA*

Abstract. The question of propagation of high power electron beam in air presents the scientific and applied interests. The high power (80 kW) CW electron accelerator "Rhodotron" with kinetic energy of electrons 5 and 10 MeV was used in the experiments. The experimental results for propagation of scanning electron beams in air are presented and discussed.

INTRODUCTION

Interest to investigations of propagation of electron beams in air has long story. The high current pulsed electron beam in air investigated for understanding of non-stability of beam and increasing of efficiency of transport in air and vacuum conditions [1,2].

The study of physical effects at during of propagation of scanning relativistic electron beams in air allows better understand the nature of the beam loses, character of interaction of beam electrons with ionized air and etc. [3,4].

The results of experimental research for propagation of high power CW scanning relativistic electron beam in air are presented in this report.

EXPERIMENT

The experiments were conducted using CW electron accelerator "Rhodotron" [5] with next main parameters:

- kinetic energy of electrons................................ 5 & 10 MeV;
- power of electron beam .... 5 – 80 kW;
- diameter of electron beam after foil window...... ... 8 cm;
- repetition of scanning for electron beam.................60-100 Hz;
- frequency of CW mode....................................108 MHz.

The scanning mode for electron beam leads to pulsed character interaction of electrons with molecules of air behind foil window of accelerator.

The current transformers (Rogowski Coils) and Faraday Cups with current shunts were used for measurements of beam current.

The standard Cellulose TriAcetate (CTA) films dosimeters (FTR-125) were used for measurements of dose distribution in cross section. The CTA films with width 100 cm

CP650, *Beams 2002: 14th International Conference on High-Power Particle Beams*,
edited by T. A. Mehlhorn and M. A. Sweeney

were fixed with step 20 cm after foil window of electron accelerator. The distance between first CTA film and foil window of accelerator was 10 cm.

EXPERIMENTAL RESULTS

The oscillograms of microbunches for electron beam (CW mode) with kinetic energy 10 MeV for beam current 2,4,6,8 mA are given on Figure 1. The distance between output foil of accelerator and current sensors was 50 cm.

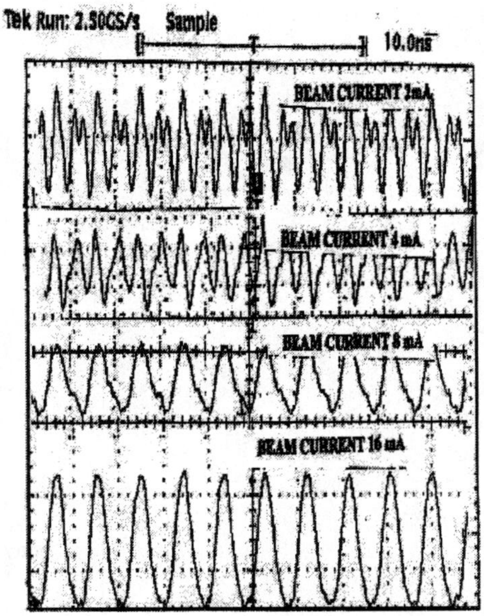

FIGURE 1. The oscillograms of microbunches for 10 MeV electron beam.

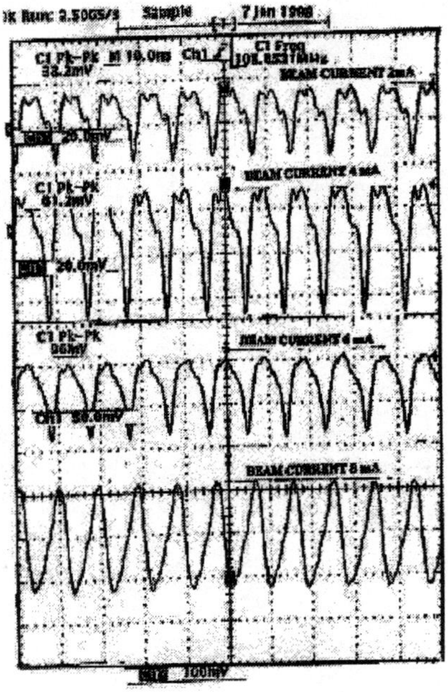

FIGURE 2. The oscillograms of microbunches for 5 MeV electron beam.

The oscillograms of microbunches for electron beam with kinetic energy 5 MeV for beam current 2,4,8 and 16 mA are presented on Figure 2. The distance between output foil of electron accelerator and current sensors was 50 cm.
The cross section distributions of absorbed doses in air are shown on Figure 3.

DISCUSSION

At during of the propagation of CW relativistic electron beam in air we can consider the next main physical processes:

- the ionization of air molecules by electrons of beam and forming air plasma;
- the heating of air and air plasma;
- the chemical reactions with produce of ozone.

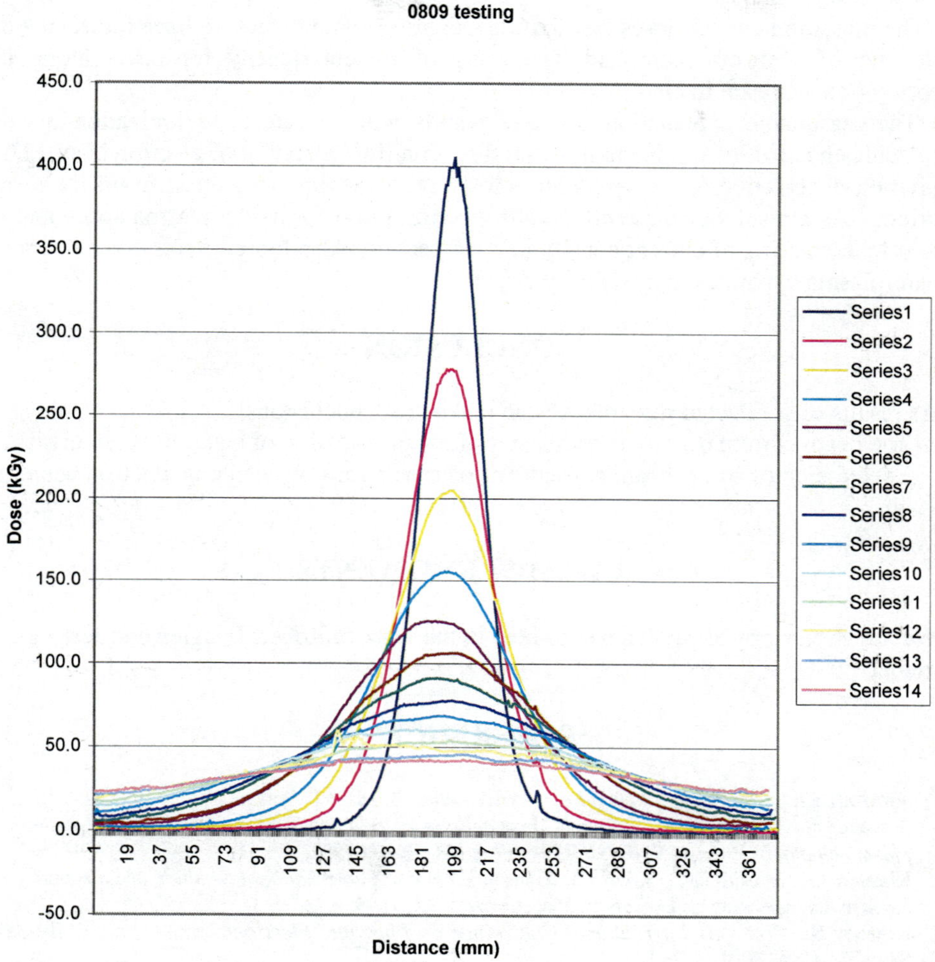

FIGURE 3. The electron beam dose distribution in air for different distance from foil window of accelerator: series 1 – 10 cm, series 2 – 30 cm and following series have quantity 20 cm.

The detail measurements of time characteristics for microbunches by beam current shown the erosion of fronts of current see oscillograms on the Figure1,2). The ionization loses for electron beam with 5 (about 250 KeV) and 10 MeV in air are small and effect of ionization of air's molecules by electron beam leads to high resistance of air plasma [3,4].

The relativistic factor $\gamma = 21$ (kinetic energy of electrons is 10 MeV) and $\gamma = 11$ (kinetic energy of electrons is 5 MeV) and parameter of ion neutralization of space charge is very small.

The measurements of doses distribution (Figure 3) show , that we have increasing of diameter of electron beam and decreasing of current density for beam along of propagation of beam in air.

The reasonable explanation of these results will be effects of ionization by air molecules by electrons in the transversal direction. It is scanning of electron beam. The variation of repletion for scanning leads to variation of time erosion of fronts by beam current. As a results of these effects the electron beam forms the plasma space and it leads to decreasing of current density in cross section. Also these effects can be lead to beam-plasma dynamics non- stability effects.

CONCLUSION

As results of conducted research we can make next conclusions:

1. The beam current density is decreasing along propagation of electron beam in air.
2. The erosion of fronts beam current was observed for CW scanning electron beam in air.

ACKNOWLEDGMENTS

We are very happy to say thanks for help in the work to Robert Langley and Jerry Kriebel.

REFERENCES

1. Graybill S.E., S.V. Nablo S.V. *Applied Physics Letter* **8,** 18-23 (1966).
2. Korenev S., Khodataev K., Rubin N. *Propagation of relativistic high current pulsed beams in plasma channels.* Preprint JINR **No R10- 417**, JINR, Dubna,1982.
3. Korenev S., Masefield J., Kriebel J. Johnson S.. *Proc. X International Seminar of Laboratory of Surface Science*, Rutgers University, Piscataway, USA, 1999, p.23.
4. Korenev S. *Proc. XII International Conference on Photonic, Electronic and Atomic Collisions.* Santa Fe, USA, 2001, p. 234
5. Pottie J. *Nuclear Instruments and Methods in Physics Research* **B40/41,** 943-947 (1989).

The Conductivity of Dielectrics during Irradiation by Scanning High Power Electron Beam

Sergey Korenev

STERIS CORPORATION, STERIS Isomedix Services,
2500 Commerce drive, Libertyville, IL 60048, USA

Abstract. The conductivity of dielectrics under electron beam irradiation is fundamental question of physics for interaction of electron beams with condensed mater. The CW electron accelerator Rhodotron with scanning electron beams was used in the experiments. The electron beam has next main parameters: kinetic energy is 5 MeV, maximum power is 80 kW, and the scanning repetition is 100 Hz. The results of measurements are discussed.

INTRODUCTION

The question of conductivity for irradiated dielectrics is very important for fundamental understanding of physics of dielectrics. The question of charging of dielectrics by propagating electrons was discussed in many papers and summarized in monograph by V. Gromov [1]. D. Vaisburg was considered the irradiation of dielectrics by nanosecond electron beams [2]. The electron and ion beam irradiation of dielectrics in the range of few hundred nanoseconds pulse duration was described in [3]. The electrical properties of dielectrics under irradiation present interest from charging of dielectrics and forming discharge inside of dielectrics [4]. The fundamental question of forming electrical discharge inside of dielectrics consists in the forming of electrical fields inside of the sample that lead to breakdown of dielectric.

This paper considers the results of measurements of conductivity of dielectrics under scanning electron beam.

EXPERIMENTAL

The experiments were conducted using CW electron accelerator “Rhodotron” [5] with next main parameters:

- kinetic energy of electrons................................. 5 & 10 MeV;
- power of electron beam .... 5 – 80 kW;
- diameter of electron beam after foil window...... ... 8 cm;
- repetition of scanning for electron beam.................60-100 Hz;
- frequency of CW mode...................................108 MHz.

CP650, *Beams 2002: 14th International Conference on High-Power Particle Beams,*
edited by T. A. Mehlhorn and M. A. Sweeney

The scanning mode for electron beam leads to pulsed character interaction of electrons with sensor with irradiated sample behind foil window of accelerator (see Figure 1).

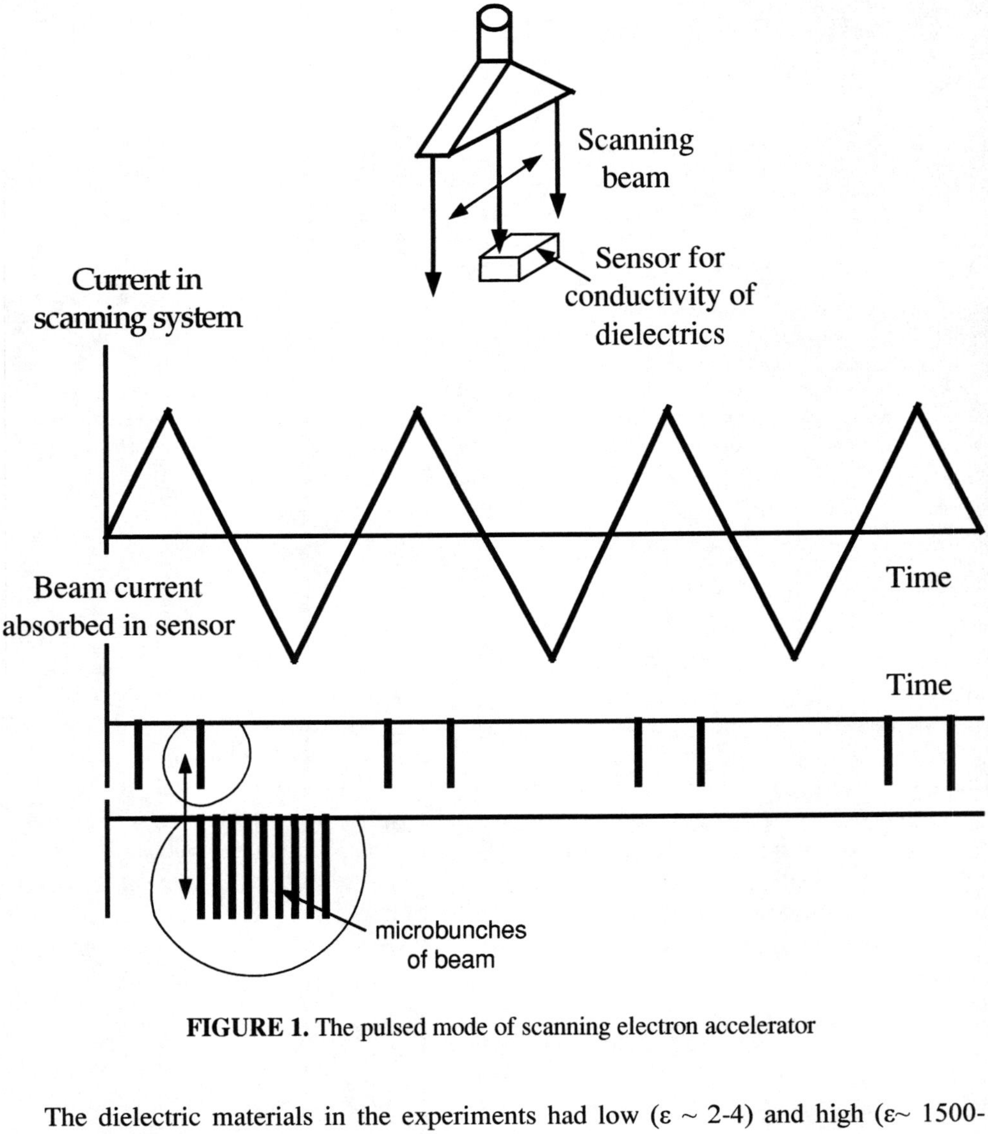

FIGURE 1. The pulsed mode of scanning electron accelerator

The dielectric materials in the experiments had low ($\varepsilon \sim$ 2-4) and high ($\varepsilon \sim$ 1500-2000) dielectric constant.

The electrical diagram for measurements of conductivity of irradiated dielectrics is given on Figure 2. The measurements of conductivity of dielectrics under irradiation were conducted using sensor described in [3,4] where used pulsed irradiation of materials [6,7] with deposited metal films for charging of capacitor by electron beam. The scanning mode of electron beam allows realizing these measurements.

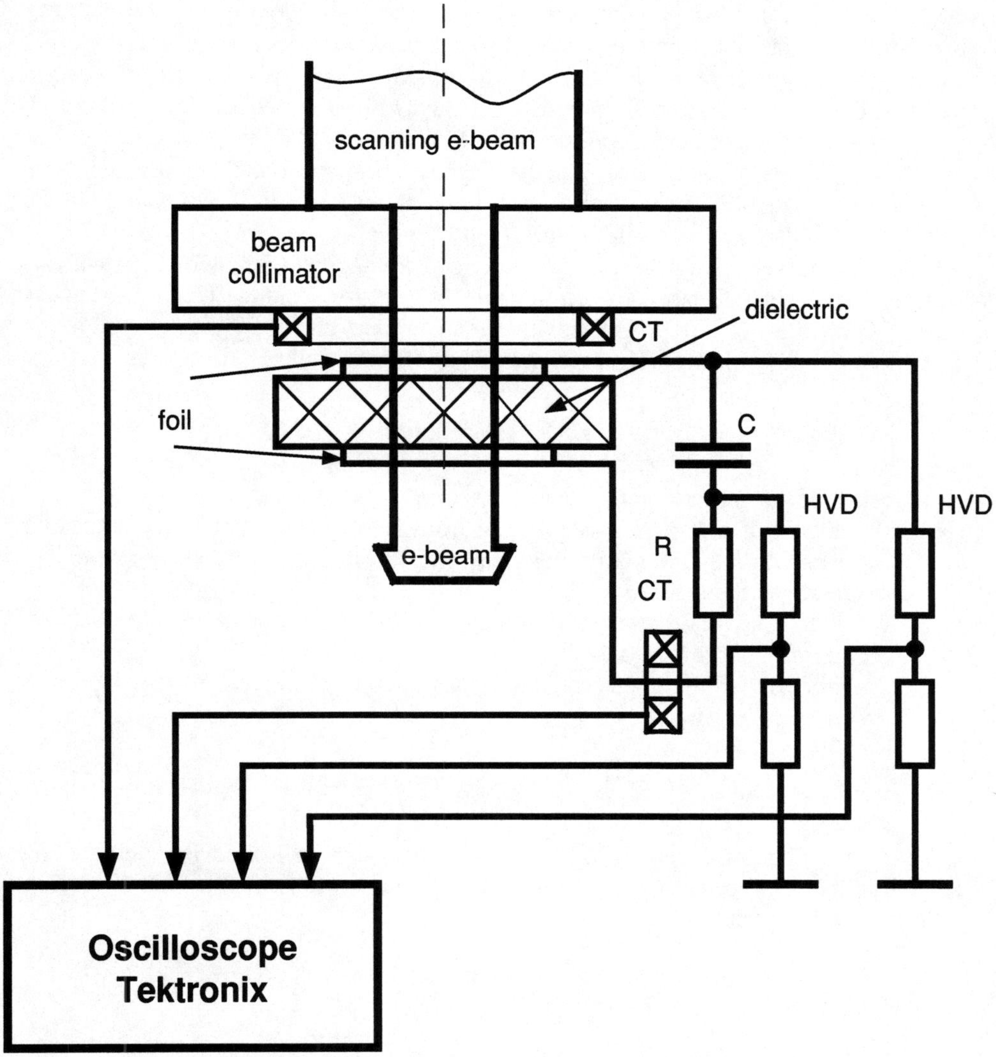

FIGURE 2. Electrical diagram for measurements of conductivity for dielectrics under irradiation by scanning electron beam

The measurements of conductivity of dielectrics with low and high value of dielectric constants were shown the general character in the kinetic and storage of electrical charge inside of dielectrics. These results confirm the model of ionized condensed mater in solid state and electrical discharge in dielectrics. Analysis of time characteristics for electrical breakdown of dielectrics gave good correlation of experimental data with experimental results considered in monograph [1].

The pulsed conductivity for dielectrics with low dielectric constant has range σ~ (0.008 – 0.3)S/m for absorbed doses 20-100 kGy. The pulsed conductivity of dielectric with high dielectric constant has range σ ~ (0.0002 – 0.01)S/m for absorbed doses 20 – 100 kGy. These experimental data have good correlation with theory [8]. It will reasonable explanation of results for comparison of absorbed electron beam doses from experiments and computer modeling [9].

The comparison of conditions for electrical breakdown for dielectrics with high dielectric constants for pulsed electron beam [4] and for scanning electron beam (for scanning electron beam for samples with dimension 1x1 cmxcm pulse duration is about 140 microseconds) confirm the same threshold electrical fields.

CONCLUSION

As results of these experiments we can make next main conclusions:

a) the effects of electrical space charge from electron of beam in dielectrics have influence on the forming of electrical conductivity of electrical discharge inside of irradiated dielectrics;
b) the pulsed conductivity of dielectrics at during of irradiation increasing with increasing of absorbed doses;
c) the time parameters of electrical discharge inside of dielectrics have dependence on dielectric constant.

ACKNOWLEDGEMENTS

I am very happy to say thanks for help in the work to Robert Langley and Jerry Kriebel and Ivan Korenev for preparation of samples for experiments and treatment of CTA film dosimeters.

REFERENCES

1. Gromov V.V. *Electrical charge in irradiated materials*. Energoizdat. Moscow, 1982
2. Vaisburg D.I., Semin B.N., Taranov E.G. et al. *Solid state high energy electronics*. Nauka, Novosibirsk, 1982.
3. Korenev S.A. Proc.: VI International Conference on Radiation Effects in the Insulators, Weimar, FRG, 1991, p. 24.
4. Korenev S.A. Proc, VIII USSR Conference on Physics of Dielectrics, **IV** , Tomsk, 1988, p. 103.
5. Pottie J. *Nuclear Instruments and Methods in Physics Research* **B40/41,** 943-947 (1989).
6. Korenev S.A. *Electron sources of diode type with cathodes anodes from high temperature superconductors*. Preprint JINR N P13-94-192, JINR, Dubna, 1994.
7. Korenev S.A., Sikolenko V.V. *Pulsed method of measuring critical current of high temperature superconductors*. Communication of the JINR, Dubna, 1993.
8. Sesler G.M. *Appl. Phys*. **43**, 405 – 408 (1972)
9. Kriviosheev O.E., Korenev S., Kolchuzkin A., Tropin I. *Distribution of Absorbed Doses in the Materials Irradiated by "Rhodotron" Electron Accelerator: Experiment and Monte Carlo Simulations,* Preprint FERMILAB-Conf-01/162-T, FNAL, Batavia, 2001.

Study of the Spatial Distribution of the Langmuir Wave Amplitude in the Turbulent Plasma using Electron Beam Shadowgraph Technique

Ritoku Ando, Hisashi Fujii, Satoshi Wada, Norio Toyosugi, Hiromasa Morishita, Naoki Okamoto, Masaru Masuzaki, and Keiichi Kamada

Faculty of Science, Kanazawa University, 920-1192, JAPAN

Abstract. New diagnostic, "electron-beam-shadowgraph", is examined for taking the spatial distribution of transient fields in a plasma. It is applied to the unmagnetized plasma, which is perturbed by an intense relativistic electron beam (IREB). Strong deflection is observed with duration of about 1 μs just after the IREB injection. The shadowgraph image is characterized by presence of aggregate holes. The size of the hole is changed with the plasma density, and corresponds to the 15 – 60 Debye lengths. Important agreements are observed between the experiment and the strong Langmuir turbulence theory.

INTRODUCTION

An electron beam excites electric field by the instabilities when it passes through a plasma. If the beam is dense enough, the power density of the electric field becomes also high comparable to the thermal energy of the plasma electron. The oscillating field with the near plasma frequency is unstable, then localized spatially and becomes bumpy. The state is called the strong Langmuir turbulence, in which many collapsing nonlinear wave packets, called the "caviton"s are produced.

Normal electron-beam probing technique employs a beam with a small cross-section. The local electric field in the plasma is estimated from the deflection angle of the beam. It is difficult to obtain the spatial distribution of the transient electric field. New diagnostic, "electron-beam-shadowgraph", with a large cross-section is examined. The shadowgraph of the unmagnetized plasma, which is perturbed by an intense relativistic electron beam (IREB), is taken.

EXPERIMENTAL SETUP

We have developed an electron beam probing technique applicable to IREB-plasma system. The "shadowgraph" technique is a kind of the beam probing with employing a beam with a large cross-section. An improvement of the detector which increases the signal-to-noise ratio, S/N, makes it possible to use the lower current density beam. The outline of the electron probing system used for this experiment is shown in Fig.1. An electron gun for the probing was set at the position z = 305 mm, where z is the distance from the anode. The probing beam propagates

CP650, *Beams 2002: 14th International Conference on High-Power Particle Beams*, edited by T. A. Mehlhorn and M. A. Sweeney

perpendicularly to the main chamber. An electron-beam has current of ~ 30 mA, energy of 50 kV, and pulse length of 100 ns. Because the probing beam is defocused, the beam diverges and the cross-section increases with the distance from the gun. The distance from the electron gun to the center of the main chamber is 220 mm, and that from the center to the phosphor screen is 346 mm. The projection magnifies the object at the center 2.6 times. The orifices are set along the path of the probing beam as shown in Fig.1 to reduce the IREB influx to the screen and the electron gun. The diameter of the probing beam is about 70 mm, which is limited by the first orifice placed at the entrance of the main chamber. The phosphor screen is set larger than the beam diameter. In this experiment, the time resolution for the measurement is set longer than the Langmuir oscillation period but shorter than the timescale of the ion movement. In this setting, the probing beam is deflected by both static and oscillating electric fields, which deflect and spread out the beam on principle.

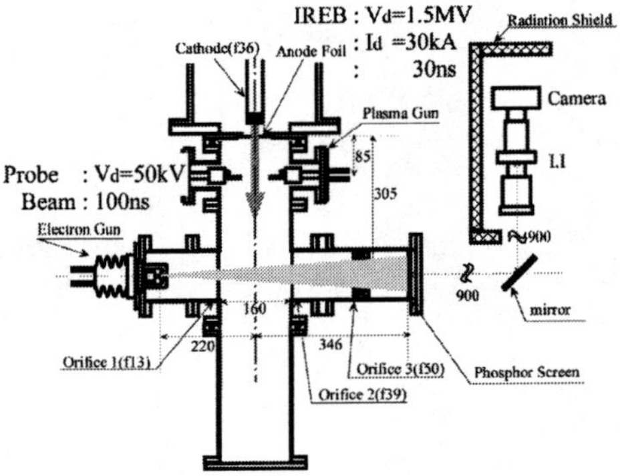

FIGURE 1. Experimental setup

An IREB with energy of 1.5 MeV, current of 30 kA, and duration of 30 ns is injected to the plasma. A drift chamber, 600 mm long and 160 mm in diameter, is used. No external magnetic field is applied. The rail-type plasma guns with carbon rods are set at the position of z = 100 mm. The plasma is produced by the rail-type guns and its after-glow plasma is utilized. The plasma density, n_p, was measured with a microwave interferometer and Langmuir probes. The n_p is a function of the time, τ, from a start of the discharge. The n_p at z = 175 mm reaches 3 x 10^{13} cm^{-3} at τ=12 μs. After the plasma reaches at the chamber wall at τ=20 μs, the n_p is distributed with a gentle slope in the chamber and decrease exponentially in time. The main data presented in this paper is obtained with the timing of τ= 50 μs, 60 μs, 80 μs, 100 μs, with plasma density of n_p= 4 x10^{11} cm^{-3}, 1 x10^{11} cm^{-3}, 4 x10^{10} cm^{-3} and 2 x10^{10} cm^{-3} respectively. The main component of the plasma is probably carbon, but it contains other components, like hydrogen or oxygen as the impurities.

EXPERIMENTAL RESULTS

The examples of the beam pattern are shown in Fig.3. The dark part on the picture is the location where the electrons hit

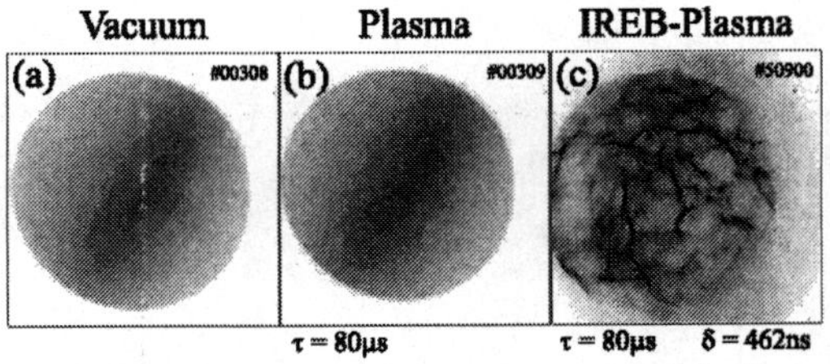

FIGURE 2. Examples of the beam pattern before and after the deflection

the screen. Fig.3(a) shows the beam pattern without deflection in the case of a vacuum. The cross-section of the beam is observed as a large dark circle in the pictures. The beam density is not perfectly homogeneous. The alteration of the spatial distribution of beam density from the original beam pattern is measured.

The probing beam is not affected by the presence of undisturbed plasma, as shown in Fig.2(b). But the pattern is changed strongly by the participation of the IREB injection. Many holes with various sizes are observed in Fig.2(c). The typical shadowgraph image is characterized by appearance of aggregate holes. From another point of view, the pattern behind the holes is a mesh. It is an irregular mesh. The mesh size from a few-millimeter to few-centimeters is observed through the experiment. It is observed that the mesh size, *b*, strongly depends on the n_p. It becomes larger when the n_p decreases, *i.e.*, as the increase of the τ, as shown in Fig.3. The deflection is observed with duration of about 1 μs just after the IREB injection. The parameter δ is taken as the time interval from the beginning of IREB pulse to the beginning of the probing beam. Following dependency on the δ is observed. Figure 4 shows the typical change of the pattern with the time transition under a fixed n_p. (1) Clear image of the hole is not observed for any shot at $\delta < 400$ ns. (2) Holes are clearly recognized at $\delta = 450 - 700$ ns. The size changes in this time duration as shown in Fig.5. (3) At $\delta = 700 - 1600$ ns, the picture of the holes becomes thinner. The holes with small circular shape are observed. (4) It becomes impossible to observe any deflection after $\delta > 1600$ ns.

The perturbation is decreased by reducing the IREB current. A masking plate with many holes is attached to the anode foil for reducing the IREB current. The beam density dependence in the case of $\tau = 100$ μs is shown in Fig.6. The transmittance rate is changed between 4 % and 100 %. The n_b/n_p is estimated about between 0.023 and 0.29. The aggregate holes gradually disappear as the IREB current decreases. A number of holes cover the whole surface in every case when it appears. The mesh size decreases with the decrease of the n_b. The perturbation is

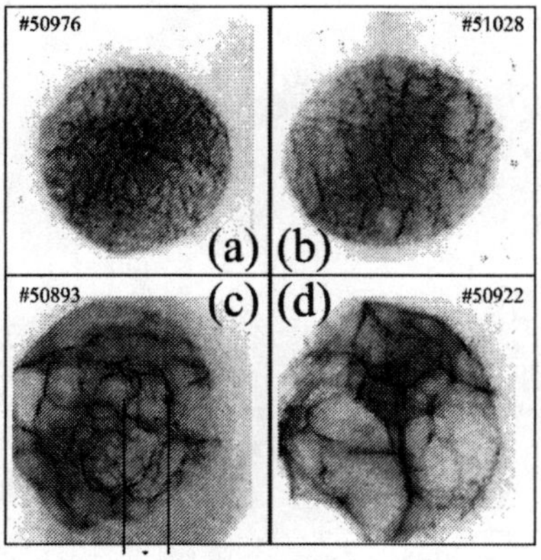

FIGURE 3. The plasma density dependence. (a) τ= 50 μs, (b) 60 μs, (c) 80 μs, (d) 100 μs. The diameter of the hole is measured from one mesh to the next mesh, which is mentioned as the mesh size, *b*, in this paper.

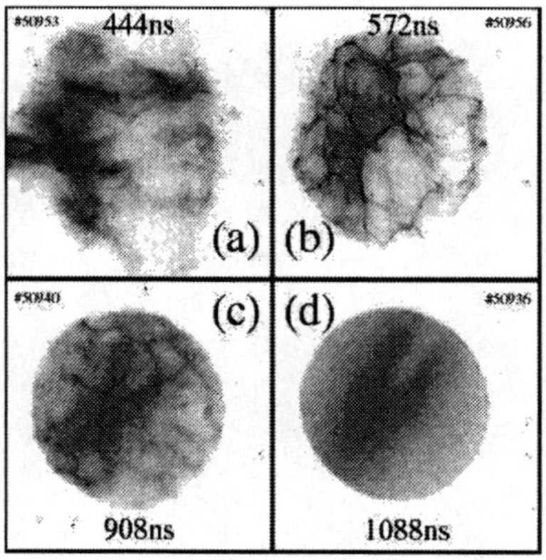

FIGURE 4. Typical shadowgraph after an IREB passed through. (a) δ=444ns, (b) 572ns, (c) 908ns, and (d) 1088ns. The plasma density is 2×10^{11}cm^{-3}, and the beam density is about 10^{10}cm^{-3}.

small enough to recognize on the image for n_b/n_p= 0.023.

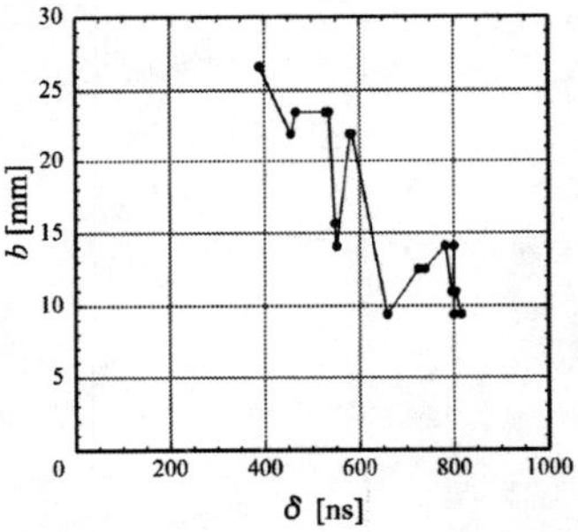

FIGURE 5. The mesh size *b* decreases with *δ*.

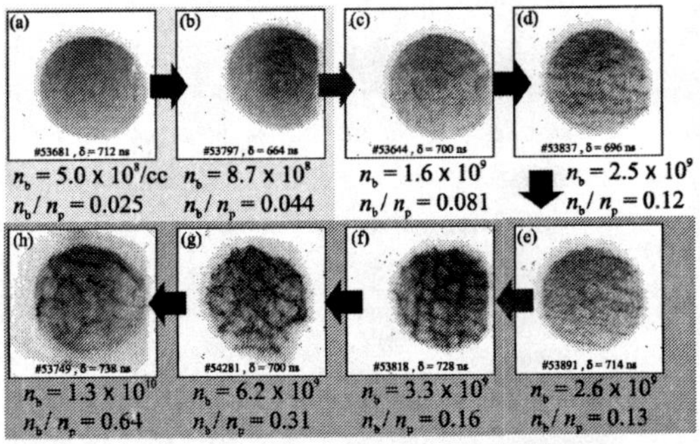

FIGURE 6. The beam density is decreased with keeping plasma density constant, τ=100 μs.

CONCLUSIONS

The electron-beam-shadowgraph image of the strongly disturbed plasma by IREB was always characterized by presence of aggregate holes. The various dependences on basic parameters were investigated experimentally. Those dependences can be explained by assuming the proper "caviton" field. A measurement of the accurate electron temperature probably justify the caviton model and improve the reliability of the diagnostic. Application of this new diagnostic to other phenomena is expected.

ACKNOWLEDGEMENTS

A part of this work is supported by a Grant-in-Aid for Scientific Research from Ministry of Education, Science, Sports and Culture, Japan.

REFERENCES

1. P. A. Robinson, Rev. Modern Phys., 69, 507-573 (1997).
2. Ando, R., et. al., IEEE trans. Plasma Sci., **27**, 1545-1547 (1999).
3. Ando, R., et. al., Proc. BEAMS2000, 893-896 (2000)
4. Ando, R., et. al., Research Report NIFS PROC-50, ISSN 0915-6348, 163-172 (2001).

The Study of Explosive Emission from Carbon Nanotubes

Sergey Korenev

STERIS CORPORATION, STERIS Isomedix Services,
2500 Commerce Drive, Libertyville, IL 60048, USA

Abstract. The carbon nanotubes (CNT) found applications for high density current electron emitters. The main interest for forming of high current electron beams using CNT is high concentration of electrical field on the nanotubes and high value of yield by electrons for field emission. The experimental results for time processes of forming cathode plasma and extraction of electron beam are presented in the report.

INTRODUCTION

The results of investigations of field emission cathodes on basis of carbon nanotubes were shown good emission characteristics with ability to form the high density beam current [1,2,3]. But the main problem of this type of field emitters is lifetime. The small sizes of nanotubes determine the high of geometry factor for increasing of electrical field on the surface of nanotubes. The investigations of carbon fiber cathode allowed to elaboration of plasma cathodes with explosive electron emission [4]. The general concept of carbon fiber cathode with explosive electron emission consists in the using of carbon fiber for forming of cathode plasma, which is pulsed emitter of electrons.

The similar properties of carbon nanotubes also allow to use this material for initiator of cathode plasma. The experimental data for time parameters and emission properties have different interpretations for case of forming of electron beams and vacuum discharge in gap of diode with cathode from carbon nanotubes [1,5].

The results of experimental investigations of explosive electron emission from carbon nanotubes are presented in this paper.

EXPERIMENT

The pulsed electron source [6,7] was used for investigation of questions for forming of cathode plasma of the cathode from carbon nanotubes. The block-diagram of electron source with cathode from carbon nanotubes is given on Figure 1. The SEM images of carbon nanotubes are presented on the Figure 2 a,b. The current transformers (Rogowski Coils) and Faraday Cups were used for measurements of emission characteristics. The high resistance divider was used for measurements of high voltage from Marx generator.

The pressure of residual gas in vacuum chamber of electron source was 10^{-5} Torr.

CP650, *Beams 2002: 14th International Conference on High-Power Particle Beams,*
edited by T. A. Mehlhorn and M. A. Sweeney

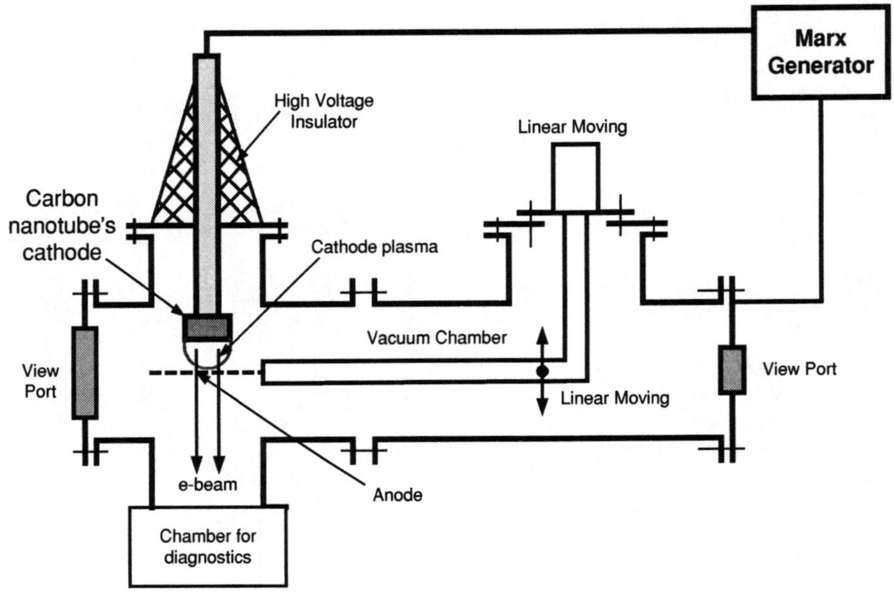

FIGURE 1. The block-diagram of electron source for investigation of explosive emission from carbon nanotubes cathode.

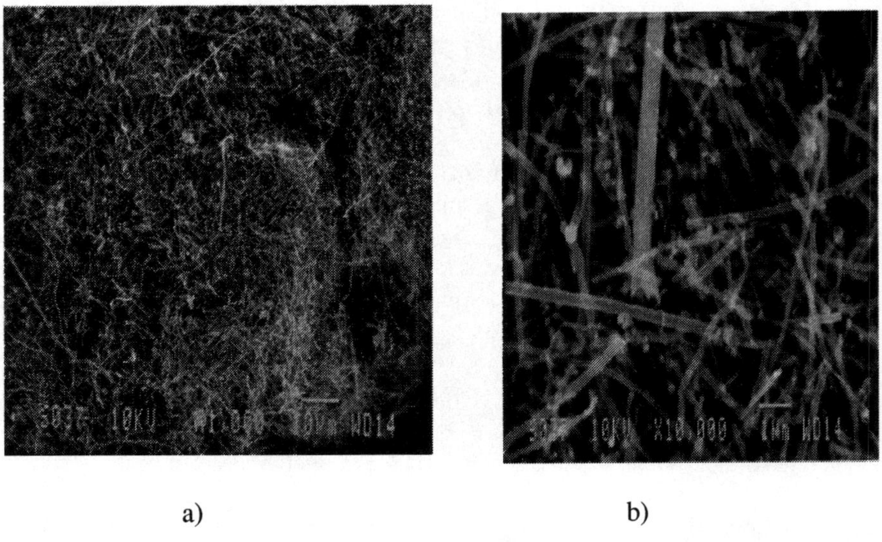

a) b)

FIGURE 2. The SEM images of carbon nanotubes for cathode.

RESULTS OF EXPERIMENTS

The typical emission characteristics for carbon nanotubes cathode with diameter 1.0 cm and distance between cathode and anode 1.0 and 2,0 cm (SEM images on Figure 2 a,b) are presented on the Figure 3. The pulse duration of accelerating voltage is 100-400 nsec.

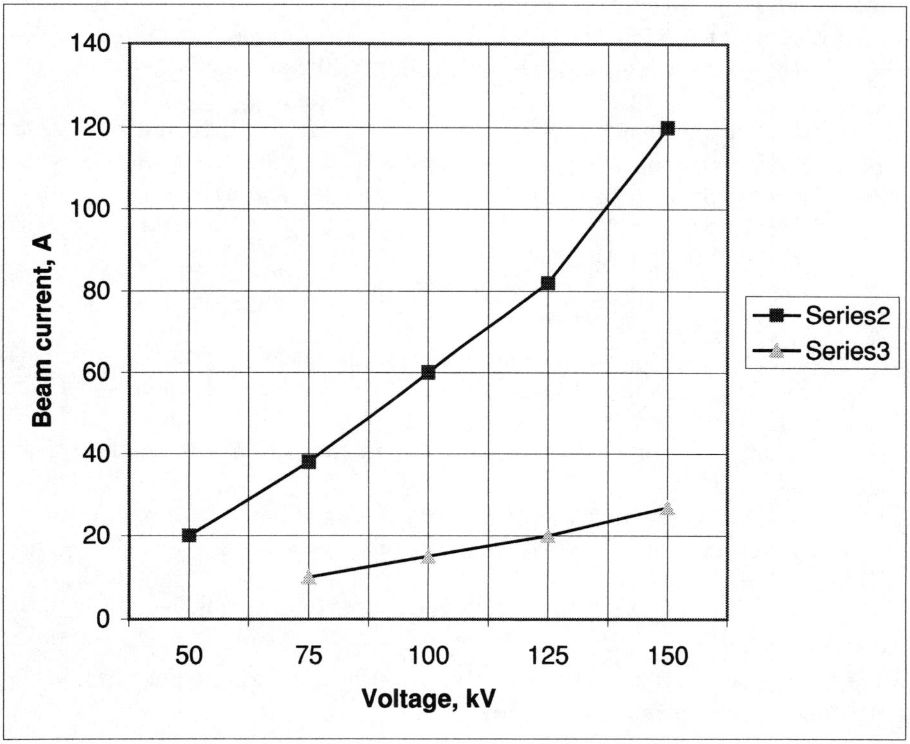

FIGURE 3. Emission characteristics of carbon nanotubes cathode with explosive electron emission: gap is 1 cm (series 2) and gap is 2 cm (series 3).

The high emission properties for carbon nanotubes for autoelectron emission can be explained by geometry factor of increasing of electrical field. For explosive electron emission mode of carbon nanotubes cathode the emission properties determines by parameters of cathode plasma. The parameters of cathode plasma depend from energy dissipated to vacuum discharge from high voltage generator. Analysis of current yield for electron beam after anode shows good correlation of experimental data with Child-

Langmuir Law. The geometry of nanotubes allows forming the stable cathode for short time (few nanoseconds).

The investigations of lifetime for carbon nanotubes for cathode with explosive electron emission shown the life time about 10^4 pulses with following decreasing of beam current. It can be explain by destroy of surface for carbon nanotubes for effect of deposition of film for anode material. In the case of forming of cathode plasma, the cathode plasma moved to direction of abode and interaction of this cathode plasma with anode leads to pulsed plasma ablation. As a result of these phenomena we have film on the surface of carbon nanotubes that decreasing the influence of effect of small sizes of nanotubes and variation for conditions for forming of cathode plasma.

The mechanism of forming cathode plasma on the surface of carbon nanotubes can be explained autoionization of gases molecules around nanotubes by high electrical field and autoemission electrons with following of evaporation of carbon by ions in primary cathode torch.

CONCLUSION

As results of these investigations we can make next main conclusions:

a) the emission properties of explosive emission cathode from carbon nanotubes determine by properties of cathode plasma;
b) the current yield for electron beam has good correlation with classical Child-Langmuir Law.
c) The lifetime of carbon nanotubes cathode has limitation and consists about 10^4 pulses in range of few hundred nanoseconds.

ACKNOWLEDGMENTS

I want to say thank you for help in the work to Ivan Korenev and Dr. Mike Kozlov for providing of carbon nanotubes material.

REFERENCES

1. Guliaev Yu.V., Chernozatonsky L.A., Kosakovskaia Z.Ya., Sinitcin N.I., Torgashev G.V., Zaharchenko Yu.F. *Vacuum Sci.& Techn.*, **B13,** 234 - 238 (1995).

2. Guliaev Yu.V., Sinitcin N.I., Torgashev G.V., Chernozatonsky L.A., Kosakovskaia Z.Ya., Zaharchenko Yu.F. *Microelectronics*, **26,** 84-88 (1997).

3. Dean K.A., Chamada B.R. *Appl. Phys. Lett.,* **76,** 375 – 377 (2000).

4. Korenev S.A., Baranov A.M., Kostychenko S.V., Chernenko N.M. "Cathode on the basis of carbon-fibrous plastic material". Preprint JINR No 13-88-292, Dubna, 1988; *J. Instrum. Exper. Technics,* **5,** 194 – 196 (1988.)

5. Shiffler D. LaCour M., Ruebush M., Howorth M., Zadar D. *Proc. IEEE International Conference on Plasma Science*, Alberta, Canada, 2002, p. 151.

6. Korenev S. Design and Investigations of pulsed electron and ion sources and their applications, Thesis of Doctor Science , JINR, Dubna, 1995.

7. Korenev S. Proc. IEEE International Conference on Plasma Science, Alberta, Canada, 2002, p. 333.

ION BEAM PHYSICS AND APPLICATIONS

Plasma Channel Transport for Heavy Ion Fusion: Investigation of Beam Transport, Channel Initiation and Stability

Andreas Tauschwitz[a], Christoph Niemann[b], Dan Penache[b], Richard Birkner[b], Dieter H.H. Hoffmann[b], Renate Kobloch[b], Stefan Neff[a], Radu Presura[a], Dave Ponce[c], Frank Rosmej[a], Simon Yu[c]

[a]*GSI Darmstadt, Planckstr. 1, 64291 Darmstadt, Germany*
[b]*Technische Universität Darmstadt, Schlossgartenstr. 9, 64289 Darmstadt, Germany*
[c]*Lawrence Berkeley National Laboratory, 1 Cyclotron Road, Berkeley, CA, USA*

Abstract. For final beam transport in an IFE reactor three alternatives are mainly discussed. These are neutralized ballistic transport, self-pinched transport, and plasma channel transport. Discharge plasma channels were investigated in the recent years at GSI Darmstadt and at LBNL Berkeley in a number of experiments. Different initiation mechanisms for gas discharges of up to 60 kA were studied and compared. In the Berkeley experiments laser ionization of organic vapors in a buffer gas was used to initiate and direct the discharge while at GSI laser gas heating and ion beam induced gas ionization were tested as initiation mechanisms. Measurements of temperature, electron density, gas density, and magnetic field distribution in the channels are compared with results of beam transport experiments at the GSI UNILAC accelerator and with MHD simulations of the 1D-fluidcode CYCLOPS, which was developed in Berkeley. Good agreement between plasma diagnostics results, measured ion optical properties and MHD simulations was found. Parameters that are required for a reactor application are a discharge current of 50 kA, a channel diameter below 1 cm, a pointing stability better than 500 µm, and MHD stability for more than 10 µs. These parameters have been demonstrated in the recent experiments. The results imply that transport channels work with sufficient stability, reproducibility and ion optical properties in a wide pressure range and for various discharge gases.

INTRODUCTION

Neutralized beam transport and focusing in gas discharge plasmas is under investigation for heavy ion beam fusion since 1995 at the Lawrence Berkeley National Laboratory (LBNL) [1] and at GSI Darmstadt [2] following extensive work on discharge channel transport for light ion beams that was done at Sandia, NRL, and Osaka mainly in the years from 1977 to 1983. The key elements of a focus and transport system for an IFE reactor are described in [3]. Beam combining is done at the entrance of an adiabatic plasma lens outside of the reactor, which reduces the beam diameter to the desired focal spot size on the target. The adiabatic lens is then extended into a laser initiated discharge channel which transports the beam at constant diameter through the reactor chamber to the target.

CP650, *Beams 2002: 14th International Conference on High-Power Particle Beams*, edited by T. A. Mehlhorn and M. A. Sweeney

CURRENT WORK AT GSI DARMSTADT AND LBNL

Recently PhD works were completed at LBNL by Ponce [4] and at GSI by Niemann [5] and Penache [6]. The LBNL work concentrates on the investigation of dynamical properties both of the neutrals and the ions of a laser induced discharge in a mixture of nitrogen and benzene. A density depression before ignition of the main discharge is created by an electrical prepulse which is mandatory in this experiment. This prepulse reduces the gas density in the channel by an order of magnitude. The stability of the high current discharge is attributed to the high gas density that surrounds the central depression channel. The current distribution in the channel was investigated by measuring the Faraday rotation of a CO_2 laser beam intersecting the discharge channel.

The work by Niemann covers both the Berkeley and the Darmstadt experiments and concentrates on the plasma diagnostics by interferometry, schlieren diagnostics, Faraday rotation, and a time resolved spectroscopic temperature determination using a rate equation model. Different channel initiation methods employing gas ionization and gas heating by a laser as well as gas ionization by an ion beam are described. First experiments on intersecting discharge channels are described which show that the discharge configuration must be symmetrical so that magnetic forces on the channels cancel. In addition these experiments show that the breakdown risk between channels is small. In a proof-of-principle experiment an ion beam was transported through one of the intersecting channels without problems.

The work by Penache describes the channel transport experiment at GSI and emphasizes the ion beam transport properties using the UNILAC accelerator. In several runs different ion species with an energy of 11.4 MeV/u were transported through the discharge channel with 50 cm length. At the time of peak current (up to 60 kA) the ions make more than one full betatron oscillation in the short channel. The transport properties are investigated spatially resolved by using a pepperpot mask in front of the channel. The resulting beam pattern of the single beamlets on a scintillator behind the channel are interpreted by a Monte Carlo simulation of the beam transport. The results are compared to magnetic probe measurements in the discharge channel.

In the following sections different channel initiation mechanisms are compared and the stability of the discharge channels is described which is a key issue to make discharge channel transport a viable concept for ion beams. With measurements of temperature, electron density, gas density, and magnetic field distribution in the channels and the comparison with results of beam transport experiments at GSI the channel transport principle is well characterized by experimental data over a wide range of discharge conditions.

DISCHARGE CHANNEL INITIATION MECHANISMS

Gas Ionization by Lasers

The obvious way to induce and to direct a gas breakdown is to ionize the gas along the desired breakdown path. There are different ways to ionize a gas with a laser. One of these possibilities is direct ionization by single photon absorption. With few exceptions, like the ionization of alkali vapors, a laser in the VUV range is needed for this scheme. A possibility is the ionization by two-photon absorption which is possible for many organic molecules using excimer lasers. This approach was extensively tested and investigated at Berkeley.

Some molecular gases like ammonia can be efficiently heated in the IR region where they have strong absorbing rotation or vibration levels of the molecules. Due to the temperature increase a small amount of thermal ionization can be expected.

Finally short pulse lasers can be used to achieve field ionization along the laser path in the gas. This last possibility has the advantage that the choice of the discharge gas is not limited by the initiation technique

Thermal Gas Expansion

Although experimentally very similar to the mentioned gas ionization by a laser, the physics of discharge initiation by thermal gas expansion is very different. For this approach the discharge gas is heated by a laser. Only few laser – gas combinations allow an efficient heating of the gas. The combination chosen for the experiments at GSI was a line tunable CO_2 laser and ammonia gas following similar work in Sandia [7]. Details of the discharge initiation for the GSI experiment are described in [8].

At GSI the breakdown initiation was investigated for different discharge delays with respect to the laser pulse and for different laser intensities. It was found that a well defined minimum delay between laser and discharge trigger exists. If a discharge is triggered before this minimum delay time it does not ignite or ignites with a very large jitter. This minimum delay is in the range of 10 to 20 µs depending on the gas pressure. With the help of hydro-simulations using the CYCLOPS code the delay is identified as the expansion time of the laser heated gas to the minimum on-axis density. The process of gas expansion was directly observed by interferometry. The interferometry shows that the expanding gas forms a steep gas wall surrounding a density depression zone along the laser path. The development of the gas depression channel was investigated with temporal resolution measuring the small angle scattering of the UNILAC heavy ion beam in this channel [9]. Both the interferometric as well as the beam scattering measurements show a density variation in agreement with simulations using the CYCLOPS code which assume a gas heating by the laser to a temperature of 500 to 800 K.

Gas Initiation by an Laser Beam

Stable discharge channels were initiated by an ion beam pulse from the UNILAC linear accelerator at GSI, which creates a trace of seed electrons along the desired path of breakdown. For these experiments an ion beam with 2 mm diameter aligned exactly to the channel axis was used to prepare the ionization channel. In this way, discharges were created in NH_3 and argon gas. The discharge was triggered 20 µs after the beginning of a 50 μs long ion beam pulse with an electric current of less than 5 μA inside the 50 cm long discharge chamber. The discharges were stable and reproducible even at very late times in both ammonia and argon at currents exceeding 50 kA. Channels in NH_3 could only be produced at pressures up to 5 mbar. The pressure in Ar discharges was only limited by the differential pumping system. Even after the current reversal of the ringing discharge current the channels show no sign of instabilities. The discharge current as well as the channel diameter evolution for discharges in ammonia and argon are compared in figure 1. The diameter of discharges in argon is considerably smaller than in ammonia, which is important for beam transport for heavy ion fusion since the channel diameter limits the beam spot size on the target.

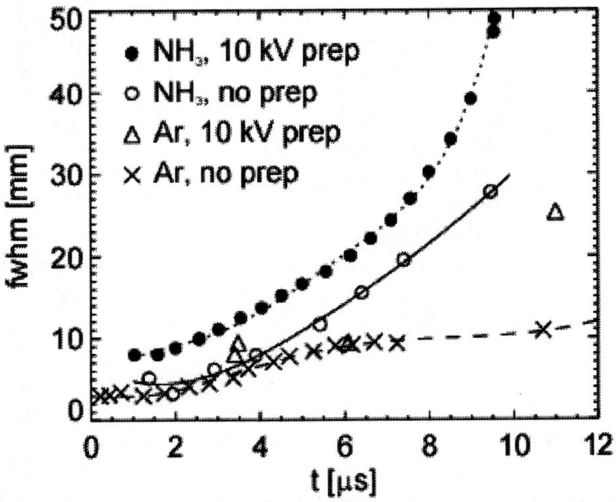

Figure 1: Channel diameter (FWHM) as a function of time for discharges in ammonia and argon with and without an electrical prepulse.

Besides typical MHD instabilities the discharge channels show a shift of the channel axis during the discharge. For laser initiated channels this shift of the axis can be several millimeters. This observation was made by fast-shutter images of the plasma light emission recorded side-on and confirmed by a deflection of the transported ion beam. Beam initiated channels in NH_3 show only a shift of about ±200 μm, for Ar discharges this shift is reduced to about ±100 μm. Discharges in other gases like N_2, He, Kr and Xe have been successfully initiated but need further investigation to determine their stability. The transverse stability of the discharge axis is important for a fusion reactor, where the beam must hit a mm-sized target.

The improved stability, the simple initiation mechanism and the possibility to create channels in any gas make ion beam initiation of the discharge especially interesting for heavy ion fusion.

DISCHARGE CHANNEL STABILITY

The MHD stability of discharge channels is crucial for ion beam transport. The dominant type of instability in all investigated channels was the kink (m=1) instability. To investigate the growth rate of this instability the channels were recorded side-on with a framing camera. Below a pressure of 15 mbar no instabilities were observed during the first half wave of the current pulse. For a pressure of 18 mbar instabilities were found in about 25% of the channels. These instabilities usually start to become visible around the time of peak current at about 4 μs. At a pressure of 21 mbar the instabilities appear between 2 and 3 μs after ignition and are observed in all discharges.

This increasing growth rate of instabilities with increasing gas pressure is unexpected, but a scaling law derived by Mannheimer [10] shows that the growth rate depends on many variables, which are all interconnected in a rather complex way.

Here Γ is the growth rate of an instability with wave number k for an azimuthal

$$\Gamma \propto k \frac{B_\Phi}{\sqrt{\mu_0 n_i m_i}} \left(\frac{\rho_{ch}}{\rho_g} \right)^{1/2}$$

magnetic field of B_Φ, an ion density n_i, and a mass of the ions m_i. If the discharge takes place in a channel with initial gas density ρ_{ch} which is surrounded by a gas of density ρ_g, the growth rate is reduced by the higher mass of the surrounding gas. With this scaling the growth rate depends on the ionization degree, the radial current distribution, in case of a molecular discharge gas like ammonia on the gas chemistry and on the gas density profile before and during the discharge. These quantities depend on the local temperature and conductivity distribution in the plasma.

In addition to the framing pictures the channel stability was investigated using the ion beam deflection. This was done by imaging the location of an ion beamlet on a scintillator behind the discharge chamber. The azimuthal magnetic field of the

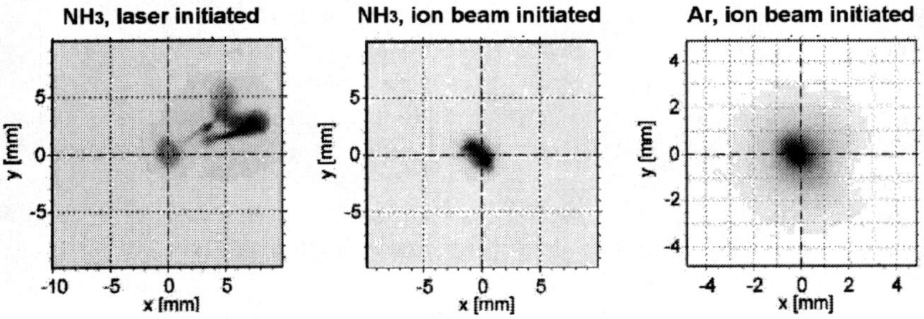

Figure 2: Scintillator images of an ion beam propagating through different channels. The motion of the beamlet on the scintillator was integrated over the first current half wave.

discharge causes betatron oscillations of the beamlet, which are centered around the discharge axis. If the discharge is aligned with the beam axis, the image of the beamlet on the scintillator changes in size but remains at a fixed position for all times during the discharge current oscillation. For an unstable channel, which moves around its initial axis, the beamlet also moves on the scintillator. Figure 2 compares time integrated scintillator images for laser- and ion beam-initiated discharges in NH_3 and for ion beam initiated channels in Ar. The exposure time of the pictures started 1 μs before breakdown and extended over the first half wave period of the discharge.

Laser initiated discharges typically exhibit shifts of the beam position on the scintillator of up to 2 mm when an electrical prepulse is used. This situation is slightly improved without the prepulse. Although the use of a prepulse reduces the growth of MHD instabilities by the factor $(\rho_{ch}/\rho_g)^{1/2}$ it facilitates the shift of the channel axis due to the reduced gas density in the gas depression channel. The same behavior is found in ion beam initiated channels in argon. Directly comparing laser- and ion beam initiated channels in ammonia shows the significantly better stability of the beam initiated discharges. An explanation for this could again be the gas density depression by the laser, which facilitates the movement of the channel axis within this density depression zone. Typically the shift of the discharge axis for beam-initiated channels is an order of magnitude smaller than the shift of laser initiated discharges, in agreement with results of the framing camera investigations.

CONCLUSIONS AND PERSPECTIVES

The recent results at LBNL and at GSI on discharge characteristics and beam transport properties of laser and ion beam initiated discharges show that discharge channel transport is a viable concept for the final beam transport in heavy ion fusion.

A number of topics needs close attention in the near future. All experiments so far were restricted to a channel length below half a meter while the required channels in a fusion reactor have to be a factor of ten longer. Experiments to extend the channels at GSI up to 2m are planned in the same set-up that is currently used. A second point is an intense study of channel instabilities. Since instabilities have shown to depend on many details of the initiation scheme and discharge conditions it would be desirable to investigate instabilities in an environment that is very similar to the anticipated reactor conditions. To achieve this goal discharge initiation by a short pulse laser has to be tested. Although field ionization by a short pulse laser is the envisaged method of choice for a fusion reactor it has so far not been tested. First experiments are planned at GSI using the 10J / 500fs pulse of the PHELIX-preamplifier. These tests will allow investigating discharges in Xenon gas and it will be possible to include the effects of a small partial pressure of other gases, like constituents of a molten salt mixture that will be present in a liquid wall reactor, on the discharge characteristics.

REFERENCES

1. A. Tauschwitz, et al., Proc. Part. Accel. Conf., Dallas, 1852 (1995)
2. A. Tauschwitz, et al., Proc. Int. Conf. on Inertial Fusion Sci. and Appl.(IFSA), Bordeaux, 521 (1999)
3. S. Yu, et al., Nucl. Instr. and Meth. **A 415**, 174 (1998),
4. D. Ponce, PhD thesis, Univ. of Calif. at Berkeley (2002)
5. C. Niemann, PhD thesis, TU Darmstadt (2002)
6. D. Penache, PHD thesis, TU Darmstadt (2002)
7. J.N. Olsen, L. Baker, J. Appl. Phys. **52**, 3286 (1981)
8. C. Niemann, et al., J.Appl. Phys. **91**, 617 (2002)
9. S. Neff, et al., submitted to J. Appl. Phys. (2002)
10. W.M. Manheimer, M. Lampe, J.P. Boris, Phys. Fluids **16**, 1126 (1973)

Preparation of $SrAl_2O_4$:Eu,Dy Thin Films on Polyethylene Substrates by Pulsed Ion-Beam Evaporation

Hisayuki Suematsu, Atsushi Kosaka, Keizo Kato*, Weihua Jiang and Kiyoshi Yatsui

Extreme Energy-Density Research Institute, Nagaoka University of Technology, Nagaoka 940-2188, Japan
**Graduate School of Science and Technology, Niigata University, Niigata 950-2181, Japan*

Abstract. Thin films of $SrAl_2O_4$:Eu,Dy were prepared by a pulsed ion-beam evaporation method on Si single crystal, soda lime glass, stainless steel and polyethylene substrates at room temperature. In the XRD patterns for the thin films, crystallized $SrAl_2O_4$:Eu,Dy phases were detected in all thin films deposited on the substrates including polyethylene. From these results, we have successfully developed a novel thin film deposition method to prepare crystallized thin films on low-melting-point substrates.

INTRODUCTION

Strontium aluminate ($SrAl_2O_4$) has been known to be a phosphorescence material. Partial substitution of europium and dysprosium for strontium in $SrAl_2O_4$ yielded a long lasting phosphorescence[1]. Using this property, $SrAl_2O_4$: Eu,Dy powders have been used as pigments in luminous paints.

Thin films of $SrAl_2O_4$: Eu,Dy have been prepared by sputtering[2]. Since the as-deposited thin films consisted of an amorphous $SrAl_2O_4$:Eu,Dy phase, which has to be crystallized after the deposition. Furthermore, to exhibit phosphorescence properties, oxidation state of Eu has to be 2+, which is formed by treating the thin films at high temperatures in reducing atmosphere[2]. From these reasons, phosphorescence of the $SrAl_2O_4$:Eu,Dy thin films was only exhibited after annealing the as-deposited thin films.

A novel thin film preparation method of ion-beam evaporation (IBE) was developed[3]. This method utilized a high power pulsed ion beam produced in a pulsed power generator[4]. The ion beam was focused on a target, from which high density plasma was formed. The plasma was deposited on the surface of substrates which are placed facing toward the target. By the IBE method, various thin films were obtained with instantaneous deposition rate of a few mm/s. Furthermore, crystallized TiFe, B_4C[5-7] and $SrAl_2O_4$[8-10] thin films were successfully prepared without substrate heating or sample annealing.

CP650, *Beams 2002: 14th International Conference on High-Power Particle Beams,*
edited by T. A. Mehlhorn and M. A. Sweeney

Because of the necessity of the annealing, $SrAl_2O_4$ thin films had to be deposited on substrates with higher melting points than the annealing temperature. Contrary to the conventional methods, the IBE process did not require annealing to crystallize the thin films. Thus, it is also expected that crystallized $SrAl_2O_4$:Eu,Dy thin films can be obtained on substrates with low melting points, i.e., glass, metals or plastics. In the present research, preparation of crystallized $SrAl_2O_4$:Eu,Dy thin films on low melting point substrates, i.e. soda lime glass, stainless steel and polyethylene substrates were attempted. Crystallinity of the as-deposited thin films were examined by powder X-ray diffraction (XRD).

EXPERIMENTAL PROCEDURE

Thin films were prepared in a pulsed ion beam generator ('ETIGO-II')[4]. Ion beam and deposition chambers are schematically shown in Fig. 1. A voltage of 1 MV (peak) with pulse width of ~50 ns and current of 60 kA was applied between the cathode and the anode, which was covered with a polyethylene sheet (flashboard). The application of the pulsed voltage produced a pulsed light ion beam which mainly consisted of protons. The ion beam was geometrically focused on a sintered $SrAl_2O_4$:Eu,Dy target. The ion beam irradiation produced ablation plasma from the surface of the target. Substrates of (100) Si single crystal, stainless steel, soda lime glass and polyethylene plates were placed at 70 mm apart from the target. The ablation plasma was deposited on the substrates. The ion beam was irradiated on the target for 25 times. The substrates were kept in a vacuum of $2x10^{-4}$ Torr and at room temperature.

Phases in the as-deposited thin films were identified by XRD.

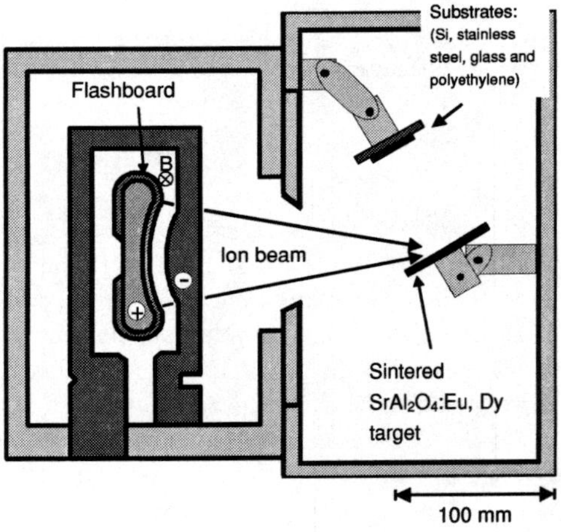

FIGURE 1. Experimental setup for thin film deposition by IBE.

RESULTS AND DISCUSSION

After the ion beam irradiations on the target, approximately 2 μm thick films were obtained on all substrates. An XRD pattern for a thin film deposited on a Si substrate is shown in Fig. 2(a). All peaks are matched with those for $SrAl_2O_4$ phase, which indicates that a single-phase-$SrAl_2O_4$:Eu,Dy thin film was prepared at room temperature by IBE. This result is comparable to the previous studies[8-10].

An XRD pattern for a thin film deposited on a glass substrate is shown in Fig. 2(b). Most peaks correspond to those for $SrAl_2O_4$ phase. A peak at 45°, which matches to the 330 reflection for $SrAl_2O_4$ phase but is stronger than the relative intensity shown in JCPDS card. This may be caused either by preferential orientation of some $SrAl_2O_4$ grains or by presence of a secondary phase. An XRD pattern for a thin film deposited on a stainless steel substrate is shown in Fig. 2(c). From Figs. 2 (b) and (c), crystallized and essentially single phase $SrAl_2O_4$:Eu,Dy thin films were deposited on

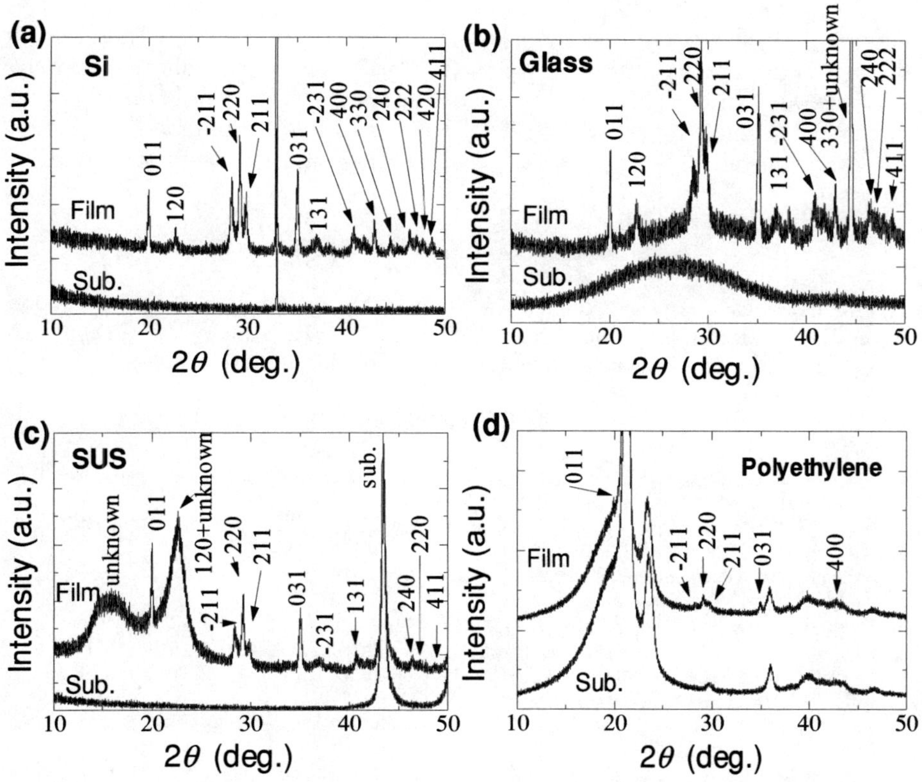

FIGURE 2. XRD patterns for $SrAl_2O_4$:Eu,Dy thin films deposited by IBE on (a) Si single cyrstal, (b) soda lime glass, (c) stainless steel and (d) polyethylene **substrates.**

the soda lime glass and stainless steel substrates. These results imply that crystallinity of $SrAl_2O_4$:Eu,Dy thin films deposited by IBE is not sensitive to the substrate materials.

Finally, a thin film was deposited on a polyethylene substrate. The XRD pattern for the thin film is shown in Fig. 2(d). Faint but apparent peaks were observed in the pattern and the diffraction angles for the peaks coincide with those of $SrAl_2O_4$ phase. A crystallized $SrAl_2O_4$:Eu,Dy thin film was successfully obtained by IBE on a polyethylene substrate.

Deposition of the $SrAl_2O_4$:Eu,Dy thin film on a polyethylene substrate is discussed here. In the IBE method, substrates were subjected to heat flux associated with dense and high temperature plasma. Temperature in the Si substrate during W thin film deposition by IBE was numerically calculated[11]. The temperature increased up to 800 °C until 30 μs after the beginning of the deposition. Then, the temperature rapidly dropd to 100 °C at ~1 ms. Thus, the substrate is heated during deposition but the duration of the heating must be restricted to approximately 1 ms.

Softening temperature of polyethylene is 80 °C, which is lower than that for soda lime glass (~550 °C) and melting point of stainless steel(~1500 °C). However, no damages were observed on the substrate[12]. The reason why the polyethylene substrate was not degraded should be the shortness in the heating duration during deposition. In the present study, a novel process to deposit crystallized thin film on low-melting-point substrates was successfully developed using the pulsed ion beam generator.

ACKNOWLEDGMENTS

This work was partly supported by a Grant-in-Aid for Scientific Research by the Ministry of Education, Culture, Sports, Science and Technology of Japan. The $SrAl_2O_4$:Eu,Dy target was supplied by Nemoto & Co. Ltd.

REFERENCES

1. Matsuzawa, T., Aoki, Y., Takeuchi, N. and Murayama, Y., *J. Electrochem. Soc.* **143,** 2670-2673 (1996).
2. Kato, K., Tsutai, I., Kamimura, T., Kaneko, F., Shinbo, K., Ohta, M. and Kawakami, T., *J. Luminescence* **82,** 213-220 (1999).
3. Shimotori, Y., Yokoyama, M., Isobe, H., Harada, S., Masugata, K. and Yatsui, K., *J. Appl. Phys.* **63,** 968-970 (1988).
4. Yatsui, K., *Laser and Particle Beams* **7,** 733-741 (1989).
5. Kitajima, K., Suzuki, T., Jiang, W. and Yatsui, K., *Jpn. J. Appl. Phys.* **40,** 1030-1034 (2001).
6. Suematsu, H., Kitajima, K., Suzuki, T., Jiang, W., Kurashima, K., Bando, Y. and Yatsui, K., *Appl. Phys. Lett.* **80,** 1153-1155 (2002).
7. Suematsu, H., Kitajima, K., Ruiz, I., Takeda, M., Suzuki, T., Jiang, W. and Yatsui, K., *Thin Solid Films* **407,** 132-135 (2002).
8. Sengiku, M., Oda, Y., Jiang, W., Yatsui, K., Kato, K., Shinbo, K. and Kaneko, F., *Jpn. J. Appl. Phys.* **40,** 1035-1037 (2001).
9. Kato, K., Ogura, Y., Sengiku, M., Shinbo, K., Kaneko, F., Oda, Y. and Yatsui, K., *Jpn. J. Appl. Phys.* **40,** 1038-1041 (2001).
10. Suematsu, H., Sengiku, M., Kato, K., Mitome, M., Kimoto, K., Matsui, Y., Jiang, W. and Yatsui, K., *Thin Solid Films* **407,** 136-138 (2002).
11. Jiang, W., Suzuki, T. and Yatsui, K., *Proc. 13th Int. Conf. Pulsed Power and Plasma Sicence,* in press (2001).
12. Suematsu, H., Kosaka, A., Honzawa, T., Kato, K., Jiang, W. and Yatsui, K., *Proceedings of 2nd International Symposium on Pulsed Power and Plasma Applicatoins,* 63-65 (2001).

Novel Applications of Intense Pulsed Ion Beams in Materials Science

K. Yatsui, W. Jiang, H. Suematsu, F. Endo, and S. C. Yang

Extreme Energy Density Research Institute, Nagaoka Univeristy of Technology, Nagaoka, Niigata 940-2188, Japan

Abstract. Using pulsed ion beam evaporation technique, experimental studies were carried out to embed metals into via holes in LSI. It will be shown that tungsten is successfully embedded in via holes with the aspect ratio of 4.3. Furthermore, light emission of blue has been observed from a poly silicon nanosize powders, which is also produced by pulsed ion beam evaporation technique.

INTRODUCTION

Rapid growth of pulsed power technology makes it possible new applications, which were not possible by conventional technologies. If we transfer the pulsed power energy to pulsed ion beams such as protons, wider applications have been possible. For example, the interaction of pulsed proton beam with target produces very high density plasma because the range of protons is very small compared with electrons. Here, the range is the distance, where the beam stops by loosing its energy due to interaction with targets. For example, the range of protons in aluminum target is estimated as 14 μm, while that of electron 1.5 mm. The short range of protons gives a huge energy density on target, resulting in high-density ablation plasma.

By the irradiation of proton beam with aluminum target, we have typically obtained the plasma density on the order of 10^{20} cm^{-3}. Depositing the ablation plasma on a substrate placed nearby, we succeeded in the efficient preparation of ZnS thin films in 1988, which was called pulsed ion beam evaporation (IBE). Later, various kinds of thin films have been prepared such as YBaCuO, $SrTiO_3$, $BaTiO_3$, $(Ba,Sr)TiO_3$, B_4C, ZrO_2, ITO, etc. Since many of them were already published elsewhere [1-3], we do not mention here only referring several articles. Using IBE to prepare thin films, furthermore, we have recently succeeded in to embed metals into *via holes* of LSIs. To embed metals into *via holes* is important in recent semiconductor manufacture, particularly with a large aspect ratio, the ratio of the depth of the hole over the diameter of the hole. Because these details were not reported, they will be introduced in this paper.

Such the ablation plasma, in addition, has been used to synthesize ultrafine nanosize powders (UNP) by the rapid cooling of the plasma with ambient gas [3,4]. Various kinds of UFP have been successfully synthesized such as Al_2O_3, AlN, TiO_2, and TiN.

CP650, *Beams 2002: 14th International Conference on High-Power Particle Beams*, edited by T. A. Mehlhorn and M. A. Sweeney

Since these details were also published, we will present recent interesting results of poly-silicon powders, which emit light emission of blue.

EMBEDDING METAL INTO VIA HOLES OF LSI

Figure 1 shows the outline of the experimental setup. The ion diode used is a magnetically insulated diode (MID), where the anode-cathode gap length is 10 mm. On the anode surface, polyethylene is pasted as a flashboard. The cathode also works as a one–turn theta-pinch coil. The current fed by the slow capacitor bank produces transverse magnetic fiel, by which the electrons emitted from the cathode cannot arrive at the anode. Ions from the anode plasma source can be extracted through a vane-structure cathode, where the protonic ratio is more than 80 %. The ion beam is irradiated on tungsten target, where the energy density on target is 40 J/cm^2 with the ion beam energy of 1 MeV (peak). Although the geometric focusing distance of the anode is 160 mm, the target was placed at 200 mm to reduce the damage by the beam irradiation. The via holes are located at d_{TS} (target-substrate distance) = 70, 80, 90 and 100 mm from the target. The number of shots irradiated was 1, 3, 5, and 10. The whole assembly of the equipment was evacuated to 2 x 10^{-4} Torr.
The cross-section was analyzed by a scanning electron microscope (SEM). In order to obtain the image due to metals, we have also diagnosed by SEM composite images.

Figure 2 shows the cross-section of the via hole of LSI. The hole is 185 nm in diameter, and 800 nm deep, yielding the aspect ratio of 4.3. In Fig. 1, TEOS and BPSG are the insulating layers, which stand for tetra ethyl ortho silicate and boro phospho silicate glass, respectively.

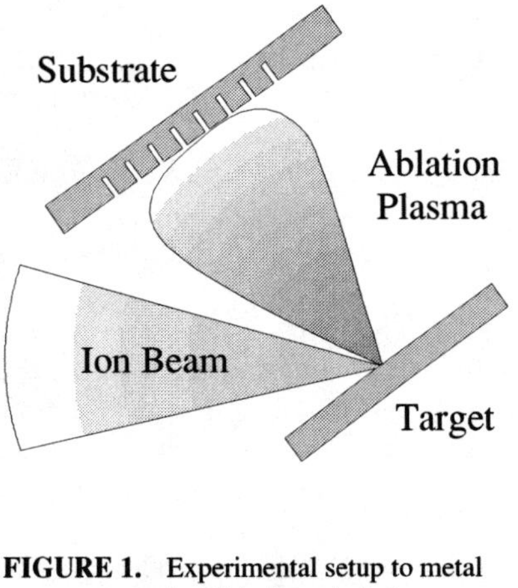

FIGURE 1. Experimental setup to metal build-in into via holes.

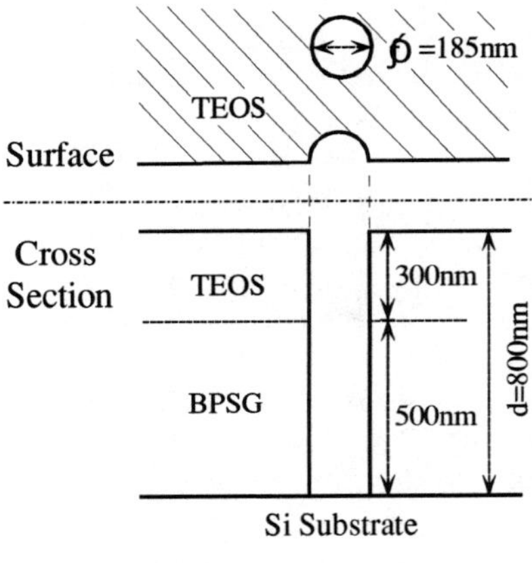

FIGURE 2. Cross-section of via hole of LSI.

Figure 3 shows the SEM images of the via hole with different d_{TS}. At the small distance, for example d_{TS} = 70 mm of Fig. 3 (a), there are too much tungsten to be embedded in the holes. At a larger distance of d_{TS}, as seen in Fig. 3 (d), we can see the smooth deposition of thin film of tungsten inside the via hole.

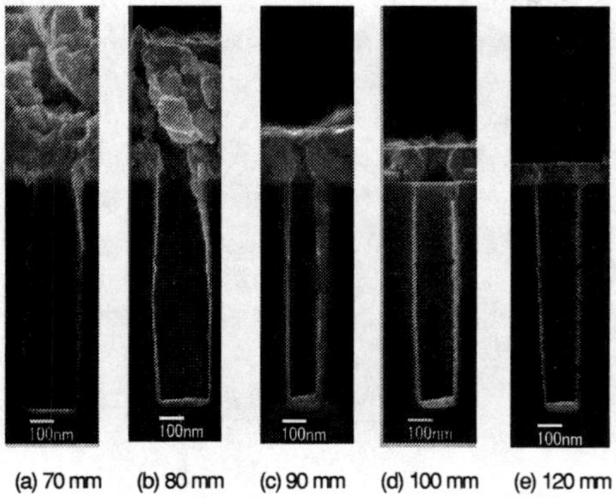

FIGURE 3. SEM images of the via hole in different target-substrate distances.

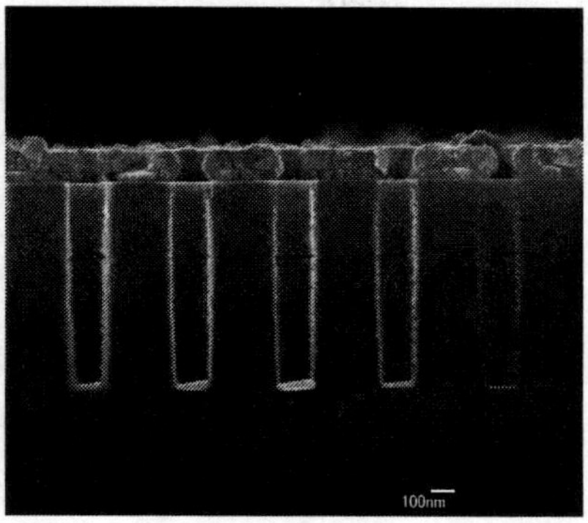

FIGURE 4. Cross-section of via holes after being fired by 5 shots.

Since we have found there exist the optimized conditions for metals to be built-in, we have studied it under the optimized conditions. Figure 4 shows the cross-section of via holes after being fired by 5 shots. It is evident that thin films are prepared uniformly inside the via holes. Since we cannot identify whether metals are prepared within the via holes by SEM images, we have studied it by SEM composites images, where images are only due to metals. Figure 5 shows the SEM composite images of the same via holes shown in Fig. 4. Clearly, we have found tungsten is uniformly embedded within the via holes.

Figure 5 shows the enlarged SEM image of single hole prepared by single shot. It is very clear that the 20-nm thick film of tungsten is prepared in the via hole, not only in the bottom of silicon, but also in the side wall.

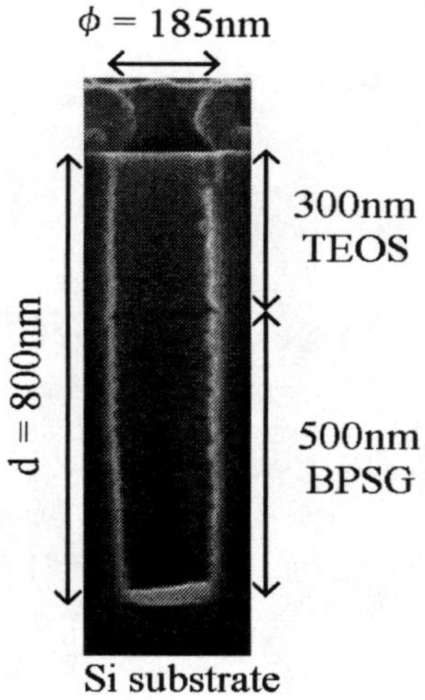

FIGURE 5. Enlarged SEM image of single hole prepared by single shot.

BLUE LIGHT EMISSION FROM NANOSIZE POWDERS OF POLY SILICON

Although silicon does not emit light because of its indirect transition, it is known that light emission might be available from nanosize powders of silicon possibly due to quantumized confinement. In fact, several authors have reported light emission. From such a viewpoint, we have attempted to obtain light emission from UNP of poly silicon. Here, UFP was produced by the rapid cooling of the ablation plasma obtained by IBE. The experiment was carried out under the almost same configuration as Fig. 1 except that the target and substrate are single-crystal silicon (100) and quartz glass in a vacuum or helium gas at 1 Torr. The distances are d_{AT} = 170 mm and d_{TS} = 90 mm.

Figure 6 shows typical data of (a) size distribution of UFP and (b) photo-luminescent light spectrum, which was obtained at 1 Torr of helium. Cleary, the UFP is dominated by 15 nm in diameter, and the dominant light emission near the wavelength of 440 nm can be observed.

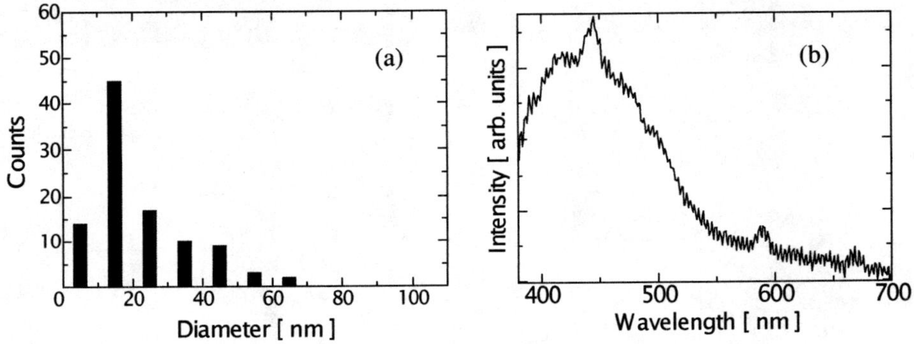

FIGURE 6. (a) Size distribution of UFP of silicon, and (b) photo-luminescent light spectrum.

CONLUDING REMARKS

High-density ablation plasma obtained by intense pulsed ion beam interaction with target has demonstrated efficient preparation of thin films into via holes in LSIs and the light emission from nanosize powders of poly silicon, both of which has been very difficult by the conventional techniques. New applications might be available using such the high-density plasma.

REFERENCES

[1] K. Yatsui, Laser and Particle Beams 7, 730 (1989).
[2] K. Yatsui et al, Phys. Plasma 1, 1730 (1994).
[3] K. Yatsui et al, Jpn. J. Appl. Phys., 36, 4928 (1997).
[4] K. Yatsui et al: Appl. Phys. Letters 67, 1214 (1995).

Preparation of Hydrogen Storage Alloy Thin Films by Intense Pulsed Ion-Beam Evaporation

Tsuneo Suzuki,[a] Takeshi Saikusa,[a] Nobuyuki Nishimiya,[b] Hisayuki Suematsu,[a] Weihua Jiang,[a] and Kiyoshi Yatsui[a]

[a]*Extreme Energy-Density Research Institute, Nagaoka University of Technology, Nagaoka, Niigata 940-2188, JAPAN*
[b]*Department of Materials Science, Toyohashi University of Technology, Toyohashi, 441-8580, JAPAN*

Abstract. Titanium iron (TiFe) is widely known as an intermetallic compound, which shows hydrogen storage characteristics. In the past, Ti-Fe thin films have been tried to prepare by sputtering methods, but crystallized thin films have not been prepared yet. The preparation of crystallized TiFe thin films has been attempted by intense pulsed ion-beam evaporation. Using the Ti-50at.%Fe target, thin films deposited on soda lime glass substrates have been found to be mixed phases of TiFe and $TiFe_2$. Thin films deposited on Si(100) substrates consist of an amorphous phase. By using the Ti-40at.%Fe target, a thin film of single phase TiFe has been successfully prepared on a soda lime glass substrate. Preparation of the crystallized, single-phase TiFe thin film is believed to be the first in the world. Even after hydrogenation treatments, the microstructures of the thin films were kept.

INTRODUCTION

Titanium iron (TiFe), widely known as an intermetallic compound, shows hydrogen storage characteristics. It reacts with hydrogen, producing metal hydride. In general, lattices of hydrogen storage alloys expand by absorbing hydrogen, and contract by desorbing hydrogen. However, there is a problem that it would be micronized in repetitive contractions and expansions. Although, several phases such as TiFe, $TiFe_2$, and β-Ti alloy were known in Ti-Fe binary equilibrium system, TiFe is the only phase that shows hydrogen storage characteristics. In the past, Ti-Fe thin films have been tried to prepare by sputtering methods [1, 2], but crystallized, single-phase TiFe thin films have never been prepared yet. Thus, new thin film preparation methods have to be developed to obtain single-phase crystallized TiFe thin films.

A new thin film preparation method of intense pulsed ion-beam evaporation (IBE) [3], has been developed. In this method, high-density ablation plasma is deposited on substrates that are placed near a target. Various kinds of crystallized thin films including intermetallic compounds such as TiAl [4] have been prepared by IBE. It is believed that the heat flux associated with the high temperature and high-density plasma tends to assist the crystallization without oxidation during the deposition on the substrates. In the present work, Ti-Fe thin films are prepared on Si and soda lime glass

CP650, *Beams 2002: 14th International Conference on High-Power Particle Beams,*
edited by T. A. Mehlhorn and M. A. Sweeney

substrates at room temperature by IBE.

EXPERIMENTS

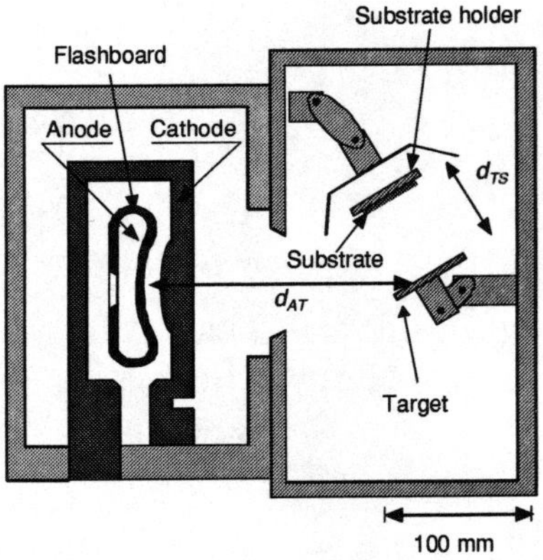

FIGURE 1. Outline of the experimental arrangement.

Figure 1 shows the schematic of the experimental arrangement for the preparation of thin films by IBE. The left-hand side represents the ion beam diode section to produce an intense pulsed light ion beam (LIB), while the right-hand side the preparation section of thin films. An intense pulsed-power generator "ETIGO-II" [5] was utilized in the experiment. Table 1 summarizes the experimental conditions. The preparation of the thin films was attempted using two kinds of targets with different compositions of Ti-50at.%Fe and Ti-40at.%Fe. The chemical composition of the targets and the thin films were determined by X-ray fluorescence analysis (XRF) and energy-dispersive X-ray spectroscopy (EDX). The phases in the thin films was identified by X-ray diffraction (XRD) using CuKα radiation of 0.154 nm. In order to obtain the hydrogen storage characteristics, the thin films were kept in hydrogen gas at high temperature. The conditions of hydrogenation treatments were described in Table 2.

TABLE 1. Experimental Conditions.

Targets	Sintered Compact (Ti-50at.%Fe, Ti-40at.%Fe)
Substrate	Si(100), Soda Lime Glass
Distance between Anode and Target (d_{AT})	170-180 mm
Distance between Target and Substrate (d_{TS})	90 mm
Number of Shots	3
Substrate Temperature	Room Temperature
Ambient Pressure	10^{-4} Torr

TABLE 2. Conditions of Hydrogenation Treatments.

Step	Holding Time (h)	Hydrogen Pressure (MPa)	Holding Temperature (˚C)	Evacuation Time (h)
1st	8	3.4	350	2
2nd	6	3.4	350	2
3rd	8	3.4	350	2
4th	6	3.4	350	2
5th	13	3.4	350	2
6th	6	3.4	350	2

RESULTS AND DISCUSSION

Ti-Fe thin films were deposited on Si(100) wafer and soda lime glass substrates.

The thickness of all thin films was about 500-600 nm. Figure 2 shows the XRD patterns of the thin films deposited by using the Ti-50at.%Fe target on (a) the soda lime glass substrate and (b) the Si(100) substrate. From the result of EDX analysis for the Ti-Fe thin film prepared on the Si(100) substrate, the chemical composition of the thin films was Ti-49at.%Fe, which was almost the same as that of the target. The thin film deposited on the glass substrate has been well crystallized (Fig. 2(a)). The strongest peak at 43° corresponds to the 110 reflection of TiFe phase. This result indicates that TiFe phase was successfully crystallized in the thin film prepared by IBE. Since small peaks for $TiFe_2$ phase also exist in Fig. 2(a), small amount of $TiFe_2$ grains co-exists in the thin film. In the meantime, since the XRD pattern in Fig. 2(b) shows a harrow peak of amorphous, the thin film deposited on the Si(100) substrate has not been crystallized. This difference in crystallinity between the thin films may come from the difference in thermal conductivity of substrates.

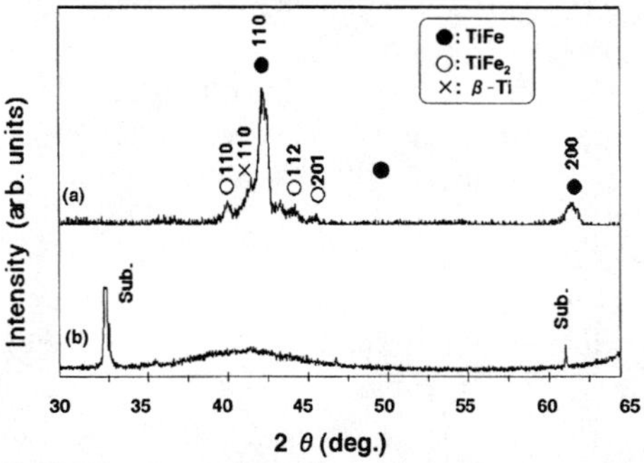

FIGURE 2. XRD patterns of the thin films deposited by Ti-50at.%Fe target at d_{AT} = 170 mm and d_{TS} = 90 mm on (a) glass substrate, (b) Si(100) substrate.

Figure 3 shows the XRD patterns of the thin films deposited by using the Ti-40at.% Fe target on (a) the soda lime glass substrate and (b) the Si(100) substrate. From the result of EDX analysis, the actual thin film composition was observed to be Ti-37at.%Fe. The thin film deposited on the Si(100) substrate (Fig. 3(b)) has not been crystallized. This is similar to the thin films prepared using the Ti-50at.% Fe target (Fig. 2(b)). From Fig. 3(a), the XRD pattern for the thin film deposited on the glass substrate has two peaks. Both peaks correspond to reflections from TiFe phase. This result indicates that the thin film of almost single phase TiFe has been successfully prepared by IBE.

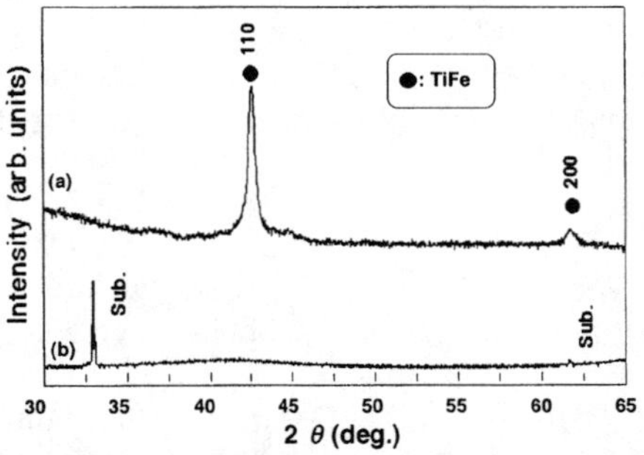

FIGURE 3. XRD patterns of the thin films deposited by using Ti-40at.%Fe target at d_{AT} = 180 mm and d_{TS} = 90 mm on (a) glass substrate, (b) Si(100) substrate..

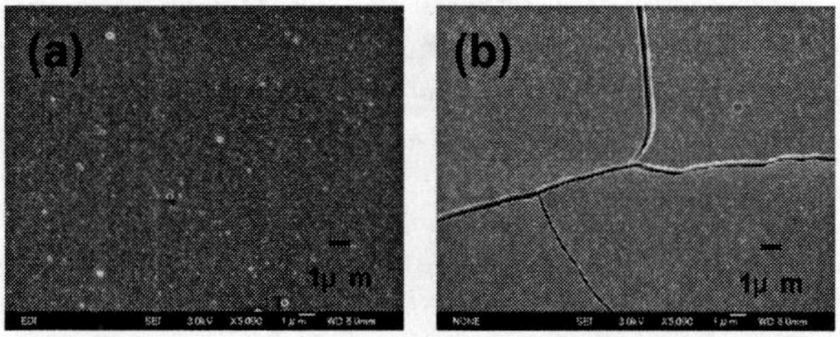

FIGURE 4. SEM images of the thin films deposited by using Ti-40at.%Fe target at d_{AT} = 180 mm and d_{TS} = 90 mm on (a) glass substrate, (b) Si(100) substrate.

Figure 4 shows the SEM

images of the thin films deposited by using the Ti-40at.% Fe target on (a) the soda lime glass and (b) the Si(100) substrate. The surface of the thin film deposited on Si(100) is smooth, whereas many cracks are seen on the surface of the thin film deposited on glass substrate.

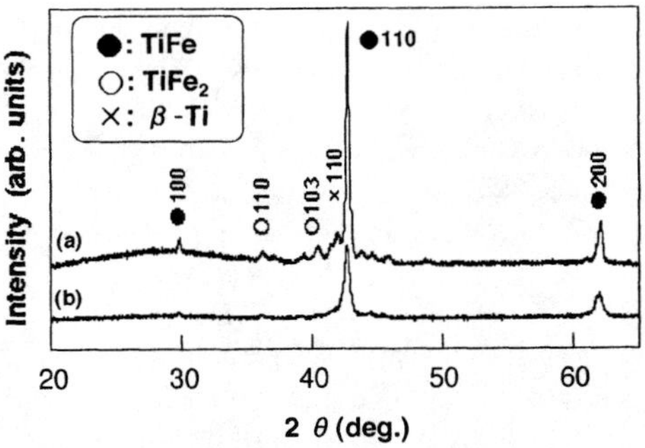

FIGURE 5. XRD patterns of the thin films (a) before hydrogenation, and (b) after hydrogen desorbing.

In order to obtain the hydrogen storage characteristics of the thin films, the hydrogenation treatments were carried out. The thin films deposited by using Ti-40at.%Fe target at d_{AT} = 180 mm and d_{TS} = 90 mm on soda lime glass substrate were kept in hydrogen atmosphere at 350 °C. The details of hydrogenation conditions were described in Table 2. The thin films could be considered to undergo hydrogen absorbing and desorbing by this hydrogenation conditions. The XRD patterns of thin films before the hydrogenation treatments and after hydrogen desorbing were shown in Fig. 5(a) and (b), respectively. The thin films kept TiFe structure as a main phase even after the hydrogenation treatments, and the change in main structure could not been observed after the hydrogenation. The second phases ($TiFe_2$ and β-Ti) which were observed in Fig. 5(a) were disappeared in Fig. 5(b). These second phases seem to be transformed to TiFe phase due to annealing effect of hydrogenation. Micronization of the thin films could not been observed after hydrogenation.

CONCLUSIONS

We have attempted to prepare the thin film of Ti-Fe intermetallic compounds by IBE. Chemical composition and phases in the thin films were investigated, and the following conclusions were drawn.

(1) Thin films of Ti-Fe intermetallic compounds could be prepared by IBE.
(2) Using a Ti-50at.%Fe target, the thin film was been found to be the mixed phases of TiFe and $TiFe_2$.
(3) Using a Ti-40at.%Fe target, the thin film of single phase TiFe was successfully prepared on a soda lime glass substrate.
(4) The thin films deposited on a Si(100) substrate consisted of amorphous phases.
(5) TiFe phase was remained in the thin films after the hydrogenation treatments.

REFERENCES

1. Nakamura, K., *Scripta Metalluragica* **18**, 793-797 (1984).
2. Sumiyama, K., Hashimoto, Y., and Nakamura, Y., *Trans. Jpn. Inst. Metals* **24**, 61-65 (1983).
3. Yatsui, K., *Laser and Particle Beams* **7**, 733-741 (1989).
4. Suzuki, T., Kishima, M., Jiang, W., and Yatsui, K., *Jpn. J. Appl. Phys.* **40**, 1042-1044 (2001).
5. Tokuchi, A., Nakamura, N., Kunimatsu, T., Ninomiya, N., Den, M., Araki, Y., Masugata, K., and Yatsui, K., *Proc. 2nd Int. Top on ICF Res., High-Power Particle Beams, Nagaoka, ed. K. Yatsui*, (Lab. of Beam Technology, Technol. Univ. of Nagaoka, 1986) pp. 430-439.

Characteristics of High Power Ion Beam Produced in Plasma Focus and Its Application to Semiconductor Processing

Katsumi Masugata, Kazuto Takao, Masaki Shiotani, Takeo Honda, Iwao Kitamura, and Takakazu Takahashi

Faculty of Engineering, Toyama University, 3190, Gofuku, Toyama, 930-8555, Japan

Abstract. To apply an intense pulsed ion beams produced in plasma focus (PF) to material processing, characteristics of the beam was evaluated. In the experiment a Mather type PF was used which was pre-filled with N_2 (5.5 Pa) to produce nitrogen ions. To clarify the mechanism of the production of impurity ions, two types of anode (type A; rod type and has an anode tip on the top, type B; hollow shape top) were used. In both type peak discharge currents of 550 kA were obtained at 1.5 μs after the rise of the current. From the measurement by Thomson parabola spectrometer large quantity of copper ions were observed when type A is used and the purity for type A and type B were evaluated to be 26 % and 90 %, respectively. To enhance the energy and the charge state of the nitrogen ions the PF with type B anode was pre-filled with mixture of nitrogen and hydrogen (N_2 20, H_2 180 Pa). From the TPS measurement, multiply ionized nitrogen ions ($N^{1\text{-}5+}$) of energy more than several MeV were observed with protons of maximum energy 1 MeV. Thin layer of amorphous silicon was irradiated by the ion beams produced in PF and we found that the layer was crystallized.

INTRODUCTION

Plasma focus (PF) is known to produce high-energy, high current pulsed ion beams (PIB) with energies several MeV [1-3]. Since that the obtainable ion current density is high and that by selecting the filling gas wide variation of ion species can be produced, it has been expected as a pulsed ion beam source for materials processing [4,5]. For example pulsed ion beam implantation process to semiconductor [6] is very hopeful since ion implantation and surface annealing is completed in the same time. For the application we have evaluated the characteristics of ion beam produced in PF. Dependence of the characteristics on the electrode shape and on the filling gas was evaluated. In addition, to evaluate the effect of irradiation the PIB was irradiated onto the thin amorphous silicon.

EXPERIMENTAL APPARATUS

A schematic of the experimental setup is shown in Fig. 1. In the experiment, a Mather-type plasma focus device was used with a capacitor bank of 43.2 μF. The charging voltage of the bank was 30 kV (19.4 kJ) giving the peak discharge current

CP650, *Beams 2002: 14th International Conference on High-Power Particle Beams*, edited by T. A. Mehlhorn and M. A. Sweeney

of 520-530 kA. The chamber is separated to a PF part and an analyzer part by a partition plate to keep the pressure difference. The plate has a pinhole of diameter 0.21 mm on the axis to introduce energetic ions into the analyzer chamber. The PF chamber was once evacuated to < 1.5 x 10^{-3} Pa and then filling gas is introduced. The analyzer chamber was evacuated by another vacuum system to keep the pressure below 1.0×10^{-3} Pa. The PF consists of an inner electrode (anode, 50 mm^{OD}, 320 mm^{L}) and an outer electrode (cathode, 100 mm^{ID}, 230 mm^{L}). The cathode has a cage like structure and it consists of 24 of 8 mm^{OD} copper rods.

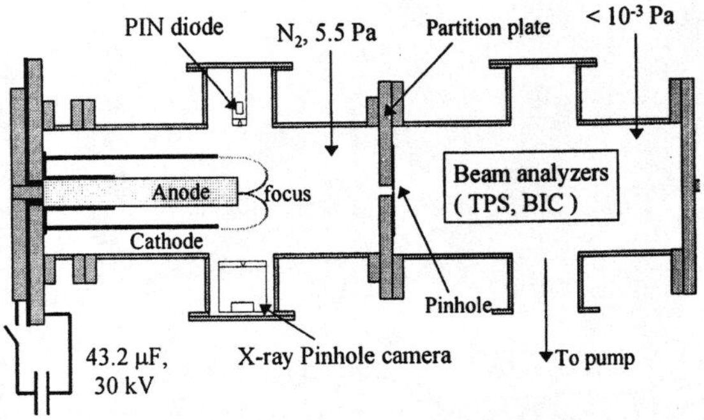

FIGURE 1. Schematic of the experimental system.

MECHANISM OF IMPULITY ION PRODUCTION

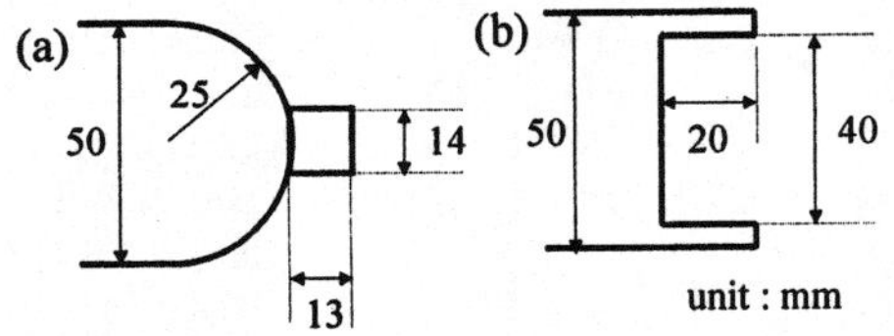

FIGURE 2. Shape of the anode top. (a) Rod type (type A), (b) Hollow type (type B).

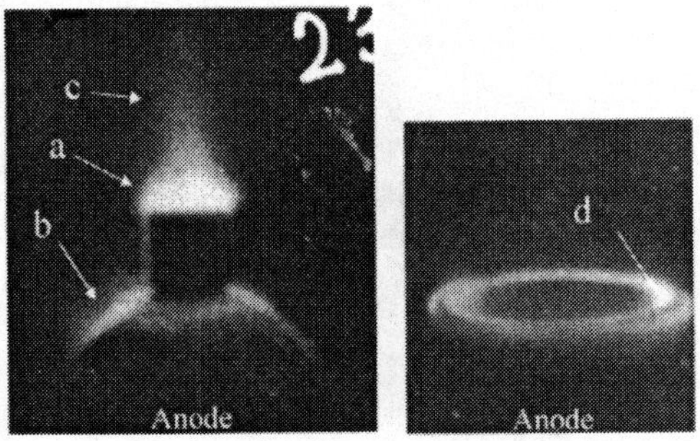

(a) Type A (b) Type B

FIGURE 3. X-ray pinhole images obtained with (a) Type A and (b) Type B.

To evaluate the mechanism of the production of impunity ions, two types of anode are used (see Fig. 2). Type A is the rod type anode and has an anode tip on the top of the anode whereas type B has a hollow shape top. In the experiment filling gas of N_2 (5.5 Pa) was used to produce nitrogen ion beam.

Figure 3 shows X-ray pinhole images obtained with both type. In type A, bright X-ray emission (indicated as a) and weaker X-ray emission (indicated as b) from the top of the rod and the hemispherical area, respectively. In addition, X-ray is observed over the electrode top (indicated as c). Here, the X-ray from the electrode surface seems to be due to the electron irradiation whereas that observed over the electrode top is from the pinch plasma. Hence high temperature pinch plasma is produced over the anode top. In contrast, in type B, X-ray is observed only form the inner edge of the hollow (indicated as d).

Figure 4 shows typical waveforms obtained with type A and type B. The discharge current (I_d), soft X-ray signal, hard X-ray signal, and the ion current density (J_i) at z = 218 mm from the top of the anode on the axis are shown for each. In type A, I_d reaches the peak of 520 kA at t = 1.25 μs and after that I_d sharply dips to 400 kA with dI_d/dt value of -0.43 kA/ns. A peak of J_i ≈1100 A/cm^2 is observed with pulse width (FWHM) of 60 ns. In type B, I_d have a peak of 530 kA at t = 1.3 μs and after that decreases with dI_d/dt of -0.24 kA/ns, hence pinch strength is much weaker than that in type A. Even though the peak intensity of the hard X-ray is almost same as in type A, the soft X-ray intensity is 40 times weaker than that of type A, hence the pinch plasma temperature seems to be lower in type B. The peak value and the pulse width of J_i were 1300 A/cm^2 and 50 ns, respectively.

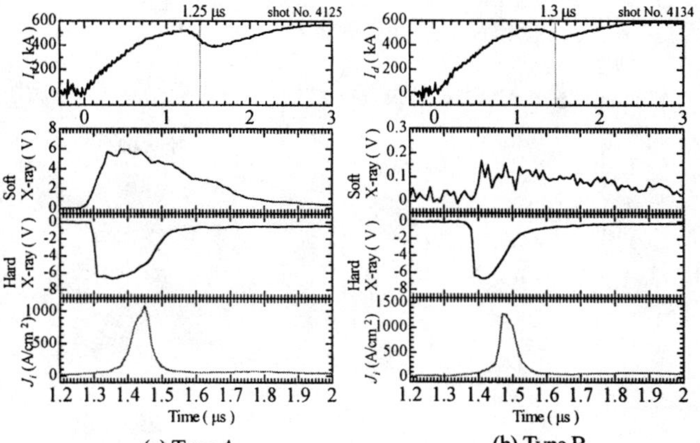

(a) Type A (b) Type B

FIGURE 4. Typical waveforms obtained with (a) type A and (b) type B.

Figure 5 shows the TPS traces obtained of CR-39 for type A and type B. In type A, impunity of $Cu^{1\text{-}2+}$, $O^{1\text{-}2+}$, C^+ are observed with nitrogen ions of $N^{1\text{-}2+}$. Here, the copper ions seem to be produced by the spattering or the ablation of the anode surface by the irradiation of energetic electrons. Energy integrated ion composition ratio for each ions is estimated by counting the ion tracks in Fig. 5(a). The result is $N^{1\text{-}2+}$: $O^{1\text{-}2+}$: C^+: $Cu^{1\text{-}2+}$ = 25: 50: 11: 14 %. In spite that the PF was filled with N_2, the ion beam is dominated by the impurity ions of oxygen.

In contrast, in type B, strong traces of $N^{1\text{-}3+}$ are observed with weak impurity traces of C^+ and $O^{1\text{-}2}$ and no copper ions. In the case ion composition ratio is evaluated to be $N^{1\text{-}3+}$: $O^{1\text{-}2+}$: C^+ = 90: 8: 2 %, hence the purity of the beam is 90 %. Considering the results of the X-ray pinhole images and the waveforms measurement, the mechanism of the production of impurity ions in type A is considered to be due to the dens irradiation of electron beam onto the anode top, which produces ablation

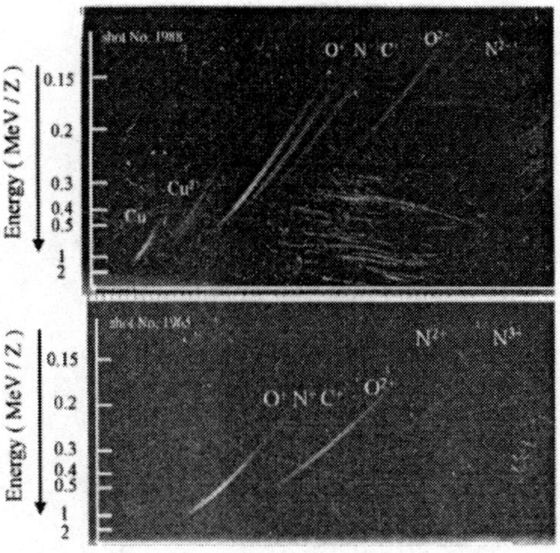

FIGURE 5. TPS traces obtained with (a) type A and (b) type B.

plasma. The ablation plasma contains cupper and des orbed molecule of O, C, etc. and the plasma diffuses to the pinch plasma.

ENHANCEMENT OF THE CHARGE STATE OF THE BEAM

To obtain highly ionized nitrogen ion beams PF was filled with mixture gas (N_2 20, H_2 180 Pa). Figure 6 shows the track pattern obtained with TPS. As seen in the figure, variety of nitrogen ions ($N^{(1-5)+}$) are observed with protons. The energy of nitrogen ion is in the range of 0.4 - 6 MeV. In addition, no impurity is observed in the condition.

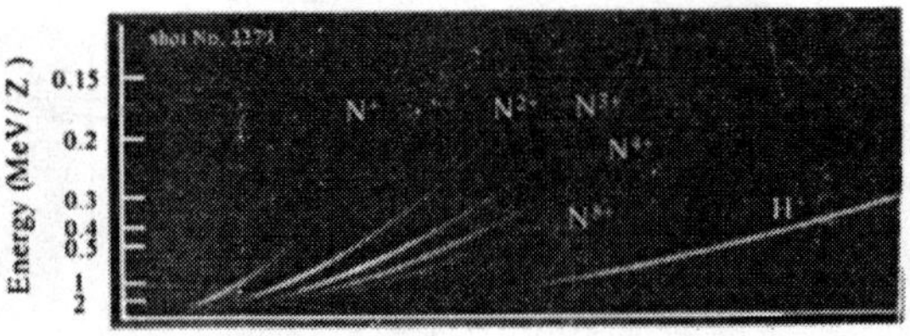

FIGURE 6. The track pattern obtained with the TPS at mixture gas.

IRRADIATION EXPERIMENT

By the irradiation of intense pulsed ion beam to materials, the surface will pulse heated to extremely high temperature and the crystal structure is expected to be modified. To evaluate the irradiation effect, amorphous silicon films of thickness 800 nm were irradiated by the proton beam produced in PF.

Figure 7 shows the X-ray diffraction patterns of the films before and after one shot of irradiation. Here, the peaks correspond to Al are due to that Aluminum holder is used in the diffraction measurement. From the figure we see that there is no crystal stricture in the target before irradiation. In contrast, on the targets at r = 0 and 30 mm, we clearly see the peaks corresponding to Si (111) and Si (220), which indicate that amorphous silicon layer was crystallized by the irradiation of the beam.

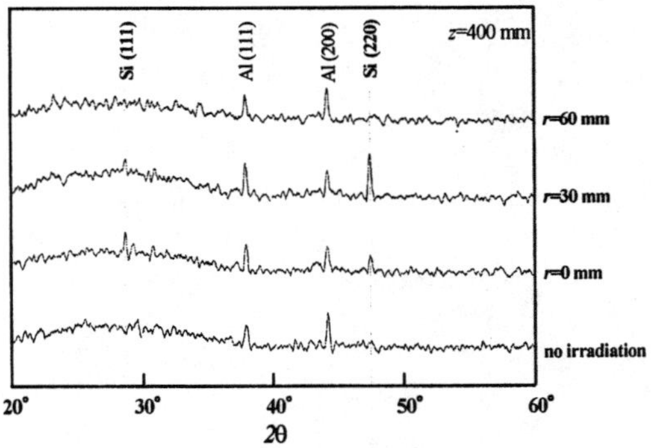

FIGURE 7. X-ray diffraction patterns of amorphous Si irradiated by pulsed proton beam.

REFERENCES

1. Zakaullah, M., Akhtar, I., Murtaza G, and Waheed A, Phys. Plasmas **6** 3188 (1999).
2. Yamamoto, T., Shimoda, K., Kobayashi, K., and Hirano, K., Jpn. J. Appl. Phys. **23** 242 (1984).
3. Takao, K., et al, Jpn. J. Appl. Phys. **40** 1013 (2001).
4. Feugeas, J.N., Llonch, E.C., de Gonzáles, C.O., and Galambos G.J., Appl. Phys. **64** 2648 (1988).
5. Sánchez, G., and Feugeas, J., J. Phys. D: Appl. Phys. **30** 927 (1997)

Development of Bipolar Pulse Accelerator for Pulsed Ion Beam Implantation to Semiconductor

Katsumi Masugata,[a] Yoshihiro Kawahara,[a] Chihiro Mitsui,[a] Iwao Kitamura,[a] Takakazu Takahashi,[a] Yasunori Tanaka,[b] Hisao Tanoue,[b] and Kazuo Arai [b]

[a] *Faculty of Engineering, Toyama University, Toyama, 930-8555, Japan*
[b] *National Institute of Advanced Industry Science and Technology, Tukuba, 305-8568, Japan*

Abstract. To improve the purity of the ion beams new type of pulsed power ion accelerator named "bipolar pulse accelerator" was proposed. The accelerator consists of two acceleration gaps (an ion source gap and a post acceleration gap) and a drift tube, and a bipolar pulse is applied to the drift tube to accelerate the beam. In the accelerator intended ions are selectively accelerated and the purity of the ion beam is enhanced. As the first step of the development of the accelerator, a Br-type magnetically insulated acceleration gap is developed. The gap has an ion source of coaxial gas puff plasma gun on the grounded anode and a negative pulse is applied to the cathode to accelerate the ion beam. By using the plasma gun, ion source plasma (nitrogen) of current density around 100 A/cm^2 is obtained. In the paper, the experimental results of the evaluation of the ion beam and the characteristics of the gap are shown with the principle and the design concept of the proposed accelerator.

INTRODUCTION

Intense pulsed ion beams (PIB) of carbon, nitrogen or aluminum, have a wide area of applications including nuclear fusion, materials science, etc. However, the purity of the conventional PIB is not sufficient for those applications [1]. To improve the purity a new type of pulsed power PIB accelerator named "bipolar pulse accelerator (BPA)" was proposed [2-3]. Figure 1 shows a conceptual drawing of the bipolar pulse accelerator. It consists of a grounded ion source, a drift tube and a grounded cathode. In the diode, bipolar pulse (V_1) is applied to the drift tube. At first the negative voltage pulse of duration τ_p is applied to the drift tube, and ions on the grounded ion source is accelerated towered the drift tube. If τ_p is adjusted to the time of flight delay of the ions to pass the drift tube, the pulse is reversed and the

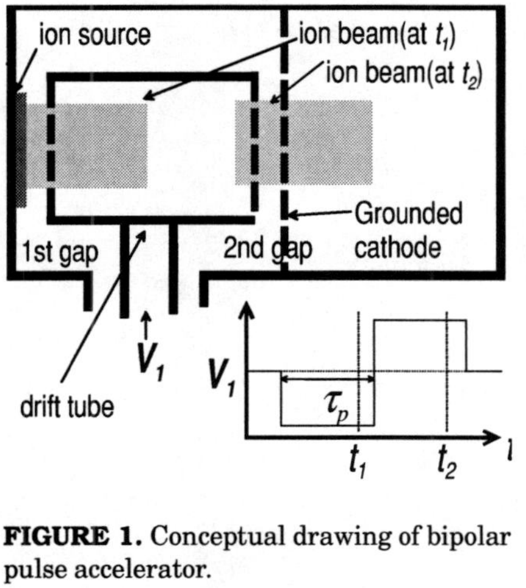

FIGURE 1. Conceptual drawing of bipolar pulse accelerator.

CP650, *Beams 2002: 14th International Conference on High-Power Particle Beams*, edited by T. A. Mehlhorn and M. A. Sweeney

positive voltage of duration τ_p is applied to the drift tube when top of the ion beam reaches the 2nd gap. In the case ions are again accelerated in the 2nd gap toward the grounded cathode. Hence by adjusting the τ_p to the aimed ion species, another ion species (impurity) can be removed. [2-3]

As the first step of the development of BPA, an experimental system was constructed to confirm the acceleration. In the paper the preliminary result of the experiment is shown.

EXPERIMENT

Figure 2 shows a cross-sectional view of the experimental system. It consists of a grounded anode and a magnetically insulated gap (MIG). The drift tube is connected to the high voltage terminal of the PFL (300 kV, 48 kA, 60 ns) and by applying negative pulse, ions on the anode are accelerated toward the drift tube.

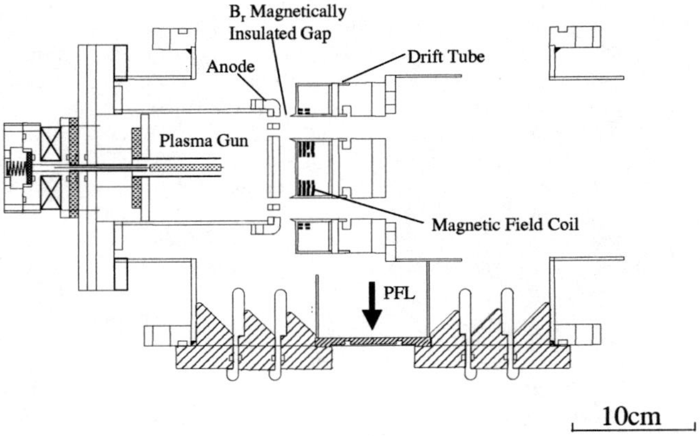

FIGURE 2. Cross-sectional view of the experimental system.

FIGURE 3. Cross-sectional view of the gas puff plasma gun.

The MIG is the B_r type and an inner and an outer coil are installed on the end of the drift tube to produces B_r field in the gap. The coils are designed to produce B_r field lines parallel to the anode surface. The rise time and the strength of the B_r field were measured to be 150 μs and 0.4 T, respectively. Since high voltage is applied to the drift tube, pulsed current is applied through an inductively isolated current feeder (IC). The IC is the helically winded coaxial cable and the outer conductor connects the grounded vacuum chamber and the drift tube with inductance of 4.6 μH. A gap length, inner and outer diameter of the effective gap area are 10 mm, 53 mm, and 93 mm, respectively. A gas puff plasma gun is used to produce anode plasma, which is placed behind the anode.

Figure 3 shows the cross sectional view of the gas puff plasma gun. It consists of a high-speed gas puff valve and a coaxial plasma gun. The gas puff valve consists of a nylon vessel, an aluminum valve and a drive coil, and the vessel is pre-filled with N_2 gas. By applying pulse current to the coil, the valve is opened and the gas expands with a supersonic velocity and reaches the gas nozzles on the inner electrode of the plasma gun.

The plasma gun has a pair of coaxial electrodes, i.e. an inner electrode of outer diameter 4 mm, length 100 mm, and an outer electrode of inner diameter 10 mm. Since it takes about a hundred μs to open the valve and several tens μs for N_2 to drift to the gas nozzle, the plasma gun is discharged with a delay time of τ_d. In the experiment capacitor banks of 5 μF and 1.5 μF are used for the gas puff and the plasma gun, respectively. Both of the banks were charged to 6.5 kV, and 28 kV, respectively.

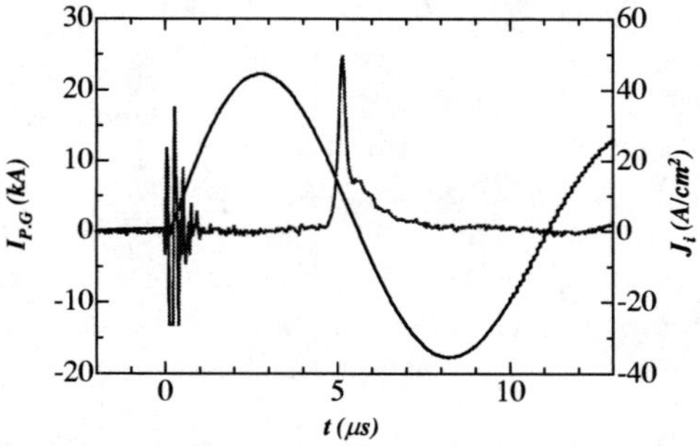

FIGURE 4. Typical waveforms of the discharge current of the plasma gun ($I_{P.G}$) and the ion current density (J_i) obtained by BIC.

Figure 4 shows the wave forms of the discharge current of the plasma gun ($I_{P.G}$) and the ion current density (J_i) obtained by a biased ion collector (BIC) when τ_d = 213 μs. As seen in the figure $I_{P.G}$ rises in 2.75 μs and have a peak of 23 kA. The peak value of J_i of 50 A/cm^2 is observed at 100 mm downstream from the plasma gun at $t = 5$ μs after the rise of $I_{P.G}$.

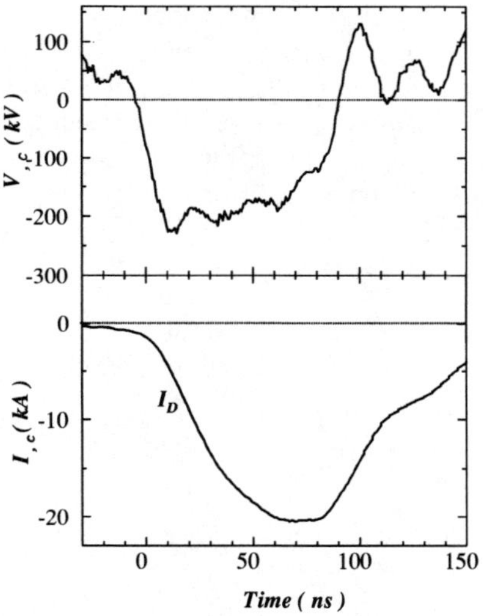

FIGURE 5. Typical waveforms of V_d and I_d.

To confirm the acceleration of the ions the system was operated at 70 % of the full charge condition of the PFL. The plasma gun was operated at $\tau_d \approx 215$ μs and high voltage negative pulse was applied to the drift tube at 6.2 μs after the rise of the $I_{P.G}$. Figure 5 shows typical waveforms of applied voltage to the drift tube (V_d) and the current (I_d). As seen in the figure V_d rises in 15 ns and have a peak value of - 210 kV, whereas I_d rises in 50 ns and have a peak of 20 kA. Considering the I_{IC}, the effective current through MIG seems to be less than 17 kA.

Figure 6 shows the pinhole image obtained by the shadow box. The shadow box consists of a pinhole plate placed 92 mm downstream from the anode and an ion track detecting plastic of CR–39 placed 10 mm downstream from the pinhole plate. The pinhole plate has 15 pinholes of diameter 0.2 mm each. As seen in the figure ion pinhole images are observed, each of which corresponds to the pinholes, suggesting uniform production of ion beam. From the track images the divergence angles of r and θ directions of the beam were estimated to be 0.1 rad and 0.2 rad, respectively.

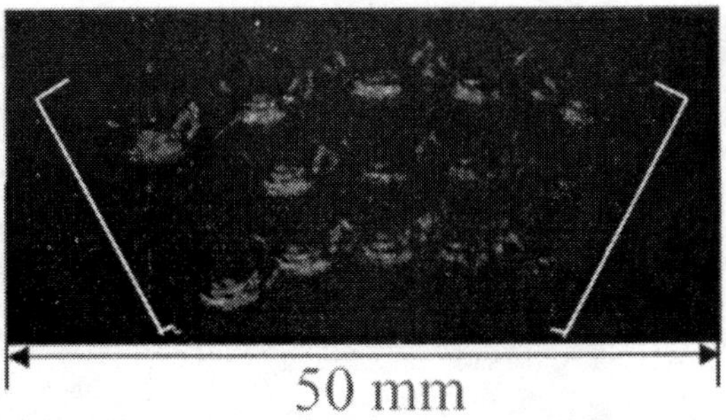

FIGURE 6. Ion pinhole images obtained by the shadow box.

CONCLUSION

To produce a pure PHIB a new type of pulsed power ion accelerator using bipolar pulse is proposed, which is called as "bipolar pulse accelerator (BPA)". As the first step of the development of BPA, an experimental system was constructed to confirm the principle of the acceleration. The system consists of a anode with ion source of gas puff plasma gun, a Br type magnetically insulated acceleration gap, and a drift tube. In the experiment the gas puff plasma gun generates a nitrogen ion flux of current density 50 A/cm^2. The ions are successfully accelerated in the gap by applying negative pulse of voltage 210 kV to the drift tube. The accelerated ions are detected by a shadow box and ion acceleration is confirmed.

REFERENCES

1. Masugata, K., Nucl. Instrum. & Methods in Phy. Res. **A 411,** 205-209 (1998)
2. Masugata, K., et al, Proc. 12th Symp. on High Current Electronics, Tomsk, Russia, Sep. 24-29 2000, 111-114 (2000)
3. Masugata, K., et al, Proc. 28th IEEE Int'l Conf. on Plasma Science and 13th IEEE Int'l Pulsed Power Conf., Las Vegas, Nevada, June 17-22 2001, O3C3 (2001)

Preparation of Crystalline $Si_{1-x}Ge_x$ Thin Films by Pulsed Ion-Beam Evaporation

Hisayuki Suematsu, Ryuma Iwashita, Tsunetoshi Arikado, Naotaka Uchitomi* S. –C. Yang, Weihua Jiang and Kiyoshi Yatsui

*Extreme Energy-Density Research Institute, *Department of Electrical Engineering, Nagaoka University of Technology, Nagaoka 940 -2188, Japan*
**Department of Electrical Engineering, Nagaoka University of Technology, Nagaoka 940 -2188, Japan*

Abstract. Thin films of single phase, polycrystalline silicon germanium ($Si_{1-x}Ge_x$) were prepared by ion-beam evaporation (IBE) using Si-Ge multi-phase targets. After the irradiation of the targets by an pulsed light ion beam with peak energy of 1 MV, 450 and 480 nm thick films were deposited on Si single crystal and quartz glass substrates, respectively. From XRD analysis, the thin films consisted of a single phase $Si_{1-x}Ge_x$, whose composition is close to those of the targets.

INTRODUCTION

Complementary metal-oxide-semiconductor structures have been and will be used in large scale and high-speed devices. For decreasing feature size and increasing clock speed, electrical properties including electron and hole mobilities in the devices has to be improved. Although silicon occupies a dominant position in materials for the devices, the required properties are reaching those for silicon. Thus, new materials have to be developed for the devices.

Silicon germanium ($Si_{1-x}Ge_x$) has been paid attention because of its high hole mobility[1]. Many researches have been performed to establish deposition processes to prepare not only epitaxially grown $Si_{1-x}Ge_x$ p-channels[2] but also polycrystalline $Si_{1-x}Ge_x$ gates[3,4]. For preparation of the $Si_{1-x}Ge_x$ gates, it is required to drive dopant s and to activate carriers in the gates by sample annealing, which could be harmful for the devices. Thus, other processes to deposit $Si_{1-x}Ge_x$ thin films have been studied.

A novel thin film preparation method of ion-beam evaporation (IBE) was developed[5]. This method utilizes a high power pulsed ion beam generated in a pulsed power generator[6]. The ion beam was focused on a target, which yields high-density plasma. The plasma was deposited on the surface of substrate s which are placed facing toward the target. By the IBE method, various thin films were obtained with an instantaneous deposition rate of a few mm/s. Furthermore, crystallized TiFe, B_4C[7-9] and $SrAl_2O_4$[10-12] thin films were successfully prepared without substrate heating or sample annealing.

Recently, polycrystalline Si thin films were prepared by IBE on Si single crystals at room temperature[13]. It is expected that polycrystalline $Si_{1-x}Ge_x$ thin films can also

CP650, *Beams 2002: 14th International Conference on High-Power Particle Beams*, edited by T. A. Mehlhorn and M. A. Sweeney

be obtained by IBE. In the present research, preparation of $Si_{1-x}Ge_x$ thin films on Si single crystal and quartz glass substrates at room temperature were attempted. Crystallinity of the as-deposited thin films was examined by powder X-ray diffraction (XRD).

EXPERIMENTAL PROCEDURE

Thin films were prepared in a pulsed ion beam generator ('ETIGO-II'). Ion beam and deposition chambers are schematically shown in Fig. 1. A voltage of 1 MV (peak) with pulse width of ~50 ns and current of 60 kA was applied between the cathode and the anode, which was covered with a polyethylene sheet (flashboard). The application of the pulsed voltage produced a pulsed light ion beam which mainly (~75 %) consisted of protons. The ion beam was geometrically focused on one of sintered $Si_{0.5}Ge_{0.5}$ and $Si_{0.8}Ge_{0.2}$ targets, which was located at 180 mm from the anode. The ion beam irradiation produced ablation plasma from the surface of the target. Substrates of (100) Si single crystal and quartz glass were placed at 70 mm from the target. The ablation plasma was deposited on the substrates. The ion beam was irradiated on the target. The substrates were kept in a vacuum of $2x10^{-4}$ Torr and at room temperature.

Thickness of the thin films was measured using cross-sectional scanning electron microscopy images. Phases in the as-deposited thin films were identified by XRD. From the peak positions, lattice parameters of $Si_{1-x}Ge_x$ phase were measured. Using known relationship between lattice parameter and composition[14], the compositions of the thin films were estimated.

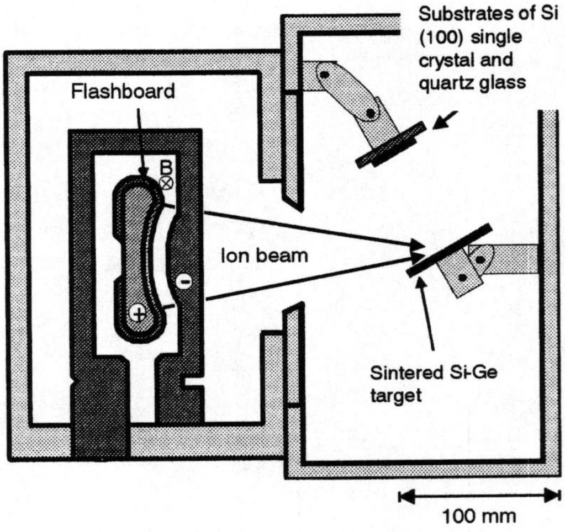

FIGURE 1. Experimental setup for thin film deposition by IBE.

RESULTS AND DISCUSSION

After one shot of ion beam irradiation on $Si_{0.5}Ge_{0.5}$ and $Si_{0.8}Ge_{0.2}$ targets, 480 and 450 nm thick films were deposited on Si single crystal substrates. An XRD pattern for a thin film using a $Si_{0.5}Ge_{0.5}$ target are shown in Fig. 2 (a). Diffraction angles of all peaks are located between those of Si and Ge phases. Peaks for other phases are not detected. Thus, it was evident that a single phase $Si_{1-x}Ge_x$ thin film was obtained by IBE. Note that the targets consisted of Si, Ge and GeO_2 phases. The germanium oxide may be reduced in the vacuum chamber. Furthermore, Si and Ge phase must be mixed in the ablation plasma after the ion irradiation of the target and form a single phase $Si_{1-x}Ge_x$ phase. An XRD pattern for a thin film prepared using a $Si_{0.8}Ge_{0.2}$ target is shown in Fig. 3 (a). Similar to the XRD pattern in Fig. 2 (a), $Si_{1-x}Ge_x$ is the only phase detected in the thin film. Comparing to the pattern in Fig. 2 (a), diffraction angles for the peaks were shifted to higher angles.

Thin films were also deposited on quartz glass substrates. XRD patterns for the thin films are shown in Figs. 2(b) and 3 (b). Similar to the results for thin films prepared

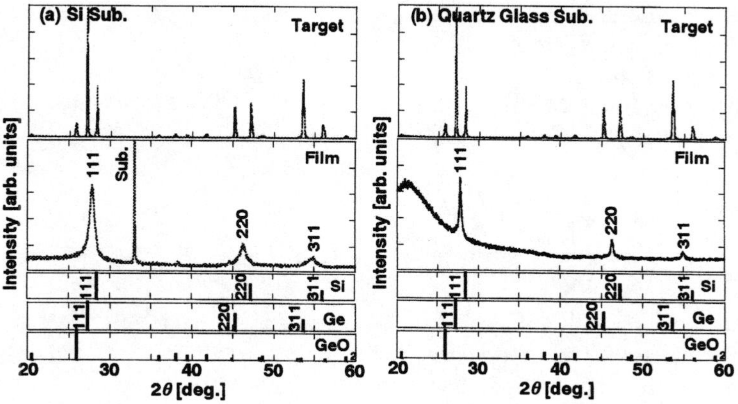

FIGURE 2. XRD patterns for $Si_{0.5}Ge_{0.5}$ thin films prepared on (a) quartz glass and (b) Si single crystal substrates. XRD pattern for the target and peak positions for Si, Ge and GeO_2 phases are also presented.

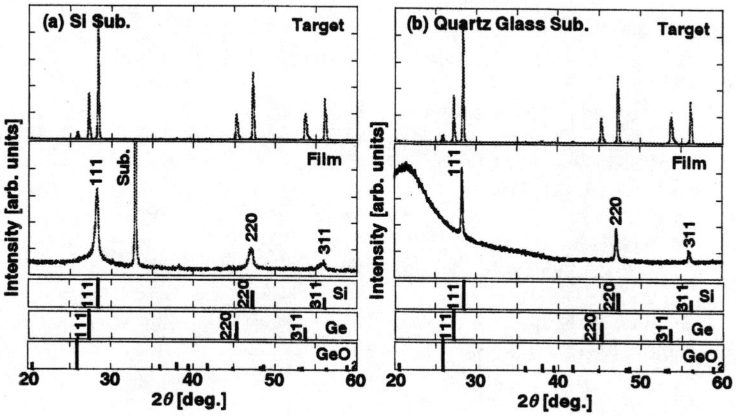

FIGURE 3. XRD patterns for $Si_{0.8}Ge_{0.2}$ thin films prepared on (a) quartz glass and (b) Si single crystal substrates. XRD pattern for the target and peak positions for Si, Ge and GeO_2 phases are also presented.

on Si single crystal substrates, all peaks are for $Si_{1-x}Ge_x$ phase. From these results, single-phase $Si_{1-x}Ge_x$ thin films were obtained on both Si single crystal and quartz glass substrates.

From peak positions for $Si_{1-x}Ge_x$ phase, lattice parameters were calculated and plotted against the composition of the target in Fig. 4. The estimated compositions of the thin films are close to those of targets. It is concluded that a novel process to prepare polycrystalline $Si_{1-x}Ge_x$ thin films with various composition were successfully developed.

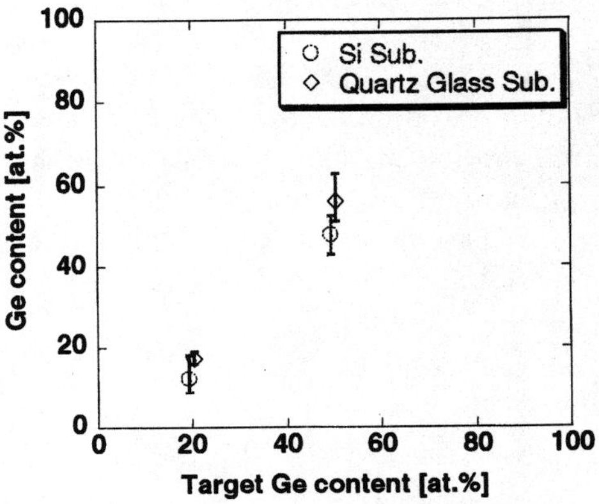

FIGURE 4. Ge content (x) estimated from lattice parameters and Ge content in the targets

ACKNOWLEDGMENTS

This work was partly supported by a Grant-in-Aid for Scientific Research by the Ministry of Education, Culture, Sports, Science and Technology of Japan. The $Si_{1-x}Ge_x$ targets were supplied by Toshiba Corporation.

REFERENCES

1. See the review by Whall, T. E. and Parker, E. H. C., *Thin Solid Films* **368,** 297-305 (2000).
2. Nayak, D. K., Woo, J. C. S., Park, J. S., Wang, K.-L. and MacWilliams, K. P., *IEEE Electron Device Lett.* **12,** 154-156 (1991).
3. King, T. J., Pfiester, J. R., Shott, J. D., McVittie, J. P. and Saraswat, K. C., *Tech. Dig. Int. Electron Devices Meet.*, **253** (1990).
4. Chang, C. L. and Sturm, J. C., *Appl. Phys. Lett.* **74,** 2501-2503 (1999).
5. Shimotori, Y., Yokoyama, M., Isobe, H., Harada, S., Masugata, K. and Yatsui, K., *J. Appl. Phys.* **63,** 968-970 (1988).
6. Yatsui, K., *Laser and Particle Beams* **7,** 733-741 (1989).
7. Kitajima, K., Suzuki, T., Jiang, W. and Yatsui, K., *Jpn. J. Appl. Phys.* **40,** 1030-1034 (2001).
8. Suematsu, H., Kitajima, K., Suzuki, T., Jiang, W., Kurashima, K., Bando, Y. and Yatsui, K., *Appl. Phys. Lett.* **80,** 1153-1155 (2002).
9. Suematsu, H., Kitajima, K., Ruiz, I., Takeda, M., Suzuki, T., Jiang, W. and Yatsui, K., *Thin Solid Films* **407,** 132-135 (2002).
10. Sengiku, M., Oda, Y., Jiang, W., Yatsui, K., Kato, K., Shinbo, K. and Kaneko, F., *Jpn. J. Appl. Phys.* **40,** 1035-1037 (2001).
11. Kato, K., Ogura, Y., Sengiku, M., Shinbo, K., Kaneko, F., Oda, Y. and Yatsui, K., *Jpn. J. Appl. Phys.* **40,** 1038-1041 (2001).
12. Suematsu, H., Sengiku, M., Kato, K., Mitome, M., Kimoto, K., Matsui, Y., Jiang, W. and Yatsui, K., *Thin Solid Films* **407,** 136-138 (2002).
13. Yang, S.-C., Fazlat, A., Iwashita, R., Suzuki, T., Suematsu, H., Jiang, W. and Yatsui, K., *Proc. 2nd Int'l Symp. Pulsed Power and Plasma Applications*, 100 - 103 (2001).
14. Okada, M., Kamioka, H., Matsuo, H., Fukuda, Y., Zaima, S., Kawamura, K. and Yasuda, Y., *Thin Solid Films* **369,** 130-133 (2000).

Combining Different Modes of High-Power Ion Beam Application for Creation of Metallic and Diamond-Like Coatings

Vasily K. Struts[a], Vasily M. Matvienko[a], Anatoli V. Petrov[a], Anatoli S. Shlapakovski[a], Valentina A. Sohoreva[a], Alexey A. Valyaev[b]

[a]*Nuclear Physics Institute at Tomsk Polytechnic University, 2a Lenina Str., Tomsk 634050, Russia*
[b]*Zelenograd Research Institute of Physical Problems, Moscow, 103460, Russia*

Abstract. Some results of the experiments on thin film deposition using pulsed high-power ion beam (HPIB) are reported. Ti, Nb, Pt, and C films were produced from ablation plasma generated under the action of pulsed HPIB on a surface of ablated materials. We are developing the combined technology that deals with cleaning a substrate surface by HPIB, thin film deposition by HPIB, and the radiation-induced mass transfer of elements of both film and substrate to increase the adhesion. The deposition of the diamond-like titanium carbide has been realized on the basis of the scheme including Ti film deposition, C film deposition, and the ion beam mixing to form TiC carbide. The main parameters of the Ti-substrate transition layers have been determined for different regimes of their creation by Rutherford backscattering (RBS). Ti, Nb, and Pt coatings were deposited on silicon cantilevers used in high-resolution scanning probe microscopes. The analysis of the composition of TiC coatings at a substrate was performed using the Auger electron spectroscopy (AES) method.

INTRODUCTION

Application of HPIBs for films deposition from ablation plasma generated through the target-beam interaction is prospective [1,2]. A primary consideration in the case of deposition of protective coatings on objects is the insufficient adhesion. This problem may be solved by using the combined method of radiation-stimulated mass-transfer of coating elements and material of substrate by the HPIB irradiation and the formation of a transition layer between the coating and the substrate [3].

The work objectives were: (i) to study the mass-transfer of elements of both the deposited film and the substrate to create a transition layer, (ii) to explore the pulsed-ion beam film deposition on silicon cantilevers, and (iii) to develop the pulsed beam technology for coating different metals with TiC in a single vacuum cycle. This technology would imply ion cleaning of a substrate surface by the HPIB, the creation of a transition layer on the substrate, the deposition of Ti and C films, and the mixing of this heterogeneous system to create TiC.

EXPERIMENTAL PROCEDURE

To deposit coatings, the "Temp-2" accelerator was used with the ion diode produced the beam of C^+ (30%) and H^+ (70%) of 350 keV energy at the ion current pulse duration of 60 ns and ion current density of 20-250 A/cm^2 (0.4 to 5.3 J/cm^2 per pulse).

CP650, *Beams 2002: 14th International Conference on High-Power Particle Beams*, edited by T. A. Mehlhorn and M. A. Sweeney

The pressure in the vacuum chamber was of about $5{\cdot}10^{-5}$ Torr. An ablation target (Ti, Nb, Pt, C) was put in the focus of the ion beam at an angle of 45° to the direction of the beam's propagation. Substrates of Si, 1020 steel, or a cantilever were established in the direction opposite to the target's surface at a distance of 50 mm.

The ground and polished samples were subjected to finish cleaning by the ion beam with the current density of 35A/cm^2 and the number of pulses n=3÷5. During the ion cleaning and the mixing, the sample was rotated by 135° to get perpendicular to the beam and moved along with targets from the focus of the magnetically insulated focusing diode to the distance of ~23.0 cm where the current density of the ion beam was ~35 A/cm^2. The deposition of the Ti + C heterogeneous layers was realized by various schemes, a holder of targets being consecutively rotated on 180°.

EXPERIMENTAL RESULTS

The film thickness was determined with the help of a Linnik interferometer and a laser-induced ellipsometer. At the deposition of Ti on the Si substrate made by a single accelerator shot, the Ti film thickness measured was of about 15 nm.

In Fig. 1, the Ti and Si concentration profiles obtained by the RBS method are presented for the following combination of the film deposition and mixing: after the 3 shots in the deposition mode, the 4 shots were made in the mixing mode, and this cycle was repeated 6 times. The profiles of impurity atoms, C and O, are not shown. It is seen from the figure that not all deposited Ti film was embedded in Si due to the mixing. The Ti concentration decreases from the maximum value of about 78 at% down to zero within the layer of the thickness of about 340 nm. At the surface of Si substrate, the Ti film has been formed; its thickness is ~200 nm.

In Fig. 2, the Ti and Si concentration profiles are shown for the case of the transition layer formation by another scheme: the cycle of 3 shots in the Ti deposition mode followed by 4 shots in the mixing mode was repeated 2 times; then, the cycle of 3 Ti deposition shots followed by 5 mixing shots was repeated 3 times; and at last, the cycle

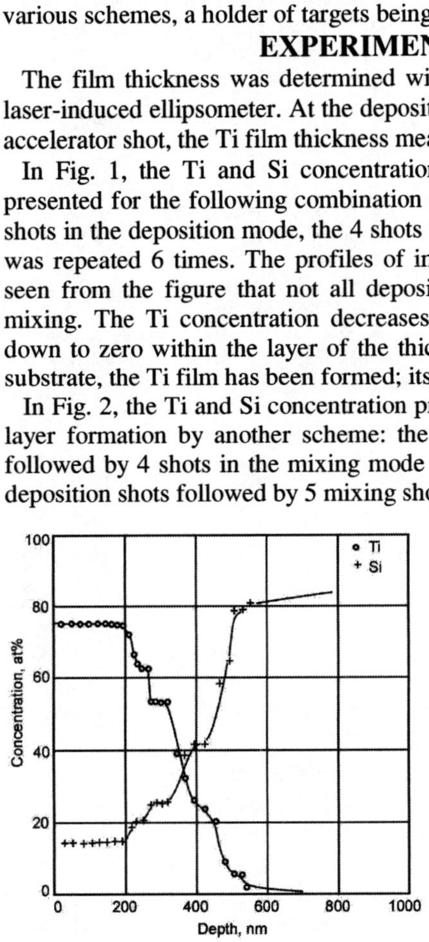

FIGURE 1. Depth distribution of Ti coating (270 nm total thickness) within the Si sample after mixing by 24 current pulses.

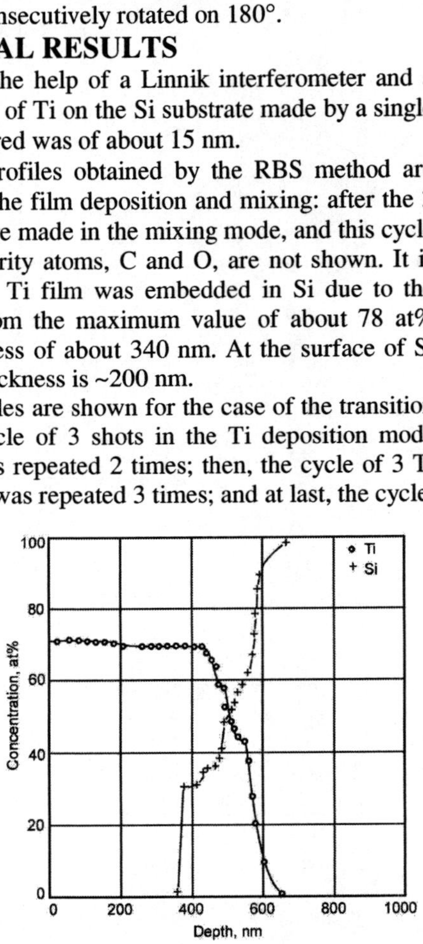

FIGURE 2. Depth distribution of Ti coating (360 nm total thickness) within the Si sample after mixing by 39 current pulses.

TABLE 1.

Summed thickness of deposited Ti, nm	180	225	336
Coated Si microhardness, GPa	12,36	11,31	10,2

of 3 deposition and 6 mixing shots was repeated 3 times. As can be seen, the transition layer thickness unexpectedly decreased down to 250 nm, in spite of increasing the number of mixing shots. At the surface, the Ti film of ~400 nm thickness has been formed. One can assume that Ti atoms are embedded in Si for a depth approximately equal to the track length of 350 keV C ions in the Ti + Si binary system. Calculations show [4] that the track length of C ions in Ti is of ~416 nm, and in Si, it is of ~470 nm. At multiple depositions of Ti films on Si and mixing under ion beam irradiation, Ti concentration at Si surface increases up to maximally possible reducing the track length of C ions in Si due to energy losses. This results in stronger decrease of Ti atoms in the volume of Si, hence, in decreasing the thickness of the transition layer.

The estimation of the transition layer influence on the coating adhesion was made by means of comparison of Ti film deposition (~650 nm film thickness) on the Si substrate with the preliminarily created transition layer (~290 nm thickness) and on the Si substrate without a layer. In the latter case, the deposited Ti film could be easily exfoliated from the substrate. As for the case with the transition layer, scratch testing with the use of a diamond needle did not show any exfoliation.

The samples' microhardness was measured by the Vickers method. Microhardness of Si with the transition layer depends on the summed thickness of Ti mixed with Si as illustrated in Table 1. The load on a diamond indenter was of ~0.3 N; the inaccuracy of microhardness measurements did not exceed 5%. Measured microhardness of initial Ti and Si was ~3.2 GPa and ~18 GPa, respectively.

We deposited metallic films (Ti, Nb, and Pt) of several nanometers thickness on cantilevers, the silicon probes for high-resolution scanning probe microscopy, in a single

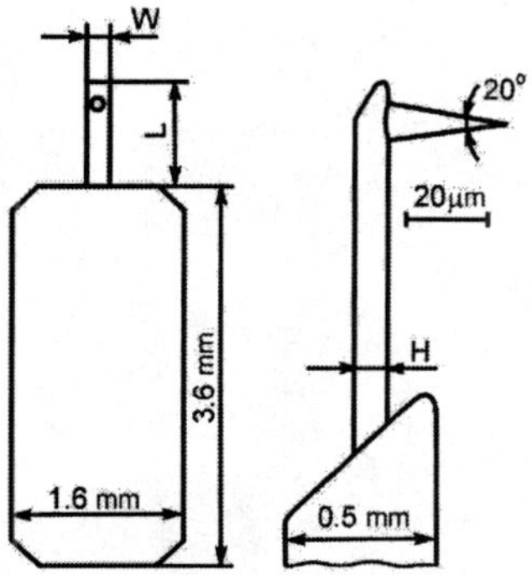

FIGURE 3. Scheme of cantilever. L=100±5μm, W=35±3μm, H=0,4mm.

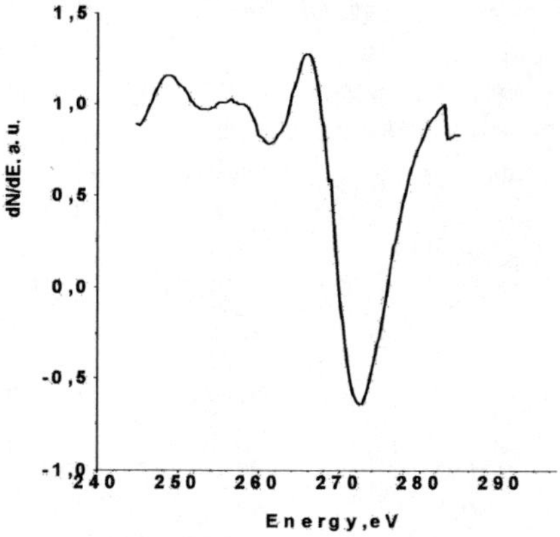

FIGURE 4. Form of carbon peak showing the presence of carbide.

TABLE 2.

Metal	Ti	Nb	Pt
Film thickness, nm	15	9	3.6
Melting temperature, C°	1609	2469	1772
Density, g/cm^3	4.505	8.75	21.45
Heat conductivity, W/mol·K	22	54	72

accelerator shot. The cantilever drawing is shown in Fig. 3. For different target materials, the thickness of films deposited on cantilevers under a single HPIB pulse is given in Table 2. It has turned out that this thickness does not correlate with melting temperature of ablated target as was noted in [5], but rather correlates (inversely proportional) with density and heat conductivity of the target material.

Coating 1020 steel with TiC was realized on the basis of the following scheme: the creation of a transition layer by 45 nm Ti deposition with mixing, the creation of 10 heterogeneous layers composed of both Ti with a thickness of 60 nm and C with a thickness of 13 nm with mixing of each heterogeneous layer by 1 pulse of ion current. The total coating thickness was ~770 nm. Under such conditions of the mixing, more than 50% of Ti and C forms Ti carbide. This estimation follows from the microhardness measurements. In this case, at the indentation load of 0.1 N, the microhardness of the sample with the coating increased by 2.9 times in comparison with the microhardness of the initial sample, and was 4.9 GPa. The form of carbon peak in the Auger-spectrum for this case is presented in Fig. 4. To enlarge the fraction of the Ti carbide formation in the coating, further studies on regimes of the mixing of the Ti-C heterogeneous systems are necessary.

CONCLUSIONS

To enlarge the Ti concentration in a transition layer, it is necessary to intermix the thin layers of Ti by the smallest possible number of pulses with the smallest possible current density of the ion beam. The presence of a transition layer increases significantly the adhesion of the titanium coating to the substrate. The thickness of Ti, Nb, and Pt films deposited on silicon under a single HPIB pulse does not correlate with the melting temperatures of the metals. The combined method of creation of the TiC coating by the deposition of Ti and C thin films with accompanying mixing by one pulse of the ion beam with a current density of 35 A/cm^2 gives formation of > 50% TiC carbides.

REFERENCES

1. Davis H.A., Remnev G.E., Stinnet R.W., Yatsui K. *MRS Bulletin*, **XXI**, No.8, pp.58-62 (1996).
2. Struts V.K., Zakutaev A.N., Matvienko V.M., Potyomkin A.N., Shulepov I.A. *Proceedings of the 5th Conference on Modification of Materials*, Tomsk, Russia, pp.464-466 (2000).
3. Yeong Yan Que, Jen Fin Lin, Chi-Fong Ai. *Thin Solid Films* **302**, 193-200 (1997).
4. Burenkov A.F., Komarov F.F., Kumahov M.A., Temkin M.M. *Tables of spatial distribution parameters of ion-implanted impurities*. Minsk, BSU Publishing, 1980 (*in Russian*).
5. Jiang W., Ide K., Kitayama Sh., Suzuki T., Yatsui K. *Jpn. J. Appl.Phys.* **40**, pp.1026-1029 (2001).

Advanced Studies of Medium-mass Ion Sources-1: Effect of Beam Neutralization on Beam Micro-divergence Angle[1]

Koichi Kasuya, Seimei Araya, Takahiro Kamiya

Department of Energy Sciences, Tokyo Institute of Technology, 4259 Nagatsuta, Midori, Yokohama, Kanagawa 226-8502 Japan

Abstract. Effect of beam neutralization on the beam micro-divergence angle was investigated for proton beam with hydrogen gas puff. About 10% reduction of the angle was observed preliminary.

INTRODUCTION

Although beam production costs for light ion-driven ICF are less than for Heavy Ion Fusion (HIF), the higher charge to mass ratio may restrict the ultimate focus ability of such a beam on an ICF target. On the other hand, the high beam production cost of a HIF driver, together with questions about whether the currents required are achievable, lead to concerns about whether heavy ion beams are the most promising-candidate for a future particle beam driver. Because of these considerations, a medium mass ion beam may represent the best focus ability and the cost compromise. So that, we recently revisited the production of such medium mass pulsed ion beams, and measurement of the beam micro-divergence angle both at the source and at a short propagation distance. The origins of this study are the earlier experiments by us [1] and [2], where pure and more intense nitrogen beams were produced from cryogenically cooled anodes.

In our recent paper [3], medium mass ion beams including nitrogen and oxygen were produced from a cryogenic diode with N_2O ice as the ion source. The nominal diode voltage was 300 - 400 KV, and the peak ion current was 240 A. The beam micro-divergence angle was measured with a 5–aperture time-integrated pinhole camera. The five camera images were analyzed to estimate the spatial distribution of the beam source micro-divergence angle along the anode radius, yielding a value of

CP650, *Beams 2002: 14th International Conference on High-Power Particle Beams*, edited by T. A. Mehlhorn and M. A. Sweeney

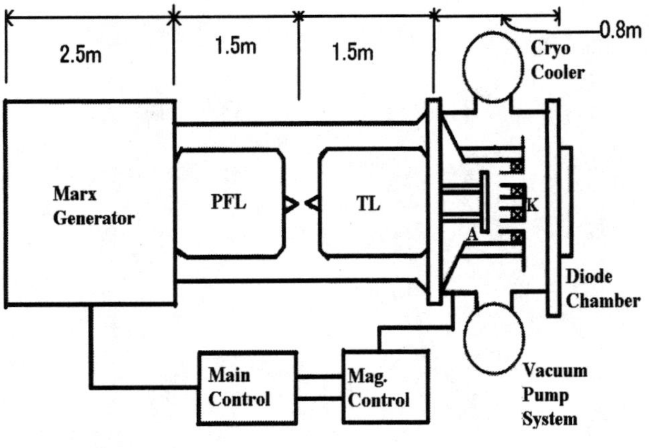

FIGURE 1. Schematic diagram of pulsed power machine and ion diode.

5-6 mrad for the average micro-divergence. This was a low enough value for this ion source without any active beam neutralization to be considered as an ion beam driver for future Inertial Fusion (IF) applications.

During the works above mentioned, one of the most important topics was the effect of the beam neutralization on the beam micro-divergence. So that, most recently, we measured the effect with our advanced methods for beam production and beam diagnostic. In this paper, we show our preliminary results concerning the effect in the case of proton beam.

EXPERIMENTAL APPARATUS

A schematic diagram of pulsed power machine and diode (PICA-3) is shown in Figure 1 and Figure 2. A gas-puff valve was used to put some amount of neutral atoms

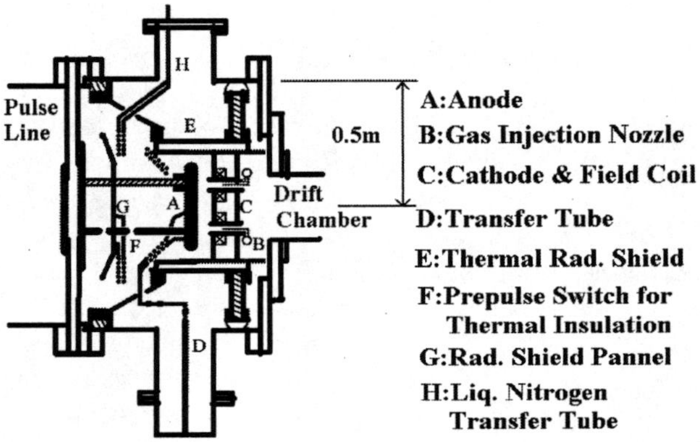

FIGURE 2. Details of diode chamber to produce and neutralize ion beams.

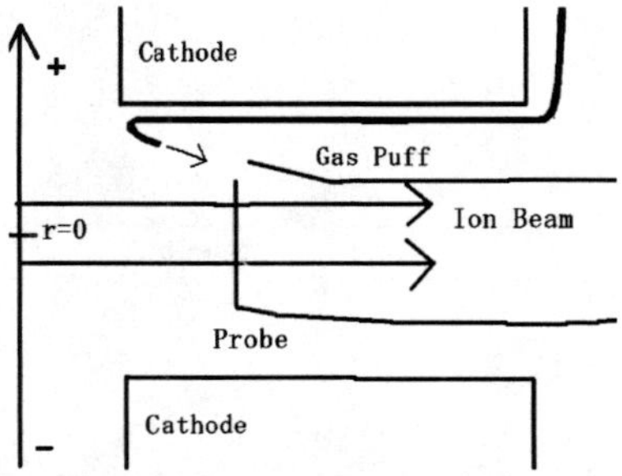

FIGURE 3. Method for pulsed gas injection.

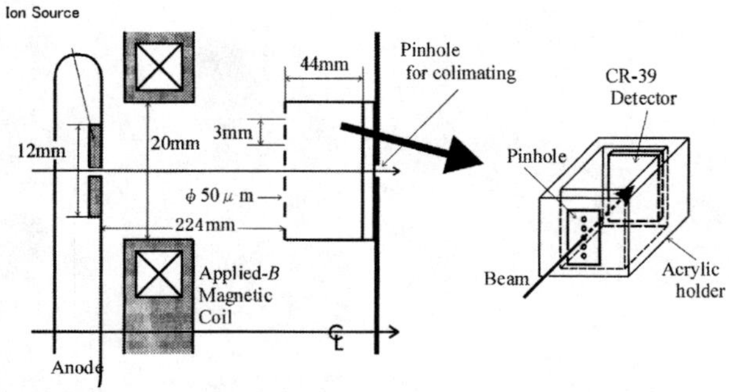

FIGURE 4. Schematic diagram of camera with arrayed (five) pinhole.

into the beam propagation region just behind the cathode, as is shown in Figure 3. We changed the gas-puff timing and the spatial distribution of the atom number density. We measured how the beam micro-divergence angle could be changed with different experimental conditions. Figure 4 is a schematic diagram of an arrayed pinhole camera with five pinholes. The time-integrated image was recorded on a plastic plate, which was etched with an alkaline liquid. The number of ion tracks per unit area on the plate after the etching was counted with a particle counter. The density distributions along the radial axis of the five images were deduced to obtain the local beam divergence angle along the same axis.

EXPERIMENTAL RESULTS

Proton beam was produced with the above apparatuses, and hydrogen gas was puffed to neutralize the beam. A self-discharge probe was used to monitor the arrival of the gas and to make a proper timing of the gas puff and the diode firing. Examples of beam micro-divergence angle on the anode surface are shown in Figure 5. There was 10% reduction of the overall angle with the neutralization, compared with the case without the neutralization, although the improvement was not so much as expected. The typical beam conditions were same with our former publication.

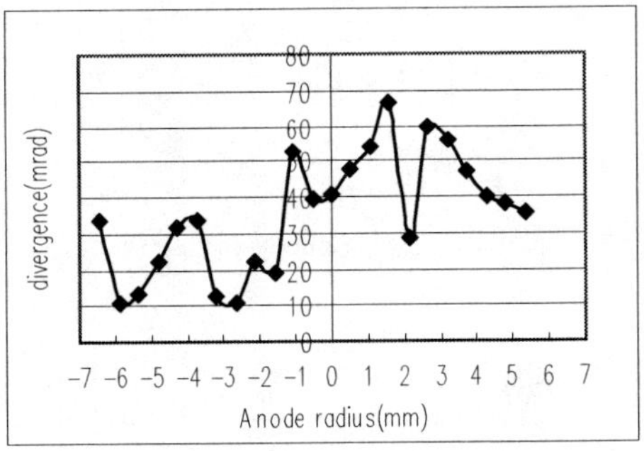

(a) Case without gas-puff neutralization.

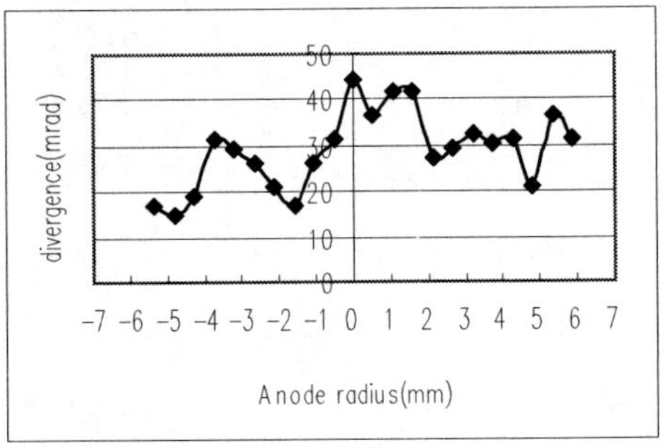

(b) Case with gas-puff neutralization.

FIGURE 5. Proton beam micro-divergence angle with and without beam neutralization.

SUMMARY

Effect of beam neutralization on the beam micro-divergence angle was measured for proton beam with hydrogen gas puff. Similar experiments are necessary to be realized for medium-mass ion beams including nitrogen beams in the near future.

REFERENCES

1. Kasuya, K., and Mitobe, K., *Proc. 11th Int. Conf. Plasma Phys. and Control. Nucl. Fusion Res.* **3**, 161-166 (1987).
2. Hanson, D., Porter, J., and Williams, R., *J. Appl. Phys.*, **70**, 2926-2938 (1991).
3. Kasuya, K., Kishi, Y., Kamiya, T., and Funatsu, M., *Laser and Particle Beams*, **19**, 309-316 (2001).

[1] Supported by the Ministry of Education, Science, Culture and Sports in Japan/ the Japanese Society for Promotion of Sciences/ ILE. Osaka University and Sandia National Laboratories.

Ablation of Various Candidate Materials for Inertial Fusion Reactor Walls-1: Ablated Surface Thickness with Intense Pulsed Proton Beams[1]

Koichi Kasuya,[a] Timothy Renk,[b] Arata Kasamatsu,[a] Craig Olson,[b]

Yohsuke Kinoshita,[a] Takahiro Kamiya,[a] Atsushi Saiki,[c]

Takayoshi Norimatsu,[d] Keiji Nagai,[d] Tatsuhiko Yamanaka,[d] Sadao Nakai[d]

[a] *Department of Energy Sciences, Tokyo Institute of Technology, 4259 Nagatsuta, Midori, Yokohama, Kanagawa 226-8502 Japan*

[b] *Sandia National Laboratories, P. O. Box 5800, Albuquerque, NM 87185-1182 USA*

[c] *Center for Advanced Material Analysis, Tokyo Institute of Technology, 2-12-1 Meguro, Tokyo 152-8550, Japan*

[d] *Institute of Laser Engineering, Osaka University, 2-6, Yamadaoka, Suita, Osaka 565-0871 Japan*

Abstract. Intense proton and nitrogen beams were focused on various sample materials to observe the surface ablation thickness. The thickness change and the surface composition change with the beam dose were measured with various analysis methods.

INTRODUCTION

To design an advanced version of inertial fusion (IF) reactor, there are lots of data to be gathered beforehand into our databases. One of these is the surface thickness of the reactor wall materials, which is ablated with intense ion beams originated from the nuclear fusion processes. As we were accustomed to produce intense proton beams with pulsed power machines, we used this beam to investigate this kind of research topic recently. An earlier example to the same direction is found in [1], where the more different topics are shown concerning the surface ablation.

Various candidate materials were collected from many sources in Japan[2]. Some parts of them were given also to Sandia National Laboratories to be irradiated with their intense X-ray (some numerical results are shown in [2]) and various kinds of intense

CP650, *Beams 2002: 14th International Conference on High-Power Particle Beams*, edited by T. A. Mehlhorn and M. A. Sweeney

pulsed ion beams. In this paper, we show our recent results with a proton beam machine at Yokohama campus of Tokyo Institute of Technology (TITech) for the cases of graphite, SiC and LiPb, and with a nitrogen beam machine at Albuquerque site of Sandia National Laboratories (SNL) for the cases of SiC and Al_2O_3.

SAMPLE MATERIALS PREPARED FOR EXPERIMENTS

Sample targets were collected from many places, including TITech, SNL, ILE-Osaka Univ. (T.Norimatsu), IAE-Kyoto Univ. (A.Kohyama and Y.Katoh) , Tokai Konetsu Co. (J.Satoh) and Atago Bussan Co. (S.Ebara). Only a part (which are listed below) of these samples could be irradiated in our experiments. The descriptions in the following list are the group number/ material/ size (vertical x horizontal x thickness, or diameter x thickness), all in mm/ supplier/ remarks.

1-a/ LiPb/ 16x19x1/ ILE/ Manufactured in Ar environment, Melting Temperature of 233C (Peak)/

1-b/ SiC/ 18x23x1 etc./ ILE/

4-b/ Carbon G650S/ 30diax5t(1)/ TITech/ Product of Nilaco Co. Ltd./

6-a/ SiC/ 100diax2/ Tokaikonetsu/ Product of Tokaikonetsu, Normal Pressure Baked, Both Surface Ground/

6-b/ SiC-Si/ 235x235x5/ Tokaikonetsu/ Product of Tokaikonetsu, RS420/

6-c/ Al_2O_3/ 135x125x5/ Tokaikonetsu/ SSA-M, Product of Nihonkagakutohgyoh, Normal Pressure Baked/

EXPERIMENTAL APPARATUS

PICA-4 machine was used for proton beam generation at TITech, while RHEPP-1 machine was used for nitrogen beam generation at SNL.

In the PICA-4 machine, a magnetically insulated diode with solid (hydrocarbon) anode was installed to extract the proton beam along the horizontal axis. The pulse length of the ion beam was 100ns, the beam energy was 200-240keV, and the ion peak current was 1.6kA. The beam was focused at 230mm from the anode, and the focused beam diameter was 23mm. The beam intensity was up to about 10J/cm^2, which was high enough to ablate most of the target materials with one shot.

In the RHEPP-1 machine, a similar diode with a gas-puff system was installed to extract the nitrogen beam along the vertical axis.

PROTON BEAM ABLATION

Sample targets were placed at the beam focal point, normal to the beam axis. The half surfaces of the targets were covered with masks for the targets not to be irradiated. The target surfaces were polished and cleaned before the beam irradiation. The beam dose was changed with the change of the supplied voltage with the pulsed power. The ablation thickness was measured with a laser displacement diagnostic tool. A series of beam irradiation with three different target materials (carbon, LiPb and SiC/ *4-b*, *1-a* and *1-b* in the former section) was tried, which results are shown in Figure 1.

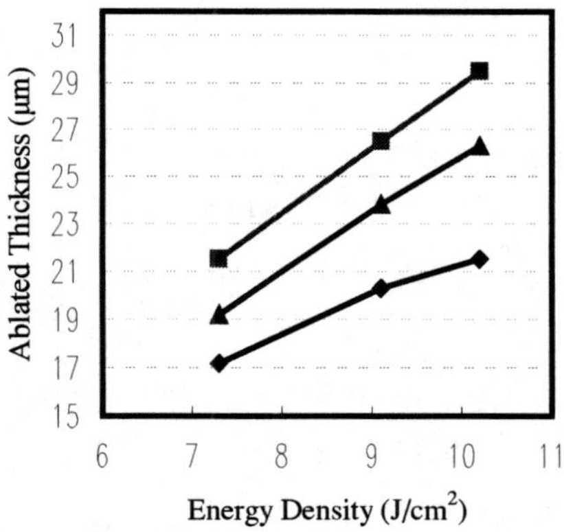

FIGURE 1. Ablated thickness vs. energy density of proton beam, SiC, LiPb and C from top to bottom.

NITROGEN BEAM ABLATION

Similar experiments were performed at SNL with the RHEPP-1 pulsed power machine to irradiate different sample targets (SiC and Al_2O_3/ *6-a* and *6-c* in the former section) with nitrogen ions. As the beam focus diameter was larger than the proton beam case, we could separately observe the different dose places on the same (larger) targets. We placed four Faraday cups along the beam diameter from the center to the outer radius. With the total number of consecutive 50 shots, the beam dose was 3.9, 3.2, 2.2 and 1.6 J/cm^2.

Two kinds of diagnostic methods were chosen to observe the target surfaces. One was a laser reflection microscope method, and the other was x-ray luminescence composition analysis method. We asked the Center for Advanced Material Analysis in the Tokyo campus of TITech.

With the multiple beam irradiations, the SiC surface condition on the microscope image was changed drastically from the original condition. The geometrical structure along the depth of the surface could be also observed with this microscope. We could estimate the mean surface ablation thickness also with this image. The beam irradiated surface looked like to have a structure after re-crystallization. On the contrary, in the case of Al_2O_3, there was no observable ablation thickness with the same diagnostic method.

Figure 2 shows the result of the x-ray luminescence analysis of SiC. The change of surface composition is shown as a function of the ion dose. With the highest dose, the carbon is lost completely from the surface after the beam irradiation. Together with the above microscope image and the luminescence result, the re-crystallized structure seems to be silicon. On the contrary, in the case of Al_2O_3, there was no observable composition change with the same diagnostic method.

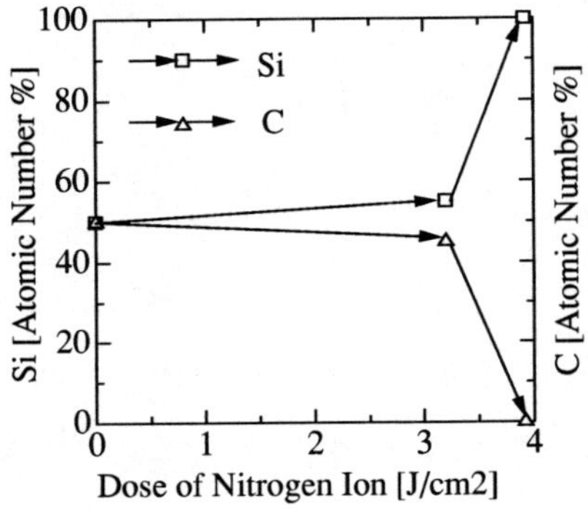

FIGURE 2. SiC surface composition change vs. nitrogen beam dose, measured with x-ray luminescence analyzer.

SUMMARY

Under the proton beam irradiation with the dose of about $10 J/cm^2$ per shot, the ablated rates were up to 20 micron for carbon, and 26 micron for LiPb and SiC. The change of the rate was investigated with the change of the dose. Under the nitrogen beam irradiation with the dose of $4 J/cm^2$, the ablated thickness of up to 6 micron was observed for SiC, while no remarkable ablation was observed for Al_2O_3. Surface conditions before and after the nitrogen beam irradiation was investigated with a microscope and x-ray fluorescence analysis. SiC surface turned into Si surface, while no discernible change was seen for Al_2O_3.

REFERENCES

1. Kasuya, K., *Report of IAEA Coordinated Research Project on Elements of Power Plant Design for Inertial Fusion Energy,* to be published on Web pages (2002).
2. Kasuya, K., Kinoshita, Y., *Bulletin of Amer. Phys. Soc.*, **46**, 237-237 (2001).
3. Kasuya, K., Kasamatsu, A., Kinoshita, Y., Kamiya, T., et al., *Proc. 2nd Int. Conf. Inertial Fusion Sciences and Applications,* to be published, (2002).

[1] Supported by the Ministry of Education, Science, Culture and Sports in Japan/ the Japanese Society for Promotion of Sciences/ ILE. Osaka University and Sandia National Laboratories

[2] The authors thank for their discussions and donations of the various samples, to Tokyo Institute of Technology (Emeritus Prof. T.Iseki), ILE-Osaka University (Prof. T.Norimatsu), Kyoto University (Prof. A.Kohyama and Y.Katoh), Tokai Konetsu Kogyo Co. Ltd. (Dr. K.Satoh) and Atago Bussan Co. Ltd. (Dr. S.Ebara).

Advanced Studies of Medium-mass Ion Sources-2: Production and Measurement of (1)Pure Carbon Plasmas as Candidate Source Plasmas, and (2) Silicon and Carbon Beams [1]

Koichi Kasuya,[a] Takahiro Kamiya,[a] Takehisa Sugimoto,[a] Waldemar Mroz,[b] Shuhji Sakabe,[c] Seiji Shimizu,[c] Sin-ichiro Okihara,[c] Tatsuhiko Yamanaka,[c] Sadao Nakai[c]

[a] *Department of Energy Sciences, Tokyo Institute of Technology, 4259 Nagatsuta, Midori, Yokohama, Kanagawa 226-8502 Japan*
[b] *Institute of Optoelectronics, MUT, 2 Kaliskiergo, Warsaw 01-485 Poland*
[c] *Institute of Laser Engineering, Osaka University, 2-6 Yamadaoka, Suita, Osaka 565-0871 Japan*

Abstract. Two kinds of source developments were tried for the generation of pulsed medium-mass ion beams. The first was the pure anode plasma formation with the fs-laser light irradiation of a (conductive) graphite target. The other was beam production from a dielectric anode, which was covered with a diamond like carbon film or silicon layer. The preliminary results were disclosed here.

INTRODUCTION

Although beam production costs for light ion-driven ICF are less than for Heavy Ion Fusion (HIF), the higher charge to mass ratio may restrict the ultimate focus ability of such a beam on an ICF target. On the other hand, the high beam production cost of a HIF driver, together with questions about whether the currents required are achievable, lead to concerns about whether heavy ion beams are the most promising-candidate for a future particle beam driver. Because of these considerations, a medium mass ion beam may represent the best focus ability and the cost compromise. So, we restarted development of new ion sources recently.

CP650, *Beams 2002: 14th International Conference on High-Power Particle Beams*, edited by T. A. Mehlhorn and M. A. Sweeney

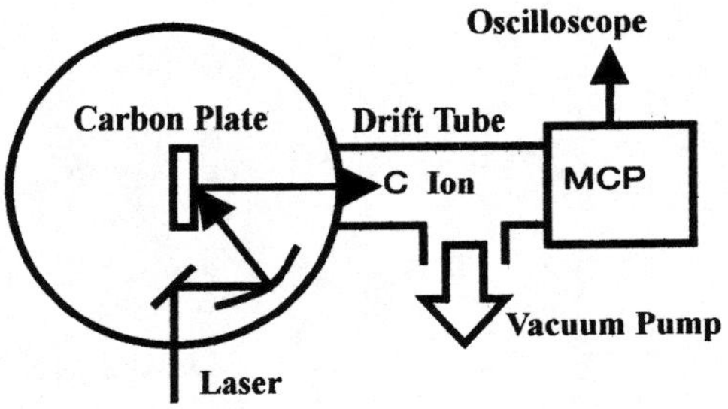

FIGURE 1. Schematic of target irradiation.

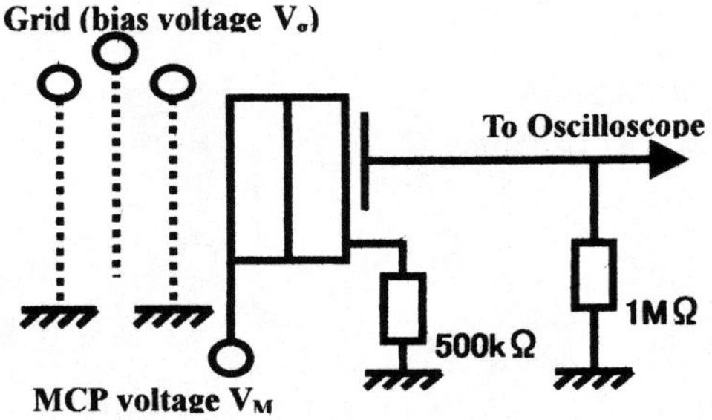

FIGURE 2. Micro-channel plate detector system.

The origin of this study is our earlier experiments [1], where we used an e-beam pumped KrF laser to produce Al plasmas. An aluminum target in a vacuum chamber was irradiated with the laser lights of 0.7J energy and 32kJ/cm^2 flux. Produced plasmas were measured with a biased ion collector, and the luminosity was observed with a fast frame camera. The plasma surface density was estimated to be 1.6x10^{16}cm^2.

In our recent experiment, we used the same laser to irradiate a carbon target. Similar measurement shown above was performed also in this case. Moreover, additional ion detector and laser were used to continue our experiments. As the third author brought an ion detector with a micro-channel plate (MCP) to Japan from Poland, this was used to measure our plasmas. As the third group (ILE, Osaka Univ.) had an ultra short pulse laser (table-top, high intensity Ti-sapphire laser [2]), a graphite target was irradiated with this laser.

As the second topic of this paper, we proposed a new method to produce pure medium-mass ion beams with a conventional type of ion diode without cooling of the anode. Diamond like carbon or silicon was the new candidate as the anode material, which was used to cover the ring-shaped anode surface. It was more convenient if we could use a room temperature anode instead of a cryogenic anode.

PRODUCTION OF CARBON ION SOURCE WITH LASER

The experimental arrangements to shoot a graphite target with a fs-laser is shown in

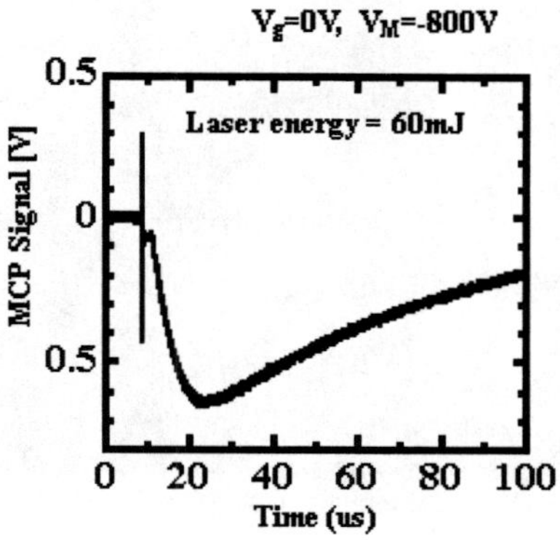

FIGURE 3. Typical ion collector signal with MCP.

Figure 1. The laser energy was 60-170mJ, and the pulse width was 130fs. The micro channel plate was placed at a distance of 150cm from the carbon target. A pinhole was put in the plasma drift tube to reduce the incoming ion flux. The background vacuum pressure was $2x10^{-5}$torr. With the different applied voltage to the detector, we observed the current signals of the detector. The more details of the MCP is shown in Figure 2.

An example of the MCP signal is shown in Figure 3. A series of similar signals were gathered with different laser light and MCP settings. We are now making analysis of these signals to disclose the characteristics of the laser-produced plasmas.

PRODUCTION OF PURE SILICON AND CARBON BEAMS

In our recent experiments [3], we used a cryogenic anode to freeze N_2O on the anode surface. Although the liquid nitrogen was useful to get the necessary anode temperature, the produced beam was a mixture of nitrogen and oxygen. On the contrary, if we use DLC (diamond like carbon) or silicon as the anode material of the similar magnetically insulated diode, we can produce pure carbon or silicon beam without any cooling. As the DLC or Silicon is dielectric (non-conductive), we need not to use the additional ionizing method as laser light to irradiate the anode surface and to produce the anode plasmas.

We used the same diode as in [3], while the anode ring surface was covered with a thin layer of DLC or Silicon. A photograph of the anode with such ring is shown in Figure 4. If the triple point configuration was realized, we could observe reasonable ion current of the pure beam. The beam specific-mass increased, when we moved from proton (1.01) to carbon (12.0), nitrogen (14.0), oxygen (16.0) and silicon (28.1). As the ion current is inversely proportional to the root of the mass, the diode impedance for silicon beam case was a lot higher than the proton beam case. As a result of this, the impedance matching between the pulsed power (5ohm) and diode became poorer. This

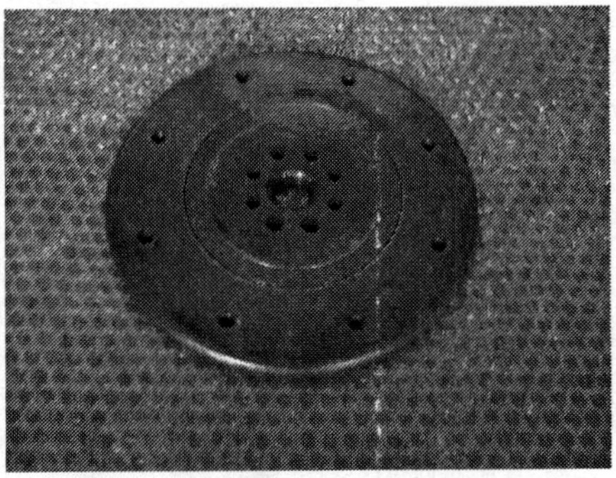

FIGURE 4. Photograph of silicon ring anode.

brought the multiple reflection of the stored energy within the pulsed power machine, and the multiple beam production was observed in silicon beam case.

SUMMARY

A graphite plate was irradiated with a fs-laser light with the aim to supply the anode plasmas for the pure carbon beam production. If possible, we wanted to produce and measure the higher charge states of carbon ions, while the obtained MCP signals were not possible to say the details of the charge state until now. Additional experiments are necessary to disclose these points.

Dielectric carbon (diamond like carbon film) and silicon were proposed to use as the anode surface materials to produce pure carbon and silicon beams. We could observe the pulsed ion signals, while more experiments were necessary to clarify the potential of these kinds of materials as the ion sources.

As the ion mass is increased with the use of heavier atomic species, we need to modify the pulsed power machine to match the load (ion diode) impedance. This is also the direction to increase the applied voltage of the ion diode, which increase the speed of the produced ions.

After the development of various ion sources with more convenience, we will be able to make various ion beam interaction experiments with various targets. More realistic applications of these kinds of beams are expected in the fields of material science and nuclear fusion, in the near future.

REFERENCES

1. Kasuya, K., Miyamoto, S., Nakai., S., Kamiya, T., Funatsu., M. et al., *Proc. 11th Int. Conf. High Power Particle Beams* **1,** 52-55 (1996).
2. Sakabe, S., Nishihara, K., Nakashima, N., et al., *Phys. Plasmas*, **8**, 2517-2524 (2001).
3. Kasuya, K., Kishi, Y., Kamiya, T., Funatsu, M., *Laser and Particle Beams*, **19**, 309-316 (2001).

[1] Supported by the Ministry of Education, Science, Culture and Sports in Japan/ the Japanese Society for Promotion of Sciences/ ILE, Osaka University and Sandia National Laboratories.

Ablation of Various Candidate Materials for Inertial Fusion Reactor Walls-2: Simple Calculation and Comparison with Experimental Result of Surface Thickness Ablated with Intense Pulsed X-ray of Sandia Z-Pinch Source[1]

Koichi Kasuya,[a] Yohsuke Kinoshita,[a] Timothy Renk,[b] Craig Olson[b]

[a] *Department of Energy Sciences, Tokyo Institute of Technology, 4259 Nagatsuta, Midori, Yokohama, Kanagawa 226-8502 Japan*

[b] *Sandia National Laboratories, P. O. Box 5800, Albuquerque, NM 87185-1182 USA*

Abstract. A simple method was proposed to calculate surface thickness of wall surface, which is ablated by intense x-ray accompanied with inertial fusion reaction. Various materials were chosen to estimate ablated depth and preliminary experiments were performed to compare the both results.

INTRODUCTION

To design an advanced version of inertial fusion reactor, there are lots of data to be gathered beforehand into our databases. One of these is the surface thickness of the reactor wall materials, which is ablated with intense X-rays originated from the nuclear fusion processes. As we were accustomed to make simple calculations of material ablation with intense electro-magnetic radiation [1-2], we tried to investigate this kind of research topic recently. An earlier example to the same direction is found in [3], where the details were not shown because of the space limit.

Various candidate materials were collected from many sources in Japan[2]. Some parts of them were given also to Sandia National Laboratories to be irradiated with their intense X-ray and various kinds of intense pulsed ion beams. In this paper, we show our recent results of simple calculations of the material thickness ablated by the intense X-ray with the spectrum and energy level of Sandia Z-Pinch Source, for the

CP650, *Beams 2002: 14th International Conference on High-Power Particle Beams*, edited by T. A. Mehlhorn and M. A. Sweeney

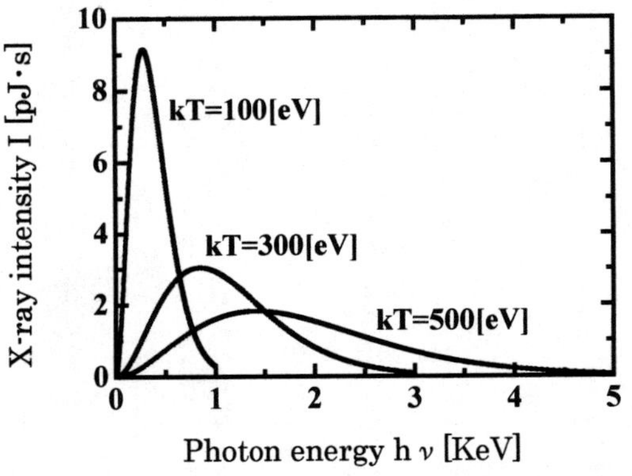

FIGURE 1. X-ray spectrum under Planck's law with conserved total energy.

cases of graphite and lithium fluoride. Preliminary experiments with PBFA-Z X-ray source were performed recently, which result was compared with the numerical one.

SIMPLE CALCULATIONS OF ABLATED THICKNESS

The ablation rates with intense pulsed x-rays with the spectrum and energy level of the PBFA-Z machine at SNL were numerically estimated with a simple model [2-3]. Under the conservation of total x-ray energy of 1MJ, we can plot the Planck's radiation law with different black body temperature (kT) of 100, 300 and 500eV in Figure 1. The ordinate is x-ray intensity, while the abscissa is photon energy. The spectral intensity of x-ray is calculated with the Planck's equation with this figure.

We chose two kinds of materials (C and LiF) to calculate the ablated depth with x-ray. The respective mass absorption coefficient for C and LiF is shown in Figure 2-1 and 2-2, as a function of photon energy. To calculate x-ray absorptions with wall materials, the

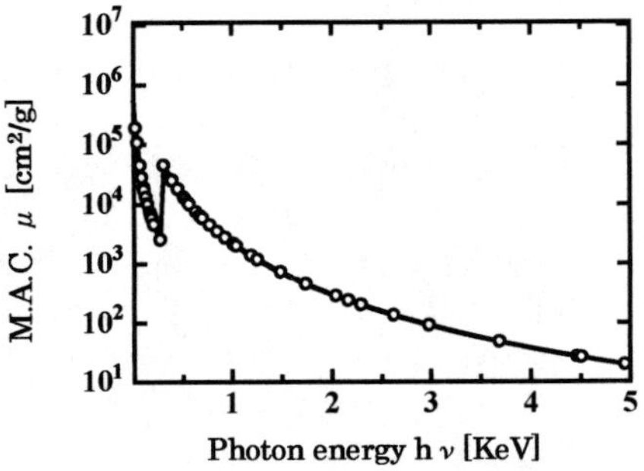

Figure 2-1. Mass absorption coefficient of C.

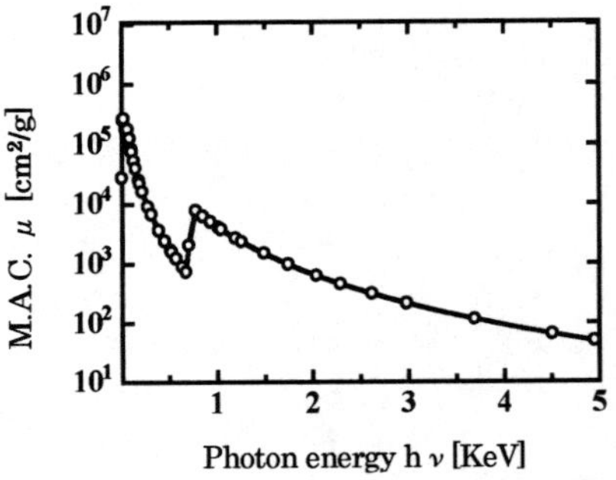

Figure 2-2. Mass absorption coefficient of LiF

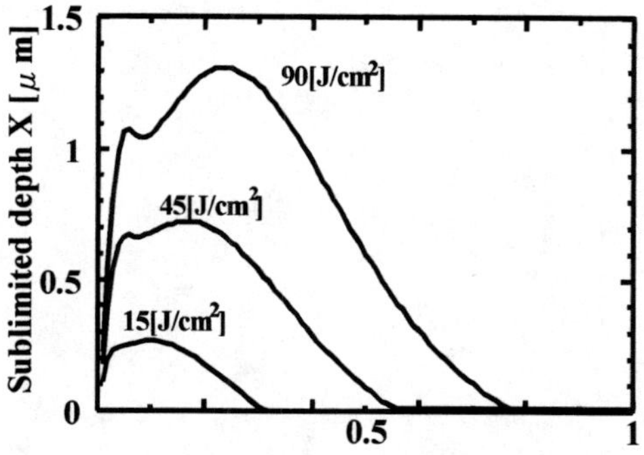

FIGURE 3. Sublimated depth of graphite vs. x-ray source temperature with three different doses.

wall thickness was sliced into very thin layers with 10nm thickness, and the x-ray absorption and transmittance were calculated layer by layer. We assumed that the layer was ablated if the absorbed energy is larger than the total energy for the layer temperature to become higher than the sublimation temperature of C, or the vaporization temperature of LiF.

We calculated the sublimation depth of C as a function of the black body temperature of x-ray under different x-ray dose (90, 45 and 15J/cm^2), which result is shown in Figure 3. The similar result for LiF is shown with three curves in Figure 4. The ablation rate was 1~7 micron in the case of LiF, while the thickness was 0.1~1.5 micron in the case of C. The maximum C ablation was obtained with the x-ray temperature of 50 ~200 eV, while the maximum LiF ablation was obtained with the x-ray temperature of 200~600 eV.

ABLATION EXPERIMENT WITH PBFA-Z X-RAY

The x-ray irradiation of a LiF sample (prepared by SNL) was performed experimentally with the SNL Z-machine. The measured ablation was about 3 microns from a single PBFA-Z shot with an estimated x-ray dose of 40 J/cm^2. SiC/Si and SiC (supplied from Tohkai Kohnetsu Kohgyo Co. Ltd.) was irradiated with x-ray of 7J/cm^2 dose. The measured ablation depth for SiC/Si was about 0.5 micron, while no discernible ablation was measured for SiC.

COMPARISON OF BOTH RESULTS

The experimental result in the case of LiF is also shown with a small white box in Figure 4 to compare with the numerical results. The black body temperature of x-ray (which corresponded to the experiment) is about 100-200eV. Similar experiments with various materials and different x-ray dose are scheduled in the near future.

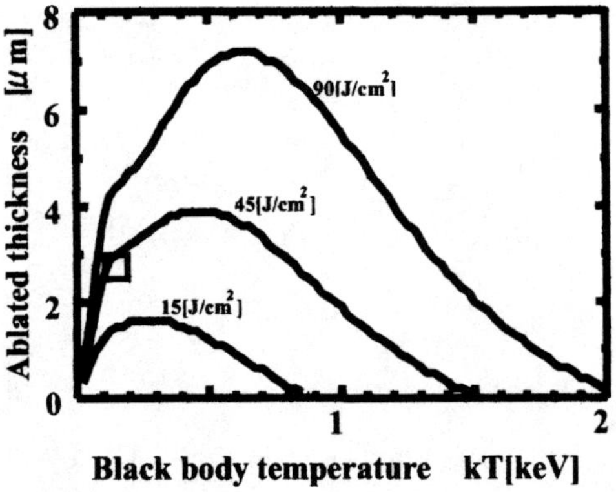

FIGURE 4. Comparison of LiF ablated thickness (rectangular box: experiments, three curves: calculations). The agreement is good although calculation is very simple, compared with other codes.

SUMMARY

We considered the mass absorption coefficient under different black body temperature and the different total energy of the X-ray. We compared the absorbed X-ray energy with the energy necessary to finish the sublimation or the evaporation of the materials. The ablation rates with the MJ level X-ray was 1~7 micron in the case of LiF, while the thickness was 0.1~1.5 micron in the case of C. The maximum ablation of C was obtained with the X-ray temperature of 50 ~200 eV, while the maximum ablation of LiF was obtained with the X-ray temperature of 200~600 eV. The obtained numerical results were compared with the experimental ones under the US Japan collaboration Project & IAEA CRP on IFE. The additional experiments with different materials from graphite and lithium fluoride will be performed in the near future.

REFERENCES

1. Kasuya, K., Funatsu, M., Saitoh, S., Kamiya, T. et al., *Proc. 18th IAEA Fusion Conf., CD-ROM Publication, IAEA-CN-77/IFP/04* (2001).
2. Kinoshita, Y., and Kasuya, K., *Proc. 2nd Int. Conf. Inertial Fusion Sciences and Applications,* to be published, (2002).
3. Kasuya, K., and Kinoshita, Y., B*ulletin of Amer. Phys. Soc.*, **46**, 237-237 (2001).

---

[1] Supported by the Ministry of Education, Science, Culture and Sports in Japan/ the Japanese Society for Promotion of Sciences/ ILE. Osaka University and Sandia National Laboratories

[2] The authors thank for their discussions and donations of the various samples, to Tokyo Institute of Technology (Emeritus Prof. T.Iseki), ILE-Osaka University (Prof. T.Norimatsu), Kyoto University (Prof. A.Kohyama and Y.Katoh), Tohkai Kohnetsu Kohgyo Co. Ltd. (Dr. K.Satoh) and Atago Bussan Co. Ltd. (Dr. S.Ebara).

Application of Flyer Acceleration by Pulsed Ion Beam Ablation to Deep Space Thruster

Nobuhiro Harada [a], Hiroshi Sakurada [a],
Kenji Kashine [c], Weihua Jiang [b], and Kiyoshi Yatsui [b]

[a] *Department of Electrical Engineering,*
[b] *Extreme Energy-Density Research Institute*
Nagaoka University of Technology, 1603-1 Kamitomioka, Nagaoka 940-2188, Japan
[c] *Department of Electrical Engineering,*
Kagoshima National College of Technology, Hayato-cho, Kagoshima 899-5193, Japan

Abstract. Flyer acceleration by ablation plasma pressure produced by irradiation of intense pulsed ion beam has been studied. Acceleration process including expansion of ablation plasma was simulated based on fluid model. In experiments, flyer velocities of 5.6 km/s and ablation plasma pressure of 15 GPa were demonstrated and these results agree well with numerical simulation. Finally, applicability of ablation propulsion was discussed. Specific impulse of 7000~8000 seconds and time averaged thrust of up to 5000~6000N can be expected.

INTRODUCTION

We have been studying on thin film flyer acceleration by ablation plasma using intense pulsed ion beam.[1-4] So far, we could achieve 5.6km/s of flyer velocity and 15GPa of ablation pressure in experiment. Although the major purpose of these studies is application of this process to materials science such as production of new material, improvement of material surface by collision of an accelerated film to another materials, and so on, another and direct possibility must be ablation propulsion.

In this paper, previous fundamental results are reviewed firstly. Then recent results on acceleration of multi-layer target and effect of waveform of incident ion beam on acceleration process are presented. Finally, capability of thrust and specific impulse of ablation propulsion is discussed.

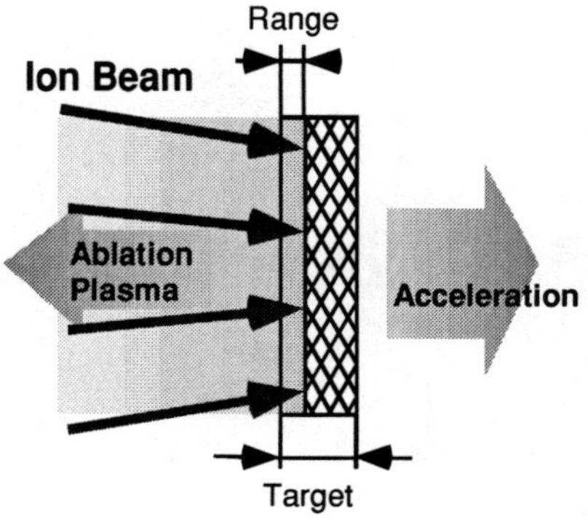

FIGURE 1. Principle of ablation acceleration by ion beam.

PRINCIPLE CONCEPT OF FLYER ACCELERATION

Figure 1 shows the basic concept of flyer acceleration by ablation plasma that

CP650, *Beams 2002: 14th International Conference on High-Power Particle Beams*,
edited by T. A. Mehlhorn and M. A. Sweeney

is produced by irradiation of an intense pulsed ion beam. Incident ion beam can penetrate into the target within a certain depth so-called "range" and deposits its energy. Deposited energy is so high that the part within the range evaporates, ionized, and finally forms high-temperature and high-pressure ablation plasma. Thus residual part of the target is highly accelerated by the reaction of this ablation plasma expansion.

EXPERIMENTAL FACILITY

Ion beam is produced by the intense pulsed ion beam facility "ETIGO-II" that is installed at the Extreme Energy-Density Research Institute (EDRI) at the Nagaoka University of Technology. We have been using two kinds of diode to produce ion beam: 1) Magnetically Insulated Diode(MID) with the energy density of up to $100 J/cm^2$ and 2) Spherically-shaped Plasma Focus Diodes(SPFD) with up to $4.3 kJ/cm^2$. Typical operating conditions are presented in Ref. 1-3.

NYMERICAL ANALYSIS

We applied 1-dimensional fluid model to this ablation process as for numerical simulation. Simple basic equations were used as follows.

Mass conservation:
$$\frac{\partial \rho}{\partial t} + \frac{\partial \rho u}{\partial x} = 0 \tag{1}$$

Momentum conservation:
$$\rho\left(\frac{\partial u}{\partial t} + u\frac{\partial u}{\partial x}\right) = -\frac{\partial (P+q)}{\partial x} \tag{2}$$

Energy equation;
$$\rho C_v\left(\frac{\partial T}{\partial t} + u\frac{\partial T}{\partial x}\right) = -(P+q)\frac{\partial u}{\partial x} + S + H \tag{3}$$

where interactions of ion beam with target and ablation plasma can be taken into account as so-called stopping power S. Term of H denotes energy loss due to thermal conduction. In addition to the above basic equations, ideal equation of state was used. Above set of equation were solved a finite difference method in the Lagrange scheme.[4]

RESULTS AND DISCUSSIONS

Accelerated flyer velocity is shown in Fig.2 against incident ion beam energy density. For measurement of flyer velocity, we adopt the Time-of-Flight method with the electric probe located at the distance of 1mm from the target. So, measured flyer velocity is averaged one over this distance of 1mm. We can see that fairly good agreement of experimental data with the numerical results over the wide range of incident beam energy density. Pressure of ablation plasma is shown in Fig.3. Experimental data are estimated under the assumption of constant acceleration, which is averaged over about 1μs in this particular case. On the other hand, results of numerical analysis show the maximum pressure for each condition. Difference between experimental data and numerical ones is about slightly more than an order.

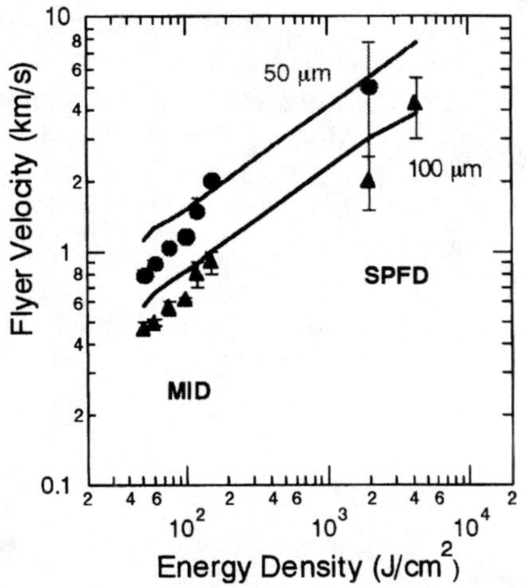

FIGURE 2. Flyer velocity against incident ion beam Energy density.

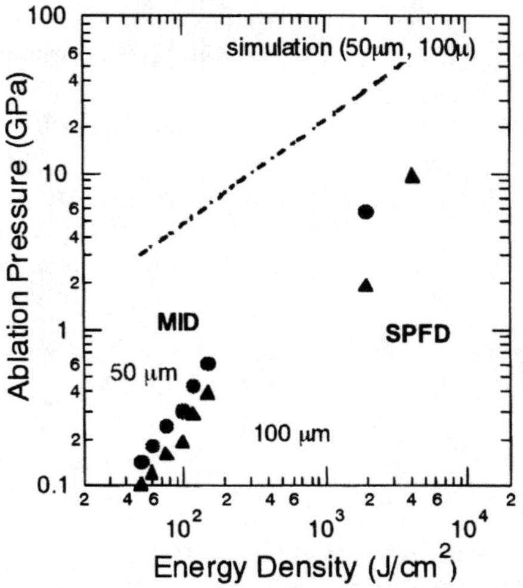

FIGURE 3. Ablation plasma pressure vs. energy density.

This fact can be understood as the difference of averaging period. For example, for the case of 100J/cm^2, integrated ablation pressure is 30GPa x 60ns from analysis, which corresponds to 0.3GPa x 1000ns from measured data. Therefore, we can understand that these differences must be reasonable.

APPLICABILITY OF ABLATION THRUSTER

We have to note that the maximum power level of the present experiments of 4kJ/cm^2, which corresponds to 40MJ/m^2 if ablation area is assumed to be 1m^2. If this process is performed once a second, required electric power becomes 40MW and this corresponds to electric propulsion power for transportation to Mars. [5]

Total momentum balance can be written as:

$$m_{ion} v_{ion} = m_f v_f - \overline{m}_{ab} \overline{v}_{ab} \tag{4}$$

where f and ab denote flyer and ablation plasma. Because of momentum of incident ion beam is negligible, ion beam only transfers its energy not its momentum. From this point of view, ion beam has the same role as lasers or electron beams. Advantages of ion beam are relatively high-efficiency to produce intense beam, short range, i.e. penetration depth, and higher-pressure and higher–temperature of ablation plasma produced by ion beam. Therefore, averaged ablation plasma velocity can be written:

$$\overline{v}_{ab} = \frac{m_f}{\overline{m}_{ab}} v_f \approx \frac{t_f - t_r}{t_r} v_f \tag{5}$$

where t_f and t_r denote initial thickness and range.

In the present experiments, flyer velocity v_f of ~8 km/s can be expected as shown in Fig.2. And the factor $(t_f\text{-}t_r)/t_r$ are around 6. Therefore, averaged ablation plasma velocity that corresponds to effective exhaust gas velocity is about 50km/s and specific impulse will be 5000 seconds. Further, it can be increased 7000~8000 seconds by optimization of waveform of ion beam as shown in the previous section.

We can expect ~20Gpa of ablation pressure as shown in Fig.3 at 4kJ/cm^2. Duration can be estimated as ~100ns. This corresponds to averaged value over 1 second of 2000N for 1m^2 of ablation area. If we use heavier material, thrust can be increased even with the same beam energy density. We just started to accelerate tungsten flyer. Comparison of energy with the same deposition energy density:

$$\frac{1}{2} m_{Al} v_{Al}^{2} = \frac{1}{2} m_W v_W^{2} \tag{6}$$

and relation between each momentum is:

$$m_W v_W = \sqrt{\frac{m_W}{m_{Al}}} \cdot m_{Al} v_{Al} \tag{7}$$

We can expect that the thrust may be increased by 2.6 times and the average thrust reaches up to 5000~6000N for the case of tungsten target.

CONCLUDING REMARKS

Flyer acceleration process by intense pulsed ion beam has been experimentally and numerically studied. We can summarize the following concluding remarks:

Numerical results of accelerated flyer velocity agreed well with measured one over wide range of incident ion beam energy density.

Flyer velocities of 5.6 km/s and ablation plasma pressure of 15 GPa were demonstrated by the present experiments.

Applicability of ablation propulsion was discussed. Specific impulse of 7000~8000 seconds and time averaged thrust of up to 5000~6000N can be expected. It will be possible to use electric thruster system for deep space explorations.

ACKNOWLEDGMENTS

For This study has been partly supported by the Grant-in-Aid for Scientific Research (B) from the Japanese Ministry of Education, Science, Sports and Culture.

REFERENCES

1. Nob. Harada, et al., 12th Int. Conf. High-Power Particle Beams, Vol.2, p.1048, 1998.
2. Nob. Harada, et al., AIAA Paper, AIAA 99-3485, 1999.
3. K.Yatsui, et al., Jpn. J. Appl. Phys. Vol.40, pp.955-959, 2001.
4. Nob. Harada, et al., Jpn. J. Appl. Phys. Vol.40, pp.960-964, 2001.
5. Ron Litchford, et al., AIAA paper, AIAA 2001-0961, 2001.

Characteristics of Cr-Al-N-O Thin Films Prepared by Pulsed Laser Deposition

Makoto Hirai, Hajime Saito, Tsuneo Suzuki, Hisayuki Suematsu, Weihua Jiang and Kiyoshi Yatsui

Extreme Energy-Density Research Institute, Nagaoka University of Technology, Nagaoka, Niigata 940-2188, JAPAN

Abstract. Chromium aluminum oxynitride (Cr-Al-N-O) thin films have been successfully prepared by pulsed laser deposition (PLD). Experiments were carried out by changing surface area ratio of the targets ($S_R = S_{AlN} / (S_{Cr_2N} + S_{AlN})$) from 0 to 100 %. The hardness of one of the Cr-Al-N-O thin films was found to be approximately 40 GPa when the aluminum content in cations (x) was 35 at. %. Elastic modulus of the Cr-Al-N-O thin films with x = 35 at. % was clarified to be 40 % higher than that of chromium oxynitride (Cr-N-O) thin film. These results implied that increasing the elastic modulus by substitution of Al for Cr in Cr-N-O grains causes the hardening of the Cr-Al-N-O thin films. In oxidation tests, the existence of oxides was not detected in the Cr-Al-N-O thin film heat-treated at 900 °C by X-ray diffraction (XRD). This fact was presumed to be due to dissolved aluminum atoms in B1 (NaCl) structure.

INTRODUCTION

A hard coating of nitrides has been prepared by various methods including physical vapor deposition (PVD). Chromium nitride (CrN), titanium nitride (TiN) and titanium aluminum nitride (Ti-Al-N) are widely used as the coating materials for cutting tools in industries. In fact, Ti-Al-N has been known to exhibit the highest hardness (30 GPa) and oxidation temperature (800 °C) among these coating materials [1]. However, to increase the lifetime of coating used under severer conditions, it is necessary to improve these properties more further.

Recently, much attention is paid for chromium aluminum nitride (Cr-Al-N). One of the reasons is that CrN has the highest solubility limit for aluminum nitride (AlN) among transition metal nitrides with B1 (NaCl) structure [2]. It can be related by the fact that solid solution hardening is effective when the solubility limit is large. Among metals and alloys, in addition, better oxidation behavior has been achieved by not only chromium but also aluminum [3]. Aluminum sesquioxide (Al_2O_3) layer, which is formed on the surface of aluminum bulks, inhibits the oxidation process. Thus, Cr-Al-N is expected to improve the oxidation characteristics of CrN.

Although Cr-Al-N thin films have been prepared by PVD [4-7], a comprehensive understanding of hardness and oxidation behavior is still poor. In addition, it is difficult to avoid completely the dissolution of oxygen into nitride as it has been reported that nitride thin films prepared by PVD methods contain the oxygen of

CP650, *Beams 2002: 14th International Conference on High-Power Particle Beams*, edited by T. A. Mehlhorn and M. A. Sweeney

approximately 10 at. %. [8,9]. It is therefore necessary to characterize the thin films in the chromium aluminum oxynitride (Cr-Al-N-O) system. The author has already prepared Cr-Al-N thin films containing oxygen by pulsed laser deposition (PLD) [10-12]. The aim of this investigation is to reveal hardness and oxidation behavior of the Cr-Al-N-O thin film.

EXPERIMENTAL SETUP

Thin films were deposited by the irradiation of a Nd: yttrium aluminum garnet (λ = 355 nm) laser beam onto a target. The laser was electrooptically Q switched by a Pockels cell to produce intense pulses of short duration (7 ns) by a repetition rate of 10 Hz. The laser beam was concentrated by a lens, and the fluence (F) was adjusted to 1 and 3 J/cm^2 on the target. Plates of chromium nitride (Cr_2N) and aluminum nitride (AlN) were used as the target. Cation composition of the thin films was controlled by changing the irradiated area of Cr_2N (S_{Cr_2N}) and AlN (S_{AlN}) plates. The surface area ratio of the targets (S_R) is expressed by the following equation: $S_R = S_{AlN} / (S_{Cr_2N} + S_{AlN})$. The chamber was evacuated to 5×10^{-6} Torr using rotary and turbomolecular pumps. Experiments were carried out under the pressure of 1×10^{-5} Torr. A single-crystal (111)-oriented silicon (Si) substrate was placed at the distance (d_{TS}) of 10 and 40 mm from the target. The substrate temperature was controlled at 400 °C by using a heater and a thermocouple. A Rutherford backscattering spectroscope (RBS) was applied to measure the composition of the thin films. The crystal structure of the thin films was identified by a X-ray diffractometer (XRD). In grazing angle X-ray diffraction (GXRD) measurements, the indent angle of X-ray was fixed at 5°. Hardness (H) and elastic modulus (E') were measured by using a microhardness indenter equipped by a Vickers diamond indenter. In oxidation tests, the thin films were heat-treated in an electrical furnace. The heat treatment conditions were as follows: temperature (T), RT to 1100 °C; gas, atmospheric air; pressure, 760 Torr; holding time, 1h.

RESULT AND DISCUSSION

Figure 1 shows cation composition of thin films as a function of surface area ratio of targets. The thin films were prepared by changing S_R from 0 to 100 % under the condition of F = 1 J/cm^2 and d_{TS} = 10 mm. In Fig. 1, the results expressed as circles falls on the dotted line, if the aluminum content in cations (x) is the same as S_R. Although the circles at S_R = 50 and 75 % were somewhat away from the line, x was found to be controlled by changing S_R.

Figure 2 shows GXRD patterns of thin films prepared by changing surface area ratio of targets. The diffraction peaks attributed to B1 200, which were observed in x ranging from 0 to 70 at %, were shifted to the higher angles with increasing x. Thus, in the thin films with x = 25 and 70 at. %, the cation sites in CrN structure was found to be partially replaced by aluminum atoms. Since the diffraction peaks attributed to AlN (B4 structure) phase appeared above x = 90 at. %, the solubility limit of AlN for chromium oxynitride (Cr-N-O) was clarified to be in x ranging from 70 to 90 at. %.

Figure 3 shows hardness and elastic modulus as a function of cation composition of thin films. The thin films were prepared by changing from $S_R = 0$ to 100 % under the conditions of $F = 1$ J/cm^2 and $d_{TS} = 40$ mm. The hardness of the thin films increased abruptly up to $x = 35$ at. % above which it decreased gradually. Moreover, the elastic modulus of the thin films showed a similar trend as hardness. Therefore, the hardness of the thin films was found to depend on elastic modulus.

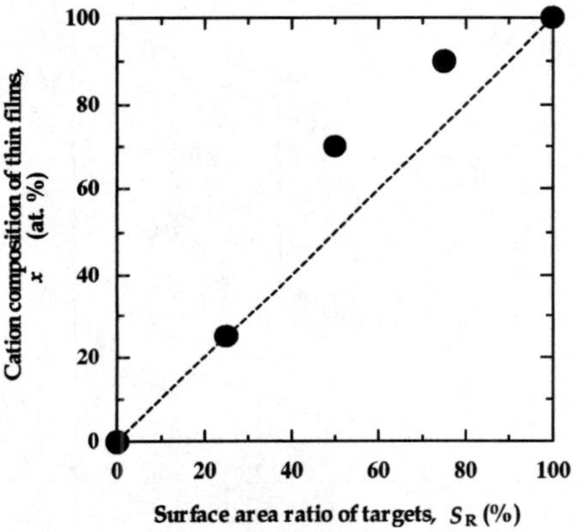

FIGURE 1. Cation composition of thin films as a function of surface area ratio of targets.

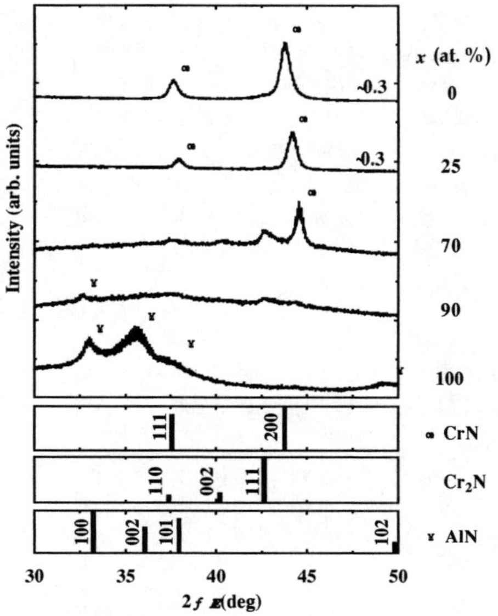

FIGURE 2. GXRD patterns of thin films prepared by changing surface area ratio of targets.

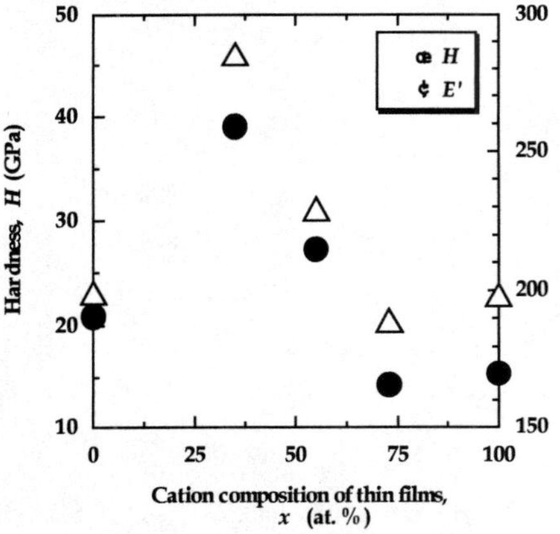

FIGURE 3. Hardness and elastic modulus as a function of cation composition of thin films.

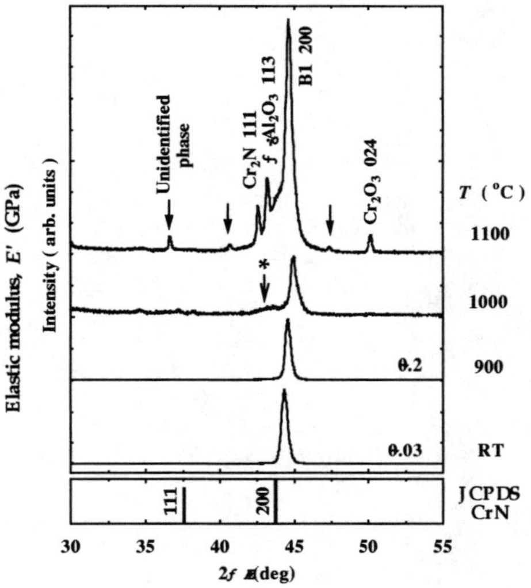

FIGURE 4. XRD patterns of Cr-Al-N-O thin films with $x = 55$ at. % after oxidation tests.

Figure 4 shows XRD patterns of Cr-Al-N-O thin films with x = 55 at. % after oxidation tests. The thin film was prepared at F = 3 J/cm^2, S_R = 25 % and d_{TS} = 10 mm. Only the diffraction peak of B1 200 was observed at T up to 900 °C, whereas the other broad peak (* in Fig. 4) was slightly detected at T = 1000 °C. The broad peak was conjectured to be due to the microcrystallinity of Cr_2N and/or α-Al_2O_3, since it changed to sharp peaks attributed to Cr_2N 111 and α-Al_2O_3 113 with increasing T from 1000 to 1100 °C. In addition, the thin film heat-treated at T = 1100 °C contained B1, chromium oxide (Cr_2O_3) and unidentified phases other than Cr_2N and α-Al_2O_3. Ti-Al-N thin films, which were known to exhibit a high oxidation resistance, have been reported to oxidize completely at 1000 °C [1]. In contrast, the Cr-Al-N-O thin film was revealed to consist mainly of B1 structure at T = 1100 °C.

CONCLUSIONS

From these experimental results, the author has obtained the following conclusions:

1) Cr-Al-N-O thin films have been successfully prepared from Cr_2N and AlN targets by the PLD method.
2) The hardness of the Cr-Al-N-O thin film has approximately 40 GPa when the aluminum content in cations (x) is 35 at. %.
3) Cr-Al-N-O thin film heat-treated at 1100 °C consists mainly of B1 structure.

ACKNOWLEDGMENTS

The authors appreciate M. Obata, Y. Sekimoto and K. Katagiri of Nagaoka University of Technology for the RBS measurements. Thanks are due to S. Ohshio and H. Saitoh of Nagaoka University of Technology for the microhardness measurements.

REFERENCES

1. Ikeda, T., and Satoh, H., J. Japan Inst. Metals **57**, 919-925 (1993).
2. Makino, Y., ISIJ International **38**, 925-934 (1998).
3. Okabe, H., Bull. of Ceramic Soc. Jpn **21**, 529-534 (1986).
4. Makino, Y., and Nogi, K., Surf. & Coatings Technol. **98**, 1008-1012 (1998).
5. Vetter, J., Lugscheider, E., and Guerreiro, S. S., Surf. & Coatings Technol. **98**, 1233-1239 (1998).
6. Ide, Y., Kishitake, K., and Nakamura, T., J. Japan Inst. Metals, **63**, 1576-1583 (1999).
7. Ide, Y., Inada, K., Nakamura, T., and Kishitake, K., Materia Japan **40**, 815-816 (2001).
8. Kools, J. C. S., Nillsen, C. J. C. M., Brongersma, S. H., Riet, E. van de, and Dieleman, J., J. Vac. Sci. & Technol. **A10**, 1809-1814 (1992).
9. Kumar, S., and Tansley, T. L., Jpn. J. Appl. Phys. **34**, 4154-4158 (1995).
10. Hirai, M., Ueno, Y., Suzuki, T., Jiang, W., Grigoriu, C., and Yatsui, K., Jpn. J. Appl. Phys. **40**, 1056-1060 (2001).
11. Hirai, M., Saito, H., Suzuki, T., Suematsu, H., Jiang, W., and Yatsui, K., Thin Solid Films **407**, 122-125 (2002).
12. Hirai, M., Saito, H., Suzuki, T., Suematsu, H., Jiang, W., and Yatsui, K., Mat. Res. Soc. Symp. Proc. **697**, P7.7.1-6 (2002).

Ablation Plasma Generated by Pulsed Ion Beam Evaporation

Weihua Jiang, Shinji Kitayama, Tsuneo Suzuki, Hisayuki Suematsu, and Kiyoshi Yatsui

Extreme Energy-Density Research Institute, Nagaoka University of Technology 1603-1 Kamitomioka, Nagaoka, Niigata 940-2188, Japan

Abstract. The ablation plasma generated by pulsed ion beam evaporation has been studied. The objective of this study is to understand the property and the behavior of the ablation plasma which is being used for thin film deposition. In the experiments reported in this paper, we have concentrated on the processes of target material ablation and ablation plasma expansion. The diagnostic results are used to analyze the process of thin film deposition.

INTRODUCTION

Pulsed ion beams have been applied to various material processes, such as ion implantation and surface treatment. In these applications, the ion-beam energy density is controlled so that the target surface is not ablated. On the other hand, when the ion-beam energy is high enough, the target material near the surface can be evaporated, producing an ablation plasma.

The ion beam generated ablation plasma is of considerable interest due to its nature of high density and high temperature. These properties have made the ion-beam generated ablation plasma a unique tool for material development, such as thin film deposition [1-4].

When an ion beam is irradiated on the surface of a solid target, it deposits its kinetic energy into the target material within a very short range. For example, this range is about 14 μm for 1-MeV proton in aluminum target. Therefore, for ion-beam current density of 1 kA/cm^2 on the target surface, an energy density of 10^5 J/cm^3 in the target material can be obtained by using pulsed ion beam of 100 ns. The thermal conduction has no time to take effect during the ion beam pulse.

The high-density, high-temperature target plasma produced by pulsed ion beam naturally expands into the vacuum area adjacent to the target. Since the ion-beam spot size is usually much larger than the depth of energy deposition, the plasma expansion is nearly one-dimensional in the direction perpendicular to the target surface. During this expansion, when the plasma touches a solid surface, it cools and solidifies itself on the surface. As a result, a thin film is obtained.

We have used this method to deposit various kinds of thin films on different substrates [1-4]. As illustrated in Fig. 1, the ablation plasma is generated by ion beam

CP650, *Beams 2002: 14th International Conference on High-Power Particle Beams,* edited by T. A. Mehlhorn and M. A. Sweeney

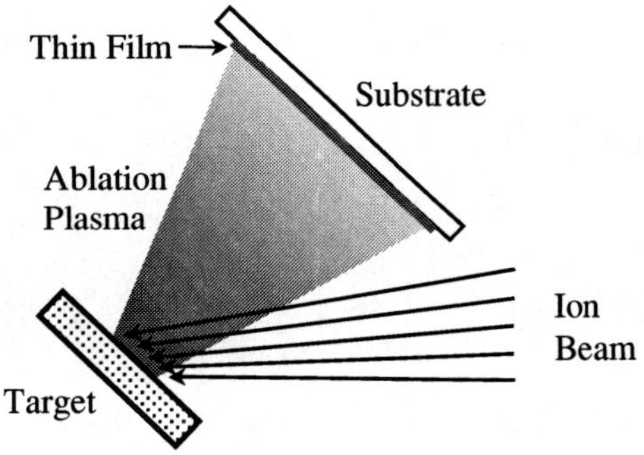

FIGURE 1. Illustrated thin-film deposition process by using ion-beam generated ablation plasma.

irradiation on the target surface and the thin film is obtained on the surface of the substrate. In this paper, we have measured the target mass that is removed by the ion beam ablation and we have studied the ablation plasma behavior by using high-speed camera.

TARGET ABLATION

The amount of the target material that is ablated by the ion beam is measured by comparing the target surface before and after the ion-beam shot. A 10-cm diameter copper mask is used to ensure the ion-beam uniformity in the ablated area. The target surface profiles obtained before and after the ion-beam irradiation are shown in Fig. 2.

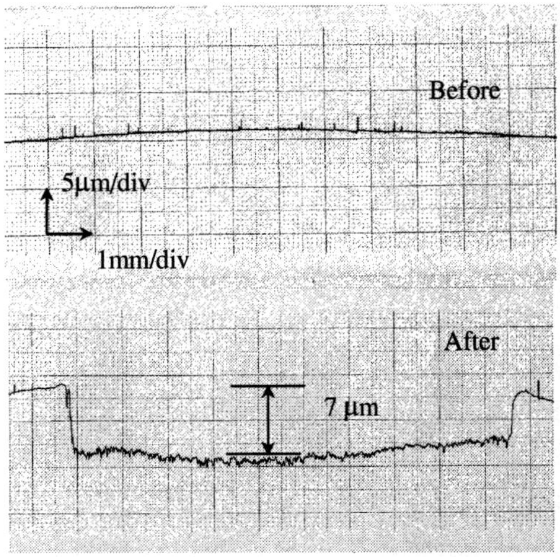

FIGURE 2. Target surface before and after ion-beam irradiation.

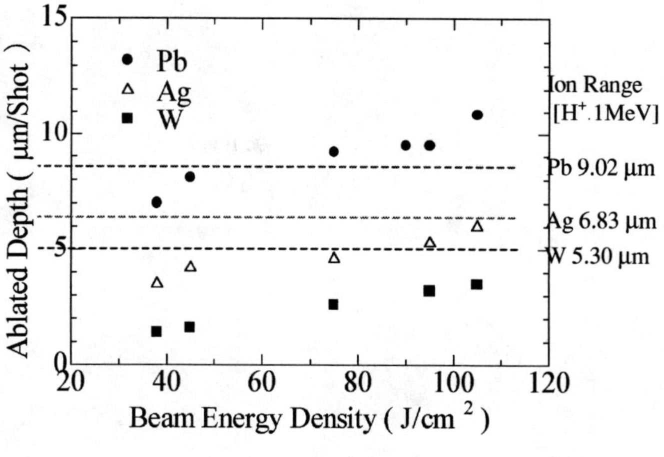

FIGURE 3. Ablated target mass versus ion-beam energy density.

For example, from the profiles shown in Fig. 2, we have obtained the average ablation depth of 7 μm. Figure 3 shows the ablation depths obtained with different ion beam energy density for different target materials. The data shown in Fig. 3 were obtained with constant ion-beam pulse width of ~ 50 ns and acceleration voltage of ~ 1 MeV. The ion-beam energy density was varied by varying the ion current density on the target surface. For comparison, the 1-MeV proton ranges in these materials are also shown in Fig. 3 by the dashed lines. It is seen from Fig. 3 that the ablation depths increase with the ion-beam energy density and, for silver and tungsten targets, they are significantly smaller than the 1-MeV proton ranges. Since most of the ions are protons in this experiment, it is concluded that the ion range is not the only factor that determines the ablation depth.

ABLATION PLASMA BEHAVIOR

The behavior of the ablation plasma has been studied by high-speed photography [5,6]. Figure 4 shows the photographs obtained for tungsten target. It is seen that, after the pulsed ion-beam irradiation (t = 0), an ablation plasma is generated on the target surface. The plasma expands very fast toward the substrate located 5 cm away from the target surface. When the plasma reaches the substrate, it stagnates and expands along the substrate surface. The plasma residence time on the substrate surface is about 20 μs, which is considered to be very important to thin-film deposition. The ablation plasma near the target lasts a longer time. However, from the intensity of its light emission, it is believed that the plasma density after t = 20 μs has significantly decreased. Therefore, from the high-speed photographs, it has been concluded that the plasma lifetime around the substrate is on the order of 20 μs.

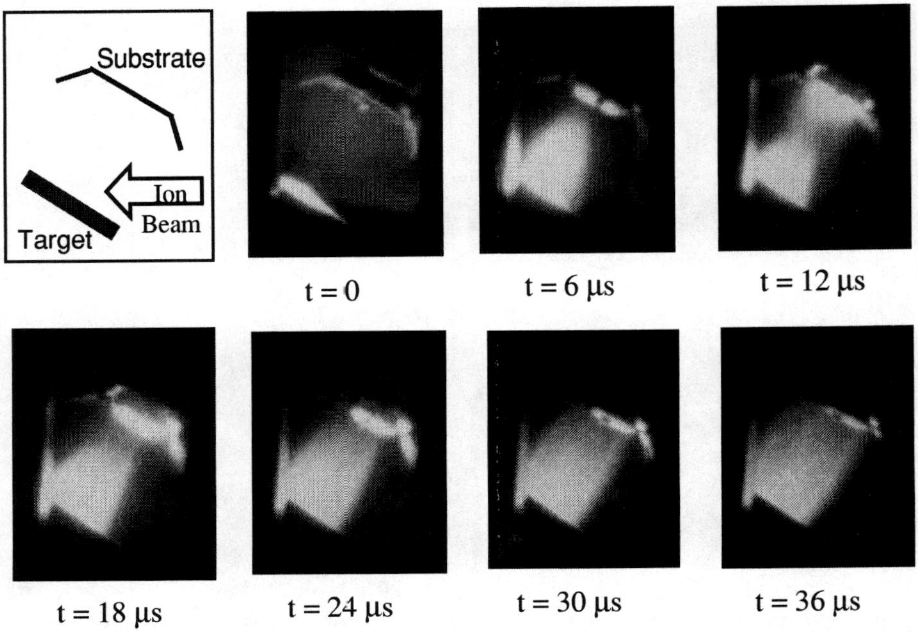

FIGURE 4. High-speed photographs of ablation plasma obtained for tungsten target.

The plasma lifetime on the substrate is of importance in two respects. First, it is believed that the thin-film deposition on the substrate happens mostly in this time. Therefore, for example, if we get the film thickness of 100 nm/shot, the deposition rate is obtained to be 5 nm/μs. Second, we have measured the thermal energy deposited by the ablation plasma onto the substrate surface. With the energy deposition of 5 J/shot, we have obtained the instantaneous substrate-surface temperature of 900K [7].

REFERENCES

1. Y. Shimotori, M. Yokoyama, H. Isobe, S. Harada, K. Masugata and K. Yatsui, *J. Appl. Phys.* **63**, 968-970 (1988).
2. K. Yatsui, *Laser and Particle Beams,* **7**, 733-741 (1989).
3. K. Yatsui, X. D. Kang, T. Sonegawa, T. Matsuoka, K. Masugata, Y. Shimotori, T. Satoh, S. Furuuchi, Y. Ohuchi, T. Takeshita, and H. Yamamoto, *Phys. Plasmas* **1**, 1730-1737 (1994).
4. K. Yatsui, T. Sonegawa, K. Ohtomo, and W. Jiang, *Mater. Chem. & Phys.* **54**, 219-223 (1998).
5. W. Jiang, N. Hashimoto, H. Shinkai, K. Ohtomo, and K. Yatsui, *Nucl. Instr. & Methods*, **A415**, 533-538 (1998).
6. W. Jiang, K. Ide, S. Kitayama, T. Suzuki, and K. Yatsui, *Jpn. J. Appl. Phys.* **40**, 1026-1029 (2001).
7. W. Jiang. S. Kitayama, T. Suzuki, and K. Yatsui, *Prof. Int'l Conf. Pulsed Power Plasma Science 2001*, Ed. R. Reinovsky and M. Newton, **I**, 376-379 (2001).

THEORY AND MODELING

Species Separation and Field-Penetration in a Multi-Component Plasma*

Stephen B. Swanekamp,[a,c] Robert J. Commisso,[a] Paul F. Ottinger,[a] Joseph W. Schumer,[a] Sean D. Strasburg,[a,d] Bruce V. Weber,[a] Yitzhak Maron,[b] and Ron Arad[b]

[a]Naval Research Laboratory, Plasma Physics Division, Washington, DC USA, [b]Weizmann Institute Of Science, Rehovot, Israel, [c]Titan/Jaycor, Mclean, VA, [d]National Research Council Associate at NRL

Abstract. Particle-in-cell (PIC) simulations confirm recent multi-component plasma-opening switch measurements in which magnetic field simultaneously penetrates the heaviest plasma components while pushing the lighter species. In these simulations a positive potential hill associated with the J×B force develops that, in the frame of the moving hill, reflects the lighter ions but is not large enough to reflect heavier ions. The simulations show that species separation can occur when the Hall speed is large compared to the characteristic speed for the heavier species but small compared to the characteristic speed of the lighter species.

INTRODUCTION

In recent plasma-opening switch (POS) experiments it was discovered that lighter ion species are pushed ahead of heavier plasma components.[1] It was also shown that the magnetic field convects with the lighter ion species leaving the more massive species behind with very little current flowing through it. As a consequence of this field penetration, there is very little pushing on the plasma remaining behind the current front leaving the center of mass of the heavier plasma components close to their original positions. In other POS experiments a slowly propagating current channel has been observed that transfers current between the POS and the load as the switch opens.[2] In these experiments most of the POS mass is left in the original location of the POS when the switch opens and it was speculated that current propagation occurred as a low density (presumably the lighter species) plasma that moved between the POS and load at 1-2 cm/μs.

The pushing of a single species plasma by magnetic field can be described by the Hall/MHD equations, which are given by

$$\begin{aligned}
&\frac{\partial n_i}{\partial t} + \nabla \cdot n_i V_i = 0 \\
&n_i M_i \frac{DV_i}{Dt} = -\nabla\left(\frac{B^2}{8\pi} + p\right) \\
&\frac{\partial B}{\partial t} = \nabla \times (V_i \times B) - \nabla \times \left(\frac{J \times B}{e n_e}\right)
\end{aligned} \tag{1}$$

where n_i and n_e are the ion and electron densities, V_i is the ion fluid velocity, B is the magnetic field strength, and J is the current density. These equations permit two types

CP650, *Beams 2002: 14th International Conference on High-Power Particle Beams,*
edited by T. A. Mehlhorn and M. A. Sweeney

of solutions depending on the electron density-gradient scale length, L_n.[3] When $L_n >> c/\omega_{pi}$ the Hall/MHD equations predict pushing of the plasma at the Alfven speed, $V_A = B/(4\pi n_i M)^{1/2}$. When $L_n >> c/\omega_{pi}$ the Hall/MHD equations predict magnetic field penetration into the plasma at the Hall speed, $V_{Hall} = cB/(4\pi e n_e L_n)$. In this case the magnetic shock moves more rapidly than the ions can respond and there is very little ion motion. Magnetic field penetration occurs in the $B \times \nabla n_e$ direction so that the penetration depends on the polarity of the magnetic field.

Analyzing magnetic field penetration into a multi-component plasma is more complicated. Assuming that electrons $E \times B$ drift in the current channel, an electrostatic space-charge separation exists of the order $\Phi = B^2/8\pi e n_e$. If the field penetrates into the plasma at a speed given by V_B, then the species can separate when the kinetic energy of the heavier species in the frame moving with the field is larger than the electrostatic potential and the kinetic energy of the lighter species is smaller than the electrostatic potential. The conditions for species separation can be summarized by $V_{AH} < V_B$ and $V_{AL} > V_B$ where $V_{AH,L} = B/(4\pi n_e M_{H,L})^{1/2}$ are the characteristic speed for the heavy and light plasma components. The characteristic speeds for the heavy and light species closely resemble the Alfven speed except that the local electron density appears in the denominator. The Alfven speed for the multi-component plasma is defined by $V_A = B/[4\pi(n_L M_L + n_H M_H)]^{1/2}$.

SIMULATION RESULTS

The simulations were performed in parallel-plate Cartesian geometry using the LSP[4] PIC code. The geometry of the simulations is shown in FIG. 1a. Plasma is located between X = 0 and X = 3 cm and between electrodes located at Y = 0 and Y = 5 cm from which space-charge-limited electron emission is allowed. The magnetic field was ramped up to 20 kG in 3 ns and then held constant for the duration of the simulations. The initial electron density profile is shown in FIG. 1b and was chosen so that the Hall speed was uniform between 1 < Y < 5 cm and equal to 0.24 cm/ns. The density was allowed to rise between 0 < Y < 1 cm to minimize electrode effects. The

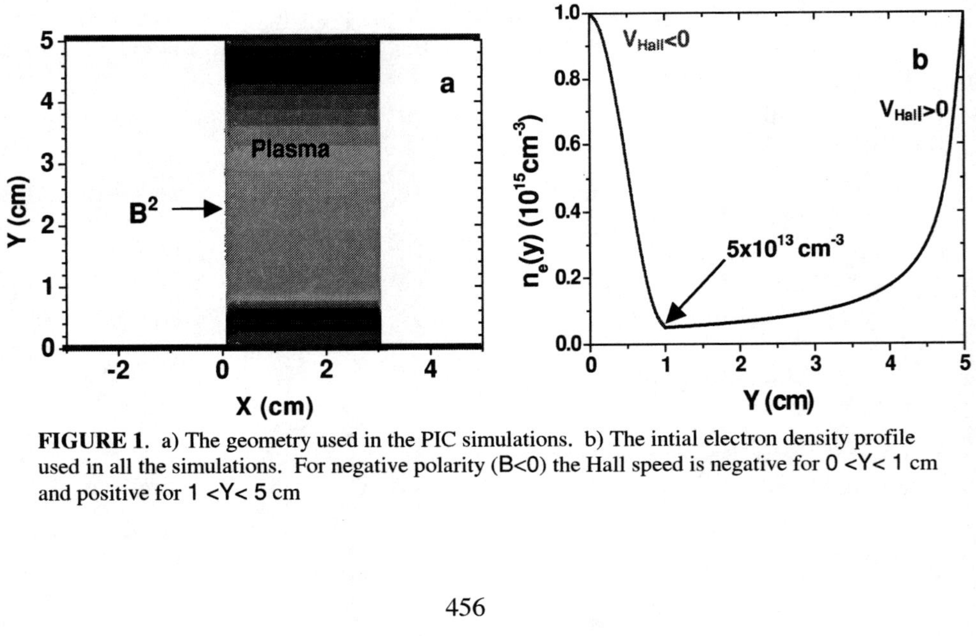

FIGURE 1. a) The geometry used in the PIC simulations. b) The intial electron density profile used in all the simulations. For negative polarity (B<0) the Hall speed is negative for 0 <Y< 1 cm and positive for 1 <Y< 5 cm

same initial electron density profile was used for all of the simulation results. The light plasma species was taken to be protons and the heavy ion species was taken either to be infinitely massive or singly charged Teflon with M_H=17 amu

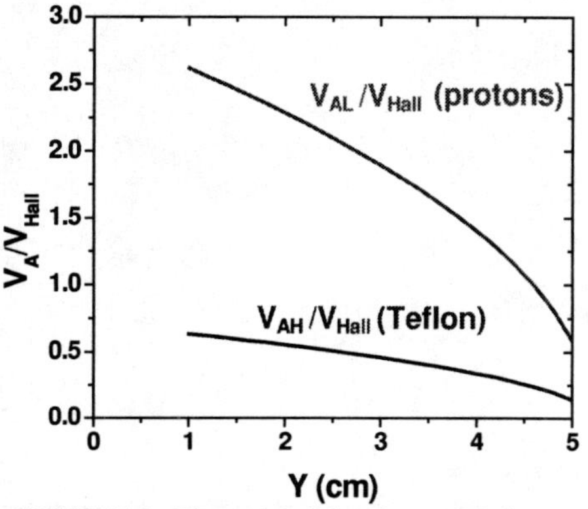

FIGURE 2. The ratio of the characteristic speeds for protons and Teflon to the Hall speed for the density profile given in FIG. 1b.

A plot of the ratio of the characteristic heavy and light ion speeds to the Hall speed for the chosen electron density profile is shown in FIG. 2. From this figure we see that V_{AH} is smaller than V_{Hall} over the entire simulation while V_{AL} is larger than V_{Hall} for Y values less than 4.5 cm. In negative polarity we expect simultaneous penetration of the heavy ions and pushing of the light ions for $1 < Y < 4.5$ cm. For $Y > 4.5$ cm we expect magnetic field penetration into both the heavy and light ions. For $Y < 1$ cm the density gradient is in the direction to exclude magnetic field from the plasma and we expect no penetration in negative polarity. In positive polarity the density gradient is in the direction to exclude magnetic field for $1 < Y < 5$ cm and we expect penetration for $0 < Y < 1$ cm.

The simulation results at t=12 ns are shown in FIG. 3. In all of the simulations shown in FIG. 3 the protons comprised 10% of the charge but nearly all of the mass was in the heavier ion species. The magnetic field and heavy and light ion density contours for an infinitely massive ion species in negative polarity are given in FIGS. 3(a), 3(d), and 3(g). These contours show that the magnetic field simultaneously penetrates the heavy ion species while pushing the protons. The penetration occurs primarily as a plane wave as predicted by the Hall/MHD equations with the exception of the appearance of paramagnetic electron vortices near Y=2 cm. These vortices have been describe elsewhere.[5] The penetration speed for this case is very close to the Hall speed. The magnetic field and heavy and light ion density contours for the case where the heavy ion species is Teflon (M_H=17 amu) in negative polarity are given in FIGS. 3(b), 3(e), and 3(h). The simultaneous penetration of the heavy ions and pushing of the protons is also seen in this case. However, the paramagnetic vortices are gone because the motion of the heavy ions reduces the space-charge separation field required to sustain the vortices. This results in a reduction of the heavy ion density. The magnetic field and heavy and light ion density contours for the case where the heavy ion species is Teflon (M_H=17 amu) in positive polarity are given in FIGS. 3(c), 3(f), and 3(i). In this case we do not see rapid magnetic field penetration into the body of the plasma. Instead we see magnetic pushing of the plasma at a speed much slower than the Hall speed. It can also be seen that the protons are pushed ahead of the Teflon ions in the body of the plasma. However, the mechanism for this separation is independent of the Hall effect. We also see the appearance of a gap at Y = 1 cm in both the heavy and light ion densities that is caused by rapid axial Hall penetration of magnetic field that occurs near Y = 1 cm and subsequent radial J×B forces.

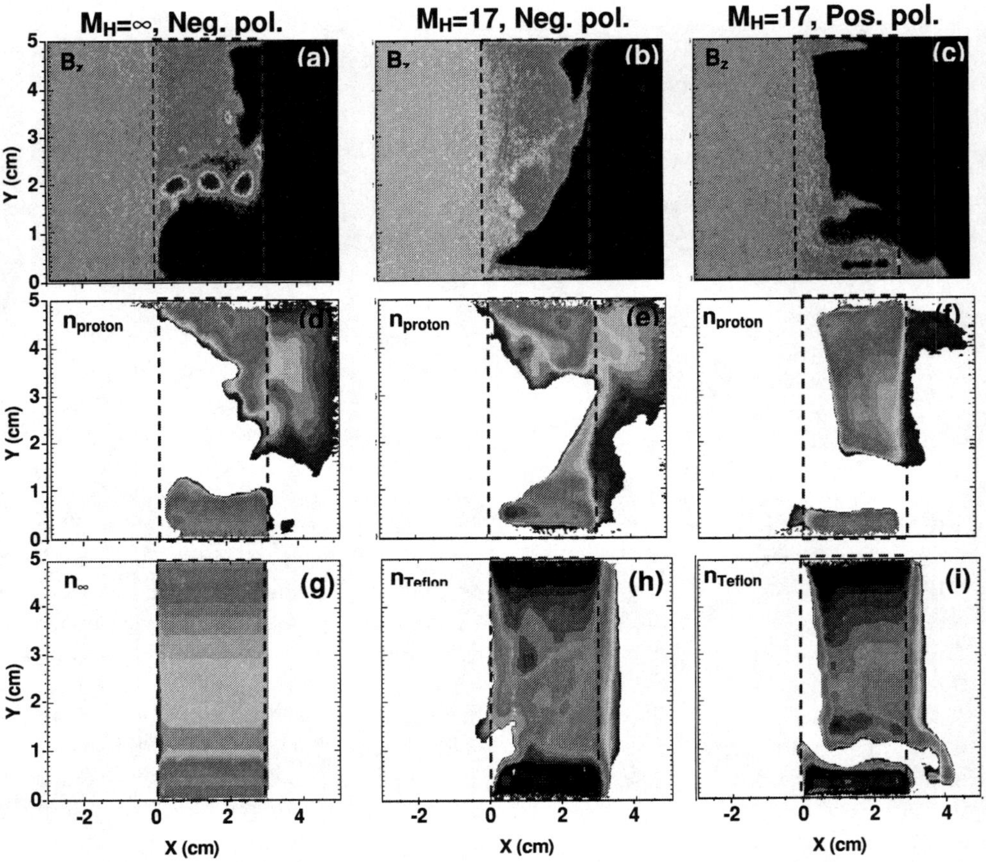

FIGURE 3. Contours of magnetic field (B_z), light ion density (n_{proton}), and heavy ion density (n_∞ or n_{Teflon}) for $M_H=\infty$, negative polarity [(a), (d), and (g)], $M_H=17$, negative polarity [(b), (e), and (h)], and $M_H=17$, negative polarity [(c), (f), and (i)] at t=12 ns. The dashed lines show the original plasma location.

1. A. Weingarten, R. Arad, Y. Maron, and A. Fruchtman, Phys. Rev. Lett. **87**, 115004 (2001).
2. B.V. Weber, P.F. Ottinger, R.J. Commisso, J.R. Goyer, D. Kortbawi, J. Thompson, J.E. Rowley, P. Filios, and M. Babineau, Proceedings of the 11th Conference on High-Power Particle Beams, K. Jungwirth and J. Ullschmied eds, p. 121, Prague (1996).
3. A.S. Kingsep, Y.V. Mohkov, and K.V. Chukbar, Sov. J. Plasma Phys. **10**, 495 (1984).
4. D. R. Welch, D. V. Rose, B. V. Oliver, and R. E. Clark, Nucl. Instrum. Methods Phys. Res. A **464**, 134 (2001).
5. S.B. Swanekamp, J.M. Grossmann, A. Fruchtman, B.V. Oliver, and P.F. Ottinger, Phys. Plasmas **3**, 3556 (1996).

*Work supported by Defense Threat Reduction Agency.

Particle-in-Cell Simulations of the Electra KrF Electron-Beam Diode

D. V. Rose*, D. R. Welch*, S. B. Swanekamp†, M. Friedman**, M. C. Myers**, J. D. Sethian** and F. Hegeler‡

**Mission Research Corporation, Albuquerque, NM 87110 USA*
†TITAN/JAYCOR, McLean, VA, 22102 USA
***Plasma Physics Division, Naval Research Laboratory, Washington, DC 20375 USA*
‡Commonwealth Technology Incorporated, Alexandria, VA 22315 USA

Abstract. Semi-analytic methods and particle-in-cell simulations are used to model large-area electron beam diodes that are used to pump KrF laser cells. The simulations include models for following the energy loss and scattering of beam particles in foils, gas and support structures. Estimates of energy deposition to the various components of the diode system are obtained allowing optimization studies to be carried out. Losses to the pressure foil separating the diode vacuum region from the laser cell are found to be of order 10% due to back-scattering from the foil and gas for 1-mil Ti foils and 500 keV operation. An upper limit on the energy deposition fraction into the gas of nearly 80% is found for the design considered here.

INTRODUCTION

The Electra facility [1] at the Naval Research Laboratory uses repetitive pulsed power systems driving large area electron-beam diodes to pump a high-power krypton-fluoride (KrF) laser. The electron beams enter the gas cell though one or more foils supported by a ribbed structure referred to as the hibachi. Maximizing the electron-beam energy deposited in the KrF gas (to maximize the pump efficiency) as well as minimizing the electron-beam energy deposited in foils and hibachi (to ensure the survivability of the foils) is a critical design element of a repetitively pulsed KrF facility like Electra. Three-dimensional particle-in-cell (PIC) simulations using LSP [2] as well as semi-analytic methods are carried out to aid in the design of the diode. These designs include a slotted cathode that provides high transmission efficiency and suppresses growth of a transit-time instability [3] that can modulate the electron-beam energy distribution. The slots are rotated with respect to the hibachi ribs so that the beam enters the hibachi parallel as it rotates in the applied and self magnetic fields.

LSP is a general-use PIC code with explicit and implicit electromagnetic field solvers. The Integrated Tiger Series (ITS) of Monte Carlo electron transport algorithms [4] have been incorporated for tracking electron energy loss and scattering in materials. The simulations presented in this paper use the ITS algorithms to model beam transport in the Kr gas where the beam charge and current are assumed to be fully neutralized in this region, and electron trajectories respond only to the applied axial magnetic field, B_o.

CP650, *Beams 2002: 14th International Conference on High-Power Particle Beams,*
edited by T. A. Mehlhorn and M. A. Sweeney

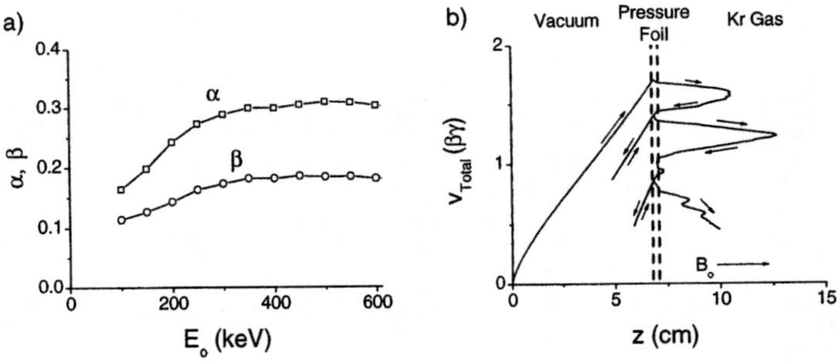

FIGURE 1. a) Back-scatter coefficients as a function of injected beam energy E_o for a 1-mil Ti foil and 30-cm of 1.3 atm Kr gas; α is the ratio of the scattered to the injected current density and β is the ratio of the scattered to the incident electron energy. b) Sample electron trajectory from a 500 keV 1-D LSP simulation demonstrating multiple passes through the 1-mil Ti pressure foil.

1-D PLANAR DIODE WITH BACK-SCATTERING

Scattering of the electron beam from the pressure foil and gas is an energy loss mechanism and must be accounted for in the diode design. Accurate determination (and minimization) of the energy deposited in the pressure foil and ribs is an important consideration in the design of these rep-rated diode systems. Previous assessments of electron back-scatter in high-power planar diodes [5, 6] have indicated reductions in the diode current density of order 10%.

In the absence of ions, the 1-D Poisson equation including electron back-scattering [6] was numerically integrated using back-scattering coefficients obtained from one-dimensional ITS calculations [see Fig. 1a)]. A reduction in the average current density below the (non-relativistic) Child-Langmiur current, J_{CL}, of $0.92J_{CL}$ was found for 500 keV operation, $B_o = 1.4$ kG, with a 1-mil Ti foil and 30-cm of 1.3 atm Kr. 1-D LSP simulations using the same diode parameters resulted in a current density of $0.89J_{CL}$. The difference in these two results is due primarily to the fact that the energy loss in repeated back-scattering events was not taken into account in the Poisson model solution. Multiple back-scattering of a single beam electron from the 1-D LSP simulation is illustrated in Fig. 1b). The electron is emitted from the cathode at $z = 0$ and accelerated to 500 keV at the pressure foil ($z = 0.68$ cm). The electron loses some energy as it passes through the foil and enters the gas, where it is scattered back into the foil and into the diode vacuum gap. The electron is reflected in the diode electric field and re-injected into the foil and gas. A second pass into the diode vacuum region occurs before the electron energy falls below the 50 keV transport threshold. Along the trajectory in the foil and gas, the lost electron energy is accumulated as a computational cell quantity.

1-D simulations at 500 keV with a solid absorber backing the 1-mil Ti foil show only 2.7% of the beam energy being deposited in the foil. With 30-cm of Kr gas at 1.3 atm backing the 1-mil Ti foil, the beam energy fraction lost to the foil is approximately 10% at 500 keV due to enhanced multiple back-scattering as discussed above.

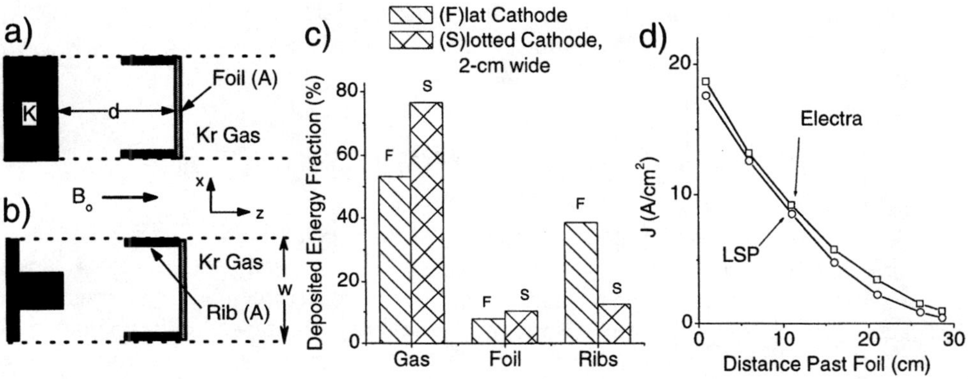

FIGURE 2. Schematics of the periodic, 3-D LSP simulation geometries for a) flat-cathode and b) slotted-cathode cases. The dashed lines represent the periodic boundaries. c) Energy deposition fractions from the periodic 3-D LSP simulations, with 1.3 atm Kr, 1-mil Ti foil at 500 keV. d) Current density J as a function of depth in 1.2 atm Kr for a flat cathode shot with $d = 7$ cm and 2-mil Ti pressure foil.

3-D PERIODIC DIODE WITH BACK-SCATTERING

In the previous section, a lower limit on foil losses was obtained, assuming 500 keV operation and a 1-mil thick Ti foil backed by 30 cm of 1.3 atm Kr. The same parameters used in the 1-D simulations are now applied to 3-D periodic simulations which include the presence of the hibachi ribs. Once again, these simulations do not include the effects of self-magnetic fields on the beam (pinching and shear), but they do allow an examination of the effects of electron back-scattering on diode operation. The hibachi ribs are 0.5 cm thick and 2.8 cm deep, with center-to-center spacing (w) of 4-cm [see Figs. 2a) and b)]. The cathode emission region extends over 20 cm in the direction perpendicular to the periodic boundaries (y). The simulations use a 6.8 cm spacing (d) between the cathode to the pressure foil.

Along with suppression of the transit-time instability [3], periodic arrays of cathode emitters, Fig. 2b), can improve the fraction of beam energy that is deposited in the gas. Figure 2c) shows the fractional energy deposited in the gas, foil, and ribs for flat and slotted cathodes. In addition, the energy deposited in a 1-mil Ti foil located at the opposite end of the gas cell is also obtained; energy deposited in this second foil is less than 1.0%. Because these simulations do not include self-magnetic field bending and shearing of the beam, the numbers represent an upper limit on gas deposition fraction for these parameters. Again, the role of back-scattered electrons is an important limiting factor in the design efficiency.

The current density as a function distance past the pressure foil, shown in Fig. 2d), agrees well with flat cathode Electra data for 1.2 atm of Kr, $B_o = 1.4$ kG, and a 2-mil Ti pressure foil at 500 keV. The simulations include a detailed model of the Faraday cups used in the experimental measurement, including filtering by a 1-mil Ti foil covering the opening of the cup. These measurements are made in the line-of-sight of one of the hibachi ribs and this was accounted for by expanding the periodic simulations in the

x-direction by one additional rib space width. The simulation results are slightly lower than the measured values at all positions but within the uncertainty due to simulation statistics.

FULL 3-D DIODE SIMULATIONS

Detailed 3-D, fully electromagnetic simulations of the Electra diode geometry [7] have also been carried out with flat and slotted cathodes. The complete details of these simulations will be discussed in a future paper. However, we note that good agreement has been found between the simulations and the total energy deposited in the gas. For a flat cathode operated at 500 keV, 1.2 atm of Kr and a 2-mil pressure foil, the fraction of beam energy deposited in the gas on Electra is 47%, compared with 44% from the simulations for a 40 ns rise, 100 ns flat top and 60 ns fall power pulse. Note that the energy deposited into the gas in slotted cathode configurations is typically much higher and detailed comparisons with 3-D simulations are underway. However, the agreement already obtained in the flat cathode comparisons demonstrates the capability for optimization and testing of new cathode/hibachi designs for the KrF program.

ACKNOWLEDGMENTS

We thank R. E. Clark at MRC for technical assistance with the simulations presented here. This work is sponsored by the US Department of Energy through the Plasma Physics Division, Naval Research Laboratory, Washington, DC.

REFERENCES

1. Sethian, J. D., Myers, M., Smith, I. D., Carboni, V., Kishi, J., Morton, D., Pearce, J., Bowen, B., Schlitt, L., Barr, O., and Webster, W., *IEEE Trans. Plasma Sci.* **28**, 1333 (2000).
2. Welch, D. R., Rose, D. V., Oliver, B.V., and Clark, R. E., *Nucl. Instum. Meth. Phys. Res. A* **464**, 134 (2001).
3. Friedman, M., Swanekamp, S., Obenschain, S., Chan, Y., Ludeking, L., and Smithe, D., *Appl. Phys. Lett.* **77**, 1053 (2000).
4. Halbleib, J. A., Kensek, R. P., Valdez, G. D., Seltzer, S. M., and Berger, M. J., *IEEE Trans. Nucl. Sci.*, **NS-39**, 1025 (1992).
5. Pereira, N. R., *J. Appl. Phys.* **54**, 6307 (1983).
6. Mosher, D., Cooperstein, G., Rose, D. V., Swanekamp, S. B., in the Proceedings of the 11th International Conference on High Power Particle Beams, Prague, Czech Republic, June 10-14, 1996, p. 1147.
7. Hegeler, F., *et al.*, these proceedings.

Improved Envelope and Centroid Equations for High Current Beams

Thomas C. Genoni, Thomas P. Hughes, and Carsten H. Thoma

Mission Research Corporation, 5001 Indian School Road N.E., Albuquerque, NM 87110, USA

Abstract.
The standard envelope equation for charged particle beams (e.g., Lee-Cooper) neglects self-field contributions from the beam rotation and the slope of the beam envelope. We have carried out an expansion that includes these effects to first order, resulting in a new equation for the edge radius. The change in beam kinetic energy due to space-charge depression as the beam radius varies is also included. For the centroid equation, we have included the "self-steering" effect due to the curvature of the beam orbit. To leading order, there is a cancellation between the self-steering effect and the space-charge depression of the beam energy, so that a more accurate centroid equation is obtained by using the undepressed value of the energy (i.e., the total beam energy) to calculate the orbit. We have implemented the envelope and centroid equations in the LAMDA code [1]. The effect of the new terms will be illustrated with calculations for the DARHT accelerators at the Los Alamos National Laboratory [2].

ENVELOPE EQUATION

We consider a cold, circular beam with zero canonical angular momentum in a cylindrical drift tube with longitudinal guide field B_0. The standard paraxial envelope equation (see, e.g., Ref. [3]) balances radial forces at the beam edge due to the applied megnetic field and the self magnetic (azimuthal) and electric fields of the cylindrical beam, i.e.,

$$a'' = -\left(\frac{k_c}{2\gamma\beta}\right)^2 a + \frac{2\nu}{\beta^2\gamma^3 a} \tag{1}$$

where a is the beam radius and

$$k_c = \frac{eB_0}{mc}, \quad \nu = \frac{Ie}{mc^3\beta} \tag{2}$$

We include below corrections to Eq. (1) which depend on the quantities ν, k_c^2, a'^2 and a'' where a' and a'' are the slope and curvature at the beam edge respectively. Assuming a rigid rotor and linear radial dependence of the radial velocity (reasonable assumptions for currents which are not too great) one can show that the beam longitudinal velocity is given approximately by

$$\beta_z^2(r) \cong \beta^2 \left[1 - \frac{2\nu}{\beta^2\gamma^3} - \frac{r^2}{a^2}\left(a'^2 + \frac{\nu_e^2}{\beta^2} - \frac{2\nu}{\beta^2\gamma^3}\right)\right] \tag{3}$$

CP650, *Beams 2002: 14th International Conference on High-Power Particle Beams*,
edited by T. A. Mehlhorn and M. A. Sweeney

where β and γ denote values at the beam edge. Again invoking the rigid rotor approximation, magnetic flux conservation, and conservation of canonical angular momentum at the beam edge we obtain the following corrections to the longitudinal magnetic field and azimuthal beam velocity:

$$B \cong B_0 - \frac{\nu v_e a}{r_w^2} \tag{4}$$

$$v_e \cong -\frac{eB_0 a}{2\gamma mc^2}\left[1 - \frac{\nu}{2\gamma}\left(1 - \frac{a^2}{r_w^2}\right)\right] \tag{5}$$

where r_w is the drift tube wall radius. Finally we consider corrections to the radial electric field by obtaining an approximate solution to Poisson's equation

$$\frac{1}{r}\frac{\partial}{\partial r}\left(r\frac{\partial \gamma}{\partial r}\right) + \frac{\partial^2 \gamma}{\partial z^2} = \frac{4I}{I_0 \beta_z(r) a^2(z)} \tag{6}$$

where $I_0 = mc^3/e$ and $a(z)$ accounts for the slope a' and curvature a'' at the beam edge. We find

$$\frac{e}{mc^2}E_r \cong \frac{2\nu}{a}\left(1 + \frac{\nu}{2\beta^2\gamma^3} + \frac{1}{4}a'^2 + \frac{1}{4}\frac{v_e^2}{\beta^2}\right) - \frac{\nu}{2}\left(a'' + \frac{a'^2}{a}\right) \tag{7}$$

Inserting the above expressions into the envelope equation, expanding and retaining terms up to 2nd order in the quantities ν, k_c^2 and a'^2 we obtain

$$\begin{aligned} a'' &= -\left(\frac{k_c}{2\gamma\beta}\right)^2 a\left[1 - \frac{\nu}{2\gamma}\left(4 - \frac{2a^2}{r_w^2} + \frac{6}{\beta^2\gamma^2}\right) + a'^2 + \left(\frac{k_c a}{2\beta\gamma}\right)^2\right] \\ &+ \frac{2\nu}{\beta^2\gamma^3 a}\left[1 + \frac{\gamma^2 a'^2}{4}\left(\frac{6}{\gamma^2} - 2\right)\right] \end{aligned} \tag{8}$$

Fig. 1 depicts a calculation of the pinching of the envelope of a DARHT-injector-like beam passing through a cusped longitudinal magnetic field. Calculation of the envelope evolution using Eq. (8) apparently agrees well with a 2-D fully electromagnetic PIC simulation using the code LSP [4]. The lower dashed curve was obtained by integration of Eq. (1) and illustrates the magnitude of the correction terms calculated here.

CENTROID EQUATIONS

We have been using the total, rather than the depressed, relativistic factor in the centroid equations in the LAMDA code [1] for some time, based on experimental measurements [5] and simple theory [6]. The following describes some more accurate calculations of the "self-steering" force. The equations of motion for the centroid $\Delta x, \Delta y$ are

$$\gamma_c \beta_c^2 \Delta x'' = -\gamma_c' \Delta x' + \frac{e}{mc^2}\left(E_x - \beta_c B_y + \beta_c \Delta y' B_z\right) \tag{9}$$

$$\gamma_c \beta_c^2 \Delta y'' = -\gamma_c' \Delta y' + \frac{e}{mc^2}\left(E_y + \beta_c B_x - \beta_c \Delta x' B_z\right) \tag{10}$$

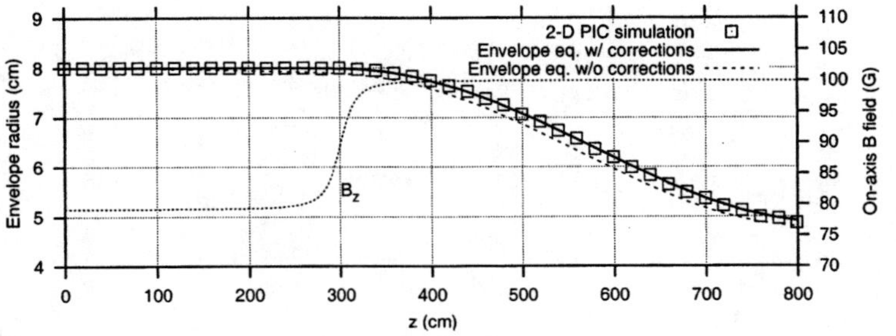

FIGURE 1. Comparison of envelope equation with and without corrections with LSP calculation for a cold, 2 kA, 3.2 MV beam with an initial radius of 8 cm. Equilibrium $B_z = 79.2$ gauss. Wall radius is 17.7 cm.

where γ_c and β_c denote values at the beam center and $E_{x,y}$ and $B_{x,y}$ are transverse wall forces for a straight beam in a smooth, circular pipe:

$$\frac{eE_x}{mc^2} = \frac{2\nu}{r_w^2}\Delta x, \quad B_y = \beta_c E_x, \quad \frac{eE_y}{mc^2} = \frac{2\nu}{r_w^2}\Delta y, \quad B_x = -\beta_c E_y \tag{11}$$

The space-charge depressed value of gamma at the center of the beam is

$$\gamma_c \cong \gamma_0 - \nu\left[2\ln\left(\frac{r_w}{a}\right)+1\right] \tag{12}$$

The above formulation neglects the so-called "hoop stress", or the transverse electrostatic and magnetostatic self forces felt by a curved beam ($\Delta x''$ or $\Delta y'' \neq 0$) in a straight cylindrical pipe. By solving Poisson's equation within the beam and matching approximately to the vacuum solution at the curved boundary we find that the radial electric field on axis is given by

$$E_r \cong -\frac{\lambda}{R}\left[\ln\left(\frac{r_w}{a}\right)+\frac{1}{4}\frac{a^2}{r_w^2}-\frac{1}{4}\right] \tag{13}$$

where the beam carries charge/unit length λ and R is the local radius of curvature. A similar magnetostatic calculation results in an on-axis nagnetic field

$$B_\theta \cong \frac{\lambda}{R}\left[\ln\left(\frac{r_w}{a}\right)-\frac{1}{4}\frac{a^2}{r_w^2}+\frac{7}{4}\right] \tag{14}$$

Combining equations (13) and (14) and making use of the correspondence between $1/R$ and $\Delta x''$ ($\Delta y''$) we arrive at an approximate expression for the total transverse self force:

$$\frac{e}{mc^2}\left(\bar{E}-\bar{B}\right)_s \cong -\nu\left[2\ln\left(\frac{r_w}{a}\right)+\frac{3}{2}\right]\left(\Delta x''\hat{\imath}+\Delta y''\hat{\jmath}\right) \tag{15}$$

(The above expression was derived under the assumption $\beta_c \cong 1$. Corrections for $\beta_c < 1$ are straightforward.) We note that the inclusion of the hoop stress force on the RHS of

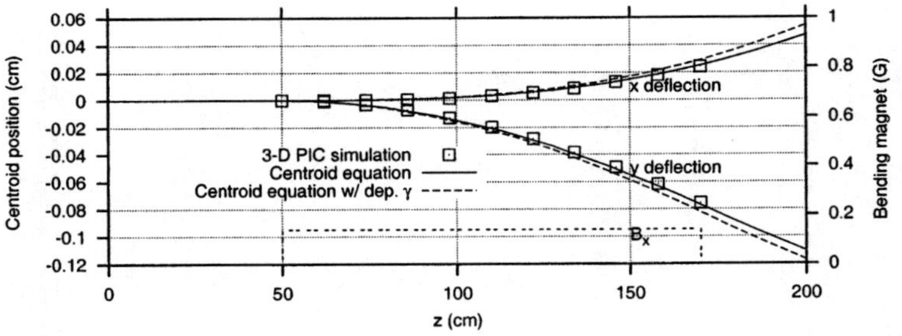

FIGURE 2. Comparison of beam centroid motion with and without corrections, with LSP calculation for a 3 kA, 3.2 MV beam. The beam is deflected by a 0.14 gauss dipole magnetic field. Equilibrium requires $B_z = 98.45$ gauss. Wall radius is 17.7 cm.

Eqs. (9) and (10) results in cancellation of the leading log term in the expression for the space-charge depressed γ_c, with the result that the "effective" value of gamma for centroid motion is

$$\gamma_c^{\text{eff}} \cong \gamma_0 + \frac{\nu}{2} \tag{16}$$

In Fig. 2 we illustrate this effect with a calculation of the deflection of a 3 kA beam in a transverse (bending) magnetic field. Use of Eq. (16) for the centroid γ_c in Eqs. (9) and (10) produces results for the transverse x and y displacements which agree very closely with 3-D PIC simulation results from LSP. The dashed curve was obtained using the space-charge depressed value of γ_c from Eq. (12) and shows the magnitude of the effect for DARHT-like beam parameters.

ACKNOWLEDGMENTS

This work was funded by Los Alamos National Laboratory. Paul Allison (LANL) developed modifications to the envelope equation similar to those discussed here, and provided copies of his calculations to us.

REFERENCES

1. Hughes, T. P., Genoni, T. C., and Clark, R. E., LAMDA User's Manual, Tech. Rep. MRC/ABQ-R-1798, Mission Research Corporation (1996).
2. Burns, M. J., et al, paper TU–02–1C, this meeting.
3. Lee, E. P., and Cooper, R. K., *Part. Accel.*, **7**, 83 (1976).
4. LSP is a software product of Mission Research Corporation, URL `http://www.mrcabq.com`.
5. Allison, P. W., Moir, D. C., Sullivan, G., and Hughes, T. P., "Observation of Self-Steering Effects on the ITS 6-MeV Linac", in *Proceedings of 1997 IEEE Particle Accelerator Conference*, IEEE, 1997.
6. Hughes, T. P., Self-steering effect in ITS experiment, Tech. Rep. MRC/ABQ-N-576, Mission Research Corporation (1996).

Magneto-Hydrodynamic Modeling of Gas Discharge Switches

P. Doiphode, N. Sakthivel, P. Sarkar and S. Chaturvedi

Institute for Plasma Research, Bhat, Gandhinagar – 382428, Gujarat, INDIA

Abstract. We have performed one-dimensional, time-dependent magneto-hydrodynamic modeling of fast gas-discharge switches. The model has been applied to both high- and low-pressure switches, involving a cylindrical argon-filled cavity. It is assumed that the discharge is initiated in a small channel near the axis of the cylinder. Joule heating in this channel rapidly raises its temperature and pressure. This drives a radial shock wave that heats and ionizes the surrounding low-temperature region, resulting in progressive expansion of the current channel. Our model is able to reproduce this expansion. However, significant difference of detail is observed, as compared with a simple model reported in the literature. In this paper, we present details of our simulations, a comparison with results from the simple model, and a physical interpretation for these differences. This is a first step towards development of a detailed 2-D model for such switches.

INTRODUCTION

Recently there has been considerable interest in high pressure gas switches. These can be used to produce high-voltage pulses with very short rise times [1]. An important use of such switches is in the development of ultra-wideband radiation systems [1-3].

One such switch is made out of a coaxial cable having an axial break in the inner conductor and dielectric [1,2]. This gap is filled with a gas at pressures ranging from a few atmospheres [2] to over a hundred atmospheres [1]. Rapid pulse-charging of the input end of the cable produces an axial voltage across the gap, leading to fast breakdown. The result is a sharpened voltage pulse at the output line.

Depending upon the parameter regime, the operation of these switches is determined by a complex interplay of radiation transport and magneto-hydrodynamics (MHD) [1]. A simple models for switch operation is available in the literature [1,2,4]. However, this model makes several simplifying assumptions, which may not always be valid.

We have simulated a switch consisting of argon gas filled in a cylindrical cavity of radius few mm, as shown in Fig. (1). Our simulations have been performed using a modified form of the RMHD code [5], which solves the multi-group radiation transport equations coupled to the MHD equations in one spatial dimension and time.

CP650, *Beams 2002: 14th International Conference on High-Power Particle Beams*,
edited by T. A. Mehlhorn and M. A. Sweeney

Equation of State and Opacity data has been generated using the IONMIX code [6]. The assumptions made in this study are listed later.

GENERATION OF MATERIALS DATA

Use of the RMHD [5] code requires input tabulated data for the equation of state, the average charge state <Z> and multi-group opacities. This data has been generated using the IONMIX code [6] over the range of densities and temperatures (0.025–20 eV) relevant to gas-discharge switches. Several checks have been made to assure the validity of the data generated by IONMIX. These include, among other things, checks of the formulas implemented in the code, a comparison of the frequency-dependent inverse-bremsstrahlung absorption coefficients with analytical expressions and a comparison of the cooling rates and average charge states at low densities with results from Ref.[7].

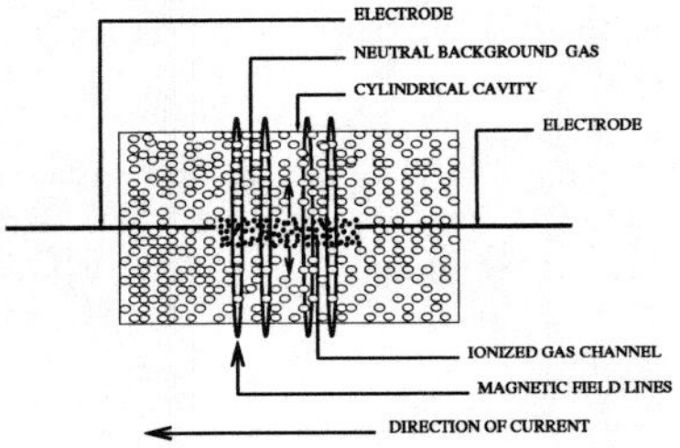

FIGURE 1. Schematic of gas switch

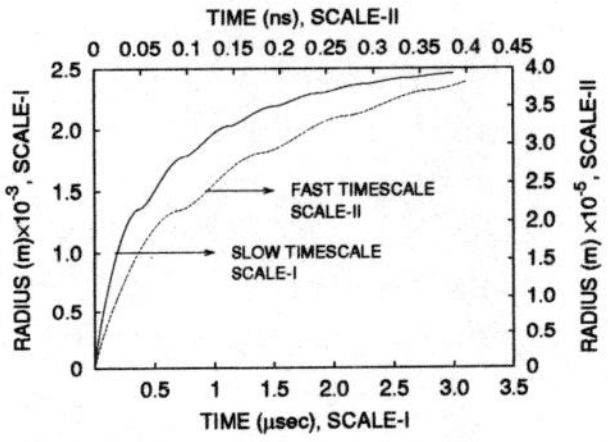

FIGURE 2. Channel radius vs time

MAJOR ASSUMPTIONS AND LIMITATIONS OF MODEL

1. Our simulation starts at the point where a "hot channel" has already formed near the axis, so that the MHD equations are valid. This means that the early stages of breakdown are not modeled, i.e., we cannot reproduce the fast-rising voltage pulse produced by such switches [1].
2. The model is one-dimensional (1-D) in space, i.e., we assume uniformity in the θ- and z-directions.
3. A modified form of Spitzer resistivity is used, which takes into account electron-ion and electron-neutral collisions. However, it does not include non-ideal effects due to high density.
4. Radiation transport is expected to modify channel expansion, by transporting energy from the hot "core" to the cooler region outside. We have made a start in this direction by generating and opacity data. However, due to numerical problems associated with radiation transport, those results are not included in this paper.

RESULTS AND DISCUSSION

A model developed by Braginskii [4] describes the development of a spark channel in a gas discharge. It provides the solution for the gas motion inside and outside the channel by considering a radially-expanding shock wave. It assumes that joule heating of the current-carrying channel heats and ionizes it. The resulting increase in pressure drives a radial shock wave, which heats and ionizes the surrounding cold gas. This is reinforced by radiation transport out of the channel, which further heats up a thin layer in the cold gas.

Braginskii has given a relation between the temporal current waveform and the channel radius [4]. Certain coefficients required for this model are available in Ref.[2] for different gases. As a first step, we have compared RMHD results with those from the Braginskii model for the temporal variation of channel radius. This has been done for helium at both high- and low pressures, with experimentally measured current waveforms taken from [1,2].

For the low-pressure switch, in the initial fill pressure is taken as 0.6 atm and the radius of the cylindrical cavity as 3 mm. Results using the Braginskii model are shown in Fig. (2). For the low pressure switch, we see that channel expansion is rapid up to t = 500 nsec, , followed by a slow expansion phase. Fig. (3) shows the radial current density profile, computed using RMHD, at different times. We see that up to 500 nsec, the current channel extends up to 2.0 mm. This value is higher than that predicted by the Braginskii model.

Fig. (4) describes the current density as a function of radius at different times for the high pressure gas switch. By 100 psec, the current channel expands up to 21.5 μm, which is less the Braginskii value shown in Fig. (2). However, given the assumptions made in the Braginskii model, the agreement is fairly good.

To estimate the importance of radiation absorption in the background gas, we have calculated the ratio of the radiation mean free path to the system radius. Fig. (5) shows results for P=0.6 atm in argon, i.e., the density is taken to be that of argon at this pressure and 300 K. The comparison has been done over a range of photon energies relevant to these switches. The plots are shown for two different plasma temperatures of 0.15 eV and 0.75 eV, which are expected to be broadly representative of initial temperatures in the current channel and the cold outer region. For a channel temperature of a few eV, we expect that most emitted photons would have energies lying below ~15 eV. For both temperatures, the computed ratio is more than unity, except at photon energies corresponding to atomic lines of neutral argon. This implies that the most of the photons which are emitted from the plasma will not get absorbed in the cold region. Beyond 15 eV, reabsorption becomes stronger due to a photo-ionization edge.

Fig. (6) shows the same plot for a fill pressure of 100 atm. Here we have assumed a channel radius of 100 μm, which is the maximum radius obtained from the Braginskii model. The ratio falls below unity even for photon energies below 15 eV. Hence we expect radiation transport to play a significant role.

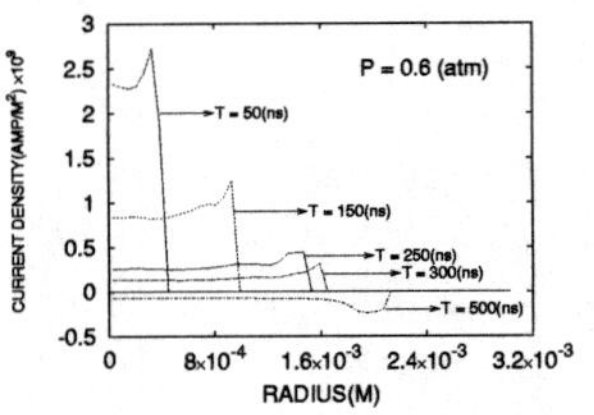

FIGURE 3. Current Density Vs. Radius

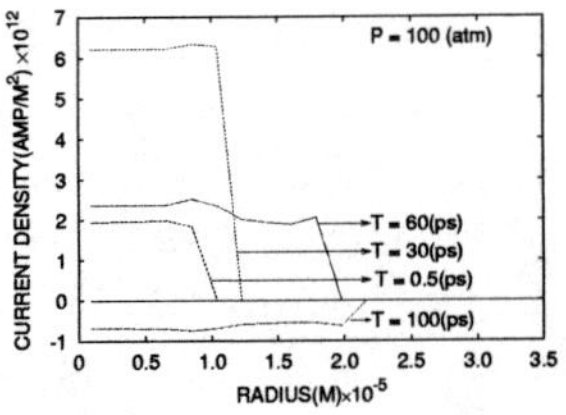

FIGURE 4. Current Density Vs. Radius

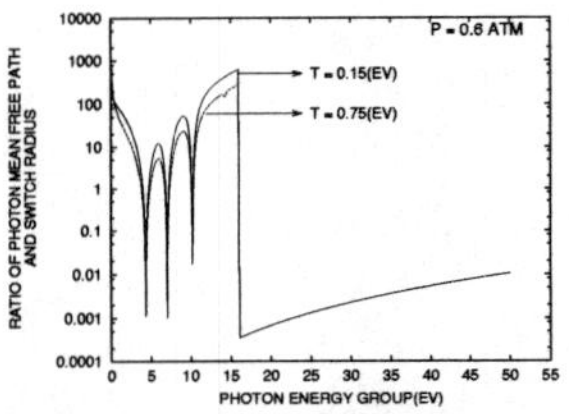

FIGURE 5. Low Pressure Comparison

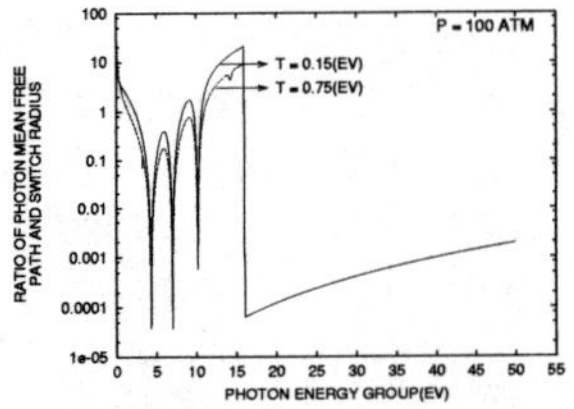

FIGURE 6. High Pressure Comparison

CONCLUSIONS AND FUTURE WORK

This MHD model reproduces the general features of the simpler Braginskii model, although there are regimes where significant mismatch is observed. The model is now being extended with the inclusion of radiation transport and non-ideal plasma effects. We also plan to set up a particle-in-cell model for the breakdown phase, in order to study the fast-rising voltage pulse associated with these switches.

REFERENCES

1. Frost C.A., et al., Proc. Ninth IEEE International Pulsed Power Conference, 491-494, (1993).
2. Martin T.H., et al., Proc. Ninth IEEE International Pulsed Power Conference, 463-470, (1993).
3. Prather W.D., et al., IEEE Transactions on Plasma Science, **28**, No.5, 1624-1630, (2000).
4. .Braginskii S. I., Soviet Physics JETP, **34**, No. 6, 1068-1074, (1958).
5. Yasar O. and Moses G. A., Comp. Phys. Commun., **69** , 439-458, (1992)
6. Macfarlane J. J., et al., Comp. Phys. Commun., **56**, 259-278, (1989).
7. Post D.E., and Jensen R. V., Atomic Data and Nuclear Data Tables, **20**,397-439, (1977).

Modeling of ETC Plasma Cartridges and their Effect on Propellant Combustion

N. Sakthivel, P. Sarkar, S. Chaturvedi

Institute for Plasma Research, Bhat, Gandhinagar – 382428, Gujarat, INDIA

Abstract. A one-dimensional, time-dependent fluid code has been developed for modeling an electro-thermal chemical (ETC) plasma cartridge. The power flow out of the cartridge is coupled to the interior ballistics code IBHVG2. For given pulse-forming network and gun parameters, the projectile velocity is maximized with respect to cartridge dimensions. In this paper, we report details of the computational model and results of optimization.

INTRODUCTION

Electrothermal-chemical (ETC) propulsion involves the use of a plasma jet to initiate and control propellant combustion. An electrothermal cartridge is a device for producing a plasma jet, which utilizes a low mass, high energy electrically generated plasma. The cartridge, typically lined with low molecular weight material, has a solid electrode at one end and a hollow electrode at the other. The two electrodes are connected by a thin wire running through the cartridge, and externally connected to a pulse-forming network (PFN). The explosion of the wire due to passage of current produces a hot, dense plasma, which then heats the cartridge walls by radiation and conduction. A complete description of ETC cartridge operation can be found in [1].

Propellant combustion is greatly affected by plasma parameters, viz., temperature, pressure and velocity at the capillary exit. These are determined, in turn, by capillary dimensions, the material lining the capillary walls and the parameters of the PFN. It is, therefore, of great interest to model this interaction self-consistently.

We have developed a one-dimensional (1-D), time-dependent fluid code for ETC cartridges. This code, which is based upon the LCPFCT code [2], solves the fluid equations using the flux-corrected transport algorithm [2]. Appropriate source and sink terms are included, along the broad lines of Ref.[1]. Only axial variation of plasma properties is taken into account. The plasma capillary acts as a resistive load for the PFN. The capacitor voltage and circuit current are evolved self-consistently in time. Any change in circuit current modifies plasma parameters such as temperature, density and velocity, which in turn modifies its electrical resistance, reflecting back on the PFN.

CP650, *Beams 2002: 14th International Conference on High-Power Particle Beams*, edited by T. A. Mehlhorn and M. A. Sweeney

Plasma flowing out of the cartridge, and into the propellant chamber, carries energy in two forms, viz., enthalpy and kinetic energy of directed flow. The 1-D cartridge code computes this power outflow as a function of time. This power is then used as an external source in the energy equation of the interior ballistics code IBHVG2 [3], replacing the primer energy. The PFN parameters are chosen such that the electrical energy is ~10% of the chemical energy in the propellant.

In this paper, we report on a study aimed at maximizing projectile velocity by varying cartridge dimensions for given PFN, propellant and projectile parameters.

COMPUTATIONAL MODEL

The computational model for ETC plasma cartridge involves solving the time dependent fluid equations in the axial dimension using the LCPFCT code [2], along with circuit equations for the capacitor bank voltage and current. LCPFCT solves coupled continuity equations of generalized form [2]. Various source/sink terms have been added to the mass continuity and energy equations. This includes Joule heating and radiation loss terms in the energy equation and mass ablation rate from the capillary wall in the continuity equation. The capillary liner material considered in this calculation is Lexan and its heat of sublimation is taken as 54 MJ/kg [1]. A table of charge-state and internal energy were generated for Lexan for the temperature and density regime covered in the ETC cartridge plasma using the Saha ionization model of the IONMIX code [4]. Bilinear interpolation has been used on these tables. The resistivity is calculated based on Spitzer's formula with contributions from electron-ion and electron-neutral contributions. The fraction of the radiation heat flux reaching the capillary walls is calculated using the vapor-shielding model of Ref.[1]. Presently, this code does not include the velocity loss due to viscous drag at the wall as well as non-ideal effects that arise in the ETC parameter regime.

We have used the choked flow boundary condition at the exit of the cartridge as described in LCPFCT manual [2] for this study. The choked flow boundary condition assumes that the fluid flows at sound speed at the capillary exit. In this study, we have used 100 computational cells. The time-step is regulated according to the Courant sound-speed criterion [2]. Conservation checks have been made routinely for mass and energy, as also a consistency check between radiation loss and mass ablation rate.

For interior ballistics, we have used the zero-dimensional interior ballistic code IBHVG2 [3]. All product gases are assumed to follow the Noble-Able thermodynamic equation of state [3]. The time dependent power output data from the capillary has been interpolated and used as an input in the energy equation, replacing primer input in the IBHVG2 code.

This model ignores axial energy transport by conduction and radiation. We also assume that plasma parameters vary significantly only in the axial direction – this has been justified by other workers. The ablation model assumes that radiation loss from the plasma causes sublimation and dissociation of the liner material (Lexan), which then flows into the plasma.

RESULTS AND DISCUSSION

This study is a linkage of three models, viz., PFN, ETC cartridge and interior ballistics. For PFN, the charging voltage, capacitance and inductance were kept as 24 kV, 7.0 mF and 10.6 μH respectively as used in [5] for driving 120mm cannon. For interior ballistics, we have used one of the sample problems provided in IBHVG2 report [2]. This consists of 105mm gun with chamber volume of 0.0062 m^3, propellant type of M30-7 perforation with charge weight and projectile weight of 5.34 kg and 10.52 kg respectively. The propellant burn rates and other gun tube parameters needed for IBHVG2 are taken as in [2]. The cartridge length is fixed as 11.0 cm and radius is varied from 2.0 mm to 6.0 mm.

In figure 1, we present the time dependent pressure profiles for 3 mm radius cartridge at three axial locations, closed end (cell no:1), middle (cell no:50) and at the cartridge exit (cell no: 100). The maximum peak pressure achieved in the first cell and at the capillary exit is 3710 Mpa and 1485 Mpa respectively. The maximum temperature and charge state reached in the capillary is 6.5 eV and 0.6 eV respectively. In figure 2, we present the time dependent power output from the cartridge. The total energy inflow in to the combustion chamber in this case is 1.8 MJ, which is 90% of the PFN energy and the peak current reached in this case is 384 kA.

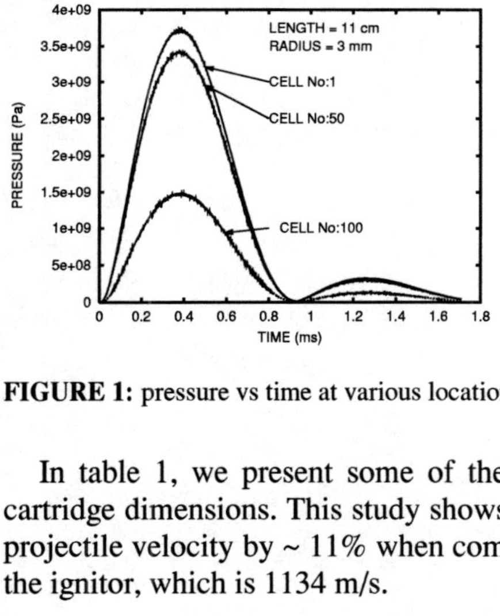

FIGURE 1: pressure vs time at various locations

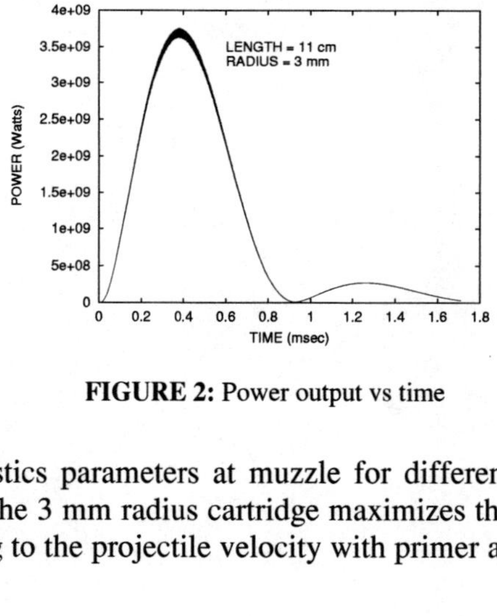

FIGURE 2: Power output vs time

In table 1, we present some of the ballistics parameters at muzzle for different cartridge dimensions. This study shows that the 3 mm radius cartridge maximizes the projectile velocity by ~ 11% when comparing to the projectile velocity with primer as the ignitor, which is 1134 m/s.

TABLE 1: Ballistic parameters at Muzzle for different cartridge radii and a fixed length of 11.0 cm.

Ballistic parameters	**Cartridge radius**						
	2.0 mm	**2.5 mm**	**3.0 mm**	**3.5 mm**	**4.5 mm**	**5.0 mm**	**6.0 mm**
Time (ms)	6.86	6.89	6.83	7.04	7.19	7.47	7.54
PV (m/s)	**1210**	**1206**	**1258**	**1167**	**1142**	**1029**	**1054**
BP (MPa)	93.3	90.6	118.7	69.9	59.2	4.4	22.4
MP (MPa)	86.9	84.3	110.6	65.1	55.2	4.0	20.9
BAP (MPa)	74.7	72.5	95.0	55.9	47.4	3.4	17.9
MT (K)	1951	1894	2468	1470	1250	95	484
PMB (kg)	5.33	5.33	5.36	5.30	5.28	5.12	5.18

[PV – Projectile velocity, BP – Breech pressure, MP – Mean pressure, BAP – Base pressure, MT - Mean temperature, PMB – Propellant mass burned].

LIMITATIONS OF THE MODEL

This study does not include non-ideal effects that arise in the ETC capillary parameter regime, which affect the resistivity and equation-of-state of the plasma. Non-ideal effects were taken into account in an earlier calculation of plasma opacity in ETC cartridges [6], and are now being added to the present code. Another important limitation is that the time-dependent breech pressure is not being taken into account in cartridge calculations; this effect is expected to be significant. Finally, viscous drag at the wall is also not accounted for in the momentum equation [1].

CONCLUSIONS AND FUTURE WORK

An end-to-end model for electrothermal chemical propulsion has been developed by coupling a 1-D fluid code to an interior ballistics code. This model has been used for optimization of plasma cartridge dimensions. This study shows that, for a given PFN, capillary dimensions play a major role in determining projectile velocity. We next plan an extension of this model to include non-ideal plasma effects, as well as the effect of breech pressure on plasma outflow from the cartridge.

REFERENCES

1. Hurley, J, D., Ph.D thesis, North Carolina State University, Raleigh, (1993), and references therein.
2. Boris, J, Landsberg, A, Gardner, J., LCPFCT – A Flux Corrected transport algorithm for solving generalized continuity equations, NRL/MR/6410-93-7192, Apr. (1993).
3. Anderson, R.D., Fickie, K.D., Interior Ballistics of High Velocity Guns, Version2 – A user's guide, Jul. (1987).
4. Macfarlane, J. J., Comp. Phys. Commun, 56, 259, (1989).
5. Katulka, G.L., 32nd JANNAF Combustion Subcommittee meeting, Volume 1, 93-101, (1995).
6. Sakthivel, N., Bhattacharya, R., Chaturvedi, S., Proc. 12th IEEE Intl. Pulsed Power Conference, Monterey, 601, (1999).

Linear Mode Computer Code MULTIWAVES-5.2 for Multiwave Cerenkov Generator Investigations

Michail P. Deichuly, Vladimir I. Koshelev, Viktor M. Pikunov, Vladimir A. Popov

Institute of High Current Electronics, RAS, 4 Akademichesky Ave., 634055 Tomsk, Russia

Abstract. The paper is devoted to experimental and theoretical investigation of two types of one-sectional generators with overmoded slow-wave structures (SWS). Numerical analysis was made by means of a computer code MULTIWAVES-5.2. Comparison and discussion of obtained results that can be used at development of multiwave Cerenkov generators were made.

INTRODUCTION

At present, intensive investigations are carried out aiming to obtain microwave radiation pulses with the power of 1-10 GW and the length of 100-1000 ns. One of the promising directions allowing investigation of the indicated problems is development of multiwave Cerenkov generators [1, 2]. However, use of overmoded SWS with $D/\lambda \approx 4$ results in arising the problem of operating mode selection and parasitic mode suppression. Here, D is the SWS diameter, λ is the radiation wavelength. A 3D computer code MULTIWAVES-5.2 [3] was used for numerical investigation of one-sectional Cerenkov generators. The code is based on the equations of a linear theory that takes into account the three-dimensional nature of electromagnetic fields and electron motion [4].

RESULTS OF EXPERIMENTAL INVESTIGATIONS

Experimental investigations were carried out at the accelerator "SINUS-7M" having the following parameters: voltage at a coaxial diode with magnetic insulation U_d = 500 kV, beam current I_b = 10 kA, pulse length τ_p = 40 ns. The hollow beam diameter was $D_b = 2r_b$ = 10-11.2 cm, magnetic field in the SWS region was B = 4-15 kG. An electron beam radius was determined by a track at a thin dielectric target. Radiation power was determined by the patterns in two polarizations measured in atmosphere. Radiation spectrum was measured by means of a bandpass filter with the bandwidth of 0.5%.

Two variants of overmoded ($D/\lambda \approx 4$) slow-wave structures of one-sectional Cerenkov generators close by the properties but different by geometry were used. The diameter of both structures was D = 130 mm, the lengths were chosen to be equal to 14d, the periods were equal to d = 13 mm, the radii of half-tores were of 3.2 mm and

CP650, *Beams 2002: 14th International Conference on High-Power Particle Beams,*
edited by T. A. Mehlhorn and M. A. Sweeney

4.8 mm, and the heights of pedestals were of 1.2 mm and 0.9 mm for the first and second structure, respectively.

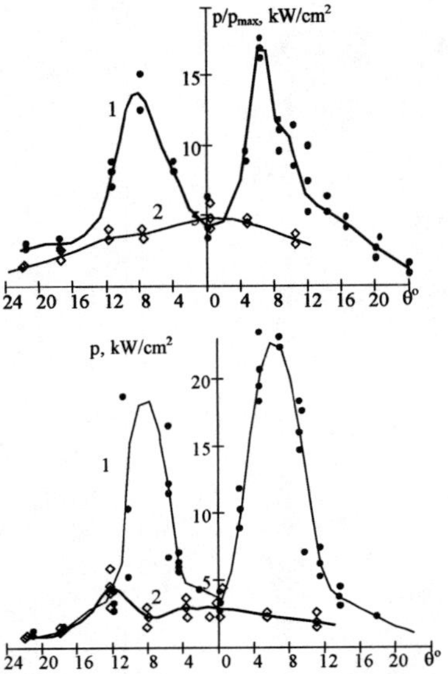

FIGURE 1. Patterns of radial (curve 1) and cross-polarized (curve 2) components of microwave radiation for the first (upper figure) and second (lower figure) generators.

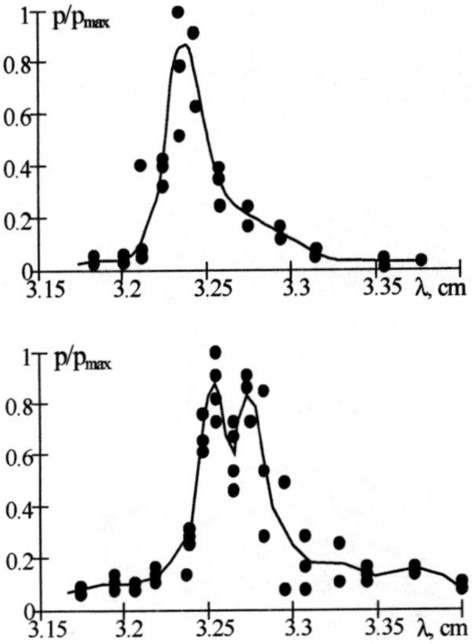

FIGURE 2. Microwave radiation spectrum for the first (upper figure) and second (lower figure) Cerenkov generators.

Figures 1 and 2 present, respectively, the pattern and the spectrum of microwave radiation of a one-sectional generator with c $L_{SWS} = 14d$.

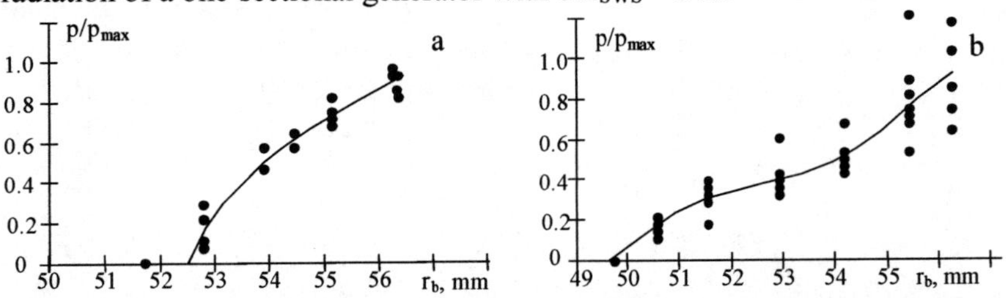

FIGURE 3. Radiation power versus beam radius at B = 7.3 kG for the first (a) and second (b) one-sectional generators.

Figure 3 presents radiation power versus beam radius. Microwave radiation power of the first generator made up 170 MW and 60 MW for the radial and cross-polarized components, respectively. Microwave radiation power of the second generator made up 200 MW and 80 MW for the radial and cross-polarized components, respectively. Microwave radiation length was limited by the beam current pulse length. The waveform of the signal from a microwave sensor was broken. Traces of electric breakdowns on the surface of the SWS No.1 were absent. Intensive traces of breakdowns are noticeable on the surface of the SWS No.2 except three end diaphragms from each side of the generator. Assymmetry of the patterns of radially polarized radiation may be related to the low ellipticity of the beam and its inadjustment in the slow-wave structure.

NUMERICAL SIMULATION RESULTS

Numerically found π-type frequencies of the operating mode E_{01} for the first and second SWS differed a little by value that was of 9.57 GHz and 9.43 GHz, respectively, and the ratio of the phase velocities to the velocity of light at these frequencies was equal to 0.83 and 0.82, respectively. Specific behavior of transmission coefficients of the symmetric E_{01} and asymmetric HE_{11} modes presented in figure 4 is availability of longitudinal resonances, i.e. frequency regions in which the behavior of the curves is resonant. Q-factors of corresponding resonances for both SWS are approximately equal but have different position on the frequency axis.

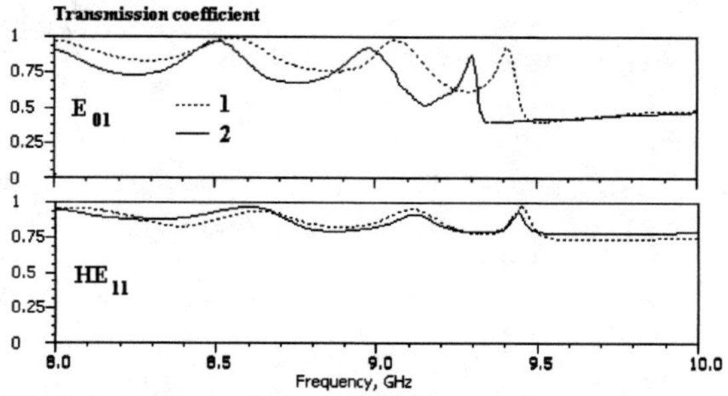

FIGURE 4. Transmission coefficients of the operating mode E_{01} and mode HE_{11} for the first (dotted lines) and second (solid lines) SWS.

Slow-down asymmetric modes $HE_{\nu 1}$ have a considerable by value longitudinal electric field. This fact can result in efficient longitudinal interaction of asymmetric modes $HE_{\nu 1}$ with electron beam (figure 5). A symmetric operating mode E_{01} has the lowest starting currents for the considered uniform structures. Note that the calculated maximum strength value of the electric field longitudinal component of the mode E_{01} between the diaphragms for the second structure exceeds the analogous value for the first structure by 1.3 times. In the first variant of a one-sectional generator only the asymmetric mode HE_{11} is excited. In the second generator the starting currents of the asymmetric modes $HE_{\nu 1}$ increase with the rise of the azimuth index $\nu = 1\text{-}5$. The difference in the starting currents and generation frequencies for the operating mode E_{01}

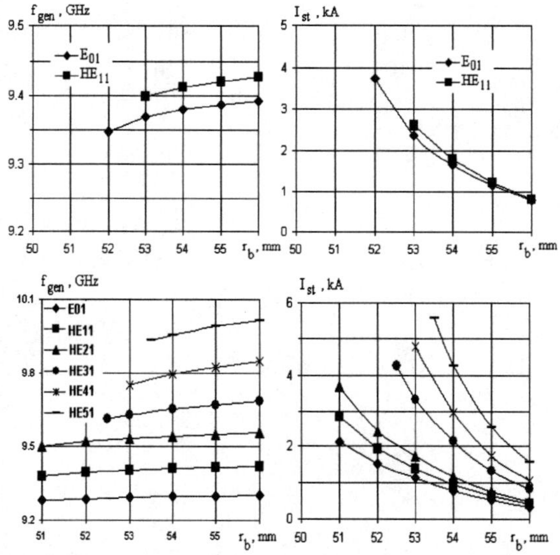

FIGURE 5. Generation frequencies and starting currents versus electron beam radius for the first (upper row) and second variant (lower row) of a one-sectional Cerenkov generator with the length of 14d.

and parasitic mode HE_{11} strongly depends on the structure geometry. Hence, if for the first generator they are equal to $\Delta I_{st}= 0.01\text{-}0.4\ kA$, $\Delta f\sim 0.4\%$, then for the second one the corresponding values are equal to $\Delta I_{st} = 0.1\text{-}0.8\ kA$, $\Delta f \sim 1.2\%$. Considerable difference in the generation frequencies of the operating and parasitic modes Δf equal to $\sim 1.2\%$ in the second case can be one of the reasons of the widening and splitting of the generation line with the frequency spacing of the order of *1%* experimentally observed for the second variant of the generator (fig.2). The used linear approximation has a limited application for explanation of experimental radiation spectra since it is valid only for the initial time for the radiation pulse when high-frequency amplitudes of the fields and convectional currents are low.

CONCLUSION

1. Results of experimental and theoretical investigations have shown that increase of the ratio of diaphragm width to the SWS period from 0.48 for the first SWS to 0.74 for the second one when the operating mode slowing-down is approximately equal results in the electric field strength increase on the SWS surface and radiation spectrum extension.
2. Results of numerical simulations of generation frequencies and electron beam starting radius agree with results of measurements.
3. Numerical calculations indicate the approximate equality of starting currents of the operating symmetric and parasitic asymmetric modes. Nevertheless, generation in the experiment is realized at the E_{01}-mode. Availability of cross-polarized radiation in the pattern indicates the availability of the HE_{11}-mode in the output radiation.

ACKNOWLEDGMENTS

This work was supported through the ISTC No.2005p / EOARD # 00-7017 Grant.

REFERENCES

1. Bugaev, S.P., Cherepenin, V.A., Kanavets, V.I., Klimov, A.I., Kopenkin, A.D., Koshelev, V.I., Popov, V.A., and Slepkov, A.I., *IEEE Trans. Plasma Sci.* **18**, 525-536 (1990).
2. Koshelev, V.I., Deichuly, M.P., "Optimization of electron beam – electromagnetic field interaction in multiwave Cerenkov generators", in *High Energy Density Microwaves,* edited by Robert M. Phillips, AIP, Woodbury, New York, 1999, pp. 347-359.
3. Koshelev, V.I., Pikunov, V.M., Deichuly, M.P., Popov, V.A., Petkun, A.A., "Cerenkov generators with overmoded one-sectional slow-wave structures", in *Proc. IEEE Pulsed Power Plasma Science Conf.,* edited by R. Reinovsky and M. Newton, 2001, pp. 1634-1637.
4. Pikunov, V.M., Chernyavsky, I.A., *Izv. Ross. Acad. Nauk, Ser. Fiz. (rus.)* **64**, 2481-2489, (2000).

Computer Model for Electrode Plasma Generation by Electron and Ion Flows*

Victor V. Ryzhov [a#], Valeri I. Bespalov [b], Alexander V. Kirikov [a], Igor. Yu. Turchanovskii [a] and Vladimir P. Tarakanov [c]

[a] *Institute of High Current Electronics, 4 Akademicheskii Ave., Tomsk, 634055, Russia*
[b] *Tomsk Polytechnical University, Tomsk, Russia*
[c] *High Energy Density Research Investigation Center, Moscow, Russia*

Abstract. A model is proposed for computer simulation of the electrode plasma generation by electron and ion flows. The distribution of the absorbed energy of particles in the electrode material is calculated by the Monte-Carlo codes. This provides a possibility to control the electrode temperature by solving the heat conductivity equation for specified distributions of thermal sources and to calculate the rate of plasma generation. The behavior of the plasma in the gap can be modeled based on simple model where the velocity, the density, and the temperature of the plasma can be given by some dependence. Within the framework of the model proposed, numerical study is performed on the effect of the plasma flows in Rod Pinch Diodes and in the Insulator Stack of the Z-accelerator.

INTRODUCTION

Complete computer modeling of the formation of plasma under the action of particles flows impinging on a surface and its motion in electric and magnetic fields is a challenging problem. This is due to the fact that the mechanism of the generation of neutral particles as well as the charged particle density and the temperature of the formed plasma change, depending on the rate of a particle flow. Therefore, to simulate these processes, different models based on the kinetic or hydrodynamic approaches are required [1, 2].

To study the influence of plasma flows on the characteristics of high-current devices, a Conductive Region Model (CRM) has been constructed. The CRM was employed in the PIC code KARAT [3] to investigate the influence of the anode plasma on the electron flow in Rod Pinch Diodes (RPD's) and in the Insulator Stack of the Z-accelerator (Z-Stack).

CONDUCTIVE REGION MODEL

To study the effect of the formation and motion of the near-electrode plasma on the processes occurring in the vacuum gap, we have developed a Conductive Region Model (CRM). In this model, the plasma is modeled by a time-variant region of homogeneous plasma (the plasma whose characteristics do not depend on coordinates).

---

* This work is being supported by Sandia under Contract #18328.
# e-mail: ryzhov@to.hcei.tsc.ru

CP650, *Beams 2002: 14th International Conference on High-Power Particle Beams*, edited by T. A. Mehlhorn and M. A. Sweeney

With this model, we can specify the characteristics of the plasma, such as the charged-particle density, the collision rate and the conduction, the permeability and permittivity, and the velocity of the free boundary of the region occupied by the plasma. In doing so, one can obtain sufficiently correct description of the behavior of the actual plasma in the gap space.

The formation and the motion of the plasma occurs on approaching a definite temperature (T_S=450°C) at the surface. In so doing, it is assumed that particles are emitted from the moving boundary of the conductive region.

RESULTS OF KARAT/CRM SIMULATION

Z - Stack

The CRM was used to investigate the influence of the anode plasma on the electron flows in a five-stage Insulator Stack of the Z-accelerator (Z-Stack), which insulated a water line and a magnetically insulated transmission line. The stack consisted of five $+45^0$ –profile Rexolite-1420 insulator rings and four hard-anodized-aluminum grading rings [4]. In model calculations, the voltage linearly increased within first ten nanoseconds, reaching V = 2-3 MV, whereupon it remained constant.

FIGURE 1 shows schematically the electron flows in the Z-Stack at one and the same moment in time t = 24 ns under like conditions with respect to voltage for two models of calculations. The first model (Fig. 1*a*) took into account the possibility of electron emission from the surface of the cathode and of the grading rings upon approaching the electric field strength E_{emis}=300kV/cm on the vacuum side. A similar model was employed in [5] in simulating the Z-Stack by the IVORY code. The second model (Fig. 1*b*), in addition to electron emission, allowed for the formation of the anode plasma under the action of an electron beam.

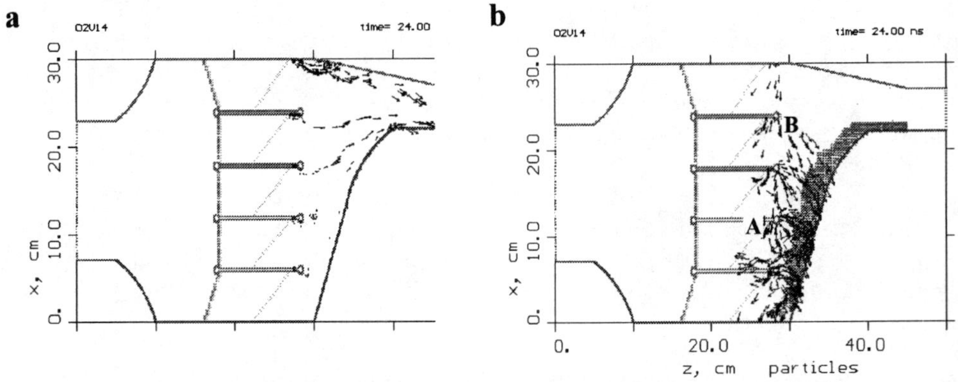

FIGURE 1. Comparison of simulation of the Z-Stack Insulator by the KARAT/CRM code with (a) and without account of plasma (b) .

The calculations have shown that account of the effect of the formation of the anode plasma changes qualitatively the pattern of the physical processes developing in the MITL. As the electric field strength in the zone of emission approaches a value higher then the threshold one, the electrons emitted from the cathode and from the grading rings start moving toward the anode. At the stage of increasing voltage, these electrons escape magnetic insulation, impinge on the anode, and heat it that results in a layer of the expanding anode plasma whose motion is modeled by the CRM. With no account of the possible influence of the plasma, as done in [5], the electrons are magnetically cut off by the external magnetic field for which the conduction current is responsible and by the self-magnetic field of the beam at a definite voltage level (Fig. 1*a*). The leakage current is therewith low and, consequently, the losses in the transfer of the electromagnetic wave energy through the insulator are small. Should the effect of the anode plasma be allowed for, the electric field may then be redistributed such that part of the electrons would strike the Rexolite-1420 insulator (Fig. 1*b*). This may most probably cause a surface discharge, i.e., short-circuiting between the rings, and then short-circuiting of the entire Z-Stack and its failure.

It should be noted that in the post-hole convolute of the Z-accelerator in the anode region AB, on which most of the electrons emitted from the cathode and from the grading rings impinge, there are technological holes whose positions have found experimentally. Therefore, no anode plasma is formed, even if shots are unsuccessful (low-rate increase in voltage, high leakage currents), and the electrons reach the region AB that precludes failure of the Z-Stack.

Rod Pinch Diode

As shown in [6,7], the motion of the near-electrode plasma strongly affects the transportation of electrons in a rod-pinch diode. This motion causes a change in effective electrode radii responsible for the value of the critical current and, consequently, for the conditions of magnetic insulation of the gap.

Using the KARAT/CRM, simulation has been performed for a RPD operating in the modes investigated experimentally on the Gamble II. FIGURE 2 shows the results of simulation for the rod-pinch diode with account of the motion of the near-electrode plasma. To compare the models, the time delays of the anode and cathode plasma, as well as their velocities for shot 7402-G II were taken from [7]. FIGURE 2 shows the dynamics of changes in the beam configuration due to the expansion of the cathode and anode conductive regions.

In calculating oscillograms of the currents, the voltage across the diode was assumed to vary with time, as in the case of shot 7402-G II [7]. The results of calculations are presented in Fig. 3 wherefrom it follows that the values of the diode current obtained in the RPD/NRL calculation with account of the plasma motion substantially disagree with experimentally measured ones. Hence it follows that the set of parameters obtained in the RPD/NRL model is self-inconsistent and requires debugging.

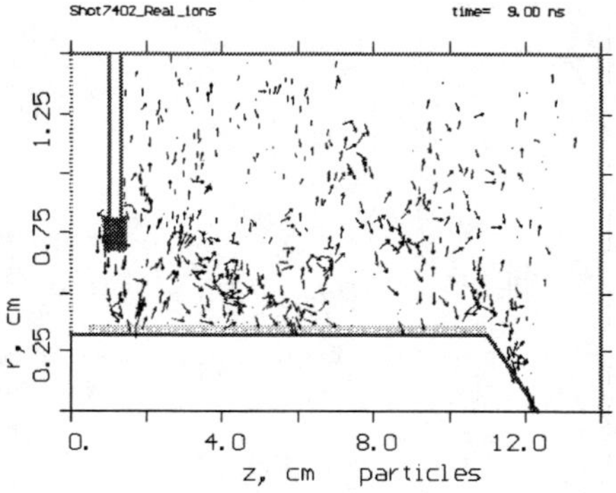

FIGURE 2. Rod-Pinch Diode (Gamble II) Simulation of the cathode and anode plasmas with CRM.

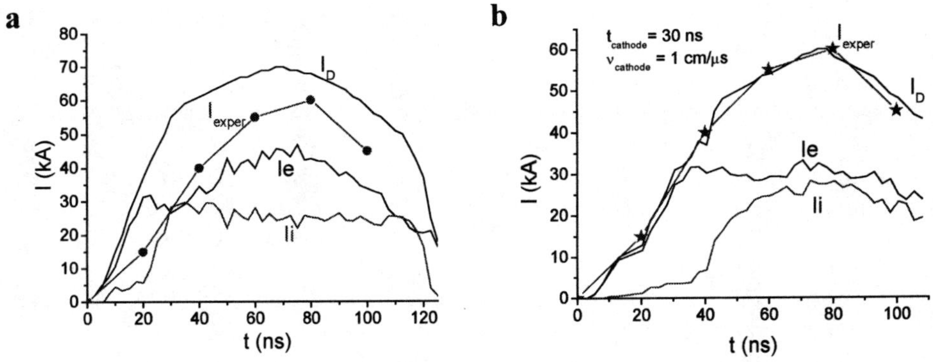

FIG 3. Calculations of the RPD currents by the KARAT/CRM code. NRL (a) and HCEI (b) models. I_e, I_i, I_D – electron, ion and diode currents, I_{exper} – experiment diode current.

Table 1 presents the delay times τ_{on} and the rate v_{AC} at which the plasma spans the anode-cathode gap, which we have obtained for Shots 7402-G II and 7497-G-II by the KARAT/CRM code. The same table cites the results obtained in [7]. It follows from comparison of the results that the effective gap span rate v_{AC} obtained by us in direct numerical simulation of the diodes is lower than the possible values of the cathode plasma velocity. This seems to be associated with the hydrodynamic spreading of the plasma out of the gap in the conditions of a complex RPD geometry.

TABLE 1

Shot	R_a	R_c	L_{Rod}	Model	v_c (cm/s)	v_a (cm/s)	v_{ac} (cm/s)	t_{aon}	α
7402	3.2	8	110	NRL	$2*10^6$	$0.6*10^6$		48	1.6
G-II				HCEI	--	-	$0.8*10^6$	52	1.26 -1.5
7497	1.6	9.8	57	NRL	$2*10^6$	$1*10^6$		46	1.85
G-II				HCEI	-	-	$1*10^6$	10	1.6 -1.75

REFERENCES

1. Welch D.R., Oliver B.V., Rose D.V., "Plasma effects in rod pinch diode for radiography", MRC/ABQ-R-2023, 2001.
2. J.W. Schumer, S. B. Swanekamp, P.F. Ottinger, R.J. Commisso, B.V. Weber, D. N. Smithe, L.D. Ludeking, "MHD-to-PIC Transition for Modeling of Conduction and Opening in a Plasma Opening Switch" IEEE Trans. Plasma Science, vol. 29, №3, pp. 479-492, June 2001
3 Tarakanov V.P. "User's Manual for Code KARAT". Springfield: BRA, 1992.
4. Stygar W.A., Spielman R.B., Anderson R.A. et. al., "Operation of a five –stage 40,000-cm^2 –area insulator stack at 158 kV/cm", Proc. 12 IEEE Inter. Pulse Power Plasma Science Conf., PPPS-1999.
5. Shoup R.W., Long F., Martin T.H., et. al., "Design validation of the PBFA-Z vacuum insulator stack", Proc. 11 IEEE Inter. Pulse Power Plasma Science Conf., PPPS-1997, pp.1608- 1613.
6. Ryzhov V.V., Belomyttsev S.Yu., Kirikov A.V., Turchanovsky I.Yu., Tarakanov V.P. "Model of a quasisteady-state electron –beam pinch in rod-pinch diode", Proc. 13 IEEE Inter. Pulse Power Plasma Science Conf., PPPS-2001, p. 1340-1343, 2001.
7. G. Cooperstein, J.R. Boller, R.J. Commisso, et. al., "Theoretical modeling and experimental characterization of a rod –pinch diode" Physics of Plasmas, vol. 8, №10, pp. 4618-4636, October 2001

BEAMS FROM ULTRA-INTENSE LASERS

Intense, High-Quality Ion Beams Generated by Ultra-Intense Lasers

Markus Roth[a], Thomas E. Cowan[b], Jean-Claude Gauthier[c], Jürgen Meyer-ter Vehn[d], Matthew Allen[b], Patrick Audebert[c], Abel Blazevic[a], Erik Brambrink[a], Julien Fuchs[c], Matthias Geissel[a], Manuel Hegelich[d], Stefan Karsch[d], Hartmut Ruhl[b], Theodor Schlegel[a]

[a]*Gesellschaft für Schwerionenforschung mbH, Planckstr.1, 64291 Darmstadt, Germany*
[b]*General Atomics, P.O. Box 85608, San Diego, California 92186-5608*
[c]*Laboratoire pour l'Utilization des Laser Intense, Palaiseau, France*
[d]*Max-Planck-Institut für Quantenoptik, Garching, Germany*

Abstract. The advent of ultra-intense short-pulse laser systems has opened the field of relativistic laser-generated plasmas in which the electron temperature exceeds several MeV. One particularly exciting topic of recent interest has been the discovery of intense, collimated, energetic beams of ions off the back surface of solid targets. The beam currents exceed several hundreds of kiloamperes at a pulse duration of only a few picoseconds.

The dominant acceleration mechanism is known as the Target Normal Sheath Acceleration (TNSA) mechanism, which is fundamentally different from the well known ion acceleration processes observed using long pulse laser over the past decades. This mechanism offers the prospect of low-emittance, laminar, directed ion beams. The laser to ion beam conversion efficiency can be as high as 10 %, which makes these beams interesting to a whole variety of applications.

The strong dependence of the ion beam acceleration on the conditions of the target back surface in agreement with theoretical predictions based on the TNSA mechanism was confirmed. Shaping of the ion beam has been demonstrated by the appropriate tailoring of the target and the laser beam. In addition, we demonstrated the use of laser accelerated protons for time resolved proton radiography, which is able provide complementary information about the properties of matter not accessible by current techniques.

INTRODUCTION

The production of energetic ions by lasers focused onto solid targets has been known for several decades. In contrast to the experiments using long pulse lasers, ions accelerated by ultra-intense, fs-laser systems are emitted predominantly from the rear, non-irradiated surface in a low divergent beam of excellent quality [1]. The dominant mechanism is understood as rear surface emission by the TNSA (Target

CP650, *Beams 2002: 14th International Conference on High-Power Particle Beams,*
edited by T. A. Mehlhorn and M. A. Sweeney

Normal Sheath Acceleration) mechanism and published in detail in [2]. Briefly, relativistic electrons generated from the laser plasma interaction, having an average temperature of several MeV, envelope the target foil and form an electron plasma sheath on the rear, non-irradiated surface. The electric field in the sheath ($E_{stat} \sim kT_{hot}/e\lambda_D$, $\lambda_D = (\varepsilon_0 kT_{hot}/e^2 n_{e,hot})^{1/2}$) can reach > 10^{12} V/m. A few monolayers of atoms at the rear surface are field-ionized and accelerated normal to the surface by E_{stat} with the most energetic electrons always extending further out into the vacuum, maintaining the accelerating field as long as the electron temperature is high. Because of the accompanying electrons, the ion beam is space charge and current neutralized. So far mainly protons have been observed from the rear surface, originating from contaminating hydrocarbon and water vapor layers which coat the target. The ions form a collimated beam with an approximately exponential energy distribution with 5-6 MeV. The conversion efficiency from laser energy to ion beam energy can be quite high and efficiencies of order of 10 % have already been measured [1]. Because of the dependence of the ion beam on the formation of the electron sheath, this process should also reveal information about the electron transport through the target. We expect details of the ion acceleration will furthermore depend on the target material and surface conditions. Therefore, we carried out experiments to investigate the influence of the target parameters on the ion beam production.

EXPERIMENTS

We have studied the influence of the target properties on laser-accelerated ion beams generated by multi-terawatt lasers. The experiments were performed using the 100 TW laser facility at Laboratoire pour l'Utilization des Laser Intense (LULI). The targets were irradiated by pulses up to 5×10^{19} W/cm^2 (~ 300 fs, λ=1.05µm) at normal incidence. To characterize the plasma conditions and ion beam parameters, we used short pulse laser interferometry, absolutely calibrated ion spectrometers, Radio-chromic film (RCF), CR-39 nuclear track detectors, nuclear activation, neutron time-of-flight diagnostics, and Thomson Parabolas to detect heavy ions with respect to their charge-to-mass ratio. Stacking the RCF and CR-39 provided spatial beam profile information at different ion energies. Details of the experimental setup and the various diagnostics were published in [3].

RESULTS

Ion Beam Properties

We performed a series of experiments to examine the properties of the laser accelerated ion beam. The majority of the accelerated ions on regular targets was found to be protons, as mentioned in the introduction. Those protons, originating from surface contaminants outrun all other ion species due to their higher charge-to-mass ratio and screen the accelerating field thereby gaining most of the energy stored in the field. The energy distribution resembles an exponential spectrum up to a sharp cut-off energy, in our experiments at about 25 MeV. The angular dependence of the energy distribution was measured with two ion spectrometers, positioned at an angle of 0° and 13° respectively. The maximum energy of about 25 MeV dropped to about 13 MeV at

13° normal to the target rear surface, consistent with the data from the RCF-stack. Details about the spectral distribution have been published in [4]. The ion beam pulse duration has been estimated to 10 ps as an upper limit due to the finfite lifetime of the accelerating electrons [5], but recent measurements using accelerated heavy ions provided more accurate data, indicating individual, much shorter pulse durations for different ion species[6], well within the limits published in [5].
The question about the ion origin has been the cause of an ongoing discussion, since there have been unambiguous experiments showing ion acceleration from the front surface[7], whereas other experiments clearly show ion emission originating from the rear surface[1,6,8].

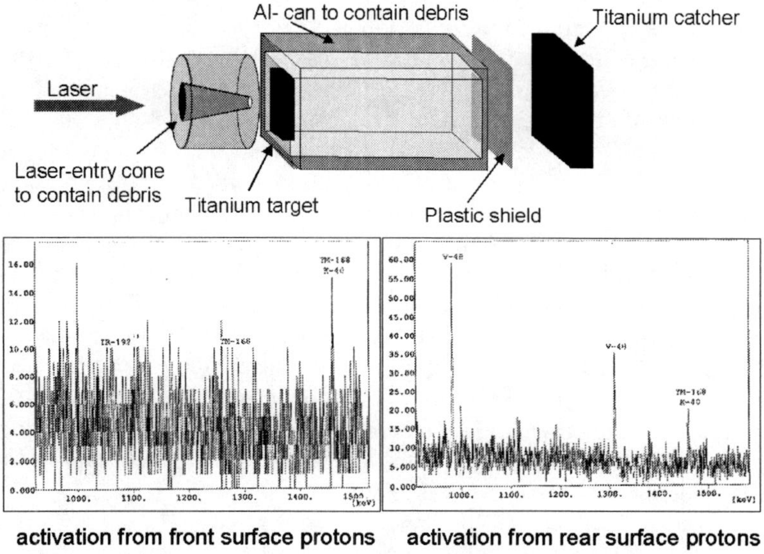

FIGURE 1. Nuclear activation of ^{48}Ti by protons from the front (left), and rear surface (right). The target was enclosed in a can to collect all the activated material from the target.

Using nuclear activation techniques, we have been able to determine the composition of the ion beam with respect to front and rear surface ions for our laser parameters. The detailed analysis will be published elsewhere. For the high energy components (> 5 MeV) we used the nuclear activation of titanium, which has a sharp energy threshold of 5 MeV for the reaction ^{48}Ti(p,n)^{48}V. The experimental setup and the results are shown in figure 1. The 50μm Ti target was enclosed inside a can to collect all the activated target debris. A second Ti foil was positioned behind the target as a catcher foil, in a distance of a few cm, shielded by a thin plastic foil. Protons above the reaction threshold emitted from the front surface through the target would cause activation of the target material, whereas protons emitted from the rear surface would cause nuclear activation of the catcher foil. Due to the lifetime of about 15 days, low background gamma detectors could be used to measure the characteristic line emission of the ^{48}V reaction products of the titanium inside the can and the catcher foil. In figure

1 the gamma ray spectra for the front surface activation and the rear surface activation are shown. There were no lines detected for the target material, indicating negligible activation caused by the protons accelerated from the front surface. In contrast, the Ti catcher target for the rear surface emission clearly shows the presence of ^{48}V. These results are consistent with the measurements using the $^{10}B(d,n)^{11}C$ reaction at lower reaction thresholds, indicating a rather soft energy spectrum for the front surface protons which we will present elsewhere.
For future applications of laser generated ion beams the beam quality is the most important characteristic. Especially for the use as an ion source or the application as an inertial fusion (ICF) ignitor beam[9], the ion beam emittance is crucial. As is apparent from the radiochromic film data the angular divergence of the beam is rather well defined and decreases with increasing proton energy. This suggests that the protons or other light ions accelerated by this mechanism may have a usefully small emittance in the sense of an actual ion beam.
We used penumbral imaging of edges at different distances to the target with the magnetic spectrometers to determine the core emittance of the proton beam. We determined the normalized emittance of the proton beam from flat gold targets to be ~0.2 π mm mrad, and a factor of at least two smaller than the resolution limited measurements we performed in [1]. Using a newly developed technique that will be published in detail in [10] (patent pending) we have been able to further improve the measurement of the beam core emittance by an order of magnitude. These measurements result in a transverse beam emittance of 0.06 π mm mrad, which is orders of magnitude lower than achieved at any conventional accelerator. Using the same technique, we also have been successful to determine the real source size of the laser accelerated proton beam. In contrast to the results obtained by penumbral imaging, the real source size of the proton beam extends to much larger areas and is strongly dependent on the ion energy, as shown in figure 2.

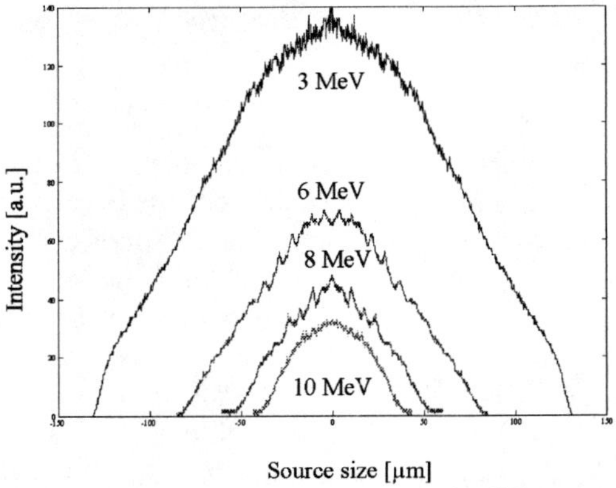

FIGURE 2. Real source size of the laser accelerated protons for different proton energies. The size at the rear surface decreases from 260 μm for 3 MeV protons to 80 μm for 10 MeV protons.

This is consistent with numerical simulations [11,12], as well as with recent measurements using imaging techniques to determine the electron distribution [13].

Dependence on target conditions

One of the interesting features of the laser accelerated ion beams is their directionality always normal to the rear surface. Structures at the target rear surface strongly influence the spatial distribution of the ion beam, as published in [3]. We have shown that beam tailoring by appropriate target design is possible, including defocusing and focusing of proton beams. In addition to the structure of the target, the imprint of the laser beam at the front surface can cause a deformation of the electron distribution at the rear surface, affecting the ion beam formation. In recent experiments we demonstrated that a strongly astigmatic focus can form an ion beam, where the orientation of the beam ellipse is perpendicular to the laser spot, consistent with numerical simulations based on the TNSA mechanism. An example is shown in figure 3, where we used an astigmatic focus to generate an elliptical proton beam from a flat target.

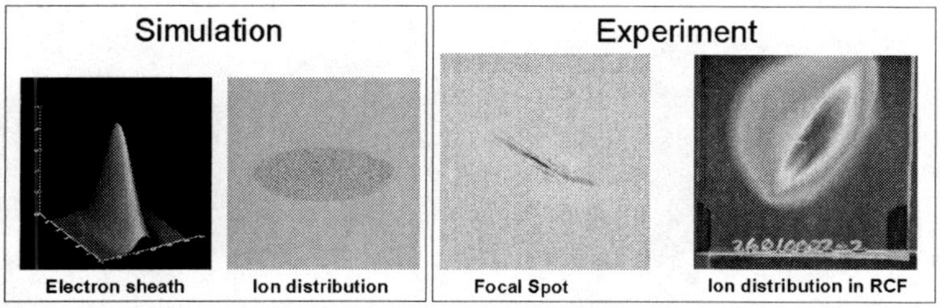

FIGURE 3. Effect of laser imprint on the ion beam shape. Computer simulations are compared to experimental results for a strongly astigmatic laser focal spot.

Proton radiography

The excellent beam quality of the ion beam is ideally matched to the requirements for imaging techniques. One scheme of particular interest is the use of laser accelerated protons to radiograph macroscopic samples to study their properties. Due to the different interaction mechanism protons can provide complementary information to techniques like x-ray backlighting. Because of the copious amounts of protons accelerated in a very short time, laser accelerated protons provide a new diagnostic quality in research of transient phenomena. We performed experiments to demonstrate the feasibility of these laser accelerated proton beams for radiography applications. In our experiments, we used a compound target of different materials to be imaged by the protons. It consisted of an 1 mm thick epoxy ring structure, several copper wires of 250 μm diameter, a hollow cylinder with 300 μm walls of steel, several Ti sheets of 100 μm thickness and a glass hemisphere of 900 μm diameter and 20 μm wall thickness. The protons were recorded in multiple layers of RCF to detect the image at different proton energies. Fig.4 shows the radiography of the target for final proton energies of 7.5 MeV. The image basically constitutes a negative image of the areal density of the target. The names of the collaborating institutes have been engraved on the epoxy ring, which results in a reduced thickness and therefore a higher energy deposition of the protons in the respective layer.

FIGURE 4. Radiography of a compound target. Details of the target are given in the text.

The areal density variation of the hollow cylinder, including a hole in the wall on the right hand side can be seen as well as a thin metal rod placed inside the cylinder. Close examination also shows the small glass hemisphere placed above the cylinder. The time of exposure in this experiment was estimated to be in the order of tens of picoseconds.

CONCLUSION

We have presented a detailed investigation of the target conditions on the proton and ion beam production from intense laser solid interactions. The observed strong dependence on the rear surface conditions in agreement with the target normal sheath acceleration mechanism. The laser spot imprint appears to have a major influence on the quality of the ion beam, and the quality of the surface finish of the target is very important for maintaining a high gradient sheath and a laminar beam. It has been shown that tailoring the ion beam (yield, shape, composition, homogeneity) by means of target shape and composition is possible.

Supported by the EU N^0 1999-0052 and in part by grant E1127 from Région Ile-de-France.

---

[1] R. Snavely, et al., Phys. Rev. Lett. **84**, 640 (2000)
[2] S. C. Wilks, et al., Phys. Plasmas **8**, 542 (2001)
[3] M. Roth, et al., Phys. Rev. ST-AB, in print
[4] M. Roth, et al., AIP Conference Proceedings, Vol. 611, ISBN 0-7354-0057-1, p. 199 (2002)
[5] Hatchett et al., Phys. of Plasmas **5**, Vol. 7, 2076, 2000
[6] M. Hegelich, et al., submitted to Phys. Rev. Lett. (2002)
[7] A. Maksimchuk, et al., Phys. Rev. Lett. **84**, 4108 (2000).
[8] A.J. MacKinnon, et al., Phys.Rev. Let. **86**, 1769 (2001)
[9] M. Roth, et al., Phys. Rev. Lett. **86**, 436 (2001)
[10] T.E. Cowan, et al., to be submitted to Nature
[11] H. Ruhl, et al., Plasma Phys. Reports **27**, 363 (2000)
[12] A. Pukhov, Phys. Rev. Lett. **86**, 3561 (2001)
[13] J. J. Santos, et al., Rapport LULI 2000, 19 (2000)

Simulations of Laser-Produced Electron Propagation in a Thin Foil

Dale R. Welch

Mission Research Corporation, Albuquerque, NM 87110 USA

Abstract. Energetic electrons are produced in the interaction of a petawatt laser with a plasma near the critical surface. The propagation of an intense electron beam through an ionized solid-density slab using a hybrid fluid-kinetic electromagnetic simulation code is examined. These calculations transport 250-1000 J of electrons with 1-5 MeV directed energy. The electrons are produced in a spot diameter of 30 microns with duration of 5 ps. The simulations ran for up to 33 ps. The background material remains fully-ionized and dissociated CH_2 (or fully-ionized Al) with an initial 5-eV temperature. All coulomb collisions are followed. All species are modeled with kinetic equations except for the fluid electrons laid down at the start of the calculation. These simulations neglect all inelastic collisions, including ionization and excitation. Results from both two and three-dimensional simulations of electron transport in solid-density plasma and in vacuum are presented.

INTRODUCTION

Energetic electrons are produced in the interaction of a petawatt laser with a plasma near the critical surface.[1,2,3] In the fast ignitor[4] concept, these electrons or secondary ions are being considered to quickly heat a relatively cool, dense hydrogen plasma to thermonuclear temperatures. Another application is the production of x-rays in a small spot for x-ray radiography. A number of authors have studied the production and transport of fast electrons using various simulation methods.[5,6,7,8,9,10] In this paper, the propagation of an intense electron beam through an ionized solid-density slab using a hybrid fluid-kinetic electromagnetic simulation code is examined. Here, electron transport in a dense plasma and the transition to vacuum can be modeled simultaneously. These calculations transport 250-1000 J of electrons with a longitudinal energy of 1-5 MeV and transverse temperature 300-1000 keV. The electrons are produced in a spot of 30-μm diameter with duration of 10 ps. The background material remains fully ionized and dissociated CH_2 (or Al) with an initial 5-eV temperature. All coulomb collisions are followed.

In this paper, we discuss the results of both 2-D cylindrical and 3-D Cartesian simulations using the LSP[11] particle-in-cell (PIC) code. The beam and plasma evolution are calculated as well as the generation of fast protons. The key features of the simulations include the propagation of the beam at roughly the Alfven current and virtual cathode formation at the vacuum interfaces. The beam is found to quickly exhibit both radial and azimuthal structure consistent with the instability of a highly

CP650, *Beams 2002: 14th International Conference on High-Power Particle Beams,*
edited by T. A. Mehlhorn and M. A. Sweeney

current-neutralized beam. At the vacuum interface, the bulk of the beam electrons are reflected back at roughly 30 degrees back and directed up along the vacuum interface. Also, a cool energetic proton beam is produced, neutralized by a co-moving electron cloud, with pulse length roughly that of the electron beam.

An electron beam produced in a laser-plasma interaction is forward-directed due to the ponderomotive force of the laser light. The beams produced in petawatt laser plasma interactions can reach > 10^{18} W/cm^2 intensities. In the problems considered here, the beam is still considered weak with respect to the roughly 10^{23} cm^{-3} plasma electron density. The beam is < 1% of this density. If the beam is to propagate, the plasma must supply not only charge neutralization but also nearly perfect current neutralization (>99%). The plasma current density supplying the neutralization resistively decays at the characteristic $\tau_d = \pi\sigma r_b^2/c^2$ decay time where σ is the plasma conductivity and r_b is the beam (or filament) radius. Assuming Spitzer resistivity and a plasma ion of charge state Z, the decay time $\tau_d\ (s) \approx 5\text{x}10^{-7} r_b^2 T_e^{3/2}/(Z\ ln\Lambda)$. For a Z=1 plasma with 10^{23} cm^{-3} electron density at electron temperature T_e=100 eV and r_b=15 μm beam radius, τ_d is roughly 600 ps. This time permits net currents roughly 1% of the beam for a 5-ps pulse. Thus, for MeV electron energy (E_b) and currents >> 1 MA, the beam will propagate through net currents at or above the Alfven limit given by $I_{net}=I_A=17\gamma\beta$ kA. The decay time for smaller filaments is much shorter.

If we assume propagation at close to the Alfven limit, we can estimate several beam conditions. By definition, the matching condition of the beam or filament temperature must be $T_b \approx 0.511\gamma\beta^2$ MeV which is roughly the beam directed energy. This value suggests that, for almost any beam conditions, sufficient net current exists to pinch the beam. The electron betatron wavelength, $\lambda_\beta=2\pi r_b(I_A/I_{net})^{1/2} \approx 2\pi r_b$. Typically $\lambda_\beta/2\pi$ ($1/k_\beta$) relates to the e-fold growth distance of resistive instabilities such as hose (m=1) and filamentation (m>1). Assuming $I_{net}= I_A$, the electron cyclotron frequency $\omega_{ce} = eB/m_ec$ is roughly $6\text{x}10^{13}\gamma\beta/r_b\ s^{-1}$.

It is also useful to consider other relevant frequencies. For the above example, the dominant frequency is the plasma electron frequency $\omega_{pe} = (4\pi e^2 n_e/m_e c^2)^{1/2}$ of order $2\text{x}10^{16}\ Z^{1/2}\ s^{-1}$. This frequency places severe constraints on explicit simulation time step. The electron-ion collision frequency $\nu=4\text{x}10^{14} Z^2 (T_e/100\ eV)^{3/2}\ s^{-1}$. Thus, we expect the plasma electrons to remain collisional ($\omega_{ce}/\nu < 1$) under most conditions.

SIMULATION TECHNIQUES

A hybrid algorithm has been developed and implemented into the 3-D parallel LSP[11] PIC code to address the issues concerning charged-particle beam interaction with high neutral and plasma densities. In particular, an energy-conserving implicit algorithm including particle collisions and gas breakdown models was added. Plasma electrons are described with PIC methods using either fluid or kinetic equations. The implicit algorithm version is particularly useful for very dense plasma where the details of electron plasma oscillations can be ignored. Coulomb scattering is modeled in Monte Carlo fashion with energy and charge-state dependent cross-sections. The collision probability is determined by the Spitzer collision frequency. This algorithm has also been employed for a number of beam transport problems including neutralized ballistic[12] and self-pinched chamber transport[13,14] of heavy ion beams in the 10–100-mTorr pressure regime.

TWO-DIMENSIONAL ELECTRON TRANSPORT SIMULATIONS

The details of our nominal 2-D (r-z) cylindrical simulation are as follows. 250 J of 1-MeV longitudinal energy electrons are injected at the front surface of a 200-µm thick (z = 0, r < 15 µm), 500-µm radius Al or plastic slab. Vacuum regions extend 300 µ behind the slab and 100 µm in front. Conducting walls bound the entire simulation box. The nominal electron beam has a 300-keV transverse temperature and 48-MA peak electron current injected in a uniform density cylinder. The pulse is 10 ps in duration (triangular 5-ps rise/fall). The plasma is initialized at 5 eV with fixed charge-state ions. Simulations with Al ions were run with charge states +1-4. The plastic slab was initialized as a fully dissociated $C^+H_2^+$ plasma. Spatial resolution is 2.5 µm radially and 5 µm axially. The temporal resolution is $c\Delta t = 0.25$ µm.

First considered is a beam produced by promoting the local plasma electrons to the nominal energy in a thin layer at the z=0 front face of an Al slab with +1 ions. The beam density and net current at 6.67 ps are shown in Figure 1. The beam initially propagates forward and filaments radially before reaching a 40-µm outer radius at the back vacuum interface (z=200 µm). The bulk of the beam rebounds off the virtual cathode and propagates backward in an annulus striking the front face at a 120-µm radius. The beam continues to rebound off alternating interfaces while it works its way up the slab. As it strikes the interfaces, a portion of the beam is also directed along the interface itself (mostly upward). The net currents push the Alfven limit for electron propagation. The net current, confining the beam annuli, is produced in thin annular regions increasing with time and propagation distance with magnitudes exceeding 100 kA, well above I_A =50 kA. The captured beam electrons have been compressed within these net current channels so that they traverse a net current < I_A. A small fraction of the beam electrons that sample a net current above I_A is diverted out of the channels.

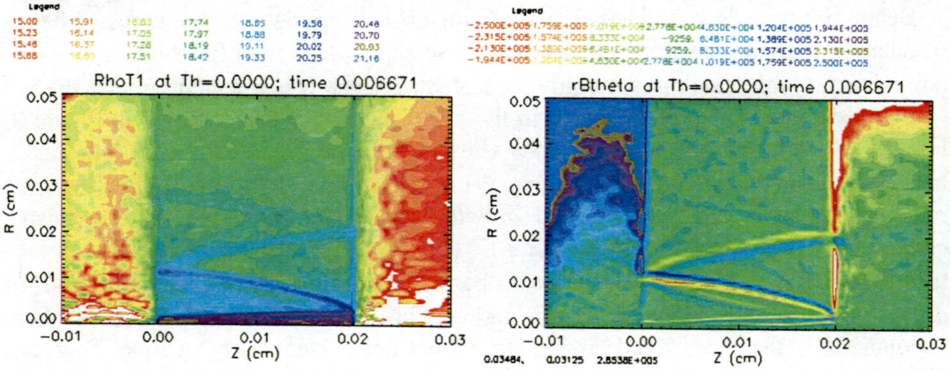

FIGURE 1. The log of the beam density (left) and the net current (amperes) within radius R are plotted 6.67 ps into the 2-D LSP simulation with an Al^{+1} plasma.

The plasma electron temperature has a peak above a keV (artificially high due to lack of inelastic losses) near the beam production region, falling off rapidly with radius. By 6.67 ps, the plasma ions barely exceed 10 eV and are not in thermal equilibrium with the rapidly heated electrons in this region. The initial backward annular path of the beam channels is heated to roughly 20 eV.

When the same beam is produced in a plastic slab, we see a hydrogen beam emerging from the back face at small radius. The motion of this material is coincident with the beam power at the vacuum interface, almost shutting off after 10 ps. The mean energy of the protons is roughly that of the beam (1 MeV) but particle energies extend out to nearly 4 MeV. The beam current right off the surface is on the order of 1 MA falling off with z due to the proton energy spread. The radius (and emittance) of the protons is quite small, $< 30\ \mu m$ at $z = 500\ \mu m$. The small change in the minimum proton radius at the two axial positions suggests the protons are well space-charge neutralized by the available electrons.

THREE-DIMENSIONAL TRANSPORT SIMULATIONS

Now presented are the results of several 3-D, 10-ps duration calculations using LSP in Cartesian coordinates. Given the collisionality of the plasma, the main focus of these simulations is the growth of the resistive filamentation instability.[15] The collisionless Weibel[16] instability requires $10^{-3}\ \mu m$ spatial resolution (plasma skin depth) which is currently beyond our computational capabilities. From the 2-D radial breakup of the beam and recent computational work,[10] the filamentation instability is expected to manifest strongly in azimuth. To speed the calculations, the size of the problem is reduced to a 200-μm radius slab (still with a 200-μm thickness) and 100-μm vacuum regions in front and behind the slab. The number of particles in the simulations still exceeds 8 million. Furthermore, only half of the Al slab is zoned up with a symmetry boundary at the y=0 plane. Otherwise, the beam production and plasma initialization conditions are the same as those of the 2-D simulations discussed in the previous section.

Beam behavior is examined in slabs with different resistance by varying the Al plasma ion charge state from +1 to +4. In these simulations with fixed beam injection parameters, ohmic heating is roughly a constant. Plasma σ is affected only by the ion charge state. Snapshots of the beam in the x-z plane at y = 0 are shown in Figure 2. For 250-J beam energy and Al^{+1} ions (Figure 2a), the beam filaments into small, barely resolved beamlets with roughly 3-μm radii. The structure of the filaments remains frozen in time after initial growth due to the rapid rise in the plasma conductivity from ohmic heating. For Al^{+2} (Figure 2b), the filament radius increases to 6 μm. For Al^{+4} (Figure 2c), the beam basically remains in one piece with a slight dipole motion and constant 15-μm radius. The key difference in these three simulations is the more rapid rise in the global net current early in time (first ps). Thus, the instability growth in the simulations appears to be reduced by the presence of self-magnetic fields with scale length of the order of the beam size. Finally, in Figure 2d, the electron energy is increased to 1 kJ (or 200-MA peak current). We can see that the instability reappears with significant filamentation and beam spreading. This re-emergence suggests that the instability threshold is a function of the current neutralization fraction or ohmic heating rate that must increase with increasing beam current.

As in the 2-D simulations, because these simulations neglect inelastic collisions, the rise in plasma temperature is overestimated. Lower temperatures (from inelastic collisions) and anomalous resistivity would decrease the plasma conductivity and, given the simulation stability scaling with ion charge state, might actually preserve the integrity and size of the electron beam as it propagates.

CONCLUSIONS

The interaction of a very intense electron beam in a solid-density slab of ionized and dissociated Al and plastic has been simulated. The key features of the simulations include filamented propagation of the beam at near-Alfven current. At the back surface of the slab, the virtual cathode formation drives several interesting phenomena. The beam electrons are mostly reflected backward or up along the vacuum interface. The growth of the filamentation instability is found to be a function of the plasma resistivity. Curiously, the more resistive the initial plasma, the better the quality of the beam spot at the back interface.

In the plastic slab simulations, a beam of protons is produced with an initial pulse length roughly that of the electron beam. These protons have nearly 1-MA current with a mean energy roughly that of the electron beam. The protons are space-charge neutralized by a co-moving electron cloud and propagate forward in vacuum. The beam is produced with low emittance and a 30-μm radius. Work is currently in progress[17] to study scaling of the ion beam production with the goal of driving fast ignition with such a beam. In future work, we intend to explore the possibility of increasing the ion beam production efficiency by manipulation of the virtual cathode sheath position as in the Ionization Front Accelerator.[18] Future simulation improvements will include time-dependent field and collisional ionization. Benchmarking LSP against experimental observations is also a high priority.

ACKNOWLEDGMENTS

Thanks should be extended to Bryan Oliver, David Rose, Robert Clark, Thomas Mehlhorn, Robert Campbell, Craig Olson, John Degroot, Max Tabak and Barbara Lasinski for their support and useful discussion. This work was supported by Sandia National Laboratories and Livermore National Laboratory.

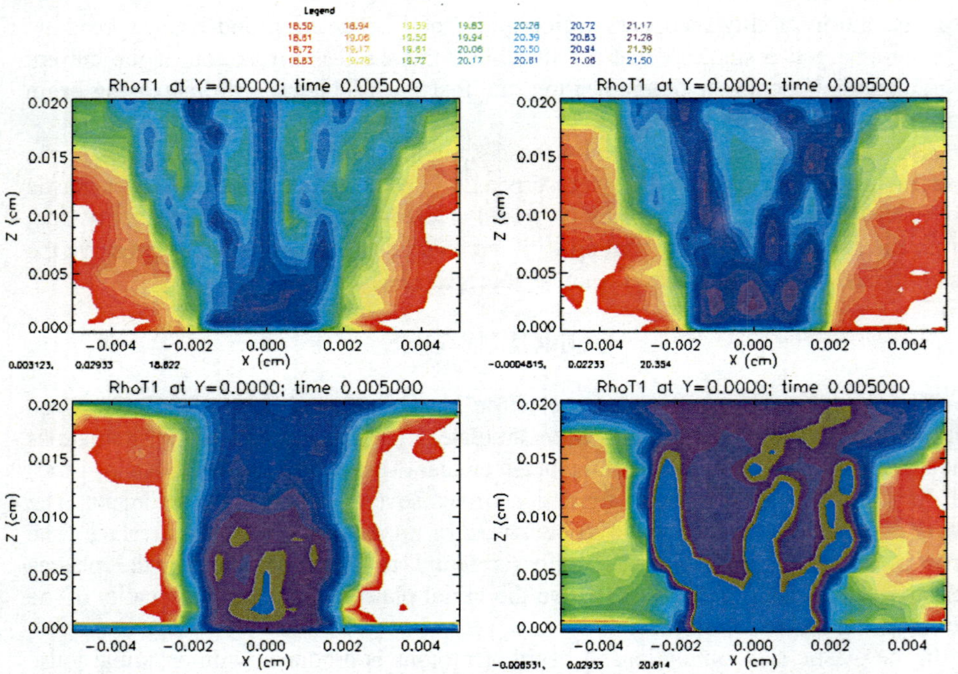

FIGURE 2. The log of the beam density (see legend above) 5 ps into the 3-D LSP simulation is plotted in the x-z plane at y=0. In (a), the total electron energy is 250 J with an Al^{+1} plasma; (b), an Al^{+2} plasma; (c), an Al^{+4} plasma; and in (d), the total electron energy is 1 kJ with an Al^{+4} plasma.

REFERENCES

[1] M. H. Key, M. D. Cable, T. E. Cowan, et al., *Phys. Plasmas* **5**, 1966 (1998).

[2] W. L. Kruer, *Phys. Plasmas* **7**, 2270 (2000).

[3] S. C. Wilks, A. B. Langdon, T. E. Cowan, et al., *Phys. Plasmas* **8**, 542 (2001).

[4] M. Tabak, J. Hammer, M. E. Glinsky, et al., *Phys. Plasmas* **1**, 1626 (1994).

[5] A. Pukhov, and J. Meyer-ter-Vehn, *Phys. Plasmas* **5**, 1880 (1998).

[6] B. F. Lasinski, A. B. Langdon, S. P. Hatchett, et al., *Phys. Plasmas* **6**, 2041 (1999).

[7] K. Witte, C. Gahn, J. Meyer-ter-Vehn, et al., *Plasma Phys. Control. Fusion* **41**, B221 (1999).

[8] M. Honda, J. Meyer-ter-Vehn, and A. Pukhov, *Phys. Plasmas* **7**, 1302 (2000).

[9] S. Weber, G. Bonnaud, and J.-C. Gauthier, *Phys. Plasmas* **8**, 387 (2001).

[10] L. Gremillet, G. Bonnaud, F. Amiranoff, *Phys. Plasmas*, **9**, (2002) 941.

[11] T. P. Hughes, S. S. Yu, and R. E. Clark, Phys. Rev. ST-AB **2**, 110401 (1999); D. R. Welch, D. V. Rose, B. V. Oliver, and R. E. Clark, *Nucl. Instrum. Meth. Phys. Res.* A **464**, 134 (2001).

[12] D. V. Rose, et al., *Nucl. Instrum. Meth. Phys. Res. A* **464**, 299–304 (2001).

[13] D. R. Welch and C. L. Olson, *Fusion Engineering and Design* **32-33**, 477 (1996).

[14] D. V. Rose, et al., *Phys. of Plasmas* **6** (1999) 4094.

[15] T. P. Hughes and H. S. Uhm, *J Appl. Phys.* **60** (1986) 577.

[16] E. S. Weibel, *Phys. Rev. Lett.* **2**, (1959) 83.

[17] R. B. Campbell, presented at the ACC2002 Conference (2002).

[18] C. L. Olson, et al, Phys. Rev. Lett., **56**, 2260 (1986).

Conceptual Design of Laser Lightning Protection System with Crossing Optical Breakdown Channel[1]

V.D.Selemir, A.V.Ivanovsky, V.I.Karelin, P.B.Repin, V.T.Selyavsky, R.V.Savchenko, T.V.Grebenev, P.S.Dadykin

RFNC-VNIIEF, Sarov, Russia

Abstract Conceptual design of laser lightning protection is developed. Creation of lightning rod is provided by long laser spark discharge, directed along two crossing channels. The discharge is powered by current source, located on the ground. Current pulse shape is optimum for lightning rod creation and has the necessary parameters. Principle possibility of the scheme realization was checked in model experiments with lightning rods of meters scales. Requirements to parameters of laser radiation, focusing system and current source for laser lightning protection realization with lightning rod of ~100 m height are formulated.

INTRODUCTION

Laser lightning protection is one of the applications of laser directed discharges (LDD) [1]. Works are being performed in this direction [2, 3]. However, its realization has a number of technical troubles. In particular, there are problems with lightning channel overcommutation to LDD channel in case of the main type of lightning discharge – negative lightning [2].

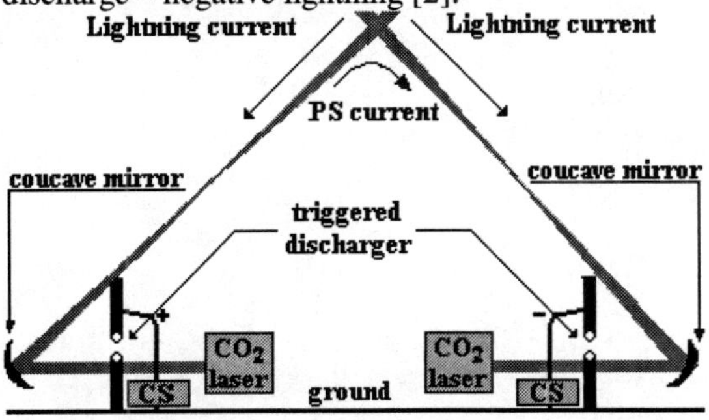

FIGURE 1. Conceptual design of laser lightning protection

In this paper we consider the method of lightning protection with lightning rod created by the discharge, powered from current source (CS), located on the ground (see fig. 1) and directed along crossing laser sparks. Protection of the CS is provided by controlled discharger, catching lightning current.

Parameters of created highly-conducting channel do not depend on lightning discharge characteristics. That is why the system can provide reliable capture of both negative and positive lightning. Principle efficiency of the scheme was tested in

[1] Supported by ISTC project № 880

CP650, *Beams 2002: 14th International Conference on High-Power Particle Beams*, edited by T. A. Mehlhorn and M. A. Sweeney

experiments with lightning rods of meter scale. Results of such experiments are presented below.

EXPERIMENTAL EQUIPMENT

Experimental bench consisted of CO_2-laser (radiation wave length is 10.6 μm). Diameter of resonator mirror was 200 mm. Laser radiation (LR) energy was varied in the range of 0.3÷2 kJ at reproducibility of ~20%. Total duration of LR was ~1 μs, divergence was not higher than $2x10^{-3}$ rad.

Capacity bank (CB) was used for LDD powering. It consisted of two five-cascades branches, connected in a way to provide different voltage pulses polarity on the output of branches relatively the central grounded point. Diversity of branches operation was not higher than 1 μs at total operation time of 1 μs. In idle regime GPV provides voltage pulses of positive and negative polarity of the amplitude of ~300 kV with decay time of ~500 μs (see fig. 2).

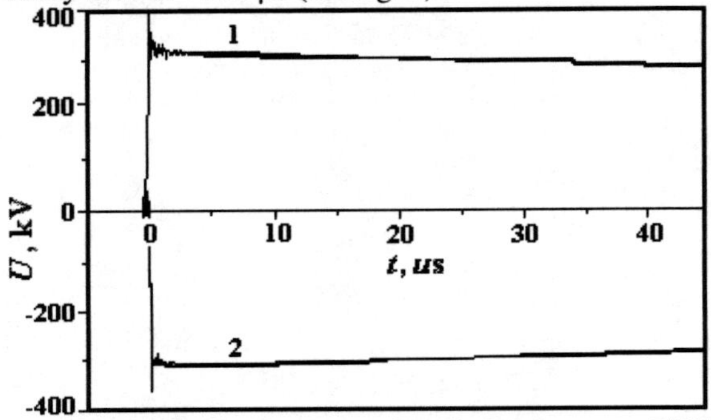

FIGURE 2. Voltage pulses oscillograms on CB output in idle regime:
1 – positive polarity pulse;
2 – negative polarity pulse.

In the experiments we controlled voltage on GPV output and LDD discharge gap with voltage attenuators, discharge current with Rogovsky coils, space-time luminescence pattern with fast photo-recorder (FPR) in the regime of slit and frame scanning.

Triggering of the equipment was performed from FPR trigger pulse. CB triggering delays relatively to LR pulse were varied from 0 up to ~200 μs.

EXPERIMENTS WITH SINGLE OPTICAL BREAKDOWN CHANNEL

Optical breakdown was created by LR focusing with spherical mirror, having focal distance of 10 m. LDD electrodes were located along the beam in the place where intense luminescence of plasma optical breakdown is recorded. CB triggering delay relatively to LR was ≤1 μs and ~30 μs.

In the experiments with single laser channel is was established that:

- at used LR parameters the reliably recorded length of optical breakdown channel increases 2 m, typical space-time scanning of laser spark nidi is presented on Fig. 3;

- at CB triggering delay relatively LR pulse of ≤1 μs it was impossible to get increase the length of air gap breakdown, which is equal to ~1 m without laser spark at voltage of ~600 kV;

- increasing CB triggering delay relatively to LR up to ~30 μs we observed stable breakdown of the gaps up to 2.1 m on the GPV voltage pulse front.

FIGURE 3. Photo-chronogram of laser spark luminescence, obtained in the regime with slit scanning.

Analysis has showed that discharge propagated better on ionization nidi, formed by LR, at voltage pulses with steep front, than at quazi-stationary voltage. At front of ~1 μs breakdown voltage is could be decreased twice. This points at necessity of creation of effective lightning protection, based on breakdown initiation with laser spark channel in quasi-stationary electrical field of thundercloud.

That is why we have developed the scheme of lightning protection with crossing channels of optical breakdown. Principle possibility of the scheme realization was checked in model experiments with lightning rods of meters scales.

EXPERIMENTS WITH CROSSING CHANNELS

Goal of the experiments is to clarify possibility of LDD creation in electrical field with force lines, which do not coincide with discharge propagation direction. Functional scheme of the focusing system is presented on Fig. 4.

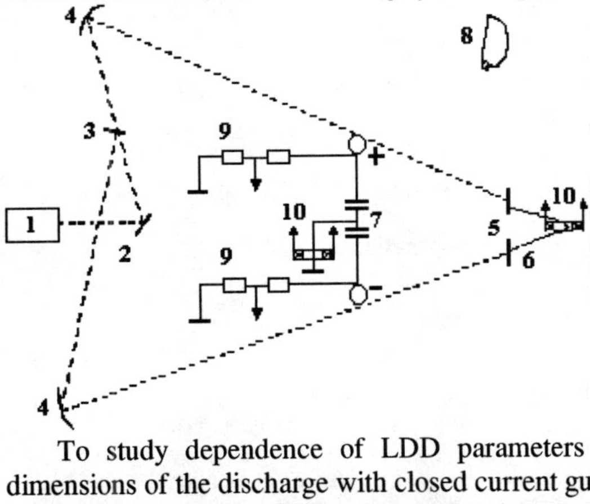

FIGURE 4. Scheme of experiments with crossing laser channels: 1 - laser; 2,3 - system of plane mirrors; 4 - focusing spherical mirrors; 5 - metal plates; 6 - current guide; 7 - bipolar generator of pulsed voltages; 8 - fast photo-recorder; 9 - resistive voltage attenuators; 10 - Rogovsky coil.

To study dependence of LDD parameters on the length we adjusted linear dimensions of the discharge with closed current guide 6 by metal plates 5.

Distance between high-voltage ball contacts of the voltage attenuators of ~1.2 m prevented breakdown between them even at GPV operation in idle regime. Distance from high-voltage contacts of voltage attenuators up to the plates (on laser beam) was varied from 0.8 up to 1.3 m, GPV triggering delay relatively LR pulse was varied from 30 up to ~200 μs.

The main result of the experiments – demonstration of ability to work of the proposed scheme. In the majority experiments we recorded breakdown on laser spark crossing channels in the studied range of inter-electrode gaps and delays of voltage supply on the gap relatively to LR. Typical fragments of frame scanning of the laser spark luminescence and discharge along its channels are presented on Fig. 5.

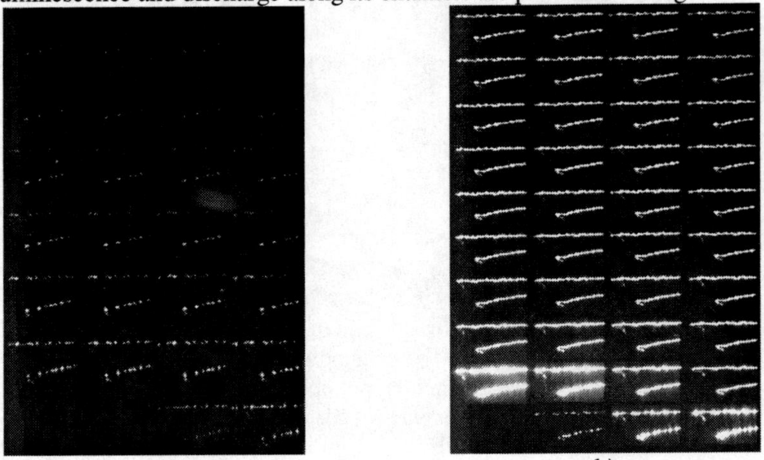

a) b)

FIGURE 5. Fragments of frame scanning of the laser spark luminescence – a) and discharge, directed along its channels – b).
Duration of each frame and motion period – 0.5 μs, frames sequence order: in rows - from left to right, rows are passing from top to bottom.

CONCLUSION

In model experiments we confirmed ability of work of laser lightning protection with crossing channels of optical breakdown.

Laser sparks of ~100 m length could be created by CO_2-lasers with energy of ~5 kJ using mirrors with focal distance of ~150 m and LR divergence improvement up to 10^{-4} rad.

Taking into account threshold of air leader breakdown of ~4-5 MV and obtained decrease of breakdown voltage up to two times, necessary for LDD formation on the base of ~100 m, CB output voltage does not exceed 2÷2.5 MV.

REFERENCES

1. L.M.Vasilyak Laser Directed Electrical Discharges // Proceedings of the Conference on Low-Temperature Plasma PLTP-98. Petrozavodsk: PSU, 1998, V.2, P.135-156.
2. M. Miki, A. Wada. T. Shindo and Y. Aihara "Basic study of laser-triggered lightning" Proceedings of 10th Int.Conf. Atmospheric Electricity, Osaka, Japan, 1996, pp. 656-659.
3. S. Uchida, Y. Shimada, H. Yasuda, K. Tsubakimoto, S. Motokoshi, C. Yamanaka Z-I. Kawasaki, T. Yamanaka, K. Matsu-ura, T. Ushio M. Adachi, Y. Ishikubo "Laser triggered lightning experiments in the field" Proceedings of 10th Int.Conf. Atmospheric Electricity, Osaka, Japan, 1996, pp. 660-663.

APPENDICES

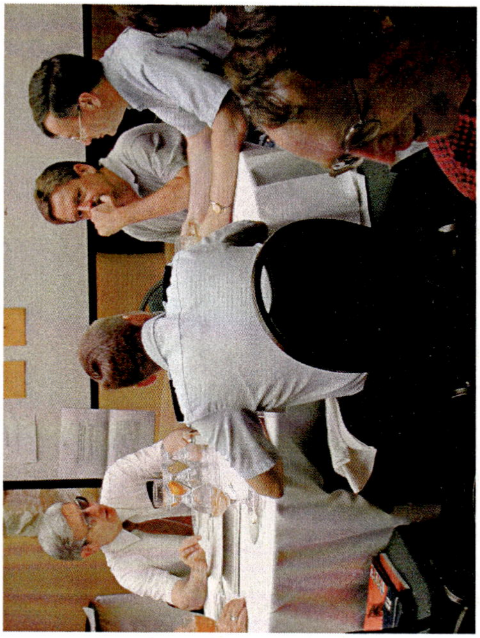

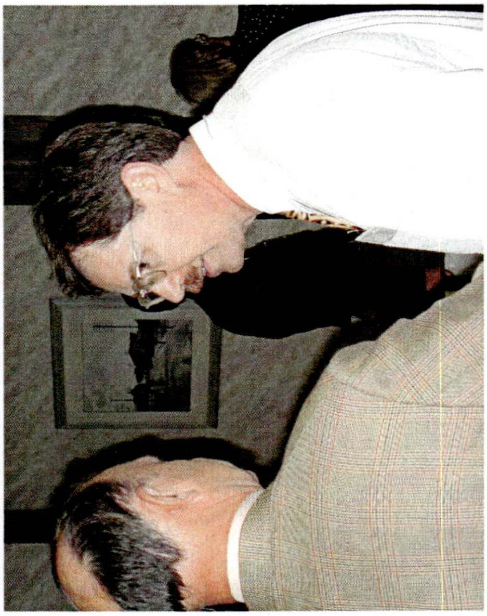

Hilton

List of Participants

Aase, Nicholas
neaase@sandia.gov

Abubakirov, Edward
edward@appl.sci-nnov.ru

Abyar Monfared, Kashani Mehrdad
kashani@phys.cst.nihon-u.ac.jp

Allen, Raymond
allen@suzie.nrl.navy.mil

Alliot, Jean-Claude
alliot@onera.fr

Ampleford, David
davidampleford@ic.ac.uk

Ando, Ritoku
ando@plasma.s.kanazawa-u.ac.jp

Andrey/Petrovich, Orlov
ivanovsky@ntc.vniief.ru

Andrey/Vladimirovich, Ivanovsky
ivanovsky@ntc.vniief.ru

Apruzese, John
apruzese@ppd.nrl.navy.mil

Armijo, Joseph
jcarmijo@lanl.gov

Arzhannikov, Andrei
arzhannikov@inp.nsk.su

Bailey, Jim
jebaile@sandia.gov

Bailey, Vernon
vbailey@titan.com

Baksht, Rina
baksht@ovpe2.hcei.tsc.ru

Barker, Robert
robert.barker@afosr.af.mil

Barnea, Gideon
barnea@actcom.co.il

Baronova, Elena
baronova@nfi.kiae.ru

Bauer, Bruno
kari@physics.unr.edu

Beezhold, Wendland
beezholt@physics.isu.edu

Bell, David
david.bell@ieee.org

Belyaev, Vadim
VadimBelyaev@mtu-net.ru

Bezdek, Milan
mbezdek@vtupv.cz

Bland, Simon
sn.bland@ic.ac.uk

Blaugrund, Abraham
fnbla@weizmann.ac.il

Bliss, David
debliss@sandia.gov

Bloomquist, Douglas
ddbloom@sandia.gov

Bluhm, Hansjoachim
bluhm@ihm.fzk.de

Brinsmead, William
kari@physics.unr.edu

Brzosko, Jan
brzoskoj@diana0hitech.com

Burtsev, Vladimir
burtsev@niiefa.spb.su

Buttram, Malcolm
mtbuttr@sandia.gov

Bystritskii, Vitaly
vbystrit@uci.edu

Bystritsky, Viacheslav
bystvm@nusun.jinr.ru

Cagliostro, Dominic

Caporaso, George
caporaso1@llnl.gov

Carey, William
carey@apelc.com

Cartwright, Keith
Keith.Cartwright@kirtland.af.mil

Cassany, Bruno
cassany@bordeaux.cea.fr

Castagno, Scott
sgcasta@sandia.gov

Cevallos, Michael
mic-dana@swbell.net

Chaikovsky, Stanislav
stas@ovpe.hcei.tsc.ru

Chalise, Priya
chalise@hotta.es.titech.ac.jp

Chandler, Gordon
gachand@aol.com

Chandler, Katherinie
kmc25@cornell.edu

Chantrenne, Sophie
schantrenne@titan.com

Chapman, Randall
Randy.Chapman@arnold.af.mil

Chen, Yanqiao
yqchen@tsinghua.edu.cn

Chernenko, Andrey
kingsep@dap.kiae.ru

Chittenden, Jeremy
j.chittenden@ic.ac.uk

Choi, Junho
choi@kingtrabu.nrl.navy.mil

Choi, Song Hun
samchoi@unm.edu

Chong, Young
chong@ppdmail.nrl.navy.mil

Chornyi, Valentin
chorny@pht.univer.kharkov.ua

Christenson, Peggy
pjc@lanl.gov

Chuaqui, Hernan
hchuaqui@fis.puc.cl

Clark, Robert
clark@ppdmail.nrl.navy.mil

Clough, Stephen
stephen.clough@awe.co.uk

Cochran, Frederick
flc@lanl.gov

Coleman, Philip
plcoleman@mailaps.org

Commisso, Robert
commisso@suzie.nrl.navy.mil

Cook, Donald
colson@sandia.gov

Cooper, Graham
graham.cooper@awe.co.uk

Cooperstein, Gerald
cooperstein@nrl.navy.mil

Corcoran, Patrick
pcorcoran@titan.com

Cotter, Timothy
tim.cotter@arnold.af.mil

Coverdale, Christine
cacover@sandia.gov

Crotch, Ian
ian.crotch@awe.co.uk

Cuneo, Michael
mecuneo@sandia.gov

Curry, Randy

Dale, Gregory
daleg@missouri.edu

Dannenberg, Korbie
korbster@umich.edu

Dasgupta, Arati
dasgupta@ppdmail.nrl.navy.mil

Davanloo, Farzin
fdavan@utdallas.edu

Davis, Jack
davisj@ppdmail.nrl.navy.mil

Davis, Harold
davis@lanl.gov

Davis, Randolph
randy.davis@dtra.mil

De Groot, John
jsdegroot@ucdavis.edu

Deeney, Chris
cdeene@sandia.gov

Deng, Jianjun
bcys@caep.ac.cn

Desjarlais, Michael
mpdesja@sandia.gov

Dickens, James
jdickens@coe.ttu.edu

Dobbins, Aaron
ahdobbi@sandia.gov

Dolgachev, Georgy
kingsep@dap.kiae.ru

Dovbnya, Anatoliy
zakutin@kipt.kharkov,ua

Droemer, Darryl
Dwdroem @sandia.gov

Duselis, Peter
pud2@cornell.edu

Dutkowski, Eugene
dutkowski_joe@crane.navy.mil

Egorov, Oleg
egorov@triniti.ru

Ekdahl, Carl
cekdahl@lanl.gov

Elegure, Folorunso
statemin@yahoo.com

Engelko, Vladimir
engelko@niiefa.spb.su

Faehl, Rickey
rjf@lanl.gov

Faretto, Harold
kari@physics.unr.edu

Favre, Mario
mfavre@fis.puc.cl

Feduschak, Vladimir
ratakhin@ovpe.hcei.tsc.ru

Filatov, Alexander
Filatov@iep.uran.ru

Fisher, Amnon
amfisher@physics.technon.ac.il

Freeman, Bruce
bruce.freeman@ne.tamu.edu

Frese, Michael
Michael.Frese@numerex.com

Frese, Sherry
Sherry.Frese@numerex.com

Frey, Wolfgang
wolfgang.frey@ihm.fzk.de

Fridman, Boris
fridman@mail.infostar.ru

Fuelling, Stephan
fuelling@physics.unr.edu

Fuks, Mikhail
fuchs@eece.unm.edu

Gahl, John
gahlj@missouri.edu

Garanin, Sergey
sfgar@vniief.ru

Garasi, Chris
cjgarasi@sandia.gov

Genoni, Thomas
genoni@mrcabq.com

Gericke, Dirk
gericke@lanl.gov

Gilgenbach, Ronald
rongilg@umich.edu

Ginzburg, Naum
ginzburg@appl.sci-nnov.ru

Giuliani, John
giul@ppdmail.nrl.navy.mil

Glazyrin, Igor
I.V.Glazyrin@vniitf.ru

Goldsack, Timothy
tim.goldsack@awe.co.uk

Goldstein, Steven
goldstsa@nv.doe.gov

Golub, Tatiana
gng_nt@sprynet.com

Juan, Carlos
jhernand@ttacs.ttu.edu

Hinshelwood, David
ddh@suzie.nrl.navy.mil

Holt, Thomas
tinyball@yahoo.com

Hoppe, Peter
hoppe@ihm.fzk.de

Horioka, Kazuhiko
khorioka@es.titech.ac.jp

Hu, Min
mh99@cornell.edu

Hughes, Thomas
tph@mrcabq.com

Ivanovsky, Andrey
ivanovsky@ntc.vniief.ru

Jennings, Chris
christopher.jennings@ic.ac.uk

Jiang, Weihua
jiang@nagaokaut.ac.jp

Jobe, Daniel
djobe@sandia.gov

Johnson, David
djohnson@titan.com

Johnston, Mark
markdj@engin.umich.edu

Jungwirth, Karel
jungwirth@fzu.cz

Kablambaev, Beissengazy
ratakhin@ovpe.hcei.tsc.ru

Kaganovich, Igor
ikaganov@pppl.gov

Kalinin, Yuri
kalinin@dap.kiae.ru

Kamada, Keiich
kkamada@plasma.s.kanazawa-u.ac.jp

Kamada, Masaki
kokamada@plasma.s.kanazawa-u.ac.jp

Kantsyrev, Victor
victor@physics.unr.edu

Karpinski, Leslaw
lekarp@ifpilm.waw.pl

Kasuya, Koichi
kkasuya@es.titech.ac.jp

Kawata, Shigeo
kawata@cc.utsunomiya-u.ac.jp

Keely, Sean
kpkeely@sandia.gov

Kemp, Mark
makemp@sandia.gov

Khakberdiev, Ibragim
khakb_i@yahoo.com

Khishchenko, Konstantin
konst@ihed.ras.ru

Kiefer, Mark
mlkiefe@sandia.gov

Kim, Alexandre
kim@oit.hcei.tsc.ru

Kim, Hyoung Suk
hyoungskim@hotmail.com

Kindel, Joseph
jkindel@lanl.gov

Kingsep, Alexander
kingsep@dap.kiae.ru

Knudson, Marcus
mdknuds@sandia.gov

Koidan, Vasili
koidan@inp.nsk.su

Kolb, Alan
alan.kolb@sbr.global.net

Konkashbaev, Isak
isak@anl.gov

Kononenko, Vladimir
bratchikov@five.ch70.chel.su

Korenev, Sergey
sergey@korenev.com

Kormilitsyn, Alexei
bratchikov@five.ch70.chel.su

Korolev, Yuri
korolev@lnp.hcei.tsc.ru

Korovin, Sergei
korovin@hcei.tsc.ru

Koshelev
Vladimir
koshelev@lhfe.hcei.tsc.ru

Koustantin, Khishchenko
konst@ihed.ras.ru

Kovalchuk, Boris
kim@oit.hcei.tsc.ru

Kovalev, Nikolay
kovalev@appl.sci-nnov.ru

Krastelev, Evgueni
e_krastelev@mail.ru

Krauz, Vyacheslav
krauz@nfi.kiae.ru

Kravarik, Jozef
kravarik@fel.cvut.cz

Krompholz, Hermann
hermann.krompholz@coe.ttu.edu

Kruecken, Thomas
thomas.kruecken@philips.com

Kubes, Pavel
kubes@fel.cvut.cz

Kubyshkin, Alexander
kub@sch2.ru

Kunze, Hans-Joachim
Hans-Joachim.Kunze@ruhr-uni-bochum.de

Kussse, Bruce
brk2@cornell.edu

Kuznetsov, Sergey
sfgar@md08.vniief.ru

Kwan, Thomas
tjtk@lanl.gov

Laroussi, Mounir
laroussi@jlab.org

Lassalle, Francis
lassallf@cegramat.fr

Lauer, Eugene
lauer1@llnl.gov

Le Galloudec, Nathalie
kari@physics.unr.edu

LeBeau, Hank
kari@physics.unr.edu

Lebedev, Sergey
s.lebedev@ic.ac.uk

Leeper, Ramon
rjleepe@sandia.gov

Lemaire, Jean-Louis
jlemaire@cea.fr

Lemke, Raymond
rwlemke@sandia.gov

LePell, Paul
pdlepel@sandia.gov

Levine, Jerrold
jlevine@titan.com

Ligachev, Aleksander
ligachev@msk.sitek.net

Lindemuth, Irvin
irl@lanl.gov

Lockner, Thomas
trlockn@sandia.gov

Loza, Oleg
loza@fpl.gpi.ru

Lu, George
lu@ao.dtra.mil

Lucero, Robert
rjlucer@sandia.gov

MacFarlane, Joseph
jjm@prism-cs.com

Maenchen, John
jemaenc@sandia.gov

Makhin, Volodymyr
kari@physics.unr.edu

Mardahl, Peter

Markovits, Meir
meirm@rafael.co.il

Maron, Yitzhak
fnmaron@wisemail.weizmann.ac.il

Masugata, Katsumi
masugata@eng.toyama-u.ac.jp

Mathuthu, Manny
mathuthu@science.uz.ac.zw

Matzen, M. Keith
mkmatze@sandia.gov

Mayes, Jon
mayes@apelc.com

Mazarakis, Michael
mgmazar@sandia.gov

McCarrick, James
mccarrick1@llnl.gov

McDaniel, Dillon

McGurn, John
jsmcgur@sandia.gov

Mehlhorn, Tom
tamehlh@sandia.gov

Mendel, Clifford
mendel@swcp.com

Merle, Eric
eric.merle@cea.fr

Mesyats, Gennady
mesyats@pran.ru

Meyer, Ryan
rmmeyer@sandia.gov

Miklaszewski, Ryszard
rysiek@ifpilm.waw.pl

Mitchell, Marc
mdm46@cornell.edu

Mix, Louis
lpmix@sandia.gov

Miyamoto, Tetsu
miyamoto@asrl.org

Molina, Isidro
imolina@sandia.gov

Mond, Michael
mond@menix.bgu.ac.il

Mosher, David
mosher@suzie.nrl.navy.mil

Mueller, Georg
georg.mueller@ihm.fzk.de

Musk, Jeffrey
jhm29@cornell.edu

Myers, Matthew
myers2@ccf.nrl.navy.mil

Navon, Itamar
navoni@barak-online.net.il

Neuber, Andreas
andreas.neuber@ttu.edu

Nielson, Dan
nielson5@covad.net

Ning, Cheng
chengnqy@public3.bta.net.cn

Ning, Ding
ding_ning@mail.iapcm.ac.cn

Novoselov, Yuri
nov@iep.uran.ru

O' Malley, John
jomalley@awe.co.uk

Oleinik, Gueorgui
oleinik@triniti.ru

Oliver, Bryan
boliver@mrcabq.com

Olson, Craig
clolson@sandia.gov

Onishchenko, Ivan
onish@kipt.kharkov.ua

Oreshkin, Vladimir
oreshkin@ovpe.hcei.tsc.ru

Orlov, Andrey
mailbox@ntc.vniief.ru

Ossakow, Sidney
ossakow@ccf.nrl.navy.mil

Ottinger, Paul
ottinger@suzie.nrl.navy.mil

Ouart, Nicholas
kari@physics.unr.edu

Oxner, Andrew
kari@physics.unr.edu

Palmer, James
james.palmer@awe.co.uk

Paraschiv, Ioana
ioana@physics.unr.edu

Pereira, Nino
pereira@speakeasy.org

Peskov, Nikolai
peskov@appl.sci-nnov.ru

Peterson, Darrell
dlp@lanl.gov

Peterson, Kyle
kyle.peterson@arnold.af.mil

Petrov, Peter
pvpetrov@snezhinsk.ru

Pikunov, Viktor
vmp@vmp.phys.msu.su

Pikuz, Sergei
clj1@cornell.edu

Plokhoim, Vladimir
V.V.Plokhoi@vniitf.ru

Pointon, Timothy
tdpoint@sandia.gov

Polevin, Sergey
polevin@lfe.hcei.tsc.ru

Powell, Charles
powellc@diana-hitech.com

Prestwich, Kenneth
krprestwich@cs.com

Presura, Radu
kari@physics.unr.edu

Price, David
price@titan.com

Putnam, Sidney
sputnam@titan.com

Quintenz, Jeffrey
jpquint@sandia.gov

Ramirez, Juan
mccasau@sandia.gov

Ratakhin, Nikolai
ratakhin@ovpe.hcei.tsc.ru

Rej, Don
drej@lanl.gov

Renk, Timothy
tjrenk@sandia.gov

Riordan, John
jriordan@titan.com

Rizakhanov, Rajudin
kercgor@dol.ru

Robledo-Martinez, Arturo
a.robledo@ieee.org

Roderick, Norman
roderick@unm.edu

Rodgers, Matthew
msrodge@sandia.gov

Rogowski, Sonrisa
strogow@sandia.gov

Rose, Evan
ear@lanl.gov

Rose, David
drose@mrcabq.com

Rosenthal, Stephen
serosen@sandia.gov

Rostomyan, Eduard
evrostom@irphe.am

Rostov, Vladislav
rostov@lfe.hcei.tsc.ru

Roth, Markus
m.roth@gsi.de

Rousculp, Christopher
rousculp@lanl.gov

Rousskikh, Alexander
russ@ovpe2.hcei.tsc.ru

Rovang, Dean
dcrovan@sandia.gov

Rudakov, Leonid
rudakovl@usam.umd.edu

Rukin, Sergei
rukin@iep.uran.ru

Ryutov, Dmitri
ryutov1@llnl.gov

Ryzhov, Victor
ryzhov@to/hcei.tsc.ru

Sachs, Bob
bob.sachs@teamsp.com

Sakamoto, Nobuhiro
nobusakamoto@mac.com

Sandoval, Joe
js_sandoval@lanl.gov

Sanford, Thomas
twsanfo@sandia.gov

Sarkar, Partha
psarkar@ipr.res.in
psarkar_70@yahoo.com

Sarkisov, Gennady
gssarki@sandia.gov

Sasorov, Pavel
sasorov@itep.ru

Savage, Mark
mesavag@sandia.gove

Schamiloglu, Edl
edl@eece.unm.edu

Schmidt, Jiri
schmidt@ipp.cas.cz

Schneider, Larry
lxschne@sandia.gov

Schoenbach, Karl
schoenbach@ece.odu.edu

Schultheiss, Christoph
christoph.schultheiss@ihm.fzk.de

Schumer, Joseph
schumer@calvin.nrl.navy.mil

Schwoebel, Paul
paul.schwoebel@sri.com

Seamen, Johann
jfseame@sandia.gov

Sethian, John
sethian@this.nrl.navy.mil

Shao, Hao
lwei@nint.ac.cn

Sheehey, Peter
pete@lanl.gov

Shelkovenko, Tatiana
clj1@cornell.edu

Sherlock, mark
mark.sherlock@ic.ac.uk

Shi, Jinshui
bcys@caep.ac.cn

Shimomura, Naoyuki
simomura@ee.tokushima-u.ac.jp

Shishlov, Alexander
ash@ovpe2.hcei.tsc.ru

Shkuratov, Sergey
sshkuratov@ppl.ee.tu.edu

Shlapakovski, Anatoli
shl@npi.tpu.ru

Shlyaptsev, Vyacheslav
slava@llnl.gov

Shlyaptseva, Alla
alla@physics.unr.edu

Shoup, Roy
roy.shoup@itt.com

Shpak, Valery
radan@ief.uran.ru

Shu, Ting
mrtingshu@yahoo.com.cn

Sinars, Daniel
dbsinar@sandia.gov

Sincerny, Peter
psincerny@titan.com

Sinclair, Mark
mar.sinclair@awe.co.uk

Sinitsky, Stanislav
sinitsky@inp.nsk.su

Smirnov, Valentin
VSmirnov@nfi.kiae.ru

Smith, John
smith@lanl.gov

Smith, Ian
ismith@titan.com

Sokovnin, Sergei
sokovnin@iep.uran.ru

Sotnikov, Vladimir
kari@physics.unr.edu

Soto, Leopoldo
lsoto@cchen.cl

Spencer, Thomas
Thomas.Spencer@kirtland.af.mil

Stallings, Charles
cstallings11@attbi.com

Steele, Jimmy
jim.steele@arnold.af.mil

Steen, Paul
psteen@titan.com

Stoltzfus, Brian
bsstolt@sandia.gov

Strasburg, Sean
stras@suzie.nrl.navy.mil

Strickler, Trevor
strickler@umich.edu

Struts, Vasily
struts@ephc.npi.tpu.ru

Struve, Kenneth
kwstruv@sandia.gov

Sturtz, Jason
jestur@sandia.gov

Sudan, Ravindra
clj1@cornell.edu

Sunka, Pavel
sunka@ipp.cas.cz

Swanekamp, Stephen
swane@calvin.nrl.navy.mil

Sweeney, Mary Ann
masween@sandia.gov

Sze, Henry
hsze@titan.com

Takasugi, Keiichi
takasugi@phys.cst.nihon-u.ac.jp

Talantsev, Yevgeniy
etalantsev@ppl.ee.ttu.edu

Tauschwitz, Andreas
A.Tauschwitz@gsi.de

Teramoto, Yusuke
tera@st.eecs.kumamoto-u.ac.jp

Terry, Robert
terry@ccs.nrl.navy.mil

Thoma, Carsten
cthoma@mrcabq.com

Thomas, Kenneth
Ken.Thomas@awe.co.uk

Thornhill, Joseph
thornhil@ppdu.nrl.navy.mil

Threadgold, Jim
jim.threadgold@awe.co.uk

Truesdale, Robert
bob.truesdale@arnold.af.mil

Turchi, Peter
peter.turchi@kirtland.af.mil

Tyo, J. Scott
tyo@eece.unm.edu

Velikovich, Alexander
velikov@ppdmail.nrl.navy.mil

Vezinet, Rene
vezinetr@cegramat.fr

Vidmar, Robert
rvidmar@unr.edu

Vikhrev, Victor
vikhrev@nfi.kiae.ru

Vladimir/Fedorovich, Ermolovich

Volkov, Nikolay
nbv@ami.uran.ru

Waisman, Eduardo
waisman@aasc.net

Walter, John
j.walter@ieee.org

Wang, Xinxin
wangxx@mail.tsinghua.edu.cn

Wang, Huacen
bcys@caep.ac.cn

Watrous, John
jwatrous@numerex.com

Weber, Bruce
weber@suzie.nrl.navy.mil

Welch, Dale
drwelch@mrcabq.com

Wemlinger, Erik
ecwemli@sandia.gov

Westenskow, Glen
westenskow1@llnl.gov

White, Roger
rogerw@titan.com

Whitney, Kenneth
whitney@ppdmail.nrl.navy.mil

Wong, Sik-Lam
swong@titan.com

Woodworth, Joseph
jrwoodw@sandia.gov

Xie, Weiping
bcys@caep.ac.cn

Yalandin, Michael
yalandin@ief.uran.ru

Yang, Libing
bcys@caep.ac.cn

Yatsui, Kiyoshi
yatsui@nagaokaut.ac.jp

Yu, Edmund
epyu@sandia.gov

Yusupov, Odil
khodjiev@online.ru

Zhong, Huihuang
hhzhong@nudt.edu.cn

Author Index

A

B

C

D

E

F

G

H

I

J

K

L

M

N

O

P

Q

R

S

T

U

V

W

X

Y

Z